The J&P
Transformer Book

J&P Books

General Editor: C. A. Worth

These two books, published originally by Johnson & Phillips Ltd, have for many years been accepted as standard works of reference by electrical engineers concerned with transformers and switchgear. They now appear under the Newnes-Butterworths imprint with the same team of contributors and editor.

The J&P Transformer Book

The J&P Switchgear Book

The J&P Transformer Book

A PRACTICAL TECHNOLOGY OF THE
POWER TRANSFORMER

S. AUSTEN STIGANT C.Eng., F.I.E.E., F.I.E.E.E.

and

A. C. FRANKLIN C.Eng., F.I.E.E.

LONDON
NEWNES-BUTTERWORTHS

THE BUTTERWORTH GROUP

ENGLAND
Butterworth & Co (Publishers) Ltd
London: 88 Kingsway, WC2B 6AB

AUSTRALIA
Butterworths Pty Ltd
Sydney: 586 Pacific Highway, NSW 2067
Melbourne: 343 Little Collins Street, 3000
Brisbane: 240 Queen Street, 4000

CANADA
Butterworth & Co (Canada) Ltd
Toronto: 14 Curity Avenue, 374

NEW ZEALAND
Butterworths of New Zealand Ltd
Wellington: 26–28 Waring Taylor Street, 1

SOUTH AFRICA
Butterworth & Co (South Africa) (Pty) Ltd
Durban: 152–154 Gale Street

First published in 1925 by Johnson & Phillips Ltd
Ninth edition published in 1961
   Second impression 1965 by Iliffe Books Ltd.
Tenth edition published in 1973 by Newnes-Butterworths,
an imprint of the Butterworth Group

ISBN 0 408 00094 5

Filmset by Ramsay Typesetting Ltd, London and Crawley

Printed in England by Fletcher & Son Ltd, Norwich
Bound in England by Richard Clay (The Chaucer Press) Ltd, Bungay, Suffolk

Editor's Foreword

Known in practically every part of the world as the transformer engineer's 'bible', *The J & P Transformer Book* is the leading treatise on the power transformer and is an essential reference book for engineers concerned with the design, installation and maintenance of power transformers. It is invaluable as a textbook for the architect and system planner as well as for the electrical engineering graduate and the student.

The book started as a series of pamphlets published by Johnson and Phillips Ltd. in 1922 entitled *Transformer Abstracts*, and an immediate and increasing demand led to extensive reprinting, until in 1924 it was decided to publish the work as a single volume. The first edition of *The J & P Transformer Book* appeared in January 1925 and this was quickly followed by a second edition in August of the same year. The continuing interest of engineers and the favourable reviews of the electrical press necessitated further editions from 1928 until the appearance of the eighth edition in 1941. A scarcity of printing materials during and immediately after World War II, and the retirement of Mr. H. Morgan Lacey who was one of the original authors, led to the book being allowed to go out of print, although the demand for it continued. In 1958, work on the ninth edition commenced with Mr. A. C. Franklin, at that time Chief Designer at Johnson & Phillips Ltd, as co-author with Mr. S. A. Stigant. This edition was published in 1961 and a revised impression appeared in 1965.

This is the tenth edition and it is almost entirely the work of Stigant and Franklin, although credit has been given by the authors to the past work of H. Morgan Lacey in the acknowledgements. It is now published by Butterworths, and represents a very considerable amount of research and study on the part of the authors in respect of the data supplied by transformer manufacturers and allied industries together with that made available by relevant institutions and associations. It contains a wealth of new technical information with tables, calculations, diagrams, line drawings and pictorial illustrations. Conversion to S.I. units has been undertaken and the whole book has been reset and repaginated.

The information contained herein will prove even more useful to engineers than that contained in past editions.

C. A. WORTH,
Editor

Preface to the 10th edition

Some time has elapsed since the appearance of the 9th edition of *The J & P Transformer Book* and every effort has been made by the authors to produce a revision that is as comprehensive and as up to date as possible before going to press.

As the book contains innumerable references to the practice and designs of many manufacturers, some difficulties arose during its production due to the rationalisation that has occurred in the transformer and allied industries, and the authors express their grateful thanks to those manufacturers mentioned in the acknowledgements for the generous assistance given by them in the preparation of new data and by the provision of up to date illustrations.

Account has been taken of the latest editions of B.S.171 and B.S.148 together with other British Standards concerned in any way with power transformers, and throughout the text and in calculations S.I. units have been included.

It is hoped that the new generation of engineers will find this edition of *The J & P Transformer Book* even more useful and informative than previous editions have been in the past.

S.A.S.
A.C.F.

Acknowledgements

The authors gratefully acknowledge the many services and assistance given to them in the preparation of this book. In particular, they offer their thanks to those manufacturers who as well as providing data also supplied photographs and line drawings to illustrate them. They are:

Alpha-Laval Co. Ltd
Associated Tapchangers Ltd
Bonar, Long & Co. Ltd
British Standards Institution
British Steel Corporation
Broadbent, Hopkinson Ltd
J. O. Buchanan Engineering Ltd
Burmah-Castrol Industrial Ltd
Controlled Heat & Air Ltd
Dawe Instruments Ltd
Ferranti Ltd
GEC Measurements Ltd
GEC Transformers Ltd
Hawker Siddeley Power Transformers Ltd
Metalectric Furnaces Ltd
National Physical Laboratories (Robinson and Dadson)
A. Reyrolle & Co. Ltd
Redman Heenan International Ltd
Telcon Magnetic Cores Ltd
Vokes Ltd (Stream Line Filters Division)
The Electrical Research Association (M. Waters)

Some parts of the work contained in the book must still be attributed to the late H. Morgan Lacey who collaborated with the present authors prior to the 9th edition.

In addition we are particularly grateful to the engineers of Bonar, Long & Co. Ltd who have spent considerable time on the preparation of technical information and certain new diagrams for this the 10th edition of *The J & P Transformer Book*.

Contents

Appendices

Chapter 1

Fundamental transformer principles

As the real interdependence of engineering theory and practice is now fully recognised, it becomes almost essential when writing a book upon one aspect of the subject to deal briefly with the other. As the present publication is intended to be a general survey of the practice concerning power transformers in Great Britain, it is therefore desirable to introduce the subject with a brief theory of transformer action, illustrated by fundamental formulae and simple phasor diagrams.

As is generally known, a transformer consists essentially of a magnetic core built up of insulated silicon steel laminations upon which are wound two distinct sets of coils suitably located with respect to each other and termed the primary and secondary windings respectively. Such a combination may be used to derive a voltage higher or lower than that immediately available, and in the former case the transformer is termed a step-up transformer, while in the latter it is termed a step-down type. The primary winding is that winding to which the supply voltage is applied irrespective of whether it is the higher- or lower-voltage winding; the other winding to which the load is directly connected is termed the secondary winding.

As the phenomenon of electromagnetic induction can only take place in static apparatus when the magnetic flux is continually varying, it is clear that static transformers can only be used in electrical circuits having such characteristics, that is, in alternating-current circuits.

If an alternating e.m.f. is applied to the terminals of the primary winding of a transformer with the secondary winding open circuited, a very small current will flow in the primary circuit only, which serves to magnetise the core and to supply the iron loss of the transformer. Thus an alternating magnetic flux is established in the core which induces an e.m.f. in both primary and secondary windings. The magnetising ampere-turns are given by the product of the magnetising current and the primary turns. The no-load current is given by the total no-load ampere-turns divided by the primary turns. As primary and secondary windings are wound on the same core, and as the magnetising flux is common to the two windings, obviously the voltage induced in a single turn of each winding will be the same, and the induced voltages in the primary and secondary windings are therefore in direct proportion to the number of turns in those windings. The formula connecting induced voltage, flux and number of turns is as follows:

1

$$E = 4K_f\Phi_mNf \tag{1.1}$$

where E = r.m.s. value of the induced e.m.f. in the winding considered
         K_f = form factor of the e.m.f. wave (1·11 for sine wave)
         f = frequency of the supply in hertz
         Φ_m = total magnetic flux through the core (max. value) in webers
         N = number of turns in the winding considered.

The above formula holds good for the voltage induced in either primary or secondary windings, and it is only a matter of inserting the correct value of N for the winding under consideration. Figure 1.1 shows the simple phasor diagram

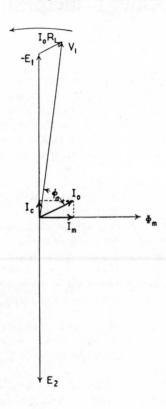

V_1 primary terminal voltage
E_1 primary induced e.m.f.
E_2 secondary induced e.m.f.
I_0R_1 resistance voltage drop due to I_0
Φ_m maximum (peak) value of magnetic flux
I_0 primary no-load current
I_c primary core loss current
I_m primary magnetising current
$\cos \phi_0$ primary no-load power factor
Magnetic leakage is negligible and is ignored.

Fig. 1.1   Phasor diagram for a single-phase transformer on open circuit. Assumed turns ratio 1:1

corresponding to a transformer on no-load, and the symbols have the significance shown on the diagram. Usually in the practical design of transformers, the small drop in voltage due to the flow of the no-load current in the primary winding is neglected.

Reverting to the eqn. 1.1 this simplifies to

$$E = 4·44f\Phi_mN$$

for a sine wave of voltage which, of course, is always assumed. In actual design calculations this formula is rearranged as:

$$\frac{V}{N} = \frac{B_mAf}{22·51 \times 10^2} \tag{1.2}$$

where $\dfrac{V}{N}$ = volts per turn, which is the same in both windings

B_m = maximum flux density in the core, tesla

A = net cross-sectional area of core, cm^2

f = frequency of supply, Hz.

In practical designing B_m and A are predetermined, whilst f is already known, so that the volts per turn are derived from the formula. It is then an easy matter to determine the number of turns in each winding from the specified voltage of the winding and the derived volts per turn.

The alignment chart of Fig. 1.2 provides the means for determining rapidly

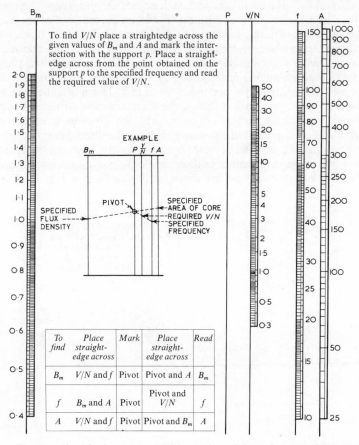

Fig. 1.2   Chart for determining V/N for single- and three-phase transformers

the value of any one of the factors in eqn. 1.2 when the others are known.

The losses which theoretically occur in an unloaded transformer are the iron losses, copper loss due to the flow of no-load current in the primary winding, and dielectric loss. In practice only the iron losses are of importance in transformers, and these losses are the sum of the hysteresis and eddy-current losses which are constant for a given applied voltage and unaffected by the load on the transformer. The dielectric losses are also functions of the primary and secondary voltages but they vary slightly with the temperature of the windings as affected by the load on the transformer. The copper loss due to the no-load current is generally negligible and it is independent of the load for a given excitation.

On open circuit the transformer acts as a single winding of high self-inductance, and the open-circuit power factor averages about 0·15 lagging.

The application of a load to the secondary side of the transformer produces a considerable change in the internal phenomena. When the secondary circuit is closed, a secondary current flows, the value of which is determined by the magnitude of the secondary terminal voltage and the impedance of the load circuit. The m.m.f. due to the secondary load current produces a certain load flux in the core which is in phase with the secondary current, but as the secondary load current is immediately balanced by a primary load current of such a value that the primary and secondary load ampere turns are equal, the secondary load flux is similarly counteracted by a primary load flux which is in phase with the primary load balancing current, and therefore in phase opposition to, and of the same magnitude as, the secondary load flux. Therefore, the core is left in its initial state of magnetisation, that is, the magnetisation corresponding to the open-circuit conditions, and this explains why the iron loss is independent of the load. The total current in the primary circuit is the phasor sum of the primary load current and the no-load current. Ignoring for the moment the question of resistance and leakage-reactance voltage drops, the conditions for a transformer supplying a non-inductive load are shown in phasor form in Fig. 1.3. The losses

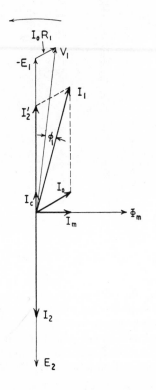

V_1 primary terminal voltage
E_1 primary induced e.m.f.
E_2 secondary induced e.m.f.
I_0R_1 resistance voltage drop due to I_0
Φ_m maximum (peak) value of magnetic flux
I_0 primary no-load current
I_c primary core loss current
I_m primary magnetising current
I_2 secondary load current
I_2' load component of total primary current
I_1 total primary current (including I_0 and I_2')
$\cos \phi_1$ = primary total load power factor
Load voltage drops ignored

Fig. 1.3 Phasor diagram for a single-phase transformer supplying a unity power factor load. Assumed turns ratio 1:1

in the transformer are now augmented by the copper losses in both primary and secondary windings, which consist of I^2R losses, losses due to eddy currents set up in the conductors, and stray losses in tank and core clamps.

Considering now the question of voltage drops due to the resistance and leakage reactance of the transformer windings, it should first be pointed out that, however the individual voltage drops are allocated, the sum total effect is manifested at the secondary terminals. The resistance drops in the primary and secondary windings are easily separated and determinable for the respective windings, but the reactive voltage drop, which is due to the total flux leakage between the two windings, is strictly not separable into two components, as the line of demarcation between the primary and secondary leakage fluxes cannot be defined. It has therefore become a convention to allocate half the leakage flux to each winding, and similarly to dispose of the reactive voltage drops. Figure 1.4 shows the phasor relationships in a single-phase transformer supplying an inductive load having a lagging power factor of 0·80, the resistance and

V_1 primary terminal voltage
E_1 primary induced e.m.f.
V_2 secondary terminal voltage
E_2 secondary induced e.m.f.
I_1R_1 primary resistance voltage drop
I_1X_1 primary reactance voltage drop
I_1Z_1 primary impedance voltage drop
I_2R_2 secondary resistance voltage drop
I_2X_2 secondary reactance voltage drop
I_2Z_2 secondary impedance voltage drop
Φ_m maximum (peak) value of magnetic flux
I_0 primary no-load current
I_c primary core loss current
I_m primary magnetising current
I_2 secondary load current
I_2' load component of total primary current
I_1 total primary current (including I_0 and I_2')
$\cos \phi_2$ secondary load power factor
$\cos \phi_1$ primary total load power factor

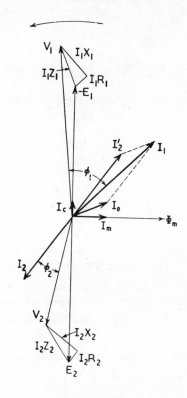

Fig. 1.4 Phasor diagram for a single-phase transformer supplying an inductive load of lagging power factor $\cos\phi_2$. Assumed turns ratio 1:1. Voltage drops divided between primary and secondary sides

leakage reactance drops being allocated to their respective windings. In fact, as the sum total effect is a reduction of the secondary terminal voltage, the resistance and reactance drops allocated to the primary windings appear on the diagram as additions to the e.m.f. induced in the primary windings.

Figure 1.5 shows phasor conditions identical with those in Fig. 1.4, except that the resistance and reactance drops are all shown as occurring on the secondary side.

5

Of course, the drops due to primary resistance and leakage reactance are converted to terms of the secondary voltage, that is, the primary voltage drops are divided by the ratio of transformation n, in the case of both step-up and

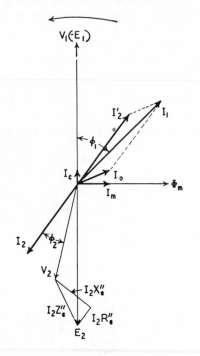

V_1 primary terminal voltage
E_1 primary induced e.m.f.
V_2 secondary terminal voltage
E_2 secondary induced e.m.f.
I_2R_e'' total resistance voltage drop
I_2X_e'' total reactance voltage drop
I_2Z_e'' total impedance voltage drop
Φ_m maximum (peak) value of magnetic flux
I_0 primary no-load current
I_c primary core loss current
I_m primary magnetising current
I_2 secondary load current
I_2' load component of total primary current
I_1 total primary current (including I_0 and I_2')
$\cos\phi_2$ secondary load power factor
$\cos\phi_1$ primary total load power factor

Fig. 1.5 Phasor diagram for a single-phase transformer supplying an inductive load of lagging power factor $\cos\phi_2$. Assumed turns ratio 1:1. Voltage drops transferred to secondary side

step-down transformers. In other words, the *percentage* voltage drops considered as occurring in either winding remain the same.

To transfer primary resistance values R_1 or leakage reactance values X_1 to the secondary side, R_1 and X_1 are divided by the square of the ratio of transformation n in the case of both step-up and step-down transformers.

The transference of impedance from one side to another is made as follows:

Let Z_s = total impedance of the secondary circuit including leakage and load characteristics.

Z_s' = equivalent value of Z_s when referred to the primary winding.

Then
$$I_2' = \frac{N_2}{N_1}I_2 = \frac{N_2}{N_1}\frac{E_2}{Z_s}$$

Therefore
$$I_2' = \left(\frac{N_2}{N_1}\right)^2\frac{E_1}{Z_s} \tag{1.3}$$

Also,
$$V_1 = E_1 + I_2'Z_1$$

where
$$E_1 = I_2'Z_s'$$

Therefore
$$I_2' = E_1/Z_s' \tag{1.4}$$

Comparing eqns. 1.3 and 1.4 it will be seen that $Z_s' = Z_s(N_1/N_2)^2$.

The equivalent impedance is thus obtained by multiplying the actual impedance of the secondary winding by the square of the ratio of transformation n, i.e. $(N_1/N_2)^2$. This, of course, holds good for secondary winding leakage reactance and secondary winding resistance in addition to the reactance and resistance of the external load.*

Figure 1.6 is included as a matter of interest to show that when the load has a

V_1 primary terminal voltage
E_1 primary induced e.m.f.
V_2 secondary terminal voltage
E_2 secondary induced e.m.f.
I_2R_e'' total resistance voltage drop
I_2X_e'' total reactance voltage drop
I_2Z_e'' total impedance voltage drop
Φ_m maximum (peak) value of magnetic flux
I_0 primary no-load current
I_c primary core loss current
I_m primary magnetising current
I_2 secondary load current
I_2' load component of total primary current
I_1 total primary current (including I_0 and I_2')
$\cos\phi_2$ secondary load power factor
$\cos\phi_1$ primary total load power factor

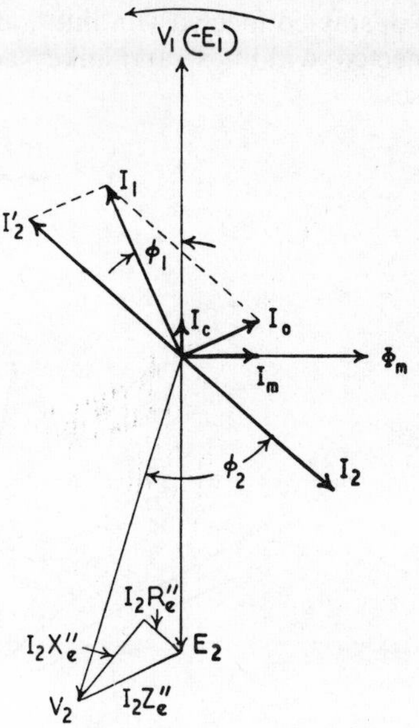

Fig. 1.6 Phasor diagram for a single-phase transformer supplying a capacitive load of leading power factor $\cos\phi_2$. Assumed turns ratio 1:1. Voltage drops transferred to secondary side

sufficient leading power factor, the secondary terminal voltage increases instead of decreasing. This happens when a leading current passes through an inductive reactance.

Preceding diagrams have been drawn for single-phase transformers, but they are strictly applicable to polyphase transformers also so long as the conditions for all the phases are shown. For instance, Fig. 1.7 shows the complete phasor diagram for a three-phase star/star connected transformer, and it will be seen that this diagram is only a threefold repetition of Fig. 1.5, in which the primary and secondary phasors correspond exactly to those shown in Fig. 1.5, but the three sets representing the three different phases are spaced 120° apart.

The output of a power transformer is familiarly expressed in kilovolt-amperes (kVA), and the fundamental expressions for determining these, assuming sine wave functions, are as follows:

Single-phase transformers

$$kVA = \frac{4\cdot44f\Phi_mNI}{1000}$$

Three-phase transformers

$$kVA = \frac{4\cdot44f\Phi_mNI \times 1\cdot73}{1000}$$

*See also Appendix 2.

7

In the expression for single-phase transformers, I is the full-load current in the transformer windings and also in the line; for three-phase transformers, I is the full-load current in each line connected to the transformer. That part of the expression representing the voltage refers to the voltage between line terminals of the transformer. The constant 1·73 is a multiplier for the phase voltage in the case of star-connected windings, and for the phase current in the case of delta-connected windings, and takes account of the angular displacement of the phases.

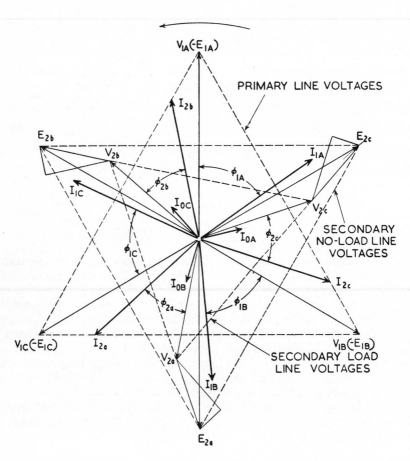

Fig. 1.7   *Phasor diagram for a three-phase transformer supplying an inductive load of lagging power factor $\cos \phi_2$. Assumed turns ratio 1:1. Voltage drops transferred to secondary side. Symbols have the same significance as in Fig. 1.5 with the addition of A, B and C subscripts to indicate primary phase phasors, and a, b and c subscripts to indicate secondary phase phasors*

Alternatively expressed, the rated kVA is the product of the rated secondary (no-load) voltage E_2 and rated output current I_2 and, in the case of polyphase transformers, by multiplying with the appropriate phase factor and by the appropriate constant depending on the magnitude of the units employed. It should be noted that rated primary and secondary voltages occur simultaneously at no load.

Single-phase transformers

$$kVA = E_2 I_2 \times 10^{-3}$$

Three-phase transformers

$$kVA = E_2 I_2 \times 10^{-3} \times 1{\cdot}73$$

The relationships between phase and line currents and voltages for star- and for delta-connected three-phase windings are as follows:

Three-phase star connection

$$\text{phase current} = \text{line current } I = \frac{kVA \times 1000}{E \times 1{\cdot}73}$$

$$\text{phase voltage} = \frac{E}{1{\cdot}73}$$

Three-phase delta connection

$$\text{phase current} = \frac{I}{1{\cdot}73} = \frac{kVA \times 1000}{E \times 3}$$

$$\text{phase voltage} = \text{line voltage} = E$$

$$E \text{ and } I = \text{line voltage and current respectively.}$$

The regulation that occurs at the secondary terminals of a transformer when a load is supplied consists, as before mentioned, of voltage drops due to the resistance of the windings and voltage drops due to the leakage reactance between the windings. These two voltage drops are in quadrature with one another, the resistance drop being in phase with the load current. The percentage regulation at unity power factor load may be calculated by means of the following formula:

$$\frac{\text{copper loss} \times 100}{\text{output}} + \frac{(\text{percentage reactance})^2}{200}$$

This value is always positive and indicates a voltage drop with load.

The approximate percentage voltage regulation for a current loading of a times the rated current at a power factor of $\cos \phi_2$ is given by the following formulae:

(i)   For transformers having an impedance voltage up to and including 4%:
percentage regulation $= a(V_R \cos \phi_2 + V_X \sin \phi_2)$

Note—for a lagging load p.f., $\sin \phi_2$ is taken as being positive; for a leading load p.f. it is taken as a negative. A positive value for the percentage regulation

9

indicates a voltage drop from no-load to the loading considered; a negative value for the regulation indicates a voltage rise.

(ii)   For transformers having an impedance voltage up to* and including 20%:

$$\text{percentage regulation} = a(V_R \cos \phi_2 + V_X \sin \phi_2) + \frac{a^2}{200}(V_X \cos \phi_2 - V_R \sin \phi_2)^2$$

where

V_R = percentage resistance voltage at full load

$$= \frac{\text{copper loss} \times 100}{\text{rated kVA}}$$

V_X = percentage reactance voltage $= \dfrac{I_2 X_e''}{V_2} \times 100$

UNITY P.F.

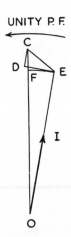

LAGGING P.F.

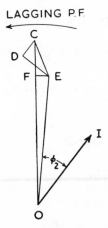

LEADING P.F.

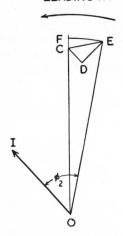

OC secondary induced e.m.f. (E_2)
OI secondary load current (I_2)
CD total resistance voltage drop $(I_2 R_e'')$
DE total reactance voltage drop $(I_2 X_e'')$
CE total impedance voltage drop $(I_2 Z_e'')$
OE total impedance voltage drop $(I_2 Z_e'')$
OE secondary terminal voltage (V_2)
$\hat{EOI} = \phi_2$
cos ϕ_2 secondary load power factor
CF = OC − OE = Regulation (usually expressed as a percentage of OE)

Fig. 1.8   Regulation diagrams

*For impedances above 20% refer to B.S. 171

V_X is obtained theoretically by calculation* or actually from the tested impedance and resistance of the windings. In the latter case

$$V_X = \sqrt{(V_Z{}^2 - V_R{}^2)}$$

where V_Z = percentage impedance voltage at full load.

At loads of low power factor the regulation becomes of serious consequence if the reactance is at all high on account of its quadrature phase relationship.

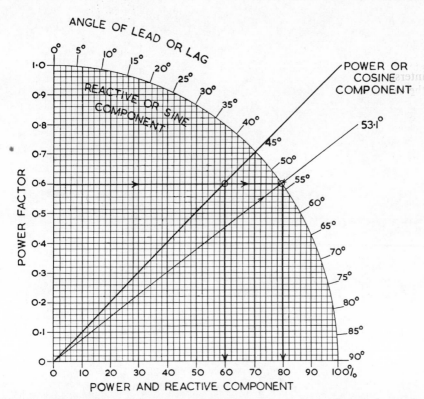

Fig. 1.9   Simple chart for determining power and reactive current components and angular displacement in an a.c. circuit

This question is, however, dealt with more fully in Appendix 4, though Fig. 1.8 shows the effect of low power factor upon the regulation. Incidentally, Fig. 1.9 indicates the disadvantages of low power as affecting the relative magnitudes of power and wattless currents.

EXAMPLE

Given that the load power factor is 0·6. Mark the point 0·6 on the scale of ordinates and project a line horizontally from this point cutting both the power

*See page 101.

component axis and the reactive component arc. Vertical projections downward give the values of the power and reactive components on the abscissa scale as percentages of the total current, these being 60% and 80% respectively. The intersection of the horizontal projection from the scale of power factors with the reactive arc also gives the angular displacement of the current — in this example 53·1°. The line from the origin to the point of intersection of the horizontal projection from the power factor scale with the reactive arc is the 100% total current in the circuit.

Chapter 2

The magnetic circuit

Steel sheets of a very inferior quality compared to present day standards were used in the very early days of transformer manufacture and magnetic ageing caused a great deal of trouble at that time. Resulting from the ageing effect, the hysteresis component of the iron loss in a transformer magnetic circuit was found to have trebled in value during the very early life of a transformer. It was

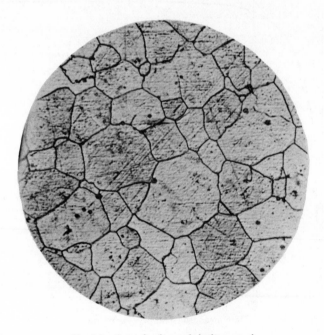

Fig. 2.1 Crystals of annealed silicon steel

subsequently found that very small quantities of silicon alloyed with low-carbon-content steel produced a material with low hysteresis losses and high permeability. In Chapter 4 iron loss/flux density curves are given illustrating values for cold rolled electrical steel available at present and it will be seen that total loss value is of the order of 1·7 watts per kilogram for sheets having a thickness of 0·33 mm at a maximum flux density of 1·6 teslas (= 1·6 Wb/m²) at 50 Hz when fabricated into laminations.

Materials used in the past were made by hot rolling. In common with other metals silicon steel is crystalline, and the hot-rolling process in conjunction

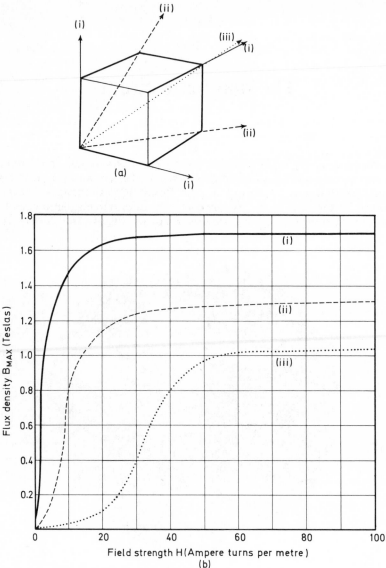

Fig. 2.2   Important directions in a silicon steel crystal. The directions of easy magnetisation are those of the cube edges

with the subsequent annealing produces a sheet in which the crystals are disposed throughout the sheet in almost a random fashion and do not take up any special alignment with respect to the direction of rolling, the sheet surface or with each other. As a result, hot-rolled electrical sheets, although not quite uniform in their magnetic properties across and parallel to the rolling direction, do not exhibit particularly strong anisotropic properties.

In the early 1930s it was discovered that cold rolling improved the magnetic properties of strip in the rolling direction. This improvement in magnetic properties is a consequence of the particularly easy magnetisation of certain directions in the crystals of silicon steel and the alignment of these crystals by certain cold-rolling and heat-treatment processes. The improved properties which were available in the rolling direction only (*i.e.* along the length of the strip) may be characterised by the iron loss which, at a maximum flux density of 1·5 teslas, approximated to 1·0 watts per kilogram for sheets having a thickness of 0·33 mm and which are now commercially available.

Annealed silicon steel sheet consists of metal crystals or grains as shown in the photograph Fig. 2.1. The magnetic properties of the sheet are derived from the magnetic properties of the individual crystals and many of these are a function of the direction in the crystal in which they are measured. This effect is shown diagrammatically in Fig. 2.2(*a*) where it will be seen that from the magnetic viewpoint, the important directions of a silicon steel crystal are those along the cube edges as these are the directions of easy magnetisation. Figure 2.2(*b*) gives a graphical comparison, in the form of B/H curves, of the magnetic properties exhibited by the cube in the three directions shown in Fig. 2.2(*a*).

Non-oriented hot-rolled sheet consists of crystals packed together in a random way so that magnetic measurements on a sheet give a result which is

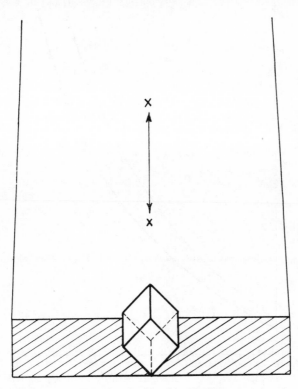

Fig. 2.3   The effect of cold rolling

almost independent of the direction in which they are taken and which is an average of the results for all directions in a single crystal. If all the crystals could be aligned then the properties of the bulk metal would approximate to those of the single crystal and further, if a direction exhibiting favourable magnetic properties in the crystal could be made to lie parallel to the rolling direction in a sheet then a much improved magnetic material would be obtained. It is fortunate that this can be achieved by cold rolling and annealing silicon steel strip. Figure 2.3 represents a cross section cut through a silicon steel strip, X-X being the rolling direction. The cube represents the orientation of the individual grains within the strip as a result of the combined effect of cold rolling and heat treatment. Since the cube edge is a direction of easy magnetisation, it is seen that in the strip this is *along* the rolling direction. Magnetic properties in the transverse direction of the strip compare very unfavourably as shown in Fig. 2.2(*b*). An important property of cold-rolled silicon steel strip is obvious from

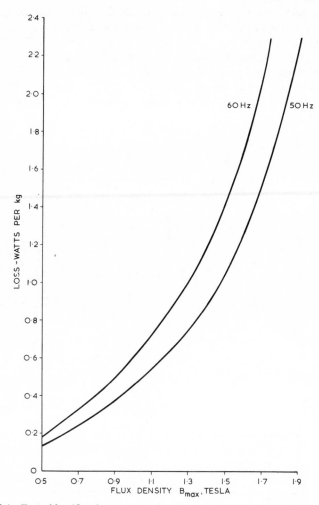

Fig. 2.4   Typical loss/flux density curves for cold-rolled silicon steel. Thickness 0·33 mm

16

these diagrams. Whilst in the rolling direction it exhibits better properties than ordinary silicon steel, in other directions the properties are not so favourable. It is for this reason that special care must be taken with the design of corners in built up cores. Not all properties are dependent on direction, exceptions being saturation flux density and electrical resistivity.

Cold-rolled silicon steel is supplied by the maker to a guaranteed maximum total loss, at a specific value of maximum flux density, usually 1·5 teslas. Figure 2.4 illustrates the total loss value for one grade of cold-rolled silicon steel for

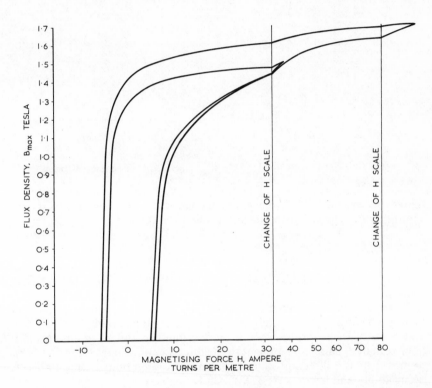

Fig. 2.5. Typical hysteresis loops in the direction of rolling for B_{max}, 1·5 and 1·7 teslas of cold-rolled silicon steel

varying flux densities at 50 and 60 Hz. Hysteresis loops for the same material are shown in Fig. 2.5.

The effect of the silicon content of oriented silicon sheet is given in Fig. 2.6. It will be appreciated that the data given in Fig. 2.6(a) are the relationship of silicon in a simple ferro-silicon alloy. Other elements also have a marked effect upon resistivity and it is generally considered that aluminium is the equivalent of silicon in its effect. In commercial grades of electrical steels resistivity must be related to the total alloy content of the steel.

Figure 2.6(*b*) represents an ideal relationship but in practice other constituents are present in electrical steels and the total alloy content affects the value of saturation flux density. Within the silicon range of cold-reduced commercial

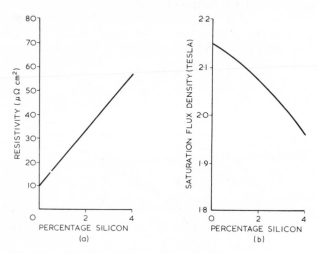

Fig. 2.6   Typical data for simple iron-silicon alloys with varying silicon content

alloys, of say, up to 3 %, the graph indicates the order of the reduction in saturation flux density with increasing silicon content.

STRIP-WOUND CORES IN GRAIN-ORIENTED SILICON STEEL

Cold-reduced grain-oriented silicon steel as we have seen earlier is a magnetic material having approximately 3·1 % silicon, manufactured by a cold-rolling and heat-treatment process which induces the constituent grains to fall into approximately parallel alignment in the direction of rolling. If the flux is made to travel in the direction of this 'preferred' orientation of the grains, high permeability and low hysteresis loss can be realised.

Strip-wound cores are manufactured by winding the material, in the form of continuous strip, upon suitably shaped mandrels, a method of construction which allows full advantage to be taken of the superior characteristics of the material, by ensuring that the direction of preferred orientation coincides with that of the flux path.

After forming, these cores are annealed for a period of three to four hours in an atmosphere that protects them against contact with, or penetration by, oxidising, sulphurous or carbonaceous gases. This annealing relieves the material from the detrimental mechanical strains imposed by the winding operation.

The illustrations shown in Fig. 2.7 give some idea of the variety of arrangements available.

The use of a single strip width results in a core of rectangular cross-section.

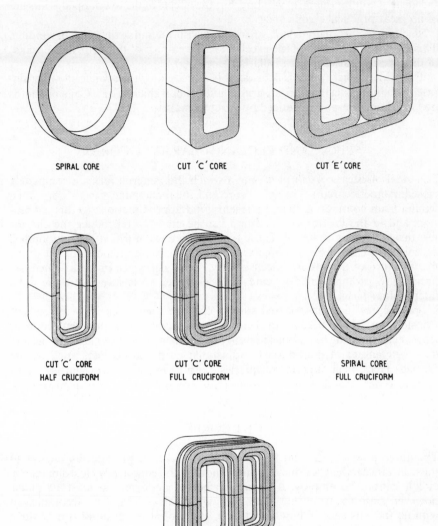

SPIRAL CORE CUT 'C' CORE CUT 'E' CORE

CUT 'C' CORE
HALF CRUCIFORM

CUT 'C' CORE
FULL CRUCIFORM

SPIRAL CORE
FULL CRUCIFORM

CUT 'E' CORE
FULL CRUCIFORM

Fig. 2.7 Forms of strip-wound cores

By employing strips of various widths, however, half or full cruciform sections can be accommodated. It may be argued that to do this means a departure from the ideal of continuous strip winding, but the effects of the overlapped joints are of no great practical significance.

Stacking factors of up to 95 % are realised without undue tightness of winding. This is possible because of the excellent surface finish of the strip, and because of the type of heat resisting insulation employed.

The usage of these cores is considerably increased by the availability of grain-oriented silicon steel in various grades and ribbon thicknesses. Coils of widths up to 800 mm are now produced by the steel makers.

SPIRAL AND RECTANGULAR TYPE CORES

The ideal physical core form is one in which the material forms a completely closed magnetic circuit. The spiral core and the rectangular type of core when wound from continuous strip, represent good approximations to this conception, and in the unstrained condition can be made to reproduce the characteristics of the raw material. These types of core are normally left unbonded, except in the case of the rectangular type core when it is intended for use in the manufacture of a cut core. Occasionally, spiral cores are given a light resin bonding to promote rigidity, care being exercised to ensure that there is no interlaminar penetration.

Spiral cores in grain-oriented silicon steel are used extensively for current transformers where, because of their superior performance, they replace cores fabricated from ring stampings in hot-rolled silicon steel. They are also used in the manufacture of continuously adjustable auto transformers such as the 'Variac'. Here, their superior properties, both as regards losses and ability to operate at high flux densities without undue exciting current, are an advantage.

CUT CORES

The development of the cut type of strip-wound core provides an answer to those instances when, for one reason or another, the completely closed magnetic circuit cannot be employed. Some sacrifice of performance must be made, however, since the provision of two solidified halves cannot be accomplished without the existence of bonding strains, and the introduction of two or three air gaps in the magnetic circuit.

Strains due to solidification of the core can be relieved to some extent during and after the cutting operation, but complete recovery to the unstrained condition is never fully realised. The characteristic most affected by the presence of the air gap is, of course, the magnetising current. Much can be done to improve the situation, however, by grinding and etching the cut faces. A further process of fine lapping can reduce the air-gap lengths to very small proportions. To a somewhat lesser degree the total iron loss is affected, but here almost complete recovery is experienced as a result of the grinding and etching operations. In this respect, the latter process is responsible for the removal of the interlaminar burring which occurs during cutting, thus reducing the eddy-current loss. Subsequent lapping to improve the mating of the cut faces tends to reintroduce this interlaminar contact, which, though of a lesser amount than before, is of some importance when frequencies higher than 50 Hz are employed. The grinding and lapping operations are not normally included in the manufacture of cut cores, since the cutting technique is now so developed that little more than an acid etch of the joint faces is required. The completed core, though having somewhat lower magnetic characteristics than the uncut spiral core, still has a remarkable performance, and in this form can allow the superior properties of grain-oriented material to be utilised in a convenient manner, with conventional windings.

A popular type of cut core is the C type core, which, as its name implies, involves the cutting of a rectangular strip wound core to form two C-shaped units. They can be manufactured to almost any combination of dimensions, and in any one of the available grades or strip thicknesses.

Because of the low values of iron loss and magnetising volt amperes, transformers of higher efficiency, smaller size and reduced weight, can be constructed for operation at power frequencies.

Two types of strip-wound core have been developed recently and these are illustrated in Fig. 2.8 (a) and (b). Both are manufactured by Telcon Magnetic

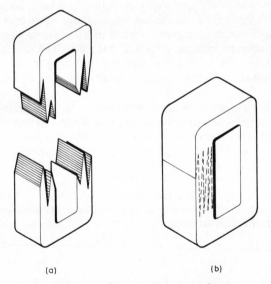

(a)   (b)

Fig. 2.8   Alternative types of strip-wound core

Cores Ltd. The Olmag core shown in (a) is a concentrically wound core with only an edge bonding, applied in the region of the cuts, to assist assembly of the core within the coil bobbin. Each individual lamination is cut and butted on itself, in both legs of the core, the butted joints of adjacent laminations being displaced from each other to give the appearance of a zigzag joint where the gap normally occurs in a C type core.

The Temcore construction shown in Fig. 2.8 (b) is also concentrically wound but is not bonded. Each individual lamination is cut and butted on itself and the joints of adjacent laminations are displaced from each other but, as none is located at the radiused corners of the core, it becomes necessary to sectionalise the radial thickness of the core into packets of from 10 to 20 laminations. The overall gap across the radial thickness of the core has a sawtooth pattern and combines the characteristics of the butt and overlapped joint because it is usual to provide a slight overlap to the last lamination of each packet. The overlap is not, however, an essential feature and the core can be obtained without it if required.

Both types of core illustrated are stress relieved by annealing in a controlled atmosphere after all the manufacturing processes are completed, and, owing to the virtual absence of bonding stresses and distribution gaps, the optimum properties of grain-oriented electrical steels are more closely approached.

CUT AND UNCUT CORES FOR SINGLE-PHASE RURAL TYPE TRANSFORMERS

In common with all other C type assemblies, these cores are manufactured initially from closed-circuit rectangular shaped units. The main difference in construction is one of a cross-sectional nature, for here, instead of confining the strip width to a fixed dimension throughout, as is commonly the case, several strip widths are employed to build up the core section in a number of distinct steps, after the manner of a full cruciform. The now accepted method of revolving tubes around the core legs, thus running the wire on to form the coils, has resulted in lightly bonded uncut cores being used in increasing numbers.

In general, the use of cut or uncut strip wound cores in grain-oriented silicon steel in distribution transformers has led to a weight reduction of the order of 10%, a saving in iron loss of as much as 30% and an improvement in magnetising current by an amount approaching 70%, against the corresponding characteristics experienced with transformers in which the cores are manufactured from laminations and built by interleaving. Figure 2.9 shows a typical single-phase rural type transformer core and windings, in which the core is of cold-rolled grain-oriented steel and the windings are interlocked to provide exceptional mechanical strength.

CUT CORES FOR THREE-PHASE APPLICATIONS

Extension of the cut-core technique to the production of cores for three-phase applications is easily achieved by winding strip round two closed-circuit rectangular cores. When the strip build-up is such as to provide a core construction having three limbs of equal cross section, forming, annealing and bonding operations are carried out, and a cutting process resolves the unit into two component E shaped parts.

Cut E type cores are manufactured in many standard and non-standard sizes, and most commonly in ribbon thicknesses of 0·33 mm and 0·10 mm respectively. They are somewhat inferior in performance to cut C type cores, however, because at high flux densities, as a result of third-harmonic flux, there is an increase in iron loss of the order of 25%. This arises from the fact that the E core construction is one in which the flux of each energising coil is divided into two separate components, a condition which can also give rise to an increase in magnetising current of approximately 30%. Even so, their characteristics

Fig. 2.9   Single-phase rural type transformer core and windings

are sufficiently outstanding to facilitate the production of smaller and lighter three-phase transformers for use in electronic and industrial equipment and installations where size and weight are limiting factors.

The application of the E core principle to distribution transformers as an alternative to the conventional in-line three-limb laminated assembly, has met

23

with little success in this country because of the disadvantages associated with cores in which there is distorted flux.

THE Y TYPE CORE

Recent developments in three-phase strip-wound core design have produced a core assembly in which low losses and exciting current characteristics can be obtained. This type of core is manufactured from two solidified rectangular shaped cruciform section cores by cutting the latter through the mid-point of the short sides and machining three of the resultant C shaped units to an angle of 120° at the cut faces. The three C pieces, when placed together, form a symmetrical Y type core in which there is no division of flux paths, and in which there is only one junction point common to all three legs. Great care must be exercised during the machining operation to ensure accurate mating of the joint faces, for otherwise some of the outstanding performance will be lost and a noisy assembly may result. Figure 2.10 illustrates a Y core.

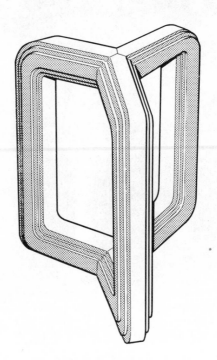

Fig. 2.10 'Y' type strip-wound core

It will be appreciated that former wound coils cannot be applied to the limbs of this core, although attempts to do this have been made by further cutting of the straight portion of the legs. This procedure is inadvisable, since it creates the difficult situation of having accurately to mate two sets of cut faces in planes at right angles to one another. The accepted methods of applying windings are either to revolve tubes round each limb, thus running the wire on, or to take each leg in turn, fix it between centres, and revolve both tube and limb together.

TYPICAL DATA

The data presented in Table 2.1 have been compiled to give some measure of comparison between the performance levels of the various types of strip-wound core. The two characteristics, viz., iron loss and magnetising volt-amperes, have been chosen as yardsticks, and all figures represent guaranteed rather than typical levels of performance. In many instances, acceptance tests

Table 2.1 MAXIMUM FLUX DENSITY 1·5 teslas

Ribbon thickness 0·33 mm. f = 50 Hz

Description	Possible ratings	Total iron loss— w/kg	Total r.m.s. excitation—VA/kg
Unimpregnated spiral and uncut rectangular cores in all grades.		1.02 1·12 1·24	1.46 1·68 2·00
Standard range cut C type cores	5 to 1600 VA	1·50	see note 2
Cores for 1-phase distribution transformers			
(a) Uncut	5 to	1·24	1·86
(b) Cut	50 kVA	1·32	3·85
Standard range cut E type cores	30 to 5000 VA	1·92	see note 4·00
Y type cores for 3-phase distribution transformers	25 to 500 kVA	1·32	3·85

would be carried out at flux densities higher or lower than those indicated here, but comparison is made easier by assuming those values used in sampling tests by the steel manufacturers.

Note: Cut C and E type cores in 0·33 mm ribbon
The effects of the air gaps are not included in the figures for total r.m.s. excitation, and it is necessary to add an allowance which varies with the core size. At the usual 50 Hz operating flux density, the magnetising VA for the core lags sufficiently behind the induced e.m.f. to enable the arithmetical addition of core VA and gap VAR to be made with negligible error.
    For cut C cores the allowance is:

$$\text{VAR/kg}_{gap} = \frac{57\,(\text{maximum flux density in teslas})^2}{\text{mean magnetic length in cm}}$$

and for cut E cores

$$\text{VAR/kg}_{gap} = \frac{(\text{maximum flux density in teslas})^2 A}{\text{weight in kg}}$$

where A is the net area of iron section in cm^2.

The sum of the effective lengths of the air gaps round the magnetic circuit is assumed to be equal to 0·032 mm.

PRODUCTION OF TRANSFORMER CORE LAMINATIONS

It will be apparent that the inherent properties of cold-rolled transformer steel will only be fully utilised by the production of core laminations fabricated to obtain the greatest possible benefits from the processing of the steel. With this in mind the remainder of this chapter is devoted to the cutting, punching and mitring of the core laminations and to the final stress relieving and insulating operations.

Figure 2.11 illustrates a hydraulically operated cutting line specifically designed to produce transformer laminations up to 5·0 m long and 0·80 m wide. Once the machine has been set to the requirements of a particular size and type of lamination, it will run automatically under the overall control of the operating console until the predetermined number of laminations have been produced. Stock, in the form of a coil of the steel, is loaded upon the mandrel of the decoiler unit, shown on the right of the machine, and fed down into the looping pit. The drive to the mandrel is actuated by a photoelectric cell in the pit. Mounted upon the main frame are a series of notching and punching tools, each independently adjustable and hydraulically operated. Beyond these tools is a gripper head which, when actuated, firmly takes hold of the strip and travels forward towards the cropping head. A stopping device fixes the exact length of each lamination before the cropping takes place. The cuts can be made at varying angles between 45° and 90° and to ensure that the length cut is accurate the sheet is fully tensioned before the final cutting takes place.

This machine has an output of 5 to 12 components per minute, the lower number being obtained at the maximum length, and will produce laminations within a tolerance of ±0·13 mm in length.

Beyond the cutting head is the unloader which can off-load laminations to either side alternately or to one side as required. Transfer to the unloader from the output side of the cutting head is effected by travelling magnetic pickups which move each plate from the cutting position to a point where automatic transfer is made to the unloading tables. These tables are lowered approximately 50 mm for every batch of 144 plates produced, thus ensuring that off-loading always takes place from the machine at a constant height from ground level.

A number of types of furnace have been designed for the final stress relieving of transformer laminations; the annealing furnace illustrated in Fig. 2.12 is of the roller-hearth type in which single laminations are loaded on to the roller conveyor and are given an annealing heat treatment cycle of 70 to 80 s at a temperature of 800°C. The roller-hearth type of furnace permits rapid heating to take place without distortion of individual laminations. It will be appreciated that core plate must be perfectly flat in order to build acceptable cores.

Each plate travels at a speed of approximately 6 m/min through the furnace, which has an overall length of 19 m, including the loading and take-off tables. The main heating section has three individually controlled zones with nickel-chromium alloy tape elements located in the roof and hearth. These zones have power ratings of 126 kW, 60 kW and 42 kW. In addition, a fourth zone, with

Fig. 2.11   Hydraulically operated automatic cutting line machine designed to produce transformer laminations. (Redman Heenan Ltd.)

Fig. 2.12   Roller hearth continuous stress-relieving furnace. (Metalectric Furnaces Ltd.)

coiled rod heating elements rated at 27 kW, is brought into operation when the furnace is started from a cold condition, to prevent the chilling of the laminations at the outlet end of the heating section before the refractory brickwork has reached the correct working temperature. This zone automatically disconnects when the furnace reaches the correct working temperature.

After leaving the fourth zone the laminations pass through an air-blast cooling section which reduces their temperature from approximately 500°C to one which will allow handling for unloading and stacking.

The roller conveyor is 1100 mm wide and, based upon an 80 % coverage, the furnace has an output of 750 kg per hour. The comparatively high speed at which annealing takes place results in a short heating and cooling cycle and it is not necessary to provide a protective atmosphere within the furnace. Installation of gas-producing plant is therefore unnecessary.

After the laminations have been annealed the final operation is to apply a surface coat of varnish or similar insulant. For small or medium transformers the insulant coating applied by the steelmaker is quite adequate, but for the larger cores it is considered necessary to supplement this initial coating after all the fabrication has been completed. Figure 2.13 shows a varnishing machine, stoving oven and cooling chamber. The laminations are fed into this through a pair of rubber-covered pinch rollers and then onto a roller chain conveyor through the heating and cooling sections.

The varnishing rollers operate in conjunction with a pressure roller and this group of rollers is adjustable so that the thickness of the varnish coating applied on each side, together with the plate thickness, can be set to the required position. All metal parts in contact with the insulant are made of stainless steel,

Fig. 2.13   Varnishing machine and stoving oven for insulating transformer laminations. (Controlled Heat & Air Ltd.)

including the storage and roller supply tanks. Both tanks are fitted with mechanical agitators to ensure that a consistent varnish is supplied to the rollers and no settlement takes place in the tanks.

Thermostatic temperature control of the gas-fired stoving section is incorporated together with electronic protection in the combustion chamber, to take care of flame or fan failure. Three heating zones are provided and each is arranged for an increasing temperature to a maximum of 350°C. The laminations take approximately 10 to 12 s to pass through each zone. After leaving the stoving chamber the laminations pass through a cooling chamber where cold air circulation reduces their temperature to one suitable for stacking.

Chapter 3

General types and characteristics

The main parts of a transformer may be listed as follows:
  (1) The magnetic circuit with its clamping structure.
  (2) The primary and secondary windings, and clamping arrangements.
  (3) Insulation of windings.
  (4) The leads and tappings from the coils, together with their supporting arrangements and the terminals.
  (5) The tank containing the transformer together with cooling surface and the insulant.

MAGNETIC CIRCUIT

The ideal shape for the section of the cores is a circle, since this would waste no space beyond that taken up by the insulation between laminations. A perfectly circular core section would, however, involve making a variation in dimensions for each successive lamination, which is possible but uneconomical. In practice, a compromise is effected by arranging the core section in steps in such a way that the net sectional area is a maximum for the number of steps employed and so that the corners of the various portions lie on a circle of predetermined diameter. Typical core sections having seven and fourteen steps respectively are illustrated in Fig. 3.1. The net sectional area is calculated from the dimensions of the various portions, and an allowance is made for the insulation and lost space between laminations. The yoke section is arranged similarly to that of the limb section, the dimensions of the various sections of the yoke being arranged so that the sectional area of each portion bears the same relation to the corresponding portion of the core. It is most important that this feature should not be overlooked, as otherwise there would be a tendency for cross fluxing to take place between the various sections of the core, which would cause excessive eddy currents to be set up. The limbs and yokes comprising the magnetic circuit are built up of laminations of specially alloyed electrical sheet steel, containing a percentage of silicon. In order that the steel should be suitable for the purpose, it should have a high electrical resistance and a high permeability at as high a flux density as possible. The hysteresis loss should also be as small as possible, while in order to reduce eddy-current loss to a minimum the laminations should be as thin as is practicable. The thickness should not, however, be reduced to such an

30

extent that the plates become weak mechanically, and consequently liable to buckle in handling. The material now used is grain-oriented cold-rolled steel having a thickness of 0·33 mm. The individual plates are insulated from one another by coating one or both sides of each plate with varnish or other insulating material, but the insulant must be completely oil proof. The limb laminations in the core are held together by stout webbing tape tightly applied on the smaller cores or by suitably spaced impregnated glass-fibre bands which are effectively secured in position by the heat treatment of the resin after application

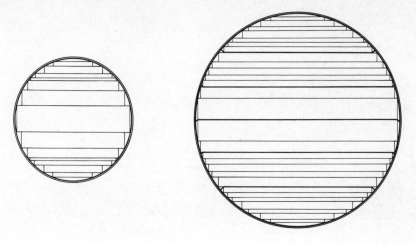

Fig. 3.1   Core sections. Seven step, taped (left); and fourteen step, banded (right)

on the core. The use of insulated bolts passing through the limb laminations has been superseded by this new method of construction, and core bolt failures on the limb section of cores are thereby eliminated. The top and bottom mitred yokes which complete the magnetic circuit are interleaved with the limbs and are clamped by steel sections held together by a number of insulated yoke bolts, the spacing and size of which are dependent upon the dimensions of the core assembly. In some cases, glass-fibre bands are also employed to clamp the yoke laminations, but this feature is determined by other design and constructional considerations.

The steel frames clamping the top and bottom yokes are held together by vertical tie bolts. These complete the clamping arrangement, and make the magnetic circuit a perfectly rigid structure.

When interleaving the yokes with the limb plates, the best arrangement, theoretically, is obtained by interleaving the plates one at a time. While this is certainly true when considering only the magnetic properties of the circuit, it is usually desirable for mechanical reasons to interleave the plates two, three or four at a time, as this minimises the risk of imperfect interleaving due to buckling of the plates.

The magnetic circuit is assembled on channel irons or supports welded to the main clamps. One yoke channel horizontal-iron is placed in position, and frame temporary plates for one side of the core are supported in their corresponding positions on channel irons suitably chosen to bring these plates to the correct level. The various bolts which ultimately hold the clamping structure together

Fig. 3.2   Three-phase mitred, taped core for a 750 kVA distribution transformer. The l.v. spiral coils are in position. (Bonar Long & Co. Ltd.)

are then placed in position, and the insulating pieces which separate the laminations from the frame are laid on. The cores and the bottom yoke are then built up from laminations cut to size. Finally, another layer of insulating board and the remaining core clamps and yoke frame are added, and the whole is bolted together by means of the bolts already in position.

Figures 3.2 and 3.3 show typical core assemblies.

PRIMARY AND SECONDARY WINDINGS

There are four types of coils used on core type transformers, these having the following designations:

(1) spiral type
(2) crossover type
(3) helical type
(4) continuous disc type.

(1) Spiral type

These coils are only suitable for windings carrying a heavy current, and are almost always used for l.v. windings.

Spiral coils are also suitable for h.v. windings when the current to be carried is of sufficient magnitude. In general, spiral coils are not normally used for currents of less than 100 A. They consist of layers wound in a continuous length from top to bottom of the coil, and the composite conductor consists of a number of square or rectangular strips in parallel. The individual strips comprising the conductor are taped together just before passing on to the mandrel, this tape serving the double purpose of holding the strips together during the process of winding and protecting them from subsequent damage before and during the course of erection. Spiral coils are former wound or are wound direct on to solid insulating cylinders, and are therefore of robust mechanical construction. Edgeblock packing consisting of tapered strips of insulation is wound into the ends of each coil at each layer, in order to form the coil into a true cylinder and to give mechanical support to the coil when assembled on the core limb. In winding, a sufficient length of conductor is left at each end for connection to the leads. This type of coil lends itself well to reinforcement of the insulation between turns, since to provide this additional insulation it is only necessary to wind strips of pressboard or other suitable material between adjacent turns.

Fig. 3.3   Three-phase mitred core of a 30 MVA, 132 kV, 50 Hz transformer showing glass-fibre banding of core limb laminations. (Bonar Long & Co. Ltd.)

The normal insulation between turns may consist of strips of insulation in addition to the paper covering on the conductor, or in some cases the latter may be augmented by taping the conductor. Where more than one conductor is arranged in the radial direction, it is necessary to introduce transpositions

Fig. 3.4   Two-layer spiral coil

throughout the length of the winding to minimise the effect of resistance and leakage reactance.

In small transformers spiral coils are sometimes wound with multilayers if the number of turns renders it difficult to arrange a satisfactory single layer coil. This type of coil might also be regarded as a special case of the crossover type. Figure 3.4 shows a typical two layer spiral coil.

(2) Crossover type

This type of winding is suitable for currents up to about 20 A, and is very largely used for the h.v. windings of distribution transformers. Crossover coils are wound on formers and each coil consists of a number of layers having a number of turns per layer, the conductor being a round wire or a strip insulated with a paper covering. Except at the ends of windings adjacent to terminals and at open ended tapping leads, it is not usual to apply any extra insulation to the conductor itself in this type of coil apart from the paper covering; between layers, however, it is customary to wind in one or two layers of some flexible insulation such as paper. This insulation between layers is usually wrapped round the end turn of the layer, thereby assisting to keep the whole coil compact. The complete winding consists of a number of such coils in series. They are spaced apart by means of insulating key sectors, and in order to avoid internal joints the inside leads are heavily insulated and brought out through the spaces between these sectors. The axial length of each coil is usually about 75 to 100 mm though it may vary from this dimension according to the voltage and radial depth of the winding, while the distance between adjacent coils is approximately 6 mm,

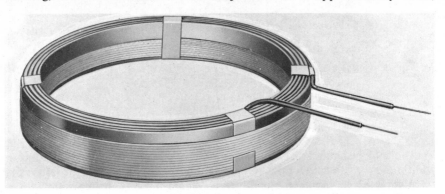

Fig. 3.5   Crossover coil

34

Fig. 3.6   Three-phase 750 kVA, 11 000/433 V, 50 Hz transformer with a mitred core. H.V. windings in delta, l.v. in star. Crossover h.v. coils; spiral l.v. coils. H.V. windings fitted with tappings brought up to an externally operated tap selector. (Bonar Long & Co. Ltd.)

unless it is desirable to increase this distance on account of the voltage consideration, or to adjust the overall length of the complete winding stack. Figure 3.5
shows a typical crossover coil, and Fig. 3.6 a complete three-phase winding
assembly including the core.

(3) Helical type

This coil, as its name indicates, is wound in the form of a helix, and consists of a
number of rectangular strips wound in parallel radially so that each separate turn
occupies the total radial depth of the winding. The helical coil, in effect, covers the

35

intermediate range of current and total winding turns between the heavy current spiral coil and the multiple conductor disc coil. Each turn is wound on a series of key spacers which form the vertical oil ducts, and each turn or group of turns is spaced by radial key sectors. A system of oil ducts is therefore provided in the horizontal and vertical directions and each separate conductor of each turn is in contact with the oil. This type of coil is eminently suitable for the lower-voltage windings of the larger transformers at voltages of 11 to 33 kV. When used in such an application, the corresponding higher voltage winding would most likely be provided with adjusting tappings for voltage variation and it would be necessary to arrange for the ampere-turn balance on the lower voltage winding. This 'thinning out' of the ampere-turn distribution opposite the higher voltage tappings is very readily achieved in the case of the helical coil and is obtained by increasing the thickness of the radial sectors in order to obtain the optimum effect. Figure 3.7 is an illustration taken during the winding stage of

Fig. 3.7   Helical coil

a helical coil. The individual parallel conductors can be seen clearly together with the method of 'thinning out' the winding at approximately one-quarter and three-quarters of the axial coil length. Transpositions are introduced at appropriate intervals to ensure uniform current distribution.

(4) Continuous disc type

These coils, as their name implies, consist of a number of discs wound continuously from a single wire or a number of strips in parallel. Each disc consists of a number of turns wound radially over one another, the conductor passing uninterruptedly from disc to disc. Whenever possible the conductor is used in such drum lengths as will be sufficient for a complete winding, or section of a winding between tappings, but such joints as are necessary in manufacture should be made at the outside of the winding and be electrically welded or brazed. The conductor may be either a single rectangular strip or a number of rectangular strips in parallel, wound on the flat. This reduces considerably the risk of the strip twisting slightly in winding and thereby making an unsatisfactory disc. With a multiple strip conductor, transpositions are made at intervals to ensure uniform current distribution. The discs are wound on to an insulating cylinder, spaced from it by strips running the whole length of the cylinder; and separated from one another by hard pressboard sectors keyed to the vertical strips. These vertical and horizontal spacers provide a system of ducts giving unrestricted circulation of the cooling oil, with which every turn of the winding is in direct contact. The keying of the spacers ensures that these cannot become dislodged in service. The whole coil structure is mechanically sound and rigid and capable of resisting the stresses set up under the most onerous short circuit conditions.

Insulation reinforcement, where necessary, is provided by increasing the

(a)   (b)

Fig. 3.8   Continuous disc coils. Taps brought out on left-hand coil for connection to an on-load tap changer

thickness of the paper covering of the conductor, or by lapping with varnished cambric tape.

Tappings, if required, are taken from the outside of the discs, the number of turns of the tapped discs being arranged to suit the number of turns between tappings.

Figure 3.8(a) shows a typical single conductor continuous disc coil wound upon its insulating cylinder, while Fig. 3.8(b) shows a similar coil but wound with two conductors in parallel.

COIL TREATMENT

The windings of power transformers and particularly those of the larger ratings are not generally impregnated before assembly on the core. Also the older practice of impregnating coils with varnish in a vacuum drying and pressure process cycle has now been largely superseded by the use of transformer oil as an impregnant.

Manufacturers now fully understand the problems and importance of adequately drying the insulation structure of transformers, and although the techniques may vary from one factory to another, the drying and pre-shrinking of coils and, later, the oil impregnation of the insulation, follow a clearly defined drying cycle. In general there are three main stages in preparing the windings for oil impregnation. These are, first, the pre-drying and shrinking of the coils; second, further drying until the required insulation resistance is obtained, and finally, the removal of vapour and gas from the assembled core and windings prior to oil impregnation.

The first two stages are usually accomplished at atmospheric pressure in an oven through which hot air is continuously circulated, but care must be taken to ensure that the oxidation of the insulation does not take place. For distribution transformers this method of a drying process cycle employing the two initial stages described is usually perfectly satisfactory and is commonly used.

For transformers of larger ratings and all high-voltage units, it is essential that stage three is included in order to ensure that gas pockets in the insulation are avoided. The processing must be carried out at a temperature of approximately 100°C and a vacuum of at least 0·5 mm.

When a transformer is being processed it is not possible to specify an exact or rigid time cycle, and it is a matter of experience gained over a period to determine when a drying cycle has been completed. It is usual to monitor the temperature and vacuum continuously whilst drying proceeds, together with the insulation resistance and power factor of the windings. In addition, when water ceases to be collected in the exhaust condenser on the vacuum system and the appropriate values of insulation resistance and power factor have been obtained, it can be taken that the drying has been effectively completed.

At this stage, and whilst the core and windings are still under vacuum, de-aerated and dehydrated transformer oil at a temperature of 65° to 75°C, is admitted to the transformer tank, and the windings are completely flooded.

In the following illustrations in this chapter many interesting examples of the various types of coils described are shown assembled on cores.

Adjustable coil supports

In the larger transformers, and especially those connected to large power systems, it is desirable to provide means for taking up any slight shrinkage of insulation that sometimes occurs. In coils of the spiral type shrinkage is negligible, but in other types of windings, even after the most exacting factory treatment, a certain amount of shrinkage may occur after prolonged service. If this shrinkage is not taken up, serious damage may be caused by movement of the coils when short circuits take place. (See Chapter 27 for detailed consideration of electromagnetic forces.)

To take up the shrinkage, substantial clamping rings are placed at the tops of the windings, being pressed down upon them by means of adjusting screws passing through the lower flanges of the core frames which clamp the top yoke of the core.

It is most important that the clamping structure should be arranged so that there is no possibility of circulating currents arising due to electrically connected portions of the structure forming completely closed turns round the magnetic cores. For this reason metal clamping rings (when used) are each provided with two gaps, while, in addition, the adjusting screws are arranged to bear directly on to the rings, and so provide effective earthing. Horizontal movement of the clamping rings is prevented by means of steel pins, which are screwed through the core frames clamping the top yoke of the core, and which are seated in cups welded onto the steel rings. As an alternative to this arrangement, insulating clamping rings are often provided. These have the advantage that they form part of the end clearance, and so do not necessitate an increase in the length of the limb, while being of insulating material, there is no possibility of circulating currents being set up in them.

As a general rule it is necessary to provide adjustable coil supports on the h.v. windings only. If, however, the voltage of the l.v. winding is such that the winding is similar in character to h.v. windings, it may be desirable to provide for taking up shrinkage in this winding also. If the h.v. and l.v. windings are similarly constructed, the adjustable coil supports may consist of single rings clamping both windings, but if unequal shrinkage of the two windings is likely, separate rings are provided. In order to reduce subsequent shrinkage to a minimum, all windings are tightened progressively during the coil drying out process in the ovens, and also when the completed transformer is finally dried out before being placed in its tank.

As a safeguard against any possible slackening of the adjusting screws due to vibration, locking devices are fitted to these screws.

INSULATION OF WINDINGS

The insulation of the windings may be divided into the two main classes, major and minor.

(a) Major insulation

This comprises the insulating cylinders between the l.v. winding and the core, and between the h.v. and l.v. windings, the insulating barrier which is inserted between adjacent limbs when necessary, and the insulation between the coils and the core yokes. The cylinders consist of a number of layers of specially selected pressboard or synthetic resin bonded cylinders. These cylinders are manufactured by winding the paper which is already coated with the varnish on to a mandrel, which is maintained at a fairly high temperature. In order to keep the paper compressed during the process of winding, it is also usual to arrange a second heated mandrel in such a way that it presses on to the paper along the line where the latter passes on to the first mandrel. The cylinders, together with the mandrel, are subsequently baked, and when finally allowed to cool, the cylinder is easily removed owing to the contraction of the mandrel. Cylinders made in this way are not only excellent insulators, but they are also very strong mechanically. The end insulation usually consists of a series of pressboard washers and spacing blocks, or of built-up insulating cylinders, having inside and outside diameters corresponding to those of the coils to be supported. This latter type of end insulation is particularly useful when the coils are of the spiral type. The continuity of the oil ducts between windings is preserved by means of suitable end washers, to which are fastened spacing blocks varying in thickness according to the size of transformer and the number and size of oil ducts for which provision is to be made.

(b) Minor insulation

This is the insulation on individual turns and between layers, as described under the various types of coils. The two principal materials used for interlayer insulation are pressboard and varnished cloth, and the methods of manufacturing them are as follows:

Pressboard

The materials for the manufacture of pressboard are jute, cotton and similar vegetable fibres or an admixture of each, the mixture being dependent upon the ultimate usage. These materials are manufactured by the 'intermittent board' machine process and are afterwards subjected to great pressure in order to remove excess water and produce a very dense material. Thicknesses up to 0·10 mm in widths up to 1·8 metres are made in continuous rolls, whilst the thicker boards are now made in sheets suitable for use on the largest transformers. Where the pressboard is not folded in the form of bent washers in the transformer assembly, it is possible to use the denser or pre-compressed board, for instance as in the case of coil sectors and washers. The advantage of this material is that it removes virtually all the axial shrinkage in the transformer winding.

Varnished cloth

Before varnishing, the cloth, a good quality cambric, is treated with alkali to protect it against fatty acids which may develop in the varnish. The cloth is then passed through a pan of varnish, and thence over a series of rollers arranged at the top and bottom of a steam heated tower. The temperature in the tower is maintained at from 75°C to 80°C, and the rate of progress of the cloth is from 25 to 150 mm per minute. By the time the cloth has passed through the tower it is sufficiently dry to be made up into rolls immediately. When this cloth is required for use as tape, it is cut into strips, and it is usual to cut at an angle of 45° to the direction of the threads. The reason for this is that if the strips are cut in the direction of the threads they are liable to break in handling, while 'bias' cut tape is much easier to wrap onto the conductor.

LEADS FROM COILS, ETC.

When the coils are assembled, the leads from the top and bottom of the windings and from such tappings as may be provided, according to the customer's requirements, are brought out to a length of a few centimetres only. The electrical connections from these leads to the terminal boards or bushings consist either of copper rod, copper strand or rectangular-section copper strip, depending upon the current they have to carry. If they consist of copper rod, they are insulated at the cleats by means of Bakelite tubes, the thickness of which is determined by the voltage of the winding to which they are connected. These tubes pass through insulating cleats which have previously been drilled to such a diameter that they hold the tubes firmly. The cleats are supported from the vertical tie rods passing through the top and bottom yoke clamps. When the connections from the coils consist of rectangular-section copper strip, which is often used on the l.v. side, it is not necessary to provide any insulation other than the insulating cleats themselves, which hold the leads in position, although it is customary to wrap each lead with linen tape or varnished cloth at the point where it passes through the cleat, in order to avoid any risk of chemical action between the cleat and the copper lead.

Leads from tappings are normally brought to a point just below the cold oil level, and so arranged that tappings may readily be changed by means of the link device provided.

Bushing insulators

These are made of porcelain, although in high-voltage transformers for indoor use, terminals built up of paper and Bakelite varnish are occasionally employed. The porcelain should be of the best electrical quality with a highly glazed surface. Generally speaking, the surface of the insulator should conform approximately to the shape of the dielectric field. In the case of transformers for indoor installation the porcelain contour should be as simple as possible. When they are for outdoor service, however, it is necessary to provide watersheds. Porcelain insulators for higher voltages are oil filled in order to take advantage of the higher permittivity and higher dielectric strength of oil, as compared with air.

When tappings are provided and the leads are brought to the outside of the tank, the bushing insulators are similar to those for the main phase leads, the only difference being that a number of tapping leads suitably insulated from one another may be brought through one insulator, as the voltages between tapping leads are relatively low.

Bushing insulators for very high voltages are sometimes of the capacitor type. These insulators consist of a number of alternate layers of paper, together with Bakelite varnish and metal foil. The various dimensions are proportioned

41

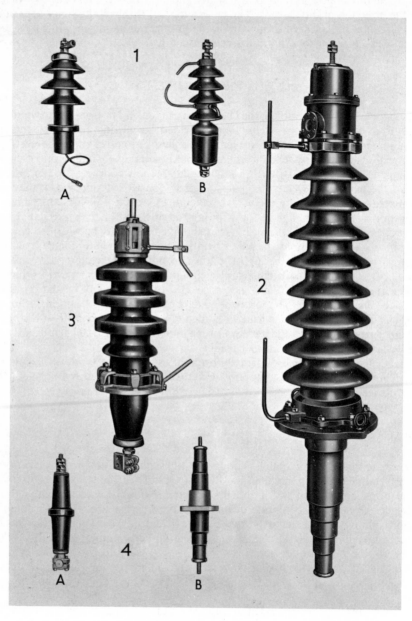

Fig. 3.9 1. (A) 11 kV outdoor pull-through type bushing
              (B) 11 kV outdoor type
            2. 88 kV outdoor capacitor type terminal
            3. 33 kV outdoor bushing
            4. (A) 11 kV standard cable box bushing
              (B) 33 kV capacitor bushing for cable boxes

so that the capacitance between adjacent layers of foil is approximately constant, and consequently the insulating layers are all equally stressed. The insulator is wound on to a brass tube which serves a triple purpose, being the mandrel for winding, the inner plate of the first capacitor, and the conductor. Insulators of this type are suitable for indoor service, but they may be rendered weatherproof by providing an outer shell of porcelain having the necessary watersheds.

A few typical terminals, both for indoor and outdoor use are shown in Fig. 3.9.

TANK AND COOLING SURFACE

For self-cooled transformers the main types of tank in modern use are:

(1) Plain sheet steel
(2) Boiler plate with external cooling tubes
(3) Radiator tanks
(4) Tanks with separate coolers.

(1) Plain tanks

These are used when the size of the tank to give the necessary electrical clearances from the transformer windings is such as to give sufficient cooling surface to dissipate the heat generated by the transformer. The output above which it is usually necessary to provide additional cooling surface is about 50 kVA for normal voltages. Up to this output, therefore, plain sheet steel tanks are almost always used. Owing to the fact that they are generally used with small transformers, the thickness of the sides need only be 3 mm. The four sides of the tank are usually made from one sheet of steel, and consequently there is only one vertical joint which is electrically welded. The base is of similar material, while at the top it is usual to form a flange as a suitable bearing surface for the cover.

(2) Boiler plate tubular tanks

This type of tank is used for all sizes or self-cooled distribution transformers except the very small ones as outlined in (1). As its name implies, the additional cooling surface is obtained by means of external tubes which are welded to the tank walls. The walls of the tank consist of boiler plate, the thickness of which varies from 5 to 10 mm, according to the tank dimensions. It is also usual to weld stiffeners to the sides of the tank at intervals of about 1 m in order to prevent bulging when the tank is transported full of oil. The tubes are arranged about 8 cm apart, centre to centre, and if there is more than one row, the rows are placed about 10 cm apart vertically. The tubes are electrically welded on the inside of the tank. All the remaining joints of the tank are electrically welded on both inside and out. The bottom of the tank consists of a sheet of boiler plate somewhat thicker than the sides, together with a suitable under base. Tank covers are formed from a sheet of boiler plate 3 to 6 mm thick, according to the size of the tank. The tubes are 1·6 mm thick and 5·0 cm in diameter.

In the larger transformers it is sometimes found impossible to accommodate on the tank walls the requisite cooling surface by using round tubes. For example, certain types of on-load tap-changing switches may utilise the whole of the h.v. side of the tank, preventing the fitting of tubes to this side. Alternatively, a low temperature rise may be specified needing greater cooling surface than otherwise would be necessary. In both cases it may be impossible to accommodate sufficient round tubes to dissipate the transformer losses within the required temperature rise limits, and then tubes of elliptical section were used. Such tubes have the same peripheral length as the corresponding round tubes, but owing to the fact that they can be spaced at smaller centres, more tubes can be used and greater cooling surface thus obtained.

Transformers, the tanks of which are fitted with elliptical cooling tubes, are shown in Figs. 3.40 and 3.44.

(3) Radiator tanks

When the rating of the transformer exceeds about 5000 kVA it is no longer possible to accommodate sufficient cooling surface, even in the form of elliptical tubes, on the tank sides without a wasteful increase in tank dimensions and oil quantity. In order to overcome this difficulty it is usual to fit detachable radiators to the tank. They consist of a series of separate elliptical tubes, or a pressed-steel plate assembly formed into elliptical oil channels, welded into top and bottom headers. The complete radiator is bolted to the tank wall and two isolating valves are fitted at the oil inlet and outlet. This arrangement permits the radiators to be assembled on site and the main tank containing the core and windings to be transported to site filled with oil. In cases where transformers are to be exported it sometimes happens that the inland transport abroad imposes a restricted rail gauge limit on the design. By adopting detachable radiators for the cooling surface it is possible to obtain smaller shipping dimensions than would be the case with a tubular tank, and a larger rating of transformer may then be delivered to site than would be the case with tubes on the tank wall. An additional advantage with detachable radiators, is that for maintenance and painting they can readily be removed by closing the appropriate isolating valves and draining the oil, when work can conveniently be undertaken at ground level. Figure 3.45 is an illustration of a typical radiator tank design.

(4) Tanks with separate coolers

Eventually there comes a time with increasing ratings when the tank is not sufficiently large to permit enough detachable radiators to be bolted to the tank side and the designer is forced to mount the coolers quite separately from the tank proper, and to pipe them to the tank by large diameter oil pipes. The coolers may be built in the form of a number of detachable radiators mounted vertically to horizontal headers. Figure 3.48 illustrates a transformer with this arrangement. The separate cooler is frequently employed for the natural cooling of oil immersed transformers, but in addition it forms an ideal arrangement for many forms of forced cooling. By the fitting of a number of fans beneath the detachable radiator bank, an appreciable increase over and above the natural dissipation is obtained. This method is known as air-blast cooling (B.S. type

ONAF). By the insertion of a flooded oil pump in the return cold oil pipe to the tank from the cooling bank it is possible also to supplement the dissipation from the radiator surface as compared with natural oil circulation. This method is of course the B.S. type OFAN. In both of these cases it will be appreciated that natural cooling of the transformer is available without the auxiliary fan or pump in operation and by suitable design it is possible to vary the rating at which the transformer will operate under natural conditions to suit varying specifications and operating requirements. Where a combination of natural and forced cooling is used the types are designated ONAN/ONAF, ONAN/OFAN, ONAN/OFAF, etc. It is common practice on the larger transformers to divide the total cooling surface into two parts, each having a separate oil circuit, independently controlled by valves. This sub-division may take the form of, say, two 50% rating coolers instead of one 100% unit. The advantage of this method is that one section of the cooler can be kept in operation and thus the transformer on part full load, in the event of a failure of either a cooler section or of the auxiliary equipment. In those cases where natural cooling is not required and the transformer is to be solely forced cooled, two other methods are recognised, namely, types OFAF and OFWF. Here, very compact external coolers are connected by oil pipes to the transformer tank. In the former the oil is pumped through a radiator which has a strong blast of air passing over the oil-filled cooling surfaces, whilst in the latter type the cooler consists of a number of internal tubes through which the oil circulates, whilst water is pumped over the outside of these tubes and the cooler acts as a heat exchanger with the heat flowing from the oil to the water. Figures 3.46 and 3.47 show external views of type OFAF cooling.

Tanks for outdoor transformers

These may be of any of the various cooling types mentioned. There are certain precautions, however, which must be taken to ensure watertightness. Although standard transformers for indoor use are practically watertight, it is not usual to take any special precautions to ensure their so being. The points at which water would tend to enter in the event of such transformers being exposed to heavy rain, are at the bolts holding down the cover and at the joints between the bushing insulators and the cover or sides of the tank. In outdoor transformers the cover bolts are closely spaced and a substantial tank flange of ample width and a Neoprene bonded cork gasket between the tank flange and the cover is provided.

The bushing insulators are dealt with in various ways, according to the size and h.v. and l.v. voltages of the transformer. Small pole mounting transformers have pockets formed on the walls of the tank, whilst larger transformers and transformers for high voltages have the insulators mounted on the cover or on side pockets. The joints are rendered watertight by means of Neoprene bonded cork gaskets.

On all high-voltage transformers and also on large units, a conservator is fitted. This consists of an expansion vessel connected to the main tank and arranged to maintain a slight head of oil in the main tank under all conditions of ambient temperature and of load on the transformer. The most important advantage of this fitting is the added protection given to the oil, by avoiding oil deterioration and the reader should refer to Chapter 17 for greater detail regarding this subject.

It is difficult to lay down specific rules for the fitting of a conservator to a given transformer but the present tendency is to consider its inclusion on units rated above 1000 kVA and for units designed for system voltages of 33 kV and above.

The remainder of this chapter is devoted to illustrations of typical transformers from the smallest to the largest sizes. These are shown with different types of tanks and with different terminal arrangements, and are typical of modern practice in the design of power transformers.

Fig. 3.10 Three-phase, outdoor platform mounting transformer. 50 kVA,
11 000/433 V, 50 Hz. H.V. windings in
delta, l.v. windings in star. Crossover
h.v. coils. Spiral l.v. coils. H.V. windings
fitted with plus and minus $2\frac{1}{2}\%$ and 5%
tappings brought up to an externally
operated off-circuit tapping selector.
View shows transformer in its tank,
looking on the h.v. side

Fig. 3.11 Single-phase
outdoor pole mounting
transformer. 15 kVA,
11 000/250 V, 50 Hz with
wound core of cold
reduced grain-oriented
steel strip. Tappings provided on the l.v. winding
to vary the secondary
voltage by plus or minus
5% are brought out to
separate bushings and
may, if required, be used
to obtain a constant out-
put with variation of the
primary voltage. View
shows the h.v. side

47

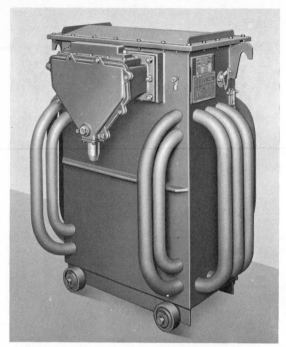

Fig. 3.12 Three-phase outdoor core-type transformer. 50 kVA, 3300/415 V, 50 Hz. H.V. windings in star, l.v. windings in delta. Crossover h.v. coils. Spiral l.v. coils. H.V. windings fitted with plus and minus 2½% and 5% tappings brought up to oil level. View looking on l.v. side. Tank fitted with cable boxes

Fig. 3.13 Three-phase indoor type transformer fitted with silica-gel breather 500 kVA, 6000/420 V, 50 Hz. H.V. and l.v. windings star connected. H.V. windings fitted with plus and minus 2½%, 5% and 7½% tappings brought up to an externally operated off-circuit tapping selector. View shows transformer in its tank, looking on l.v. side. Tank fitted with h.v. cable box and bushings on the l.v. side

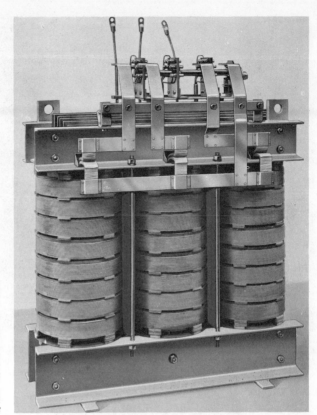

Fig. 3.14 Three-phase core-type transformer. 750 kVA, 11 000/433 V, 50 Hz. H.V. windings in delta, l.v. windings in star. Crossover h.v. coils. Spiral l.v. coils. H.V. windings fitted with tappings brought up to an externally operated off-circuit tapping selector. View shows transformer out of its tank, looking on l.v. side

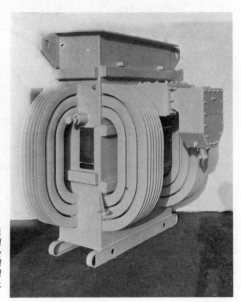

Fig. 3.15 Three-phase 6600/440 V, 750 kVA hermetically sealed transformer with a welded tank cover. Cable boxes and disconnecting chamber fitted with on h.v. and l.v. sides and off-circuit tapping selector providing plus and minus $2\frac{1}{2}\%$ and 5% variation. (Bonar Long & Co. Ltd.)

Fig. 3.16   Three-phase 33 000 V, 50 Hz earthing transformer designed for an earth fault current of 1050 A for 30 seconds and having a star connected 415/240 V auxiliary winding continuously rated at 220 kVA. This winding has all ends brought to a link board to enable the l.v. phasor group to be changed.
(Bonar Long & Co. Ltd.)

Fig. 3.17   Three-phase core-type transformer. 1750 kVA, 50 Hz, 11 000/3300 V, connected delta star. Disc h.v. and l.v. coils. H.V. winding fitted with tappings brought up to an externally operated tap selector (Bonar Long & Co. Ltd.)

Fig. 3.18 Three-phase 750 kVA factory assembled substation with h.v. and l.v. equipment directly connected to a 750 kVA, 11 000/433 V transformer. The view shows the transformer and the l.v. feeder pillar mounted on a common base. (Bonar Long & Co. Ltd.)

Fig. 3.19 Three-phase 1000 kVA, 11 000/433 V factory assembled substation with transformer, high voltage switch fuse and Bonar Long feeder pillar. The unit is designed to be transported to site ready for h.v. and l.v. cabling. (Bonar Long & Co. Ltd.)

Fig. 3.20   Substation described in Fig. 3.19 with one section of its enclosure removed to give access to the h.v. switch fuse. The operating platform can be clearly seen. (Bonar Long & Co. Ltd.)

Fig. 3.21   Substation with its external cladding in position. The complete unit can be shipped to site for installation. (Bonar Long & Co. Ltd.)

Fig. 3.22   Three-phase core and windings for a 90 MVA, 132 000/33 000 V, 50 Hz transformer. Connected star/delta with continuous disc h.v. and l.v. windings fitted with plus 10% to minus 20% tappings in 18 steps of 1·67% at the neutral end of the h.v. winding. View shows l.v. side and tapping connection. (Bonar Long & Co. Ltd.)

Fig. 3.23   View from the h.v. side of the 90 MVA, 132 000/33 000, 50 Hz core and windings in Fig. 3.22 (Bonar Long & Co. Ltd.)

53

Fig. 3.24 Magnetically shielded 15 MVA, three-phase, 31·5 kV feeder reactor with half the shields removed to show the continuous disc coils. The illustration shows the reactor out of its tank. Tank fitted with conservator, breather, Buchholz protective device, explosion vent and cable boxes on incoming and outgoing sides.

Fig. 3.25 High-voltage, oil testing set. 3 kVA, 235/50 000 V, single-phase, 50 Hz testing set consisting of a main testing transformer, panel carrying main switch, voltmeter and change-over switch indicating lamp, regulator hand wheel and fuses

Fig. 3.26   Three-phase 240 MVA, 400/132 kV auto-transformer with 132 kV tapping winding having a range of 20%. The tappings are controlled by a fully insulated 132 kV tap changer. (Hawker Siddeley Power Transformers)

Fig. 3.27   Three-phase outdoor-type power transformer. 1000 kVA, 11 000/400 V, 50 Hz. Fitted with conservator, cable boxes and silica-gel breather. View shows transformer in its tank looking from l.v. side

Fig. 3.28 Three-phase indoor core-type power transformer. 500 kVA, 6300/400 V, 50 Hz. H.V. windings in delta, l.v. windings in star. H.V. crossover coils. Spiral l.v. coils. H.V. windings fitted with plus and minus 2½% and 5% tappings brought up to oil level. View shows the transformer in its tank, looking on the h.v. side

Fig. 3.29 Three-phase kiosk core-type power transformer. 500 kVA, 6300/440 V, 50 Hz. H.V. windings in delta, l.v. windings in star. H.V. crossover coils. Spiral l.v. coils. H.V. windings fitted with plus and minus 2½% and 5% tappings brought up to oil level. View shows the transformer in its tank, looking on the l.v. side

Fig. 3.30 Single-phase outdoor core-type power transformer. 2·5 MVA, 50 000/11 000/√3 V, 50 Hz. Continuous disc h.v. coils. Spiral l.v. coils. L.V. windings fitted with taps to compensate for plus 6% to minus 14% variation in h.v. voltage, controlled by an on-load tap changer. This transformer is one unit of a 7·5 MVA, 50 000/11 000 V, delta/star connected bank

Fig. 3.31   Three-phase core-type interbusbar transformer for a C.E.G.B. power station. 13 MVA, 11 000/3300 V, 50 Hz. H.V. windings in delta, l.v. windings in star. Continuous disc h.v. coils and helical l.v. coils. View shows l.v. risers and h.v. tappings controlled by a high-speed resistor type on-load tap changer

Fig. 3.32 Three-phase outdoor core-type power transformer. 1000 kVA, 66 000–33 000/11 000 V, 50 Hz. H.V. and l.v. windings in star. Crossover h.v. coils. Continuous disc l.v. coils. H.V. windings in series-parallel for 66 000 or 33 000 V. L.V. windings fitted with plus and minus $2\frac{1}{2}\%$, 5%, $7\frac{1}{2}\%$ and 10% tappings. View shows two transformers in their tanks. Weatherproof bushings are fitted in the covers and the transformers are complete with dial type thermometers, conservators, and Buchholz protective devices

Fig. 3.33   Three-phase outdoor core-type power transformer. 5·6 MVA, 10 500/3150 V, 50 Hz. H.V. windings in star, l.v. windings in delta. Continuous disc h.v. coils. Helical l.v. coils. H.V. windings fitted with plus and minus $2\frac{1}{2}\%$ and 5% tappings brought up to an externally operated off-circuit selector. View shows the transformer out of its tank, looking at the l.v. side

Fig. 3.34 Three-phase outdoor core-type power transformer. 6 MVA, 100 000/11 400 V, 50 Hz delta/star connected. View on h.v. side

Fig. 3.35   Three-phase outdoor core-type power transformer. 10 MVA, 33 000/3300 V, 50 Hz. H.V. windings in l.v. windings in star. Continuous disc h.v. coils, spiral l.v. coils, h.v. windings fitted with plus and minus 2½% and 5% tappings brought up to an externally operated off-circuit selector. View shows the transformer out of its tank, looking at the l.v. side

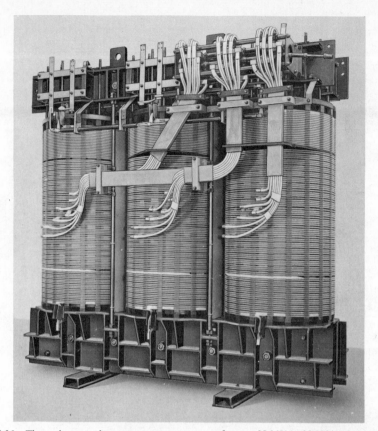

Fig. 3.36 Three-phase outdoor core-type power transformer. 25 MVA, 22 000/11 000 V, 50 Hz. H.V. windings in star, l.v. windings in delta. Helical h.v. and l.v. coils. H.V. windings fitted with tappings to give 20 000 V to 22 500 V, in 500 V steps. Tappings brought up to an externally operated off-circuit selector. View shows the transformer out of its tank, looking at the h.v. side

Fig. 3.37 Three-phase flameproof dry type mining transformer. 300 kVA, 3300/565–1130 V, 50 Hz. An air-insulated flameproof circuit breaker is fitted to the h.v. incoming side, and the l.v. chamber contains earth leakage protective equipment. The flanged rollers are designed for passing over uneven rail tracks with small-radius bends (Bonar Long & Co. Ltd.)

63

Fig. 3.38   Three-phase 15 MVA, 22/11–6·6 kV, system transformer. The h.v. winding has a tapping range of 20% and the tappings are arranged for connection to a linear tap changer. (Hawker Siddeley Power Transformers Ltd.)

Fig. 3.39 Three-phase 33/11 kV continuous emergency rated transformer fitted with high-speed resistor tap changer. The ONAN/OFAF cooling bank is connected to the tank via valves seen at right of illustration. (Ferranti Ltd.)

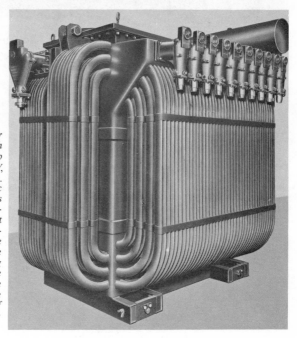

Fig. 3.40 Three-phase outdoor core-type power transformer with oil conservator and on-load tap changing gear, 3 MVA, 10750/550 V, 50 Hz. H.V. windings in delta, l.v. windings in star. Continuous disc h.v. coils. Spiral l.v. coils. Tappings provided on the h.v. winding covering plus and minus 5% in eight 1¼% steps. Terminal outlets consist of a trifurcating box on the h.v. side and of single-core cable boxes on the l.v. side for taking three single-core cables per phase and one single-core cable for the neutral. Tank cooling tubes are elliptical in cross-section. View of transformer inside tank, looking on l.v. side

65

Fig. 3.41 Three-phase outdoor core-type power transformer. 20 MVA, 33 000/11 000 V, 50 Hz. H.V. windings fitted with plus and minus $2\frac{1}{2}\%$ and 5% tappings brought up to an externally operated off-circuit tapping selector. View shows transformer out of its tank looking from h.v. side. This transformer is also rated as 25 MVA with ONAF cooling

Fig. 3.42   Three-phase outdoor core-type power transformer. 20 MVA, 33 000/11 000 V, 50 Hz. H.V. windings in delta, l.v. windings in star. Continuous disc h.v. coils. Helical l.v. coils. H.V. windings fitted with plus and minus 2½% and 5% tappings brought up to an externally operated off-circuit tapping selector. View shows tappings and l.v. connections

Fig. 3.43   Three-phase outdoor core-type power transformer. 30 MVA, 132 000/33 000 V, 50 Hz.
H.V. windings in star, l.v. windings in delta. Continuous disc h.v. and l.v. coils. H.V. windings fitted with
plus 10% to minus 20% taps in eighteen steps of 1·67% brought up to a resistor type on-load high-speed
tap changer. View shows transformer out of its tank looking from the h.v. side

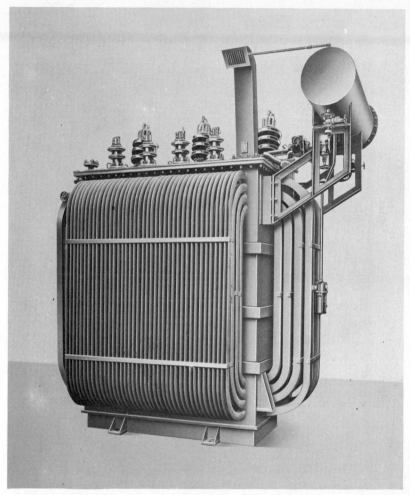

Fig. 3.44 Three-phase outdoor-type power transformer. 3 MVA, 33 000/11 000 V, 50 Hz delta/star connected. View shows transformer in its tank looking at l.v. side

Fig. 3.45   Three-phase outdoor core-type power transformer. 6 MVA, 100 000/11 400 V, 50 Hz. H.V. windings in delta, l.v. windings in star. Continuous disc h.v. and l.v. coils. H.V. windings fitted with plus and minus $2\frac{1}{2}\%$, 5%, $7\frac{1}{2}\%$, 10% and $12\frac{1}{2}\%$ tappings brought up to an externally operated off-circuit selector. Tank fitted with detachable radiators, conservator, silica-gel breather, explosion vent and Buchholz protective device. View shows transformer in its tank looking at h.v. side

Fig. 3.46   Three-phase 30 MVA, 132/33–11 kV, 50 Hz transformer designed as a system emergency spare unit. Connected star/delta, it is fitted with a high-speed resistor tap changer having a range of 30%. It can be transported by road or rail with minimum dismantling and with the OFAF coolers, tap changer and control equipment in position. (Bonar Long & Co. Ltd.)

Fig. 3.47   Another view of the three-phase, 30 MVA, 132/33–11 kV system emergency transformer illustrated in Fig. 3.46. This shows the two separate 50% coolers, pumps and tap changer control equipment in position and ready for transportation to site. (Bonar Long & Co. Ltd.)

Fig. 3.48 *Three-phase 750 MVA, 400/275 kV auto-transformer installed at a C.E.G.B. substation.*
(Ferranti Ltd.)

Chapter 4

Transformer inquiries and tenders

INQUIRIES

In the initial stage of a transformer inquiry there is nothing so important as a full and explicit statement of the total requirements that, from the user's point of view, have to be met, and from the maker's standpoint have to be considered. Frequently, inquiries are received giving insufficient information concerning the relevant details, so with a view to saving both time and trouble when issuing an inquiry the following tables have been compiled to show precisely what information should be given to the maker to enable him to prepare a quotation in accordance with the user's exact requirements.

Note

The information marked with an asterisk in Table 4.1 is required only after an order has been placed. Data listed under (a) should be given where applicable; such extras as may be required should be specified in accordance with (b).

The following general information should be given where applicable when issuing transformer inquiries:

Power factor of the load if this is to be taken into account.

Highest system voltage for the purpose of selecting the insulation level of both h.v. and l.v. windings.

Limits of variation of the supply voltage and/or frequency.

If any point of the transformer windings or system will be permanently earthed, and if so whether solidly earthed or through a current-limiting device or arc-suppression coil.

Whether the windings are to be fully insulated or designed for graded insulation.

Altitude if in excess of 1000 m.

Whether the transformer is to be installed in an electrically protected or exposed situation (see B.S. 171).

Special conditions with particular reference to the installed location, atmospheric pollution, increased bushing clearances necessary to avoid flashover from birds and vermin.

Any transport limitations of weight, dimensions, or access to site.

Table 4.1 POWER TRANSFORMERS

			Examples		
		Standard indoor transformer	*Weatherproof transformer*	*Pole-mounted transformer*	*Underground mining transformer*
(a) Essential data					
Number of units required	.. ..	2	3 for connecting as a three-phase group	6	1
Rated outputs in kVA	..	1000	3333	20	300
Single or polyphase units	..	3	1	3	3
Voltage ratio at no-load	..	11 000/433	66 000/(11 000/√3)	6600/433	3300/565
Frequency, Hz	.. ..	50	25	50	50
§Type of cooling	.. ..	ONAN	ONAN	ONAN	AN
Auto or double wound	.. ..	Double	Double	Double	Double
†Interphase connections	.. ..	Delta/star	Delta/star group	Delta/star	Delta/star
If series-parallel connections required on either winding; state voltages and accessibility required for such connections	..				
Neutral point terminal requirements	..	l.v. neutral brought out to porcelain insulator	—	l.v. neutral out	l.v. neutral out
†Overloads	.. .. ..	To C.P.1010	To C.P.1010	To C.P.1010	To C.P.1010
†Temperature rise at rated output	..	55°C by oil 65°C by resist.	50°C by oil 55°C by resist.	55°C by oil 65°C by resist.	150°C by resist.
†Maximum ambient temperature	..	30°C	20°C	30°C	30°C
Type of load, *i.e.*, lighting, power, furnace, etc.	Mixed power in paper mills	Typical power and lighting loads	Farm power and lighting loads	Coal cutters, power and lighting

§The majority of transformers are oil-immersed, naturally cooled, and the type of insulating medium should be left to the manufacturer to specify whenever possible. If synthetic insulant specify L. See page 81.

†See remarks under 'Tenders'.

Table 4.1 POWER TRANSFORMERS—CONT.

	Examples			
	Standard indoor transformer	Weatherproof transformer	Pole-mounted transformer	Underground mining transformer
(a) *Essential Data* (continued				
†Location, i.e., indoor, weatherproof, pole or platform mounting, mining—underground or surface, etc.	Indoor	Weatherproof	Weatherproof—pole-mounted	Underground mining flameproof
†Voltage tests	To B.S. 171	To B.S. 171	To B.S. 171	To B.S. 171
If packing to be suitable for home or export	Home	Export shipment	Home	Home
If required to operate in parallel with any other transformer or transformers; if so, give the following particulars of the existing transformers	Yes	—	—	—
Rated output in kVA	750	—	—	—
Interphase connections	Delta/star	—	—	—
*††Polarity	See diagram attached	—	—	—
Impedance at rated output and normal voltage, and at a specified temperature§	4·75% at 75°C	—	—	—
*Copper loss at rated output at a specified Temperature	9500 watts at 75°C	—	—	—
*Turns ratio per limb—at all tappings	1156 1128 1100 1072 1044 / 25	—	—	—

Table 4.1 POWER TRANSFORMERS—CONT.

	Examples			
	Standard indoor transformer	Weatherproof transformer	Pole-mounted transformer	Underground mining transformer
For Scott-connected transformers, state if the transformers are to be interchangeable or not	—	—	—	—
(b) Extras as Required—*h.v. or l.v. voltage adjusting tappings, and if be arranged:—				
(a) For constant kVA output	±2½% and 5% h.v. tappings arranged for constant kVA output	±10% h.v. tappings in 16 steps arranged for constant kVA output	±5% h.v. tappings arranged for constant kVA output; to oil level only	−5% and −10% h.v. tappings arranged for constant kVA output
(b) For constant current output				
(c) Available to oil level only				
Tapping selector: state off circuit or on-load	Off circuit	On-load		
Rollers: plain or flanged, swivelling or non-swivelling	Four plain non-swivelling rollers	Four flanged non-swivelling rollers to suit 1·5 metres rail gauge		Rollers to B.S. 355
Skids	Yes	Yes		—
Oil conservator		Yes	—	—
Breather, state type	—	Sealed type silica gel	—	h.v. and l.v. transwitch unit
h.v. and l.v. terminal arrangements	h.v. and l.v. trifurcating boxes	h.v. and l.v. bushings	h.v. and l.v. bushings	

*See p. 86.

77

Table 4.1 POWER TRANSFORMERS—CONT.

		Examples		
	Standard indoor transformer	Weatherproof transformer	Pole-mounted transformer	Underground mining transformer*
(b) *Extras as Required*—(continued)				
Thermometer pocket	Thermometer pocket	—	Thermometer pocket	—
‡Winding temperature indicator	—	Dial type	—	—
‡Thermometers—standard indicating, maximum reading, or remote indicating type	—	Dial type with max. indicator	—	—
Class of oil, state spare quantity	—	To B.S. 148 plus 25% spare	—	—
Spare porcelain or complete h.v. and/or l.v. bushings	—	One h.v. and l.v. bushing	—	—
Spare h.v. and/or l.v. coils and insulation	—	1 limb h.v. and l.v. coils in oil filled container	—	—
If drying out and erection is to be included	Include erection and dispatch in oil	Dispatch in oil	—	—
If tests on site are to be included; if so, specify them	—	—	—	—
If factory heat run will be required	—	Yes	—	—
Limiting dimensions	—	—	—	Height not to exceed 100/110 cm. Width not to exceed 90 cm

* Type AN dry type construction. Class C insulation, rise by resistance 150°C.
‡ If required for remote indication give distance and also state whether repeater dial is required.

78

Figure 4.1 has been prepared to assist the inquirer in specifying accurately the type of terminal outlet to be included in the quotation. Precise constructional details are not shown.

Water-cooled transformers—Type OFWF

Source and nature of cooling water supply

Head of water supply (it is recommended that the water cooler should discharge to atmosphere)

Temperature of the cooling water if in excess of 25°C

Full details of the auxiliary electricity supply for oil and pump motors

External pipework, water pump, motor and starting gear are not normally included in a quotation unless specified. Quotations usually include for oil coolers, oil pump and motor. If more than 100% cooler capacity is required this should be specifically stated (e.g., two 100% or three 50% coolers).

Static balancers for providing a d.c. neutral

Generator speed and number of poles

Generator line voltage

Total out of balance d.c. current

Number of slip rings.

Static balancers for providing an a.c. neutral

Normal line voltage

Total out of balance current

If, in connection with three-phase balancers, data are required regarding resultant loading of cables supplying the balancer, the load on each phase should be given.

Auto transformer starters

Rated output of apparatus to be started

Normal line voltage

Starting tappings required

Estimated starting kVA at different starting tappings

Whether to be started at full load or no-load

Duty cycle (refer to B.S. 587).

High voltage testing transformers

Whether one end only or possibly either end of the h.v. winding will be earthed

Whether the centre point of the h.v. winding will be earthed permanently

Whether the h.v. winding will be operated unearthed

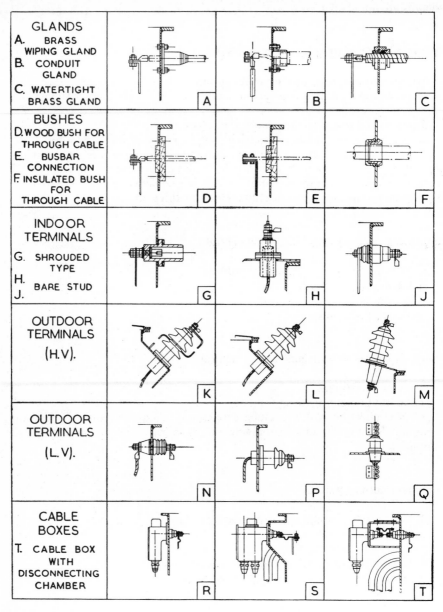

Fig. 4.1 Terminal arrangements. This figure has been prepared to assist the inquirer in specifying accurately, the type of terminal outlet to be included in the quotation. Precise constructional details are not shown

Apparatus or material to be tested

Voltage tests — magnitude and duration

If to be used with a rectifier or similar specialised equipment.

It is imperative that the following exact details be supplied with an order when h.v. and/or l.v. cable boxes are specified:

Type of cable box filling medium

Number of cables and cable cores to be terminated, *i.e.* two, three, four

Shape and section of each core

Type of core insulant, *i.e.* rubber, cambric, paper, etc.

Type of sheathing, together with the inner and outer sheath diameters

Details of servings and armourings

Direction of cable entry to the cable box: *i.e.* whether the cable approaches the box vertically upwards or downwards, horizontally from the left or right, or normal to the transformer tank.

In Table 4.2 the various methods of cooling are given whilst Table 4.3 specifies the limits of temperature rise associated with each different type of cooling. Nowadays the majority of power transformers are rated in accordance with the temperature recommendations embodied in B.S. 171 and these standards should be employed whenever possible.

Table 4.2

Dry type transformers

Natural cooling, type AN
Forced cooling, type AF

Oil immersed transformers

Oil circulation	Cooling method	Abbreviation type letter
Natural thermal head only	Air natural	ONAN
	Air blast	ONAF
Forced oil circulation by pumps	Air natural	OFAN
	Air blast	OFAF
	Water forced	OFWF

Oil immersed transformers are built with mixed natural cooling and forced cooling; the abbreviations for such mixed arrangements are ONAN/ONAF, ONAN/OFAN, ONAN/OFAF, ONAN/OFWF where O is used to indicate that the insulating liquid is mineral oil to B.S. 148. Where a synthetic insulant is employed substitute L, e.g. when specifying a chlorinated diphenyl refer to type LN.

The useful life of a transformer is dependent on the life of its insulation which ages rapidly at elevated temperatures. It is estimated that an increase of 8°C in the operating temperature of normal Class A insulation doubles the rate of deterioration. To ensure a reasonable expectancy of life it is essential that the transformer is loaded according to the prevailing ambient temperature and also to the temperature of the windings before loading. Under conditions where the maximum total temperature obtains and when the transformer has been operating continuously at full output, no continuous overload is permissible. If, however, these maxima do not occur simultaneously certain overload can be carried safely by the transformer.*

Where high ambient air temperatures prevail or adequate ventilation is not available, one of the following alternatives should be adopted:

(i) A transformer of a B.S. rated output, at the standard temperature rise, greater than that corresponding to the required output should be specified.

(ii) A transformer of the required rating but having a temperature rise of windings by resistance lower than standard.

For all practical purposes a reduction of 10°C in the temperature rise can generally be obtained by loading at 85% of the B.S. rating.

When specifying requirements it is far better to give only the essential points in the inquiry, leaving the details to the manufacturer who has expert engineering staff to deal with such matters.

Table 4.3 LIMITS OF TEMPERATURE RISE BY RESISTANCE

Type	Insulation class	Temperature rise °C
AN	A	60
AF	B	80
AN	C	150
ONAN	A	65
ONAF	A	65
OFAN	A	65
OFAF	A	65
OFWF	A	65

Temperature rise in oil
(i) conservator or sealed tanks 60°C
(ii) non-conservator or non-sealed tanks 55°C.

Note. – In the case of mixed cooling, *i.e.* types ONAN/OFAF and ONAN/OFWF, the temperature rise on the ONAN rating is the same as for an ONAN transformer and on the ONAF, OFAN, OFAF or OFWF rating as for types ONAF, OFAN, OFAF or OFWF transformers respectively, in accordance with the values in Table 4.3.

TENDERS

The submission of comprehensive and informative tenders is obviously so extremely desirable from all points of view that it behoves all parties concerned to facilitate the duties of others to the greatest possible extent. In addition, therefore, to the price, manufacturing period, terms of payment and of maintenance, and essential data and extras as specified in the inquiry according to the foregoing tables, quotations should give the following particulars regarding the specification and performance:

Type of transformer
Type of tank
Type of cooling, *i.e.* self cooled, water cooled, forced oil cooled, etc.
Core loss at normal voltage and frequency

*See C.P. 1010.

Copper loss at rated load

Percentage regulation at u.p.f.

Percentage regulation at 0·80 lagging p.f. (or at any other specified p.f.)

Percentage impedance and reactance

Efficiency at u.p.f. at 5/4, 1/1, 3/4, 1/2 and 1/4 full load

Performance reference temperature, *i.e.* 75°C

Ambient air temperature or assumed temperature of cooling medium

Maximum temperature rise by thermometer and resistance at rated load

Permissible overloads — magnitudes and durations

Oil quantity, and if included in the price

Net overall dimensions

Net weight of core and windings, tank and oil

Shipping specification (for export tenders)

Insulating medium between windings and tank, *i.e.* air or oil

Table 4.4 is a reproduction (in blank) of a typical technical data sheet such as would accompany tenders for transformers, and this, as can be seen, gives the fullest information concerning the electrical performance and weights and dimensions of the plant offered, the data being arranged to facilitate easy comparison.

If the data marked † under the heading of 'Inquiries' (p. 74) are not specified, it is usual to quote as follows:

Primary and secondary connections —
   Delta/star for three-phase transformers
   Scott connection for three- to two-phase transformers
Temperature rise, ambient temperature, and voltage test* — in accordance with B.S. 171
Overloads — In accordance with C.P. 1010
Location — Indoor
Class of oil — to B.S. 148.

Because, in the past, misunderstandings have arisen regarding exactly what is meant by 'plus' and 'minus' tappings for voltage adjustment, Table 4.5 has been prepared to show the relative location and effect of tappings when these are placed (*a*), on the primary or (*b*), on the secondary windings, to obtain variable primary or secondary voltages. As applied to tappings, the word 'plus' indicates that the tapping is located on the winding concerned to introduce more turns into the active portion of the winding than correspond to the principal tapping of the transformer. The word 'minus' conversely indicates that less turns are introduced into the active portion of the winding than correspond to the principal tapping. (See also Chapter 11.)

Tappings can be used for the following purposes:
(1) Primary tappings to vary the primary voltage
(2) Secondary tappings to vary the secondary voltage
(3) Primary tappings to vary the secondary voltage
(4) Secondary tappings to compensate for variations in primary voltage.

---

*High-voltage testing transformers are given special h.v. tests depending upon the design and method of earthing the h.v. winding. Each case is treated individually.

Table 4.4

TRANSFORMER SPECIFICATION

Offered to...

Your reference................................. *Our reference*.................................

Rated kVA at maximum daily average ambient
   temperature of...°C...
Number of phases...
Frequency...Hz...
Voltage ratio of no-load...
Winding connections...

                                                     ....
Tappings..

                                                   ....
Neutral...
Phasor diagram..

No-load loss at normal voltage ratio............watts....................................
Load loss at rated current ....watts....................................
Regulation at u.p.f. and rated current..............%.....................................
     ,,   0·8 p.f. ,,   ,,   ,, ...%.....................................
               p.f. ,,   ,,   ,, ...%.....................................
Impedance voltage at rated current ....%.....................................
Reactance voltage ,,   ,,   ,, ...%.....................................

Efficiency at u.p.f. 125% rated current ....%.....................................
    ,,   ,,   ,,   100%   ,,   ,, ...%.....................................
    ,,   ,,   ,,   75%   ,,   ,, ...%.....................................
    ,,   ,,   ,,   50%   ,,   ,, ...%.....................................
    ,,   ,,   ,,   25%   ,,   ,, ...%.....................................

Temperature rise at rated kVA by
   thermometer in oil...
Temperature rise at rated kVA by
   resistance of windings...

Permissible overload... In accordance with the B.S. Code of
                                                             Practice C.P.1010

                                                             N.B.—Performance reference tempera-
                                                             ture—75°C. Tolerances on guaranteed
                                                             performance as B.S.171

Approximate weights and dimensions
Transformer core and windings....................kg...
Tank and fittings......................................,,..
Oil..,,..
Weight of on-load tap changer......................,,..
Total weight...,,..
Quantity of oil.....................................litres...

                                                        ....
Overall length.......................................mm...
   ,,   width..,,..
   ,,   height...,,..

REMARKS

For each of these operating conditions the transformer may safely be excited on any particular tapping at the *full output* associated with that tapping and at any voltage within plus or minus 5% of its rated voltage.

In comparing tenders, the best procedure is to tabulate the data submitted by different manufacturers, on one sheet, so that a comprehensive survey can be made. The following brief remarks draw attention to some of the chief points that should be examined thoroughly.

Iron loss

As such, the iron loss mainly becomes important in cases where a lighting load is being supplied and in which the transformer itself remains excited, even though not actually supplying any load. It is also important in cases where a transformer is working on a low load factor. Obviously a low iron loss is desirable, and in those cases where it represents a continuous 24 hour loss, it may, in comparing tenders, be capitalised at a figure corresponding to the coal and other generating costs holding good in the district, such capitalisation being added to the quoted price of the transformer.

Copper loss

It is usual to keep the iron loss as low as possible since it may represent a 24 hour load, so that the copper loss is correspondingly high in proportion, as the total losses for any given kVA output and at a specified frequency, voltage ratio, and temperature rise are approximately the same for any well designed commercial transformer. In the case of a B.S. rated naturally cooled transformer, no overload is permissible,* that is, with a full-load temperature rise in oil of 60°C, and at a maximum ambient temperature. Consequently, in such transformers the ratio of copper to iron loss may be fairly high, and this ratio ranges from 4/1 to 7/1, according to the output, voltage ratio and frequency. If the ambient temperature is below the maximum, certain overloads are permissible as previously stated, but the ratios given above are such as would allow the specified overloads at the lower ambient temperatures to be reached without excessive heating of the windings. If the transformer is to be designed for a sustained overload, the ratio of copper to iron loss at normal full-load would drop, that is, the iron loss would probably be higher and the copper loss lower than in the case of B.S. rated transformers.

In general it may be said that, for a given temperature rise, the transformer with the lower losses is the more liberally designed one. As a rule it is much safer to run the core at high density rather than the windings as the insulation of the coils is much more susceptible to damage resulting from high temperatures than is the insulation of the core. It should be borne in mind, however, that a high ratio of copper loss to iron loss does not necessarily indicate a liberally designed iron circuit and a high current density in the windings, as this condition may be obtained by suitable proportioning of the weights of copper and iron. Theoretically it is possible to obtain any ratio of copper loss to iron loss by this

---

*With the overload commencing with the transformer at the maximum B.S. total temperature.

Table 4.5

POSITION AND EFFECT OF VOLTAGE ADJUSTING TAPPINGS

Variation of voltages from the normal ratio		If variation is obtained by means of:		Tappings and voltages at which B_m is greatest when using:			
				(a) Primary tappings		(b) Secondary tappings	
Primary voltage	Secondary voltage	(a) Primary tappings	(b) Secondary tappings	Tapping	Voltage equivalent	Tapping	Voltage equivalent
constant	constant	—	—	normal winding	normal	normal winding	normal
constant	above the normal	(a) minus	—	lowest minus tapping	maximum secondary voltage	—	—
		—	(b) plus	—	—	constant at all tappings	constant at all secondary voltages
constant	below the normal	(a) plus	—	normal winding	maximum secondary voltage	—	—
		—	(b) minus	—	—	constant at all tappings	constant at all secondary voltages
above the normal	constant	(a) plus	—	constant at all tappings	constant at all primary voltages	—	—
		—	(b) minus	—	—	lowest minus tappings	maximum primary voltage
below the normal	constant	(a) minus	—	constant at all tappings	constant at all primary voltages	—	—
		—	(b) plus	—	—	normal winding	maximum primary voltage

proportioning of the weights of active materials, but practically, the limits are set by considerations of mechanical design, heating in the winding, etc. For a given flux density and current density the product of the losses will be approximately constant, *and assuming that oil ducts and electrical clearances are kept constant*, it may be said that the smaller this product the more liberal the design. At the same time it should not be overlooked that by cutting down oil ducts, insulation, and clearances, it is possible to obtain a design which, while *appearing* liberal from the point of view of losses, may result in the production of a transformer which is ultimately liable to breakdown through local overheating or insufficient insulation.

Impedance

If the transformer is to be supplied from a system having a large fault capacity (and the interconnection of power networks tends towards this condition very frequently), the inherent impedance should be sufficient to protect the transformer against the mechanical forces which will be produced within the windings during an *external* short circuit.* Alternatively, values of impedance higher than normal may be specified in order to reduce the available short circuit MVA on the secondary side of the transformer and to permit the use of circuit-breakers having rupturing capacities lower than would otherwise be possible.

Tables 4.6 and 4.7 give the standard values of impedances for ONAN type single- and three-phase transformers. These values represent a compromise between the various features which contribute to a sound and economical design but, if necessary, they can be varied to suit particular requirements.

Temperature rise

It is important that the inquirer should satisfy himself that the temperature rise specified meets all requirements exactly. The national standards of temperature rise, both of oil and of windings by resistance are defined in B.S. 171 and these standards should be employed whenever possible. For power transformers the temperature rise should correspond to the rated output on the normal tapping. In the case of a transformer designed for operation at an elevated altitude, a correction will be necessary when tested at sea level due to the differing air density on the test bed. Reference should be made to B.S. 171 and Chapter 13 for more detailed information regarding the method of conducting temperature rise tests and to the interpretation of the results obtained during such tests.

Tolerances

In order to allow for inevitable variation in the qualities of materials and in the various shop processes, guaranteed performance figures are subject to the following tolerances:

Voltage ratio at no-load: 1/200 or 1/10 of the percentage impedance voltage,

*B.S. 171.

whichever is the smaller on the principal tapping

Individual iron and copper losses can each vary by $\frac{1}{7}$ provided the total losses do not vary by more than $\frac{1}{10}$

Impedance voltage for principal tapping: (a) $\pm\frac{1}{10}$ of the guaranteed value; (b) multiwinding transformers, $\pm\frac{1}{7}$ to $\frac{1}{10}$ of the guaranteed value

Regulation: in accordance with tolerances on impedance and load losses. No-load current is $\frac{3}{10}$ of the declared value.

Parallel operation

Two-winding transformers cannot be guaranteed to divide the load within closer limits than those which permit an individual variation from rated output of any transformer amounting to $\pm 10\%$ of such rated output, when the total loading on the group is equal to the sum of the rated output of all the transformers so connected.

Note.—When a transformer is ordered for operating in parallel, the rated output of the smallest transformer should not be less than one-third of the rated output of the largest transformer in the group.

The foregoing remarks have indicated briefly how quoted performance guarantees should be considered in relation to the soundness of design and manufacture, and to the all important question of continuity of service.

It is desirable to consider next, in somewhat more detail, how the electrical performance of a transformer is influenced by features of bad design and manufacture, and to show that guarantees of low losses and high efficiencies do not necessarily ensure the production of a transformer which is able to meet successfully the demands that may be made upon it by the needs of continuous operation upon a large system.

LOSSES, EFFICIENCY, AND REGULATION

In designing a transformer there are three distinct circuits to be considered, the electric, the magnetic and the dielectric circuits.

In each of these, losses occur, which may be subdivided as follows:

(1) Losses in the electric circuit
   (a) I^2R loss due to load currents
   (b) I^2R loss due to no-load current
   (c) I^2R loss due to current supplying the losses
   (d) eddy current loss in conductors due to leakage fields.

(2) Losses in magnetic circuit
   (a) hysteresis loss in core laminations
   (b) eddy current loss in core laminations
   (c) stray eddy current loss in core clamps, bolts, etc.

(3) Loss in the dielectric circuit
   This loss is small for all voltages up to 50 kV, and it is consequently included in the no-load losses, alternatively described as the iron losses.

When measuring the various losses,* they are automatically grouped as follows, the lettering being retained as above for ease of reference:

No-load losses (commonly termed iron losses)
(1) (b) I^2R loss due to no-load current
(2) (a) hysteresis loss in core laminations
(2) (b) eddy current loss in core laminations
(2) (c) stray eddy current loss in core clamps, bolts, etc.
(3) loss in the dielectric circuit.

Load losses (commonly termed copper losses or short circuit losses)
(1) (a) I^2R loss due to load currents
(1) (c) I^2R loss due to current supplying the losses
(1) (d) eddy current loss in conductors due to leakage fields.

In order that a transformer may have a high efficiency all these losses must be reduced to a minimum. It is therefore of interest to consider, firstly, the features which determine their magnitude, and secondly, the steps which should be taken to reduce them to a minimum.

Consider first the losses in the electric circuit:

1(a) I^2R loss due to load currents

This loss, as its name implies, is equal to the sum of the squares of the currents multiplied by the resistances of the various windings. As the currents are fixed by the rating, it is evidently impossible to reduce their values in order to reduce the I^2R loss. The only factors, therefore, which may be varied by design in order to reduce the loss to a minimum are the resistances of the various windings. The resistances should be reduced to a minimum, and to do this it is evident that the total sections of the conductors should be as large as possible, and their total lengths as small as possible. While to increase the section of the conductors, certainly reduces the resistance and consequently the I^2R loss, it also tends to increase the frame size, and therefore the loss in magnetic circuit. The only factors, therefore, which may reasonably be varied are the total lengths of the windings, and these may be reduced to a minimum by suitably proportioning the frame dimensions.

1(b) I^2R loss due to no-load current

This loss is very small, since the no-load current in well-designed and well-constructed transformers does not usually exceed 5% of the full-load current, and in larger transformers may even be as low as 1 to 2%. It is evident, therefore, that losses which are proportional to the square of a current of this order will, in most cases, be extremely small, and provided steps be taken, both in design and manufacture, to keep the no-load current within reasonable limits, these losses may be, and usually are, neglected. In practice the no-load current is kept to a reasonable figure by designing for a flux density below the critical saturation

*See Chapter 13.

Table 4.6

STANDARD PERCENTAGE IMPEDANCES OF SINGLE-PHASE TRANSFORMERS
FREQUENCY 50 Hz, STANDARD FLUX DENSITY

kVA	H.V. Winding (kV)										
	3·3	6·6	11	15	22	33	44	55	66	88	110
5	4·5	4·5	4·5	5·25							
7·5	4·5	4·5	4·5	4·75	4·75	4·75					
10	4·5	4·5	4·5	4·75	4·75	4·75					
15	4·5	4·5	4·5	4·5	4·5	4·5					
20	4·5	4·5	4·5	4·5	4·5	4·5					
25	4·5	4·5	4·5	4·5	4·5	4·5					
30	4·5	4·5	4·5	4·5	4·5	4·5					
40	4·5	4·5	4·5	4·5	4·5	4·5					
50	4·5	4·5	4·5	4·5	4·5	4·5	5·0	5·5			
60	4·5	4·5	4·5	4·5	4·5	4·5	5·0	5·5			
75	4·5	4·5	4·5	4·5	4·5	4·5	5·5	5·5	5·5		
100	4·5	4·5	4·5	4·5	4·5	4·5	5·5	5·5	5·5	6·0	
125	4·5	4·5	4·5	4·5	4·5	4·5	5·5	5·5	5·5	6·0	
167	4·75	4·75	4·75	5·0	5·0	5·0	5·5	5·5	5·5	6·0	6·5
200	4·75	4·75	4·75	5·0	5·0	5·0	5·5	5·5	5·5	6·0	6·5
250	4·75	4·75	4·75	5·0	5·0	5·0	5·5	5·5	5·5	6·0	6·5
333	4·75	4·75	4·75	5·0	5·0	5·0	5·5	6·0	6·0	6·5	7·0
500	4·75	4·75	4·75	5·0	5·0	5·0	5·5	6·0	6·0	6·5	7·0
667	4·75	4·75	4·75	5·0	5·0	5·0	5·5	6·0	6·0	6·5	7·0
833		5·5	5·5	6·0	6·0	6·0	6·5	7·0	7·0	7·0	7·5
1000		5·5	5·5	6·0	6·0	6·0	6·5	7·0	7·0	7·0	7·5
1333		6·0	6·0	6·0	6·0	6·0	6·5	7·0	7·0	7·0	7·5
1667		6·0	6·0	6·0	6·0	6·0	6·5	7·0	7·0	7·0	7·5
2500			6·0	6·5	7·0	7·0	7·0	7·5	7·5	8·0	8·0
3333			6·0	6·5	7·0	7·0	7·0	7·5	7·5	8·0	8·0
5000			7·0	7·5	8·0	8·0	8·0	8·5	8·5	9·0	9·0
6667					9·0	9·0	9·0	9·0	9·0	10·0	10·0
8333					9·0	9·0	9·0	9·0	9·0	10·0	10·0
10 000					10·0	10·0	10·0	10·0	10·0	10·0	10·0

Table 4.7

STANDARD PERCENTAGE IMPEDANCES OF THREE-PHASE TRANSFORMERS
FREQUENCY 50Hz: STANDARD FLUX DENSITY, STAR OR DELTA H.V. AND L.V.
CONNECTIONS

kVA	H.V. Winding (kV)										
	3·3	6·6	11	15	22	33	44	55	66	88	110
5	4·75	4·75	4·75	5·5	5·5	6·0					
7·5	4·75	4·75	4·75	5·25	5·25	5·25					
10	4·75	4·75	4·75	4·75	4·75	4·75					
15	4·75	4·75	4·75	4·75	4·75	4·75					
20	4·5	4·5	4·5	4·5	4·5	4·5					
25	4·5	4·5	4·5	4·5	4·5	4·5					
30	4·5	4·5	4·5	4·5	4·5	4·5					
40	4·5	4·5	4·5	4·5	4·5	4·5					
50	4·5	4·5	4·5	4·5	4·5	4·5	5·0	5·5			
60	4·5	4·5	4·5	4·5	4·5	4·5	5·0	5·5			
75	4·5	4·5	4·5	4·5	4·5	4·5	5·0	5·5			
100	4·75	4·75	4·75	5·0	5·0	5·0	5·5	5·5	5·5		
125	4·75	4·75	4·75	5·0	5·0	5·0	5·5	5·5	5·5		
150	4·75	4·75	4·75	5·0	5·0	5·0	5·5	5·5	5·5		
200	4·75	4·75	4·75	5·0	5·0	5·0	5·5	5·5	5·5	6·0	
250	4·75	4·75	4·75	5·0	5·0	5·0	5·5	5·5	5·5	6·0	6·5
300	4·75	4·75	4·75	5·0	5·0	5·0	5·5	5·5	5·5	6·0	6·5
400	4·75	4·75	4·75	5·0	5·0	5·0	5·5	5·5	5·5	6·0	6·5
500	4·75	4·75	4·75	5·0	5·0	5·0	5·5	6·0	6·0	6·5	7·0
600	4·75	4·75	4·75	5·0	5·0	5·0	5·5	6·0	6·0	6·5	7·0
750	4·75	4·75	4·75	5·0	5·0	5·0	5·5	6·0	6·0	6·5	7·0
1000	4·75	4·75	4·75	5·0	5·0	5·0	5·5	6·0	6·0	6·5	7·0
1250		5·0	5·0	5·5	5·5	5·5	6·0	6·5	6·5	6·5	7·0
1500		5·5	5·5	6·0	6·6	6·0	6·5	7·0	7·0	7·0	7·5
2000		6·0	6·0	6·0	6·6	6·0	6·5	7·0	7·0	7·0	7·5
2500		6·0	6·0	6·0	6·6	6·0	6·5	7·0	7·0	7·0	7·5
3000			6·0	6·5	7·0	7·0	7·0	7·5	7·5	8·0	8·0
4000			6·0	6·5	7·0	7·0	7·0	7·5	7·5	8·0	8·0
5000			6·0	6·5	7·0	7·0	7·0	7·5	7·5	8·0	8·0
6000			7·0	7·0	7·5	7·5	7·5	8·0	8·0	8·5	8·5
7500			7·0	7·5	8·0	8·0	8·0	8·5	8·5	9·0	9·0
10000					9·0	9·0	9·0	9·0	9·0	10·0	10·0
12500					9·0	9·0	9·0	9·0	9·0	10·0	10·0
15000					10·0	10·0	10·0	10·0	10·0	10·0	10·0
20000					10·0	10·0	10·0	10·0	10·0	10·0	10·0
25000					10·0	10·0	10·0	10·0	10·0	10·0	10·0
30000					10·0	10·0	10·0	10·0	10·0	10·0	10·0

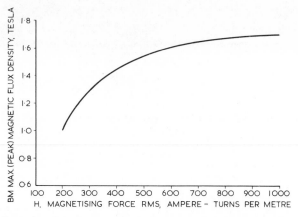

Fig. 4.2   Typical a.c. magnetisation curve for cold-rolled transformer steel 0·33 mm thick

point, and by careful building up of the laminated core structure so as to avoid abnormal air gaps. From a study of Fig. 4.2, which shows a representative B/H curve of cold-rolled, electrical steel for transformers, it will be seen that for values of B_m above 1·6 teslas (which can be taken as being the critical value) the field strength H, upon which the magnetising current directly depends, begins to increase rapidly, and consequently the normal flux density should approximate to this value for transformers continuously rated in accordance with the B.S. Specification. This feature is more important in the smaller sizes of transformers on account of the fact that the smaller the transformer the greater will be the magnetising ampere-turns corresponding to the mean length of the magnetic circuit, as compared with the total full-load ampere-turns in the primary winding.

This fact may be demonstrated by a consideration of the formulae used for calculating the magnetising current for single-phase transformers:

$$H = \frac{I_m N_1 \sqrt{2}}{10^{-2}\ell_m}$$

where H = field strength in ampere-turns per metre

            I_m = r.m.s. value of the magnetising current in amperes

            N_1 = number of turns in the primary winding

            ℓ_m = mean length of the magnetic circuit in centimetres.

Re-writing this,

$$I_m = \frac{10^{-2}\ell_m H}{N_1 \sqrt{2}}$$

Normal full-load primary current $= \dfrac{\text{kVA} \times 1000}{V_1}$

where V_1 = primary voltage.

Therefore the percentage magnetising current

$$= \frac{10^{-2}\ell_m H}{N_1 \sqrt{2}} \frac{V_1 100}{\text{kVA} 1000} = \frac{\ell_m H V_1/N_1}{1414\,\text{kVA}}$$

where V_1/N_1 = volts per turn.

92

In this expression

H is solely dependent upon the value of the flux density B_m,

I_m varies approximately as the fourth root of the kVA,

V_1/N_1 varies approximately as the square root of the kVA.

Consequently it will be seen that the percentage magnetising current varies approximately inversely as the fourth root of the kVA for constant values of H and therefore of the flux density B_m.

The no-load current, as stated earlier is but a few per cent of the normal full-load current and is therefore a relatively unimportant characteristic of the transformer. It is today, more usual to estimate the no-load current from the type of curve given in Fig. 4.3. From this curve it is possible to obtain the exciting volt amperes per kilogram of core laminations at any given value of flux density, and hence the no-load current. The value so obtained for a three-phase transformer is the arithmetic average of three line currents which differ in practice due to the dissymmetry of a three-phase, core-type magnetic circuit.

1(c) I^2R loss due to current supplying the losses

In any transformer the input exceeds the output by an amount equal to the sum of all the losses, and consequently the primary current is proportionally greater than the value obtained by calculation from the rated output and primary

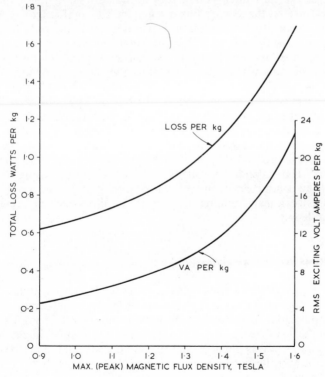

Fig. 4.3 Typical iron loss and exciting volt ampere characteristics for cold-rolled transformer laminatious 0·33 mm thick

voltage. It is the additional I^2R loss due to this increased primary current that is described as the I^2R loss due to the current supplying the losses. This loss, as in the preceding case, is usually negligible, and it is evident that in a design in which the other losses are reduced to a minimum, this loss would also similarly be reduced.

1(d) Eddy current loss in conductors due to leakage fields

This loss is caused by the eddy currents in individual conductors which are set up by stray magnetic fields. They are the most difficult losses to calculate with any degree of accuracy, but at the same time they may be of very considerable magnitude if due care is not taken in the design. Various formulæ have been propounded from time to time for calculating this loss, but there are so many factors which enter into the calculation that it is more usual and practical to add a percentage on to the I^2R loss rather than to attempt to calculate it by means of formulæ. The percentages which it is customary to add are based upon experience with the particular type of transformer under consideration. In order that this loss may be reduced to a minimum, it is first necessary to consider the various factors which tend to increase or decrease it. In general it may be said that it is approximately proportional to the square of the leakage flux density, to the total weight of copper in the transformer, and to the square of that dimension of the individual conductors normal to the path of the leakage flux. Of these factors the leakage flux density and the total weight of copper are usually fixed from other considerations. Consequently the only factor which may be varied is the dimension of the individual conductors normal to the path of the leakage flux, and this dimension should be kept as small as possible by subdividing the conductors and insulating the various portions from one another.

Another form of loss in transformers which is usually considered as part of the eddy current loss—although, strictly speaking, it is not such a loss—is the additional loss caused by the various portions or layers of a subdivided conductor not sharing the load equally. This is due to the fact that their lengths may not be equal, and also that they may not be similarly situated with regard to the leakage flux. The difficulty is usually overcome by transposing the various portions of the conductors during the winding process in such a way that each portion in its path from terminal to terminal occupies each different position for an equal length.

2(a) Hysteresis loss in core plates

This loss is dependent upon the quality of the core plates used, and is calculated from curves supplied by the manufacturers of the core plate material. It is, of course, proportional to the weight of material used, and varies according to the value of the flux density. In order to minimise this loss the weight of material should be kept as low as possible, and the flux density should not be excessive. It should, however, be borne in mind that to decrease the flux density with a view to reducing the hysteresis loss entails either the use of a greater quantity of iron, which in turn increases the length of the mean turn of the windings, or else of an increased number of turns in the windings, thus in either case increasing the loss in the windings.

2(b) Eddy current loss in core plates

This loss is dependent upon the flux density employed, the quality of the core plate material, the thickness of the core plates, and upon the efficacy of the insulation between the core plates.

Most of the remarks regarding the hysteresis loss also apply to the eddy-current loss in the core plates, and in fact the two losses are usually calculated together from the curves supplied by the manufacturers of the core plate material. Figure 4.4 gives curves showing, for 50 and 60 Hz, the approximate relation between flux density and loss per kg of 0·33 mm laminations of the type used for transformers. The loss for a given weight of material may be reduced by decreasing the thickness of the laminations, but decreasing this thickness unduly, presents the following disadvantages:

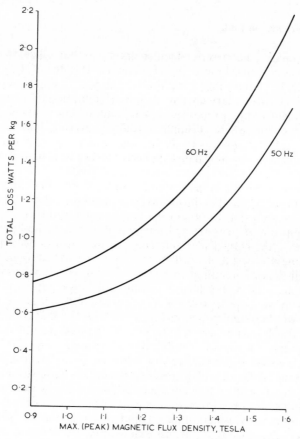

Fig. 4.4   Typical loss curves at 50 and 60 Hz for cold-rolled transformer laminatious 0·33 mm thick

(i) If the thickness of the laminations is greatly reduced, the sum total thickness of the insulation between them becomes greater and therefore more important, inasmuch as a poor space factor is obtained.

(ii) The thinner the laminations are made, the more difficult is the assembly of the transformer core, and if the thickness is too greatly reduced the advantage one endeavours to gain in iron loss may be nullified by bad interleaving in the yokes. Further, a core built of very thin laminations will not be so strong mechanically as one built from thicker plates.

2(c) Stray eddy current loss in core clamps, bolts, etc.

This loss is very difficult to predetermine and, as in the case of eddy-current losses in conductors, is usually allowed for by adding a percentage, which is determined from experience, to the magnetic circuit loss as calculated from the curves supplied by the steel plate manufacturers. In order to keep these losses to a minimum it is necessary to exercise great care in designing the clamping structure.

2(d) Stray eddy loss in tank

This loss is similar to that described under 2(c) except that it increases the copper loss and is usually allowed for in the same manner. It is due to stray flux cutting the tank structure. Under load conditions it can reach quite high proportions, particularly if the secondary current is high. Careful design of the secondary lead arrangement reduces this loss to a minimum. The same effect can also become appreciable where terminal bushings carrying high currents pass through the tank walls or cover. By the use of non-magnetic material in the neighbourhood of the bushing flange the stray loss can be kept within reasonable proportions.

From the above outline of the methods employed for calculating the various losses in a transformer, and of the means employed for reducing them to a minimum, it will be seen that to obtain the best possible results it is necessary to achieve a compromise. Features which tend to decrease the no-load or iron loss tend to increase the copper loss and, consequently, before deciding definitely along what lines the best design will be, it is necessary to determine which of the two principal losses is required to be a minimum. In connection with this it should be noted that a transformer having an exceptionally low iron loss as compared with the copper loss has its maximum efficiency at a much lower output than would a transformer having an iron loss high as compared with its copper loss. It will thus be seen that transformers which operate on a low load-factor should have the iron loss reduced to a minimum, even though this may result in a somewhat higher copper loss. A further point which should be borne in mind in deciding the ratio of the losses, is that unless a transformer is switched off from the supply when not in use, there will be a continuous loss equal to the iron loss, while the copper loss is only effective when the transformer is actually on load. From these considerations it would appear *on the surface* that in most cases it is very desirable that a transformer should be designed with the minimum possible iron loss, the copper loss being relatively unimportant. Such a transformer, however, has certain disadvantages, of which the following are the most important:

(i) Since the copper loss varies as the square of the current flowing, a transformer having a high copper loss will of necessity have a smaller emergency overload capacity than one in which the two losses are approximately equal.

(ii) A transformer which has a high ratio of copper to iron loss has in general a high impedance. This feature in itself is not necessarily undesirable, and in cases where a transformer is liable to be subjected to short circuits would even be an advantage. If, however, the impedance is excessively high it will cause the transformer to have a very high regulation, especially at the lower power factors.

(iii) In practice the current and magnetic flux densities are approximately constant so that the two principal losses are in general proportional to the weights of active material used. A transformer having a high ratio of copper loss to iron loss will also have a high ratio of copper weight to iron weight, and unless great care is taken in the design of oil ducts and coils such a transformer would be much more liable to develop hot spots in the windings than one in which the ratio of the losses is normal.

EFFICIENCY

The efficiency of a transformer expressed as a percentage is equal to

$$\frac{(\text{output in watts}) \times 100}{\text{output in watts} + \text{total losses in watts}}$$

It will be apparent from the previous remarks on the subject of losses that the losses in the magnetic circuit are constant at all loads, consequently these losses expressed as a percentage of the transformer output will vary inversely as the load. The copper losses, on the other hand, vary as the square of the load the transformer is carrying, and therefore when expressed as a percentage of the load, will vary directly as the load. By means of simple mathematics it can be proved that a transformer has its highest efficiency at a load at which the two principal losses are equal. From this it follows that the load at which the efficiency is highest will be

$$\text{percentage load} = \sqrt{\left(\frac{\text{iron loss}}{\text{copper loss}}\right)} \times 100$$

Thus, a transformer in which the ratio of copper loss to iron loss is four to one, will have its maximum efficiency at half load. Generally speaking, a close approximation of the ratio of the losses should be determined from a consideration of the kind of load to be supplied, though in fixing this ratio the previous remarks concerning the losses individually should be borne in mind.

For rapid calculations of the efficiency of a transformer the nomograms which are reproduced in Figs. 4.5 and 4.6 will be found useful. From these nomograms it will be seen that at the extreme left is a scale showing percentage copper losses, and at the extreme right a corresponding scale showing percentage iron losses. If a straight edge be laid across these nomograms so that it cuts these two scales at points corresponding to the percentage losses in the transformer under consideration at normal full load rating, the efficiencies at the various loads shown on the nomograms may be read off direct.

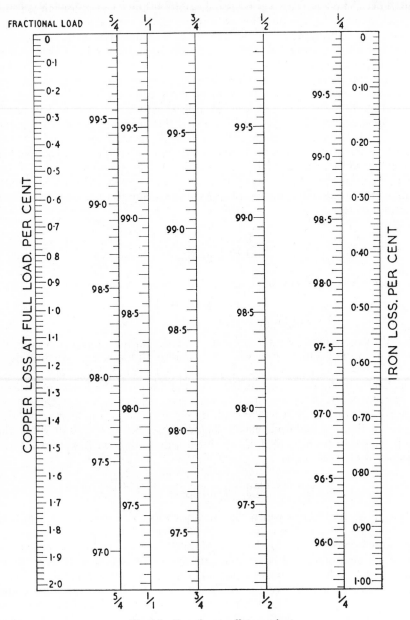

Fig. 4.5 Transformer efficiency chart
To obtain the efficiency at a given load, lay a straight edge across the iron and copper loss values and
read the efficiency at the point where the straight edge cuts the required load ordinate

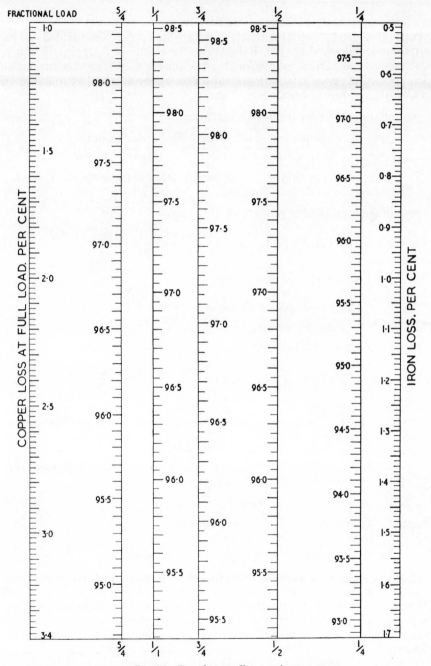

Fig. 4.6   Transformer efficiency chart.
To obtain the efficiency at a given load, lay a straight edge across the iron and copper loss values and read the efficiency at the point where the straight edge cuts the required load ordinate

99

REGULATION

In a normal power transformer having a relatively low reactance, the percentage regulation at u.p.f. is approximately equal to the copper loss divided by the output and multiplied by 100. If, however, the reactance is appreciable, it is necessary to make an allowance for this. For the range of reactances normally incorporated in power transformers, the following formula will be found to be sufficiently accurate (for impedances above 20% see B.S. 171):

percentage regulation at unity power factor

$$= \frac{\text{copper loss} \times 100}{\text{output}} + \frac{(\text{percentage reactance})^2}{200}$$

To obtain the regulation at power factors other than unity, the following formula should be used:

percentage regulation at power factor $\cos \phi_2$

$$= V_R \cos \phi_2 + V_X \sin \phi_2 + \frac{(V_X \cos \phi_2 - V_R \sin \phi_2)^2}{200} \qquad (4.1)$$

where V_R = percentage resistance voltage *i.e.* $\dfrac{\text{copper loss} \times 100}{\text{output}}$

V_X = percentage leakage reactance voltage

The formula may also be expressed as follows:

percentage regulation at power factor $\cos \phi_2 = m$

$$= V_R m + V_X \sqrt{(1 - m^2)} + \frac{\{V_X m - V_R \sqrt{(1 - m^2)}\}^2}{200}$$

The nomograms given in Figs. 4.7 and 4.8 provide an easy means for rapidly determining the percentage regulations of a transformer under given conditions at different power factors.

In commercial testing it is always the impedance voltage drop and not the reactance drop which is measured direct. The percentage reactance may easily be calculated, however, from the formula

$$V_X = \sqrt{(V_Z^2 - V_R^2)}$$

where V_X = percentage leakage reactance voltage
V_Z = percentage impedance voltage
V_R = percentage resistance voltage

It will be found from these formulæ that the reactance of a transformer has a very appreciable effect upon the regulation at the lower power factors. It is of interest to note that the worst regulation occurs at a power factor equal to the percentage copper loss divided by the percentage impedance, and at this power factor the regulation is equal to the percentage impedance. Typical curves showing the relation between regulation and power factor at full load are shown in Fig. 4.9. In preparing these curves various percentage copper losses and leakage reactances have been taken. Where low regulation is desirable it will be seen that a transformer having a low reactance would be required. Such a transformer certainly meets the requirements with regard to regulation, but it

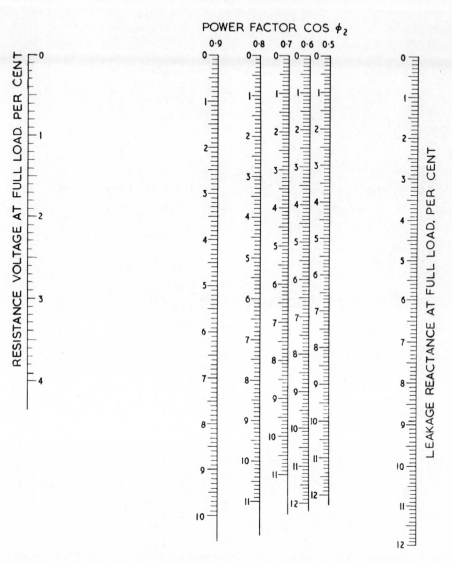

Fig. 4.7   Transformer regulation chart

To obtain the percentage regulation at a given power factor it is necessary to calculate the values of V_R and V_X and then lay a straight edge across these values reading the regulation from the appropriate power-factor scale. Figure 4.7 gives the regulation for that part of eqn. 4.1, $(V_R \cos \phi_2 + V_X \sin \phi_2)$, whilst Fig. 4.8 gives the last term, $(V_X \cos \phi_2 - V_R \sin \phi_2)^2/200$. The total regulation is the sum of these two components

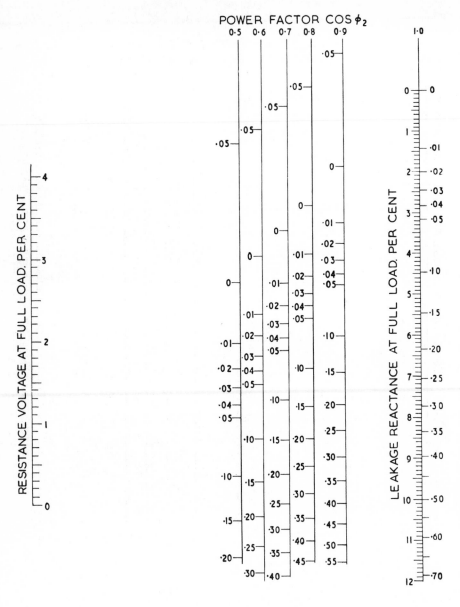

Fig. 4.8   Transformer regulation chart
To obtain the percentage regulation at a given power factor it is necessary to calculate the values of V_R and V_X and then lay a straight edge across these values, reading the regulation from the appropriate power-factor scale. Figure 4.7 gives the regulation for that part of eqn. 4.1, $(V_R \cos \phi_2 + V_X \sin \phi_2)$, whilst Fig. 4.8 gives the last term, $(V_X \cos \phi_2 - V_R \sin \phi_2)^2/200$. The total regulation is the sum of these two components

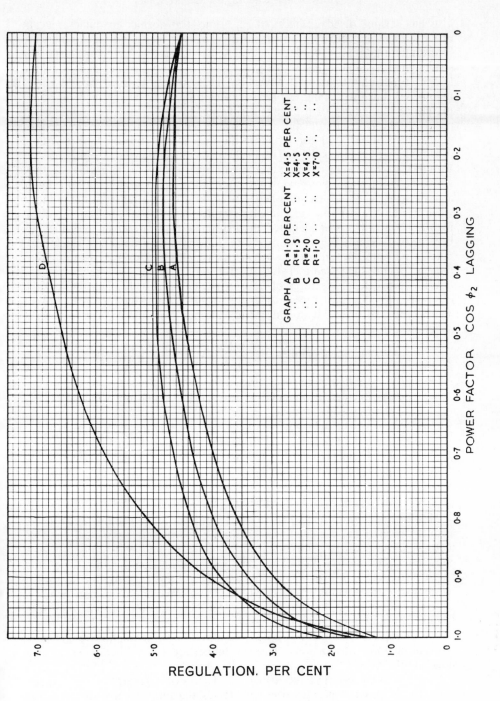

Fig. 4.9 Curves showing transformer regulations at various power factors for varying values of percentage impedance

also has serious disadvantages. At one time, transformer manufacturers, in order to meet the demands for transformers having very good regulation, were in the habit of designing for leakage reactances as low as 2%, but a transformer so designed is liable to be severely damaged if a short circuit occurs on the system, especially if the total power on the system is large. It should be remembered that the mechanical stresses set up in a transformer vary approximately as the square of the current flowing. In the event of a short circuit occurring the short circuit current varies inversely as the percentage impedance of the transformer, and consequently the mechanical stresses due to a short circuit vary inversely as the square of the percentage impedance. Such stresses would be over six times as great in a transformer having 2% impedance as they would be in a transformer having 5% impedance. In deciding what impedance a transformer should have, it is better to look at the matter from the point of view of safety rather than to consider simply the question of obtaining good regulation at all power factors. This aspect of transformer engineering has become increasingly important owing to the very large extensions which have taken a place in the total power of various systems and, as a guide, minimum impedances are suggested in Tables 4.6 and 4.7.

With regard to the features of transformer design which affect the leakage reactance, it is interesting to consider the formula from which the reactance of a core type transformer with concentric windings is calculated. The formula is as follows:

$$\text{Percentage reactance} = \frac{7 \cdot 9 I N M_w f}{V/N \ell} \left(a + \frac{b_1 + b_2}{3} \right) \times 10^{-6} \qquad (4.2)$$

where IN = total ampere-turns of l.v. or h.v. winding
  M_w = average mean length of turn of the l.v. and h.v. windings
  f = frequency in hertz
  V/N = volts per turn of l.v. or h.v. windings
  ℓ = axial length of the shorter winding*
  a = radial distance from bare copper to copper between the h.v. and l.v. windings
  b_1 = radial width over bare copper of the l.v. winding
  b_2 = radial width over bare copper of the h.v. winding

This formula holds good when all the dimensions are in centimetres. For the sake of clearness the principal dimensions are shown diagrammatically in Fig. 4.10, which also shows a flux distribution diagram, and, in addition, what is sometimes referred to as a linkage diagram. The linkage diagram shows the product of leakage flux density and turns linked at any point on the h.v. or l.v. winding. It is the area under this curve, assuming a maximum ordinate of unity, which is shown in the bracket at the end of the reactance formula (eqn. 4.2). Figure 4.11 illustrates the typical flux condition.

It is often convenient to be able to see at a glance the qualitative effects of changes in frame proportions upon reactance, and the following expressions illustrate these:

For a given kVA output and primary and secondary voltages:

(1) IN of either winding varies directly as the number of turns N in the winding

---

*For both mechanical and electrical reasons the two windings should be of equal axial length, and the reference to the shorter winding is only intended to apply to *small* unavoidable differences.

$$N \propto \frac{1}{V/N} \propto \frac{1}{d^2}, \text{ where } d \text{ is the core circle diameter}$$

(2) M_w does not change very much unless the frame size is varied considerably

(3) $V/N \propto d^2$

(4) ℓ actually is the axial winding length, but for qualitative purposes it may be taken as the core length

(5) $\left(a + \dfrac{b_1 + b_2}{3}\right) \propto (c - d)$, where c is the distance between core centres

This proportionality assumes that the relative dimensions of major insulation, oil ducts and coils do not vary greatly.

From the foregoing it is seen that, provided there is no very great change in frame proportions,

(6)
$$X \propto \left\{ \frac{1}{d^2} \frac{1}{d^2} \frac{1}{\ell} (c - d) \right\}, \text{ i.e., } X \propto \frac{c - d}{d^4 \ell}$$

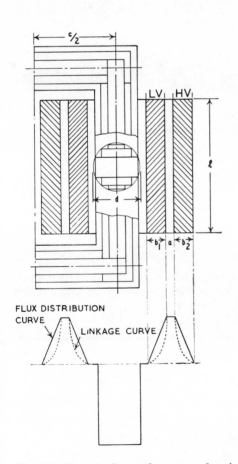

FLUX DISTRIBUTION CURVE

LINKAGE CURVE

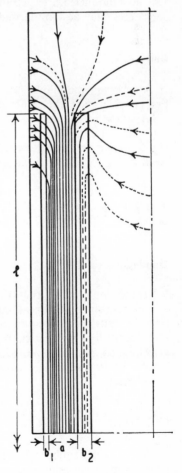

Fig. 4.10   Dimension diagram for reactance formulae

Fig. 4.11   Leakage flux diagram

105

(7) When the frame size is varied to such an extent that the mean turn length is affected considerably, the expression $M_w \propto c + d$ may be taken as an approximation.

(8) Under the circumstances cited in (7) the reactance proportionality is then,

$$X \propto \left\{ \frac{1}{d^2}(c+d)\frac{1}{d^2}\frac{1}{\ell}(c-d) \right\}, \ i.e., \ X \propto \left(\frac{c^2 - d^2}{d^4 \ell} \right)$$

Note: If the winding copper factor current densities, and magnetic flux densities remain constant, c, d and ℓ may vary individually, but the product $d^2 \ell \ (c-d)$ must remain constant.

TRANSFORMER DESIGN

The outstanding features of transformer design which should always be carefully considered may be summarised as follows:

Reliability

In both design and manufacture the reliability of a transformer should be given first consideration, the principal factors being:
  (a) sound mechanical construction
  (b) liberal oil ducts and electrical clearances
  (c) current and flux densities such as to avoid local heating
  (d) liberal radiating surfaces
  (e) good oil.
All these factors tend to increase the initial cost of a transformer, but any disadvantages are outweighed by the saving in cost of repairs and the advantage of ensuring continuity of service.

Iron loss

It has been pointed out that the iron loss of a transformer should be reduced to a minimum since this loss is constant and continuous during the whole of the time the transformer is connected to the supply. While, however, a low iron loss is undoubtedly a great advantage, it should not be obtained at the expense of a very high copper loss, and thereby the probable sacrifice of reliability, and the whole question should be considered in the light of the previous remarks on this subject.

Copper loss

It should be remembered that the operation of a commercial transformer of given proportions and for a given output entails the loss of a certain and fairly constant total amount of power, consisting generally of iron and copper losses.

Acceptance of this statement implies, therefore, that if the iron loss of a commercial transformer is very low, the copper loss will be correspondingly high and vice versa. A high copper loss may represent a distinct menace to the coil insulation on account of the possibility of excessive undetermined local heating occurring in the interior of the coils, and therefore it is far sounder practice to design for a closer agreement between copper and iron losses than to aim for a very low iron loss at the expense of the copper loss or of the electrical or thermal design of the transformer.

Impedance

The effect of impedance upon regulation has already been discussed, and it has been shown that a high impedance entails a high regulation at the lower power factors. High regulation as such is certainly an undesirable feature, but it must not be overlooked that a reasonably high impedance is the best protection against the damage a transformer is liable to sustain due to external short circuits. In deciding, therefore, what impedance a transformer should have, reliability and safety should be given preference. In this connection the question of parallel operation is of importance, since the percentage impedance of a transformer which has to operate in parallel with an existing transformer should be equal to the percentage impedance of the latter. When, however, such an impedance is too low for satisfactory service under modern conditions, it is better to insert additional reactance in series with the existing transformer, than perpetuate a design which is no longer in accordance with sound modern practice.

FORMULAE FOR THE DETERMINATION OF THE PERFORMANCE OF A TRANSFORMER

Efficiencies

Percentage efficiency at any load and any power factor $\cos \phi_2$

$$\eta = \left\{ 1 - \frac{\text{percentage iron loss} + \text{percentage copper loss}}{(\cos \phi_2 \times 100) + \text{percentage iron loss} + \text{percentage copper loss}} \right\} 100$$

The percentage losses to be inserted in the above formula are first determined for the load under consideration, as indicated in the following Table 4.8.

Table 4.8

Load	5/4	1/1	3/4	1/2	1/4
Percentage iron loss	$4/5'P_f$	$'P_f$	$4/3'P_f$	$2'P_f$	$4'P_f$
Percentage copper loss	$5/4'P_c$	$'P_c$	$3/4'P_c$	$1/2'P_c$	$1/4'P_c$

Where $'P_f$ = percentage iron loss at normal voltage

               $'P_c$ = percentage copper loss at full load

Example.—1000 kVA, 3-phase, 50 Hz, 6600/433 volts. Iron loss, 1770 watts. Copper loss, 11 640 watts at full load at 75°C. Impedance voltage, 4·75% at full load at 75°C.

(*a*) At full load, unity power factor

$$\text{percentage iron loss} = \frac{1770 \times 100}{1000 \times 1000} = 0.177\%$$

$$\text{percentage copper loss} = \frac{11\,640 \times 100}{1000 \times 1000} = 1.164\%$$

$$\cos \phi_2 = 1.0$$

$$\text{percentage efficiency} = \left\{ 1 - \frac{0.177 + 1.164}{(1 \times 100) + 0.177 + 1.164} \right\} 100 = 98.67\%$$

(*b*) At 5/4 load, unity power factor

$$\text{percentage iron loss} = \frac{4}{5} 0.177 = 0.142\%$$

$$\text{percentage copper loss} = \frac{5}{4} 1.164 = 1.455\%$$

$$\cos \phi_2 = 1.0$$

$$\text{percentage efficiency} = \left\{ 1 - \frac{0.142 + 1.455}{(1 \times 100) + 0.142 + 1.455} \right\} 100 = 98.43\%$$

(*c*) At 3/4 load $\cos \phi_2 = 0.8$

$$\text{percentage iron loss} = \frac{4}{3} 0.177 = 0.236\%$$

$$\text{percentage copper loss} = \frac{3}{4} 1.164 = 0.874\%$$

$$\text{percentage efficiency} = \left\{ 1 - \frac{0.236\% + 0.874}{(0.8 \times 100) + 0.236 + 0.874} \right\} 100 = 98.90\%$$

Leakage reactance

Percentage leakage reactance voltage at full load,

$$V_X = \sqrt{(V_Z^2 - V_R^2)}$$

where
$\quad V_Z$ = percentage impedance voltage at full load
$\quad V_R$ = percentage resistance voltage at full load

The percentage resistance voltage at full load may be taken as the percentage copper loss at full load.

The percentage leakage reactance voltage and the percentage resistance voltage for loads other than full load can be obtained by multiplying the full-load values by the factors given in Table 4.9.

Table 4.9

Load	5/4	1/1	3/4	1/2	1/4
Factor	5/4	1	3/4	1/2	1/4

Example.—1000 kVA, 3-phase, 50 Hz, 6600/433 volts. Iron loss, 1770 watts. Copper loss, 11 640 watts at full load at 75°C. Impedance voltage 4·75% at full load at 75°C.

(*a*) To obtain percentage reactance voltage at full load from the full-load percentage impedance voltage and the full-load percentage resistance voltage

full-load percentage impedance voltage = $4·75\%$

$$\text{full-load percentage resistance voltage} = \frac{11\,640 \times 100}{1000 \times 1000} = 1·164\%$$

full-load percentage reactance voltage = $\sqrt{(4·75^2 - 1·164^2)} = 4·604\%$

(*b*) To obtain percentage reactance voltage at 5/4 load

$$\text{percentage reactance voltage at 5/4 load} = \frac{5}{4}4·604 = 5·755\%$$

(*c*) To obtain percentage resistance voltage at 3/4 load

$$\text{percentage resistance voltage at 3/4 load} = \frac{3}{4}1·164 = 0·874\%$$

Regulation

Percentage regulation at any load and any power factor

$\cos\phi_2 = (\text{percentage resist. voltage} \times \cos\phi_2) + (\text{percentage react. voltage} \times \sin\phi_2)$

$$+ \frac{\{(\text{percentage react. voltage} \times \cos\phi_2) - (\text{percentage resist. voltage} \times \sin\phi_2)\}^2}{200}$$

The percentage resistance voltage and the percentage reactance voltage to be inserted in the above formula must be determined for the load under consideration from Table 4.10.

Table 4.10

Load	5/4	1/1	3/4	1/2	1/4
Factor	5/4	1	3/4	1/2	1/4

Table 4.11
COSINES AND SINES

$\cos\phi_2$	1·000	0·950	0·900	0·850	0·800	0·750	0·700	0·650	0·600	0·550	0·500
$\sin\phi_2$	0	0·312	0·436	0·527	0·600	0·661	0·714	0·760	0·800	0·835	0·866

Example.—1000 kVA, three-phase, 50 Hz, 6600/433 volts. Iron loss, 1770 watts. Copper loss, 11 640 watts at full load at 75°C. Impedance voltage, 4·75% at full load at 75°C.

(*a*) At full load, unity power factor:

$$\text{percentage resistance voltage} = 1·164\%$$
$$\text{percentage reactance voltage} = 4·604\%$$
$$\cos\phi_2 = 1·0 \qquad \sin\phi_2 = 0.$$

percentage regulation

$$= (1 \cdot 164 \times 1 \cdot 0) + (4 \cdot 604 \times 0) + \frac{\{(4 \cdot 604 \times 1 \cdot 0) - (1 \cdot 164 \times 0)\}^2}{200} = 1 \cdot 27\%$$

(b) At 5/4 load, unity power factor:

$$\text{percentage resistance voltage} = 1 \cdot 455\%$$

$$\text{percentage reactance voltage} = \frac{5}{4} 4 \cdot 604 = 5 \cdot 755\%$$

percentage regulation

$$= (1 \cdot 455 \times 1 \cdot 0) + (5 \cdot 755 \times 0) + \frac{\{(5 \cdot 755 \times 1 \cdot 0) - (1 \cdot 455 \times 0)\}^2}{200} = 1 \cdot 62\%$$

(c) At 3/4 load, $\cos \phi_2 = 0 \cdot 8$

$$\text{percentage resistance voltage} = 0 \cdot 874\%$$

$$\text{percentage reactance voltage} = \frac{3}{4} 4 \cdot 604 = 3 \cdot 453\%$$

$$\sin \phi_2 = 0 \cdot 6.$$

percentage regulation

$$= (0 \cdot 874 \times 0 \cdot 8) + (3 \cdot 453 \times 0 \cdot 6) + \frac{\{(3 \cdot 453 \times 0 \cdot 8) - (0 \cdot 874 \times 0 \cdot 6)\}^2}{200} = 2 \cdot 80\%$$

The performance of a transformer can be obtained from the preceding formulæ *for the temperature at which the copper loss and impedance voltage figures are given.* The necessary data for the calculation of the performance at any other temperature must be obtained from the manufacturer.

Note. Provided the applied voltage and frequency do not alter, the *watts* iron loss of a transformer does not vary with load or temperature. The *percentage* iron loss therefore varies inversely as the load.

The *watts* copper loss of a transformer varies as the square of the load, and therefore the *percentage* copper loss varies directly as the load.

The copper loss consists of an I^2R loss, a stray loss dependent upon the frame size, and a loss due to eddy currents in the conductors themselves. The I^2R loss increases by approximately 0·4% per degree Celsius increase in temperature, but the eddy-current component decreases in the same ratio. The stray loss is independent of frequency and temperature.

CONTINUOUS EMERGENCY RATED (CER) SYSTEM TRANSFORMERS

The continuous emergency rated system transformer is a B.S. rated transformer when operating under ONAN cooling conditions and would thus have a temperature rise of 60°C in oil and 65°C by resistance, in a standard ambient temperature. Under emergency conditions it has a continuous rating, when operating with OFAF cooling, of 200% of its ONAN rating. When so operating, it is designed to have a hot-spot temperature not exceeding 115°C with an ambient air temperature of 5°C.

A typical rating would be 12/24 MVA, with a voltage ratio of 33/11·5 kV and having an impedance of approximately 1·0% per MVA. The practice of installing two transformers in a substation with 100% spare capacity at British Standard rating is now considered unnecessary as definite economic advantages can be obtained by the use of CER system transformers. Under normal system operating conditions, two such units would be operated in parallel with each taking half of the load and, in the event of a failure of one transformer, the healthy transformer would be capable of carrying the substation load for the duration of the emergency. Such emergencies are considered to occur so infrequently that the standard operational temperatures recommended in C.P. 1010 can be exceeded during the emergency. It is assumed that the small amount of transformer life expectancy that may be lost is justified and that a maximum hot-spot temperature of 115°C is permissible.

This type of transformer has three distinct thermal ratings:

(i) An output rating of 50% of its continuous emergency rating with a B.S. ONAN temperature rise of 60°C in oil and an average winding rise of 65°C.

(ii) An output rating of 80% of the continuous emergency rating with a B.S. OFAF temperature rise of 60°C in oil and an average winding rise of 65°C.

(iii) 100% continuous emergency rating (CER) with a maximum hot-spot temperature rise of 110°C when operating in an ambient temperature of 5°C.

Chapter 5

Transformer efficiencies

It is generally known that the maximum efficiency of any transformer occurs at that load at which the iron loss is equal to the copper loss, but the significance of this statement is often entirely misunderstood.

Consider first the construction of an ordinary efficiency curve from the data usually available, namely, the output and the percentage iron and full load copper losses. The efficiency of any u.p.f. load is calculated from the usual expression, output divided by output plus total losses, or

$$\text{percentage efficiency} = \frac{100}{100 + ''P} 100 \tag{5.1}$$

where $''P$ is the sum of the percentage iron and copper losses *at the output under consideration*. The iron loss in watts (or kW) being constant at all loads (provided the excitation remains constant), varies inversely as the load when expressed as a percentage, while the copper loss in watts (or kW) varying as the square of the load, is directly proportional to the load when expressed as a percentage. Equation 5.1 can therefore be written

$$\text{percentage efficiency} = \frac{100}{100 + ''P_f + ''P_c} 100 \tag{5.2}$$

where $''P_f$ and $''P_c$ are the percentage iron and copper losses *at the output under consideration*. The application of eqn. 5.2 is shown in Table 5.1.

In order to calculate these percentage efficiencies accurately to two places of decimals on a 10 inch slide rule, the formulæ may be written as follows, where $'P_f$ and $'P_c$ are the percentage full-load losses

$$\text{percentage efficiency at 5/4 load} = 100 - \frac{100\left(\dfrac{4'P_f}{5} + \dfrac{5'P_c}{4}\right)}{100 + \dfrac{4'P_f}{5} + \dfrac{5'P_c}{4}}$$

$$\text{percentage efficiency at 1/1 load} = 100 - \frac{100('P_f + 'P_c)}{100 + 'P_f + 'P_c}$$

112

Table 5.1

Load	5/4	1/1	3/4	1/2	1/4
Percentage iron loss $''P_f$	$4'P_f/5$	$'P_f$	$4'P_f/3$	$2'P_f$	$4'P_f$
Percentage copper loss $''P_c$	$5'P_c/4$	$'P_c$	$3'P_c/4$	$'P_c/2$	$'P_c/4$
Percentage total loss $''P$	$4'P_f/5 + 5'P_c/4$	$'P_f + 'P_c$	$4'P_f/3 + 3'P_c/4$	$2'P_f + 'P_c/2$	$4'P_f + 'P_c/4$
Percentage efficiency	$\dfrac{100 \times 100}{100 + \dfrac{4'P_f}{5} + \dfrac{5'P_c}{4}}$	$\dfrac{100 \times 100}{100 + 'P_f + 'P_c}$	$\dfrac{100 \times 100}{100 + \dfrac{4'P_f}{3} + \dfrac{3'P_c}{4}}$	$\dfrac{100 \times 100}{100 + 2'P_f + \dfrac{'P_c}{2}}$	$\dfrac{100 \times 100}{100 + 4'P_f + \dfrac{'P_c}{4}}$

For a transformer having an iron loss and full-load copper loss each of 1 %, the calculations become:

Load	5/4	1/1	3/4	1/2	1/4
Percentage iron loss $''P_f$	$4/5 = 0.8$	1.0	$4/3 = 1.33$	2.0	4.0
Percentage copper loss $''P_c$	$5/4 = 1.25$	1.0	$3/4 = 0.75$	$1/2 = 0.5$	$1/4 = 0.25$
Percentage total loss $''P$	$0.8 + 1.25 = 2.05$	2.0	$1.33 + 0.75 = 2.08$	$2.0 + 0.5 = 2.5$	$4.0 + 0.25 = 4.25$
Percentage efficiency	$\dfrac{100 \times 100}{100 + 2.05} = 97.99$	$\dfrac{100 \times 100}{100 + 2.0} = 98.04$	$\dfrac{100 \times 100}{100 + 2.08} = 97.96$	$\dfrac{100 \times 100}{100 + 2.5} = 97.56$	$\dfrac{100 \times 100}{100 + 4.25} = 95.92$

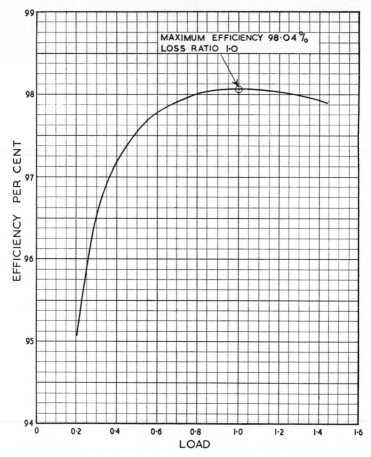

Fig. 5.1   *Efficiency curve for a transformer having iron and full-load copper losses each of* 1.0%

$$\text{percentage efficiency at 3/4 load} = 100 - \frac{100\left(\dfrac{4'P_f}{3} + \dfrac{3'P_c}{4}\right)}{100 + \dfrac{4'P_f}{3} + \dfrac{3'P_c}{4}}$$

$$\text{percentage efficiency at 1/2 load} = 100 - \frac{100\left(2'P_f + \dfrac{'P_c}{2}\right)}{100 + 2'P_f + \dfrac{'P_c}{2}}$$

$$\text{percentage efficiency at 1/4 load} = 100 - \frac{100\left(4'P_f + \dfrac{'P_c}{4}\right)}{100 + 4'P_f + \dfrac{'P_c}{4}}$$

Figure 5.1 is the corresponding efficiency curve which shows that the *maximum efficiency occurs at full load*, at which the iron and copper losses are equal.

If now, for the purpose of comparison, efficiencies are calculated in the manner shown above for other similar transformers having the same total full-load losses, *i.e.*, 2%, but different loss ratios, a family of curves can be derived, as shown in Fig. 5.2, each curve having a different maximum efficiency which in turn occurs at a different load in each case.

In the instances taken the maximum efficiency in each case is well defined, and a curve of 'maximum efficiencies' can be drawn through the maxima of the efficiency curves, as shown in Fig. 5.2

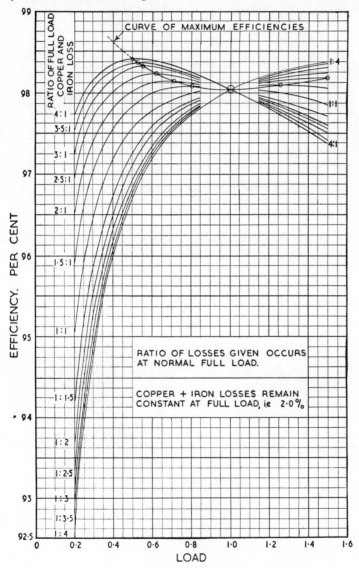

Fig. 5.2   *Effect of variation of ratio of full-load losses upon transformer efficiencies*

115

We now have to determine if the curve of 'maximum efficiencies' cuts the maxima of the different individual efficiency curves at such loads in each case as make the iron and copper losses equal at those loads. It can be said at the outset that it does, and if the full-load losses be corrected in each instance for the load giving maximum efficiency, it will be found that the iron and copper losses are equal at that load.

It is, however, more satisfactory to prove the statement mathematically, especially as the proof is so simple.

For any transformer the efficiency is a maximum when the total losses are a minimum, and mathematically the total losses are a minimum when the differential of their sum is equal to zero, that is, when

$$\frac{d(''P_f + ''P_c)}{dW} = 0$$

From eqn. 5.2 we have

$$\text{percentage efficiency} = \frac{100}{100 + ''P_f + ''P_c} \, 100$$

where $''P_f$ and $''P_c$ are the percentage losses *at the loads under consideration*. If $'P_f$ and $'P_c$ are inserted as percentage *full-load losses*, eqn. 5.2 becomes

$$\text{percentage efficiency} = \frac{100}{100 + \dfrac{'P_f}{W} + 'P_c W} \, 100 \qquad (5.3)$$

where W is the fraction of full load for maximum efficiency. Therefore for minimum losses

$$''P = \frac{'P_f}{W} + 'P_c W, \text{ is a minimum and } \frac{d''P}{dW} = -\frac{'P_f}{W^2} + 'P_c$$

for minimum losses, $$\frac{d''P}{dW} = 0$$

therefore, $$-\frac{'P_f}{W^2} + 'P_c = 0, \text{ and } W^2 = \frac{'P_f}{'P_c} \qquad (5.4)$$

and, $$W = \sqrt{\frac{'P_f}{'P_c}}$$

We thus find that the fraction of full load at which the efficiency of any given transformer is at maximum, is

$$W = \sqrt{\frac{'P_f}{'P_c}}$$

where $'P_f$ and $'P_c$ are the percentage iron and full-load copper losses respectively.

From eqn. 5.4 we have,

$$W^2 = \frac{'P_f}{'P_c}$$

from which, dividing both sides by W and multiplying by $'P_c$,

$$W'P_c = \frac{'P_f}{W}$$

As $'P_f/W$ and $'P_cW$ are the percentage iron and copper losses respectively at the load giving maximum efficiency, we thus have the proof of the statement made earlier.

It is interesting to consider next the curve of 'maximum efficiencies' and to see whether an expression can be derived for it, which at the same time is applicable to the determination of the maximum efficiency of a single transformer of any given electrical characteristics. For maximum efficiency, eqn. 5.3 may be rewritten in the following form:

$$\text{maximum percentage efficiency} = \frac{100}{100\dfrac{'P_f}{\sqrt{('P_f/'P_c)}} + 'P_c\sqrt{('P_f/'P_c)}}\, 100$$

multiplying both numerator and denominator of the right-hand side throughout by $\sqrt{('P_f/'P_c)}$

$$\text{maximum percentage efficiency} = \frac{100\sqrt{('P_f/'P_c)}}{100\sqrt{\left(\dfrac{'P_f}{'P_c}\right)} + 'P_f + \dfrac{'P_c'P_f}{'P_c}}\, 100$$

$$= \frac{100\sqrt{('P_f/'P_c)}}{100\sqrt{('P_f/'P_c)} + 2'P_f}\, 100$$

$$= \frac{100W}{100W + 2'P_f}\, 100$$

$$= \frac{50W}{50W + 'P_f}\, 100 \tag{5.5}$$

Equation 5.5 is, therefore, the expression for the curve of 'maximum efficiencies' and applies to any family of efficiency curves for a transformer with different full-load loss ratios. If the *product* of the percentage iron and full-load copper losses remains constant, the curve becomes a straight line. Figure 5.3 shows the curve for a transformer where the sum of the total full-load losses is 2% throughout, and it is interesting to note that the minimum point of the 'maximum efficiencies' curve occurs for a full-load loss ratio of 1·0. Figure 5.3 also shows as a matter of interest the percentage losses and the loss ratios corresponding to the curve of 'maximum efficiencies' plotted against the percentage loads. In order to plot the curve of 'maximum efficiencies' from eqn. 5.5 it is necessary to calculate the value of $'P_f$ for each point on the curve, as $'P_f$ is a variable. It is possible, however, to obtain an expression for eqn. 5.5 which contains only one variable, namely, the load. Let the *constant* total percentage full-load loss be $'P$,

then,
$$'P = 'P_c + 'P_f \tag{5.6}$$

also,
$$W = \sqrt{\frac{'P_f}{'P_c}} \tag{5.7}$$

117

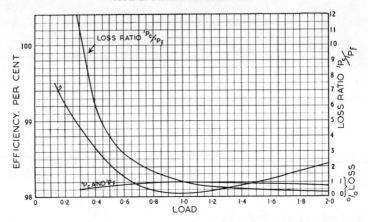

Fig. 5.3 Curves of maximum efficiencies, loss ratio and percentage losses for transformers having percentage iron and full-load copper losses each of 1·0%

from eqn. 5.6
$$'P_c = 'P - 'P_f$$

from eqn. 5.7
$$'P_c = \frac{'P_f}{W^2}$$

therefore,
$$'P - 'P_f = \frac{'P_f}{W^2}$$

and,
$$'P_f = \frac{'PW^2}{(1 + W^2)}$$

Equation 5.5 thus becomes,

$$\text{maximum percentage efficiency} = \frac{50W100}{50W + \dfrac{'PW^2}{1 + W^2}} = \frac{100(W^2 + 1)}{W^2 + 1 + \dfrac{'PW}{50}}$$

In all the foregoing calculations it has been assumed that the load power factor is unity. All the calculations may, however, be applied to transformers operating on loads of other power factors if the figures used for the percentage losses are increased throughout in inverse proportion to the power factor. For instance, a transformer having 1·0% iron loss and 1·4% full-load copper loss will have the same efficiencies at all loads at 0·8 power factor, as a transformer having 1·25% (1·0/0·80) iron loss and 1·75% (1·4/0·80) full-load copper loss has at unity power factor.

It is a well known fact that the copper loss, and consequently the efficiency, of a transformer varies according to the temperature of the windings. For an investigation of this character it is only necessary to assume that all losses and efficiencies are based on the same temperature. It is neither necessary nor desirable to assume different temperatures for different loads, because in practice load fluctuations are such that a transformer may at times be operating at full load with a low temperature or at a light load with a high temperature.

A MAXIMUM EFFICIENCY CHART

It has been shown in the foregoing that the maximum efficiency of any transformer occurs at that load at which the variable copper loss becomes equal to the fixed iron loss. That is, the percentage load at which the efficiency of any given transformer is a maximum is

$$\text{percentage load} = (100W) = \sqrt{\left(\frac{\text{iron loss}}{\text{copper loss}}\right)}\,100 \qquad (5.8)$$

In this equation the losses may be expressed in watts or as percentages, since only their *ratio* is involved. The copper loss is the full-load copper loss, and both losses, when expressed as percentages, are percentages of full-load output.

It will be realised from a study of eqn. 5.8 that the load giving maximum efficiency is not dependent upon the absolute values (*per se*) of the losses, but only upon their relative values, so that it is possible to plot a curve to the equation without any reference whatever to absolute values of the losses or to any other characteristic. The curve is shown by that marked W on Fig. 5.4, and it applies to any and all transformers.

From the curve it will be noticed that maximum efficiency occurs at full load, for instance, when the iron loss and full-load copper loss are equal. Similarly, maximum efficiency occurs at half load when the full-load copper loss is four times the iron loss.

Often it is desirable to be able to ascertain rapidly not only the load at which the efficiency is a maximum but also what the efficiency is at that load. The actual efficiency of any given transformer depends, however, upon the absolute values of the losses, and, therefore, it is not possible to determine the former from a single curve, which is of general application, such as was found possible in the case of the curve for percentage loads giving maximum efficiency.

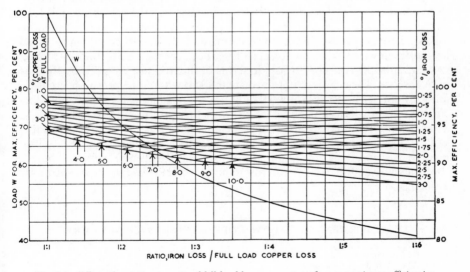

Fig. 5.4   *Effect of variation in ratio of full-load losses upon transformer maximum efficiencies*

119

For any given transformer the maximum percentage efficiency at the percentage load $100W$ of eqn. 5.8 is,

$$\text{maximum percentage efficiency} = \frac{100W}{100W + 2(\text{iron loss})} \, 100 \qquad (5.9)$$

which may be written in the alternative form,

maximum percentage efficiency =

$$\frac{100W}{100W + 2\left[\dfrac{\text{total full-load losses} \times W^2}{1 + W^2}\right]} \, 100 \qquad (5.10)$$

In these equations the load $100W$, the iron loss, and the total full-load loss are all expressed as percentages of full-load output. W is the load expressed as a fraction of the full-load output.

It will be seen that these equations do not involve simple ratios only, but that their solutions depend upon absolute values.

It is possible, however, to draw a series of efficiency curves based upon definite percentage iron losses for the loss ratios shown on Fig. 5.4, and which embrace the range of losses commonly encountered. This has been done on the present chart, for iron losses ranging from 0·25 to 3% in 0·25% steps. Smaller increments may be included as desired. The efficiencies corresponding to eqns. 5.9 and 5.10 were calculated for each of the loss ratios marked on the abscissa of the chart, and for the values of W derived from the curve so marked and for the iron losses selected.

These 'iron loss' efficiencies (so-called in order to distinguish them from the 'copper loss' efficiency curves next referred to) are shown plotted against the loss ratios, and they are marked with the corresponding absolute values of percentage iron losses.

From these curves it is also easy to draw another series showing the relationships between loss ratios and maximum efficiencies for the percentage full-load copper losses which correspond to the stated loss ratios and percentage iron losses. These curves are derived by locating and joining up points on the 'iron loss' efficiency curves which, for each loss ratio, correspond to the stated iron loss multiplied by the copper loss in terms of the iron loss as given by the loss ratio abscissa. The curves so obtained are marked with the respective percentage full-load copper losses, as shown in Fig. 5.4. For example, the 3% copper loss curve is obtained as follows. At a 1:1 loss ratio the corresponding iron loss is 3%, and the point required at this ratio is the intersection of the 1:1 abscissa value and the 3% 'iron loss' efficiency curve; at 1:2 loss ratio the corresponding iron loss is 1·5% and the point required at this ratio is the intersection of the 1:2 abscissa value and the 1·5% 'iron loss' efficiency curve; and so on, as shown by Table 5.2.

Table 5.2
THREE PER CENT 'FULL-LOAD COPPER LOSS' EFFICIENCY CURVE

(1) *Loss ratio*	1:1	1:2	1:3	1:4	1:5	1:6
(2) *Corresponding iron loss*	3%	1·5%	1%	0·75%	0·6%	0·5%

The required points for drawing the 'copper loss' efficiency curve then lie at the intersections of the abscissa values (1) in the above table, and the corresponding 'iron loss' efficiency curves (2).

The 'copper loss' efficiency curves may be omitted from the chart without impairing its usefulness, if so desired.

The maximum efficiency can be determined, therefore, from the full-load copper loss or the iron loss. The double series of efficiency curves, however, completes the presentation of the interdependence of all the factors involved in the study.

From Fig. 5.4, therefore it is easy to determine the percentage load giving maximum efficiency for any transformer and the maximum efficiency corresponding to that load.

As a case in point, suppose we have a transformer in which the iron loss is 0·75% and the copper loss 2·25%. This gives a loss ratio of 1:3, and by reference to curve W of the chart load W, giving maximum efficiency, is seen to be 57·7%. The efficiency at the load W is found by tracing vertically upwards from the loss ratio of 1:3, to intersect the efficiency curve marked 0·75% iron loss, projecting horizontally to the right to the efficiency ordinate, from which is read the figure of 97·4%. Thus, for the case cited, the maximum efficiency of the transformer is 97·4%, and it occurs at a load of 57·7% of full load.

An alternative example is where the iron loss is 1·0% and the copper loss, 5·0%. In this case the loss ratio is 1:5, and from curve W the load giving maximum efficiency is seen to be 44·7%. Tracing vertically upwards from the loss ratio of 1:5 to intersect the efficiency curve marked 5% copper loss, and projecting horizontally to the right to the efficiency ordinate, gives a figure of 95·75%. In this case, therefore, the maximum efficiency of the transformer is 95·75%, and it occurs at 44·7% of full load.

This chart is a useful supplement to the nomogram* which is commonly used for determining the percentage efficiencies at given fractional loads. Second decimal place accuracy is attainable by adopting sufficiently open scales.

*See pages 98 and 99.

Chapter 6

The effect of load factor upon the value of transformer losses

It has already been shown how the respective values of transformer iron and copper losses are influenced by the transformer load. The present study provides diagrams showing graphically (a) the relationship between transformer load factors and the load factors of the losses; (b) copper-loss correction curves for temperature for all loads from zero to full load; (c) curves showing the relationship between total copper loss and temperature; and (d) accumulated compound interest and annuity curves which are useful in determining the capital value of the annual cost of the losses.

In order to obtain a true appreciation of the capital value of the annual cost of transformer losses, proper account should be taken of the transformer load factor which thus involves a consideration of the load factors of the separate losses. For this purpose it is only necessary to consider the iron loss and the total load copper loss, as other losses, such as dielectric loss and copper loss due to the no-load current, are of negligible value.

For all practical purposes the iron loss of a transformer is constant for a given excitation, that is, for a given primary voltage, and the iron-loss load factor has a constant value of 100% during the entire period for which the transformer remains connected to the supply. The iron loss is practically independent of the load and, therefore, the iron loss load factor has the same value from no-load to full load, assuming the primary voltage is maintained constant.

The total load copper loss, on the other hand, is dependent upon the square of the load current, the proportion of true load I^2R loss to eddy current loss and the average temperature of the windings. The load factor of the total load copper loss is dependent upon the square of the load and upon the shape of the transformer load curve. Figure 6.1 shows the curves representing the extreme upper and lower limits of the relationship between transformer load factor and loss load factors which correspond to the cases where (1) the maximum load is sustained for a fraction of the total time equal to the value of the load factor, and (2) the maximum load is reached momentarily only and a fractional load equivalent to the value of the load factor is sustained over the whole period covered by the time under consideration. Typical load curves for 50% load factor are shown in Diagrams 1 and 2 of Fig. 6.1 to illustrate these conditions. Curves (a) and (b) illustrate the upper and lower limits of the relationship

122

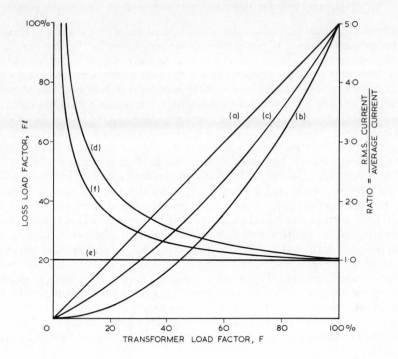

LOAD FACTOR = AVERAGE LOAD/MAXIMUM LOAD

Losses for any load factor produced as in case (a) are proportional to the load factor.

Losses for any load factor produced as in case (b) are proportional to the square of the load factor.

Losses for any load factor produced as in case (c) are proportional to one half the sum of the load factor and the square of the load factor.

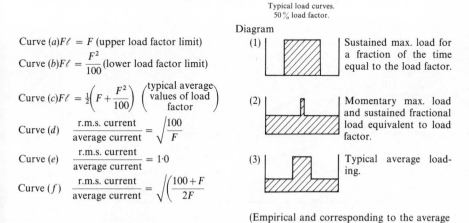

Typical load curves.
50 % load factor.

Diagram

Curve (a)$F\ell = F$ (upper load factor limit)

Curve (b)$F\ell = \dfrac{F^2}{100}$ (lower load factor limit)

Curve (c)$F\ell = \frac{1}{2}\left(F + \dfrac{F^2}{100}\right)$ $\begin{pmatrix}\text{typical average}\\ \text{values of load}\\ \text{factor}\end{pmatrix}$

Curve (d) $\dfrac{\text{r.m.s. current}}{\text{average current}} = \sqrt{\dfrac{100}{F}}$

Curve (e) $\dfrac{\text{r.m.s. current}}{\text{average current}} = 1.0$

Curve (f) $\dfrac{\text{r.m.s. current}}{\text{average current}} = \sqrt{\left(\dfrac{100+F}{2F}\right)}$

(1) Sustained max. load for a fraction of the time equal to the load factor.

(2) Momentary max. load and sustained fractional load equivalent to load factor.

(3) Typical average loading.

(Empirical and corresponding to the average Curve (c) only)

Fig. 6.1 Curves showing effect of transformer load factor upon loss load factor and current form factor

123

referred to and the respective equations to the curves which are given, are exact. Between these limits it is possible to obtain any value for the loss load factor $Fℓ$ for a given value of the transformer load factor F. Curve (c), which lies mid-way between curves (a) and (b), shows the relationship between transformer load factor and loss load factor for the typical average loading curve shown by Diagram 3 of Fig. 6.1. The equation corresponding to curve (c) is empirical only and would be different for ratios of transformer load factor to loss factor differing from those shown by curve (c). For any given maximum loading a given load factor can be obtained with a load curve having the maximum load sustained for a greater fraction of time and a fractional load of smaller magnitude sustained for a correspondingly shorter time as compared with the typical average load curve. Similarly the same load factor can be obtained with a load curve having the same maximum value sustained for a shorter period of time compared with the typical average and a fractional load greater than that of the typical average but sustained for a correspondingly greater length of time.

In the first mentioned case the loss load factor will be greater than that shown by curve (c) of Fig. 6.1, while in the second case the loss load factor will be less than the value corresponding to curve (c). The curves (d), (e), and (f) show the corresponding relationship of transformer load factor to the ratio of r.m.s. load current to average current (i.e., to the form factor of the load curve) and are useful in evaluating the total load copper loss for a given average load since the total copper loss varies as the square of the r.m.s. current and not the average current.

Any shape of transformer load curve occurring in practice will give loss load factors for any given transformer load factor lying between the limiting values of curves (a) and (b) of Fig. 6.1, the same applying to the values of the ratios of r.m.s. load current to average current.

The loss load factor is obtained by squaring all the selected ordinates of the load curve and by evaluating the ratio of the average to the maximum values of the loss load curve thus obtained.

In calculating the cost of the total load copper loss it is important to take due account of the true total copper loss at all loads involved at every point of the

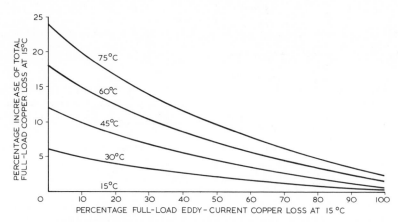

Fig. 6.2   Curves showing the percentage increase of total full-load copper loss at 15°C at temperatures up to 75°C, and for different percentage eddy-current copper losses.
The eddy-current copper losses are given as percentage of the I^2R copper loss at 15°C

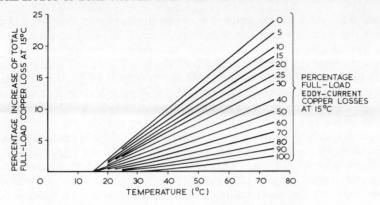

Fig. 6.3 Curves showing the percentage increase of total full-load copper loss at 15°C at temperatures up to 75°C for different percentage eddy-current copper losses. The eddy-current copper losses are given as a percentage of the I^2R copper loss at 15°C

load curve on account of the fact that the ratio between true I^2R loss and eddy-current loss is not constant at all temperatures. Moreover, the variations of I^2R loss and eddy-current loss from their respective values at a given total load and temperature are in opposite directions. That is, the I^2R loss increases with increasing temperature and the eddy-current loss decreases with increasing temperature. With increasing temperature the resistance of the windings increases at the rate of 0·4 % for every 1°C increase in temperature from 15°C, so that if the true I^2R load copper loss at 15°C is P_{cr} watts, the load I^2R load copper loss at a higher temperature t is,

$$P_{cr}' = P_{cr}\left[1 + \frac{0\cdot4(t-15)}{100}\right] \text{watts} \tag{6.1}$$

The eddy-current copper loss, on the other hand, decreases because of the increase of resistance of the windings with temperature. If the eddy-current copper loss at, say 15°C is P_{ce} then the eddy-current loss at a higher temperature t is

$$P_{ce}' = \frac{P_{ce}}{\left[1 + \frac{0\cdot4(t-15)}{100}\right]} \text{watts} \tag{6.2}$$

Figures 6.2 and 6.3 are given to show the influence of various percentage full-load, eddy-current copper losses upon the total full-load copper loss (*i.e.,* I^2R loss + eddy loss) at 15°C for various total temperatures. The curves are strictly accurate and apply to any transformer having a percentage full-load, eddy-current copper loss at 15°C within the total range covered by the curves.

Tendered guarantees for transformers usually state the total copper loss at full load and at 75°C. This is a reference temperature, and does not necessarily bear any relationship to the maximum or average operating temperature of the transformer. Thus, in calculating the overall value of any given tendered transformer, that is, in determining the total actual value obtained by adding the capital cost of the transformer to the capitalised value of the annual cost of the losses, it is necessary to assume that the transformer will be fully loaded for some portion of the time during which it is in operation, that is, that the maximum value of the load curve corresponds to the normal full-load continuous rating of the transformer; or, if the normal full-load rating is not likely to be reached by the duty imposed upon the apparatus, the guaranteed full-load copper loss should be corrected to correspond with the anticipated maximum loading, in which case due account should be taken both of the lower temperature corresponding to the load concerned and the different variations of I^2R load copper loss and eddy-current copper loss with the load as already mentioned.

Figure 6.4 shows the values of the percentage additive corrections which should be made to the total load copper loss at 15°C for temperature at all loads from zero to full load. These particular curves apply to transformers having eddy-current copper losses at 15°C of 10% of the value of the true I^2R copper losses at the same temperature, and to transformers having an oil temperature rise of 50°C, a temperature rise of the windings by resistance of 60°C at continuous full loading and working in an ambient temperature of 30°C. The figure gives four sets of correction curves corresponding to ambient temperatures of 10°, 20°, 30° and 40°C, each set containing four curves which apply to transformers having ratios of copper to iron losses of 1, 2, 3 and 4 to 1. Suppose, for instance, the total full-load copper loss of a transformer having a copper to iron loss ratio of 2 to 1 is guaranteed not to exceed 2000 watts at 15°C, then at full load and at the corresponding actual temperature rise and at an ambient temperature of 30°C the actual total copper loss would be $2000 \times 1\cdot228 = 2456$ watts. At half load and at the same ambient temperature the copper loss would be equal to $2000 \times 0\cdot5^2 \times 1\cdot128 = 564$ watts. It should be noted that the total copper loss at the appropriate load should first be determined by multiplying the guaranteed full-load copper loss by the square of the load concerned before applying the correction factor from Fig. 6.4.

An approximate estimate of the copper loss at 15°C may be obtained by dividing the given total load copper loss at 75°C by 1·2. If, on the other hand, the value of the percentage eddy-current loss is known, the two component parts can immediately be separated and the true total values of each at 15°C determined by means of the expressions given in eqns. 6.1 and 6.2.

The foregoing notes deal with loss load factors and it is interesting to consider also how these affect considerations of capital cost and provision of additional plant capacity in anticipation of a growing load.

In determining the specification of a transformer required to operate under certain known conditions, an economic comparison of machines having the same rating, but with different costs and losses, may be made by considering either the total annual running costs, or the total investment cost including the capitalised value of the transformer losses.

The overall annual running cost is the sum of the annual capital charge and the cost of the power tariff, which are as follows:

The annual capital charge $= \dfrac{(r_i + r_t + r_d)C}{100}$

where r_i = annual rate of interest

r_t = annual rate of insurance (which may be negligible)

r_d = annual rate of depreciation, which is determined by the sinking fund method, i.e., for an assumed transformer life of n years, and an interest rate on the deposits of r_c per cent

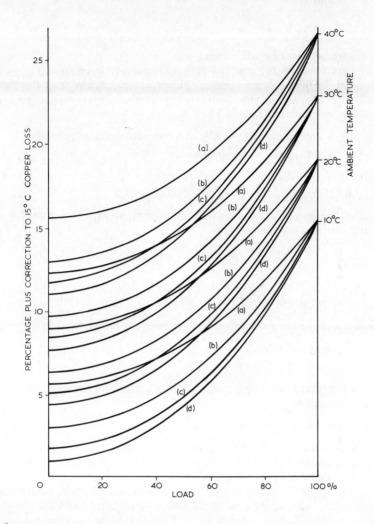

Basis of curves:
  (1) Eddy-current losses at 15°C = 10% of I^2R losses at 15°C
  (2) Oil temperature rise 50°C and temperature rise of windings by resistance 60°C at full continuous load and ambient temperature of 30°C
  (3) Total full-load copper loss guarantee given at 15°C.
Curve (a) corresponds to transformers with a ratio of copper loss to iron loss of 1/1.
Curve (b) corresponds to transformers with a ratio of copper loss to iron loss of 2/1.
Curve (c) corresponds to transformers with a ratio of copper loss to iron loss of 3/1.
Curve (d) corresponds to transformers with a ratio of copper loss to iron loss of 4/1.

Fig. 6.4   *Copper-loss correction curves for temperature for all loads from zero to full load*

$$= \frac{r_c}{\left(1+\dfrac{r_c}{100}\right)^{n-1}}$$

and C = capital cost of transformer.

The power tariff charge consists of two components, the annual costs of the iron losses and of the copper losses. The iron loss, P_f kW, and the full-load copper loss, P_c kW, are obtained from the tender, the copper loss being corrected, if necessary, as indicated earlier. Then, if

p = annual charge per kW of maximum demand (pounds)

q = charge per unit (new pence)

annual cost of iron loss $= P_f(p+8760q/100)$ (6.3)

annual cost of copper loss $= P_cD^2(p+8760qF_\ell/100)$ (6.4)

where D = the demand factor ($=$ maximum demand/full-load rating)
Therefore, the annual running cost

$$= \frac{C(r_i+r_t+r_d)}{100}+P_f\left(p+\frac{8760q}{100}\right)+P_cD^2\left(p+\frac{8760qF}{100}\right)$$

$$= \frac{Cr}{100}+K_1P_f+K_2D^2P_c \tag{6.5}$$

where $\quad r = r_i+r_t+r_d$

$K_1 = (p+8760q/100)$

$K_2 = (p+8760qF_\ell/100)$

Similarly, the total investment cost including the capitalised value of the losses may be expressed as

$$C = (K_1P_f+K_2D^2P_c)100/r \tag{6.6}$$

Equations 6.5 and 6.6 give the alternative expressions which may be used to determine the characteristics of a transformer required to operate at a certain point in a given system.

It will be realised that should the transformer not be excited continuously throughout the year, or the cost of the losses be required for a shorter period, the figure of 8760 in the constants K_1 and K_2 should be modified accordingly.

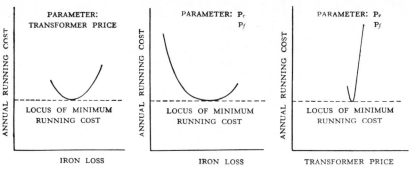

Fig. 6.5   *Curves showing the annual running cost of a transformer*

For a given tariff, a given rate of interest and a known demand factor the curves shown in Fig. 6.5 can be derived.* In each case, the curve shown is the locus of minimum points of a family of curves plotted for a range of values of the parameter stated. It will be seen that the loci have a common minimum, which represents the lowest annual operating cost, and which indicates the parameters (cost and loss values) of the most economic transformer design.

The total annual cost of the losses of a transformer is a minimum when the annual cost of the iron loss and the copper loss are equal, *i.e.* when

$$P_f(p+8760q/100) = P_c D^2(p+8760qF_r/100)$$

$$\frac{P_c}{P_f} = \frac{(p+8760q/100)}{D^2(p+8760qF_r/100)} = \frac{K_1}{D^2 K_2}$$

Now consider the 'yearly economic efficiency'. This is concerned with the *cost* of output and input, the latter being the smaller by the cost of transformation.

$$\eta_e = \frac{\text{annual cost of supply before transformation}}{\text{annual cost of supply after transformation}} 100\%$$

$$= \frac{C_A}{C_A+C_T} 100\%$$

where C_A = annual cost of supply before transformation

$$= DK(p+8760qF_r \cos\phi/100)$$

$$= DKK_3$$

Where K = rated full load output in kVA

     $\cos\phi$ = average power factor of load

     $K_3 = p+8760qF_r \cos\phi/100$

and C_T = annual cost of transformation

$$= Cr/100 + K_1 P_f + K_2 D^2 P_c$$

$$\eta_e = \frac{DKK_3}{DKK_3 + Cr/100 + K_1 P_f + K_2 D^2 P_c} 100\%$$

$$= \frac{1}{1+\dfrac{Cr/100 + K_1 P_f + K_2 D^2 P_c}{DKK_3}} 100\%$$

This is a maximum when $\dfrac{Cr/100 + K_1 P_f + K_2 D^2 P_c}{DKK_3}$ is a minimum

$$\frac{d}{dD} \frac{Cr/100 + K_1 P_f + K_2 D^2 P_c}{DKK_3} = 0$$

i.e. when $Cr/100 + K_1 P_f = K_2 D^2 P_c$

For further interest, Fig. 6.6 and Fig. 6.7 have been prepared, their use being explained as follows:

The curves in Fig. 6.6 show the accumulated value of £1 at various rates of

---

Szwander, W. 'The valuation and capitalization of transformer losses', Journal I.E.E. 1945, Vol. 92, Part II, p. 125.

compound interest against periods of years. They make it possible to determine whether the immediate expenditure of capital is justifiable on the ground that a greater expenditure at a later date will thereby be avoided. As an instance, assume that a transformer is to be provided to deal with a certain load which is likely to increase, and that a transformer capable of giving twice the output could be obtained by expending a further sum of £100, but that the extra output obtainable would not be required for five years. Assuming a rate of interest of 6%, it will be seen from the appropriate curve in Fig. 6.6 that the present expenditure of £100 is equivalent to the expenditure of £134 (i.e., £(1·34 × 100)) five years later. Hence, it follows that it is worth purchasing the larger transformer if a second transformer capable of providing the extra capacity would cost more than £134. In this connection it must be remembered that the larger transformer, which would be underloaded during its first five years of service, would have a higher iron loss and, at the reduced loading, a lower copper loss than the smaller transformer. Consequently the difference, if any, between the annual costs of the losses of the two transformers should be capitalised over the five years as shown

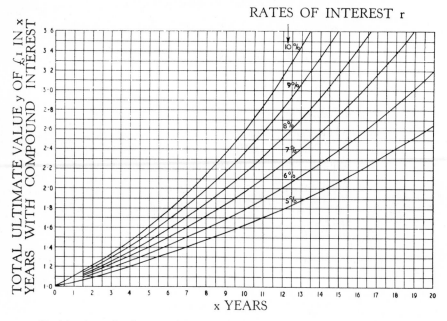

Fig. 6.6   Accumulated compound interest curves (Interest payable at the end of each year)

To determine the total value of y of £1 after x years, when x is greater than the highest value given by the curves, divide x into any number of parts x_1 and x_2 etc. (i.e. $x_1 + x_2 + etc. = x$) and multiply together their corresponding values of y, i.e. y_1, y_2, etc., i.e. $y = y_1 y_2$ etc.

Equation to the curves is $y = \left(1 + \dfrac{r}{100}\right)^x$

130

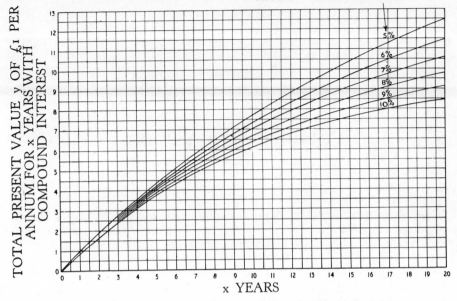

Fig. 6.7   Annuity curves (Interest payable at the end of each year)

$$\text{Equation to the curves is } y = \frac{100}{r}\left[1 - \frac{1}{\left(1 + \dfrac{r}{100}\right)^x}\right]$$

by the method covered by the description of Fig. 6.7. This capital sum should be added to or subtracted from the sum of £134 according to whether the smaller or larger transformer has the lower annual cost of losses.

The curves in Fig. 6.7 give the present value of an annual payment or charge over a number of years for various rates of interest. They are useful for converting an annual charge, such as the cost of losses, into its equivalent capital value. Suppose, for instance, that the annual cost of the losses of a certain transformer is £23, that the useful life of the transformer may be taken as fifteen years, and that the rate of interest is 6%. From the curve for 6% interest the present value of an annual payment of £1 for fifteen years is £9·7. Hence the capitalised value of transformer losses in the case under consideration is £(9·7 × 23) = £223·1. Therefore, if the annual cost of the losses can be reduced by the expenditure of a further capital sum, such expenditure will be justified if it does not exceed £22 for every 10% reduction in annual cost of losses.

Chapter 7

Polyphase connections

In deciding the particular method to be adopted for interconnecting the phases of polyphase transformers or transformer groups there are so many considerations, often conflicting, to be taken into account that the final conclusions must necessarily be of the nature of a compromise. As a consequence, it is not always such a simple matter as at first it appears to choose the most satisfactory combination. It is therefore important to study in detail the various features of the different connections together with the local conditions under which the transformers are to operate. While a large amount of investigation has been carried out and the results published at various times in the different technical journals, such results have considerably more utility if co-ordinated under a single heading, and this chapter is presented with the intention of fulfilling such a function and with the hope that it will acquaint those who have not the time available to study the different aspects in detail, with the reason for the ultimate choice of connections.

The chapter deals with those connections of double-wound transformers which have come to be recognised as the more usual ones for two, three, and six-phase service, and these have been split up into six classes in which all the connections in each class will operate in parallel with one another, though the connections of one class will not operate in parallel with those of any other class.

In addition, polyphase auto transformers are dealt with in a similar manner, except that the question of parallel operation is ignored, as such transformers, having different connections, can seldom so be operated.

The combinations reviewed are as follows:

DOUBLE-WOUND TRANSFORMERS

3 to 3 phase

Star/star	⎫
Star/star with tertiary delta	⎪
Delta/delta	⎬ Class 1
Delta/interconnected star	⎪
Vee/vee	⎪
Tee/tee	⎭

$$\left.\begin{array}{l}\text{Star/delta} \quad \dots \quad \dots \quad \dots \quad \dots \quad \dots \\ \text{Delta/star} \quad \dots \quad \dots \quad \dots \quad \dots \quad \dots \\ \text{Interconnected star/star} \quad \dots \quad \dots \quad \dots \\ \text{Star/interconnected star} \quad \dots \quad \dots \quad \dots\end{array}\right\} \text{Class 2}$$

3 to 6 phase

$$\left.\begin{array}{l}\text{Star/double star or diametric} \dots \quad \dots \quad \dots \\ \text{Delta/double delta} \quad \dots \quad \dots \quad \dots \quad \dots\end{array}\right\} \text{Class 3}$$

$$\left.\begin{array}{l}\text{Star/double delta} \quad \dots \quad \dots \quad \dots \quad \dots \\ \text{Delta/double star or diametric} \quad \dots \quad \dots\end{array}\right\} \text{Class 4}$$

3 to 2 phase

$$\left.\begin{array}{l}\text{Scott} \quad \dots \quad \dots \quad \dots \quad \dots \quad \dots \quad \dots \\ \text{Le Blanc} \quad \dots \quad \dots \quad \dots \quad \dots \quad \dots\end{array}\right\} \text{Class 5}$$

2 to 6 phase

Double Scott ...   . Class 6

AUTO TRANSFORMERS

3 to 3 phase
    Star
    Delta
    Vee
    Tee
    Star/interconnected star
    Interconnected star/star

2 to 3 phase
    Scott and Le Blanc

In the subsequent notes the advantages and disadvantages of some of the combined connections are discussed from the general aspect, and also as applied to particular types of transformer construction. This has only been done in such cases where the general features are modified by the type of transformer to which the connections are applied. In those cases where there is no discrimination between the general and particular features, the type of transformer does not influence the general deductions.

In the following phasor diagrams the phasor shall represent terminal voltages, the 180° phase displacement between primary and secondary terminal voltages shall be disregarded, and the counter-clockwise direction of rotation of the phasors shall be used.

DOUBLE-WOUND TRANSFORMERS

Star/star (Fig. 7.1)

Advantages

General (1) The cross-sectional area of both primary and secondary conductors is a maximum, the number of turns and total quantity of coil insulation per

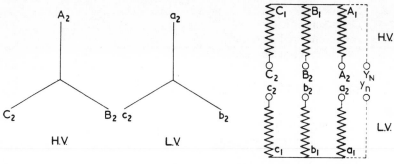

Fig. 7.1   Star/star

phase is a minimum, producing therefore a high copper factor and mechanically robust windings.

(2) Most economical connection for small output, high-voltage transformers.

(3) Both neutrals available for earthing or for giving a *balanced* four-wire supply.

(4) One of the easiest connections to phase-in for parallel operation.

(5) Owing to the relatively large conductors the electrostatic capacitance between turns is high, so that the severity of the stress, due to transient voltage waves striking the windings, is lessened.

(6) If one phase on either side becomes faulty, the remaining two phases can be operated to give a single-phase transformation. The load supplied would be $1/\sqrt{3}$ of the normal three-phase rating.

Particular

Three-phase, core-type units   (6a) If one phase fails, both its primary and secondary windings must be disconnected from the other two phases, and the sound winding (*i.e.*, primary or secondary) of the faulty phase must be short circuited.

(7) Under normal operating conditions the maximum voltage to earth on each phase is only $1/\sqrt{3}$ of the line voltage, the voltage grading down practically to zero at the neutral point. The average voltage to earth is $1/(2\sqrt{3})$ of the line voltage. Insulation stresses are consequently a minimum.

(8) Third-harmonic voltages are small and, as a rule, do not exceed a small percentage of the fundamental.

(9) Either or both neutrals may be earthed, according to the system conditions.

(10) Unbalanced four-wire loads may be supplied without undue deflection of the neutral points.

Three-phase, shell-type units   (6a) If one phase fails, both its primary and secondary windings must be disconnected from the other two phases, and the sound winding (*i.e.*, primary or secondary) of the faulty phase must be short circuited.

(7) Under normal operating conditions the maximum voltage to earth on each phase is $1/\sqrt{3}$ of the line voltage, the voltage grading down practically to zero, *only provided the maximum flux density does not exceed the value obtaining at the commencement of the 'knee' bend of the B/H curve.*

(8) Either or both neutrals may be earthed *only if the* flux *density value is in accordance with* (7).

Three single-phase, core- or shell-type units (6a) If one fails, the unit is simply disconnected and removed.

(7) Same as (7) for three-phase shell-type units.

(8) Same as (8) for three-phase, shell-type units.

(9) Differences in ratio and impedance of the individual units do not cause any appreciable unequal division of load or any circulating current.

Disadvantages

General (1) The neutrals are inherently unstable unless solidly earthed.

(2) Three-phase units or groups of opposite polarity will not operate in parallel unless the interconnection of either the primary or secondary phases of one transformer or group be reversed.

(3) A fault on one phase renders a three-phase unit or group inoperative to give a three-phase supply until repaired.

(4) Coil construction difficulties are greater and costs are higher with heavy line currents.

Particular

Three-phase, core-type units No further disadvantages.

Three-phase, shell-type units (5) Third-harmonic voltages as high as 30 to 60% of the fundamental may be present at maximum flux densities which are quite satisfactory for three-phase, core-type transformers. With isolated neutral the stress from the neutral point to earth is consequently increased considerably as compared with the stress at the neutral of a three-phase, core-type transformer. If an earthing device is fitted to the neutral the insulating medium must be proportioned according to the estimated third-harmonic voltage at the neutral.

(6) Earthing the neutral transfers the third-harmonic voltage phasors to each line terminal, and the third-harmonic voltages may be magnified by the resulting capacitance current flowing through the inductance of the transformer windings. Telephone interference may occur if the neutral is earthed, particularly as both third and ninth harmonics of appreciable magnitude may be present.

(7) Generator and transformer primary neutrals should not be connected together, as the third-harmonic currents which would flow may also cause telephone interference.

(8) Unbalanced four-wire supplies cannot be given unless the primary and generator neutrals are connected together.

Three single-phase, core- or shell-type units

(5)
(6)
(7) } Same as for three-phase, shell-type units.
(8)

Applications Star/star connected transformers find some scope as three-phase, core-type units for supplying relatively small-power loads. In practice it is generally difficult to ensure a three-phase, four-wire lighting load being always balanced and therefore this connection only finds a very limited field for such loads. For transmitting or distributing power in bulk the connection is quite suitable from an operation standpoint, providing three-phase, core-type transformers are used, but three-phase, shell-type transformers and groups of single-phase transformers may introduce disturbances due to harmonics.

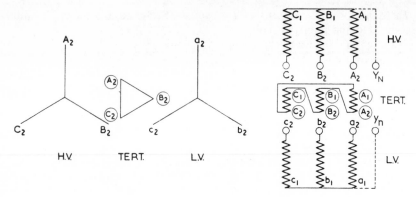

Fig. 72   Star/star, tertiary delta

Star/star with tertiary delta (Fig. 7.2)

The tertiary delta is an additional, auxiliary winding used under certain conditions with three-phase transformers or transformer groups, and it is separate and distinct from both primary and secondary main windings, though wound upon the same core, or cores. The auxiliary connection consists of a single winding per phase, the three being connected to form an ordinary closed delta circuit which may be isolated entirely from any external circuit, or from which a load may be taken for purposes referred to in the following text.

Advantages

These are best understood by reference to the applications of the connection, which it will be seen are, in actual practice, somewhat limited in extent.

Disadvantages

Such as they are, the disadvantages of the connection consists of:

(1) Additional windings depending in size upon the purpose for which they are required, which increase the frame size and the initial cost of the transformer. The characteristics of the tertiary windings are the same as those of the ordinary delta connection.

(2) If used for supplying an external load, the auxiliary circuit may, in transformers wound for high voltages on both sides, reach a dangerous voltage above earth, due to electrostatic induction, unless the circuit is earthed either at one terminal or through a three-phase neutral earthing transformer. If the delta is isolated the same abnormal voltage may be present, but as it would be confined to the auxiliary windings only, provision could more easily be made to safeguard against it.

(3) A failure of the auxiliary delta may render the transformer or group inoperative on account of triple harmonic phenomena, or if the main windings became damaged as a result of an original breakdown in the delta.

136

Applications

(1) Used in conjunction with the star/star, star/interconnected star, and inter-connected star/star methods of connecting three-phase, shell-type transformers or three-phase groups of single-phase, core- or shell-type transformers, the isolated tertiary delta provides a short-circuit path for the flow of third-harmonic components of the magnetising current which eliminates third-harmonic voltages from the main windings. The neutral points of such windings are there-fore stable, and can be earthed without any ill effects to the transformers or the system. In this case the tertiary delta is designed to provide the m.m.f. correspond-ing to that required for the elimination of the third-harmonic voltages. Three-phase, core-type transformers having the previously mentioned connections do not require a tertiary delta, as the third-harmonic voltages are negligible.

(2) In addition to providing a short-circuit path for the flow of third-harmonic magnetising current, this additional winding may be utilised for supplying an external load, such as motors or even for general distribution purposes. If the motors are of the synchronous type, a very considerable improvement in the transformer primary power factor may result, and the current loading on the primary side may be reduced appreciably. For supplying an external load, the tertiary delta may be applied to both core- and shell-type three-phase trans-formers or groups, and it would be of particular utility in cases where the voltage required for the auxiliary load is considerably lower than the voltage of the main secondary winding.

It will be appreciated that a tertiary winding must, in practice, be designed for a greater rating than that obtained from a consideration of the circulation of the third-harmonic component of the magnetising current only. The tertiary winding must safely carry the current which may flow under the most onerous fault conditions.

As compared with the delta/star connection, the tertiary delta assists in the limitation of fault currents in the event of a short circuit from line to neutral, this depending upon the impedances between the tertiary and main windings.

In general a tertiary winding should have a short-time thermal capacity at least equal to one third that of the main windings.

Where a winding is intended to fulfil the function of a tertiary winding in regard to the suppression of harmonics and is connected to external terminals for external loading, it should be considered as one winding of a multiple winding transformer. The reason for this distinction is that it is the external loading, including line fault currents, which usually determines the ultimate design limit of the winding.

Delta/delta (Fig. 7.3)

Advantages

General (1) If one phase on either side becomes faulty, the remaining two phases can be operated in open delta or vee to give a three-phase output equal to $1/\sqrt{3}$ of the total normal output. If, however, a number of delta/delta units or groups operate in parallel, the advantage as regards total output becomes less as the

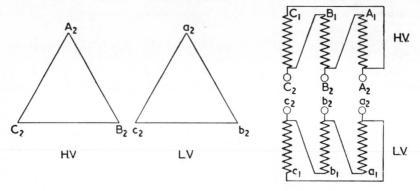

Fig. 7.3   Delta/delta

number of units or groups increases, since the sound groups do not supply their normal full load.

(2) Most economical connection for high-current, low-voltage transformers.

(3) Third-harmonic voltages are eliminated by circulation of third-harmonic currents in the deltas.

(4) One of the easiest connections to phase-in for parallel operation.

(5) With balanced line voltages no part of the windings can normally be at any excessive potential above earth, unless due to static charges.

(6) Large *unbalanced* three-wire loads may be given which only produce unbalanced voltages proportional to the internal impedances of the winding.
Particular

Three-phase, core-type units   (1a) If one phase fails, due to an open circuit or l.v. earth only, both its primary and secondary windings must be disconnected from the other two phases and open circuited. If the fault is due to a short circuit between turns, however, the transformer must not be operated in vee, as the damage is likely to extend to adjacent coils and possibly to the adjacent phase.

Three-phase, shell-type units   (1a) If one phase fails, both the primary and secondary windings must be disconnected from the other two phases, and the sound winding (*i.e.*, primary or secondary) of the faulty phase must be short circuited.

Three-phase, single-phase, core- or shell-type units   (1a) If one fails, the unit is simply disconnected and removed.

(7) Any three similar single-phase units may be connected delta/delta to form an emergency group, providing their voltage ratios are the same.

Disadvantages

General   (1) The cross-sectional area of both primary and secondary conductors is a minimum, the number of turns and total quantity of coil insulation per phase is a maximum, producing therefore a low copper factor and generally the least mechanically robust winding in the case of transformers of small ratings.

(2) No neutral points are available unless additional auxiliary apparatus be provided.

(3) A four-wire supply cannot be given unless additional auxiliary apparatus be provided.

(4) Three-phase units or groups of opposite polarity will not operate in parallel unless the meshing of either the primary or secondary phases of one transformer or group be reversed.

(5) Coil construction difficulties are greater and costs are higher with high line voltages.

(6) Under normal operating conditions the maximum voltage to earth on each phase is $1/\sqrt{3}$ of the line voltage, the minimum voltage is $1/(2\sqrt{3})$. Insulation stresses are therefore somewhat higher than with the star connection.

Particular

Three-phase, core-type units   (1a) While theoretically it is possible to operate the unit in open delta with one phase damaged, it is very seldom practicable to do so.

Three-phase, shell-type units   No further disadvantages.

Three single-phase, core- or shell-type units   (7) If the voltage ratios of the individual units are different, circulating currents will flow in the primary and secondary windings, the current being limited only by the impedances of the windings.

(8) If the impedances of the individual units are different the load distribution between the units will be unequal.

Applications

The delta/delta connection has found very little application in Great Britain, probably for the reason of the general feeling that a faulty transformer should be removed at once and the supply temporarily interrupted in preference to giving a reduced supply under abnormal conditions. In Great Britain three-phase units are by far the most common, and the advantages of the delta/delta connection are therefore of considerably less practical importance. Even with groups of single-phase transformers the policy of the larger users at least is to purchase one or more complete spare transformers, so enabling the full supply to be resumed under normal conditions with as little delay as possible should one unit fail.

Delta/interconnected star (Fig. 7.4)

Advantages

(1) Third-harmonic voltages are eliminated by the circulation of third-harmonic currents in the primary delta.

(2) The secondary neutral may be earthed, or it may be utilised for loading purposes or for providing the d.c. neutral for three-wire d.c. systems.

(3) An *unbalanced* four-wire supply may be given, and the resulting unbalanced voltages are relatively small, being proportional only to the internal impedance of the windings. Balanced and unbalanced loads may therefore be supplied simultaneously.

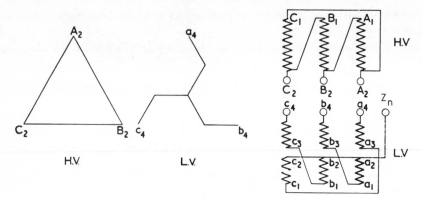

Fig. 7.4   Delta/interconnected star

Disadvantages

(1) No primary neutral is available for earthing. This is not necessarily a disadvantage, as the system on the primary side of the transformer is generally earthed at the generator or at the step up transformer secondary.

(2) A fault on one phase renders a three-phase unit or group inoperative until repaired.

(3) The delta winding may be weak mechanically in the case of a step-down transformer with a very high primary voltage, or in the case of a transformer of small rated output.

(4) On account of the phase displacement between the halves of the windings which are connected in series to form each phase, the interconnected star winding requires $15\frac{1}{2}\%$ additional copper with a corresponding increase in the total insulation. The frame size may therefore be larger and the cost of the transformer increased.

Application

The chief application of this connection is for stepping down to give a supply to three-phase synchronous converters, and at the same time providing, on the interconnected star side, a neutral point from which a d.c. connection can be taken for the purpose of providing a d.c. neutral. On account of the interconnection on the secondary side, considerable d.c. out-of-balance current can be taken without it having any ill effect on the magnetic characteristics of the transformer.

Note.—This connection only becomes desirable for three-phase, shell-type transformers and for groups of three single-phase transformers. The interconnection on the secondary side is not necessary for three-phase, core-type transformers, as if a straight star winding is used continuous current flows along the magnetic circuit in the same direction in all three limbs, and as the corresponding continuous flux must find its return path through the air or through the oil and the transformer tank, its magnetic effects are practically negligible.

140

Vee/vee (Fig. 7.5)

Advantages

(1) The only practical advantage is that some initial saving in cost may be effected when installing a transformer group to supply a load which, though likely to grow, may not do so for some time. The transformer units must be single-phase, either core- or shell-type, but two only need initially be installed instead of three for giving a three-phase supply.

(2) Third-harmonic voltages and currents are usually very small and of no practical importance.

(3) For the given vee/vee group output the windings will be more robust than with the delta/delta connection for the same group output, as the conductors must have a cross-sectional area proportioned for the full line current.

Disadvantages

(1) While only two single-phase transformers need be installed, their rated kVA output must be $15\frac{1}{2}\%$ greater than the total kVA load initially supplied. This is due to the $86 \cdot 6 \%$ internal power factor which occurs even at u.p.f. load.

(2) When the group is converted to a closed delta one, the winding characteristics become identical with the delta/delta connection for the same group output.

(3) No neutral points are available unless additional auxiliary apparatus is provided.

(4) Under normal operating conditions the maximum voltage to earth on each phase is $1/\sqrt{3}$ of the voltage, the minimum voltage is $1/(2\sqrt{3})$. Insulation stresses are therefore somewhat higher than with the star winding.

(5) The connection is electrostatically unbalanced, and is therefore not suitable for high-voltage systems.

(6) Even with balanced three-phase load the load voltages become unbalanced to an extent depending on the impedance of the transformers and the load p.f.

(7) Due to the currents in the windings leading the voltage by 30° in one phase and lagging by the same amount in the other, the power factors of the different

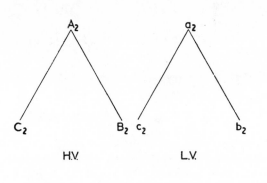
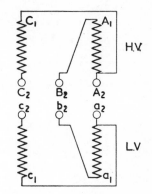

Fig. 7.5   Vee/vee

141

phases of the system will become unbalanced. This will result in a flow of wattless current, the magnitude of which will largely depend upon the load power factor.

(8) Parallel operation of vee with delta connected groups is uneconomical.

Applications

The only practical application that this connection has is enumerated under 'Advantages, (1)'. That is, under certain conditions the connection may be desirable on account of the saving in initial cost which may be effected when installing a group to supply a load which, though likely to grow, may not do so for some time. The connection has, however, not found much favour in Great Britain.

Tee/tee (Fig. 7.6)

Advantages

(1) Winding characteristics and voltage stresses are practically the same as for star/star connection.

(2) For initially supplying a load which will ultimately be greater, two inter-changeable transformers with suitable tappings may be installed to facilitate an ultimate change-over to the delta/delta connection. There would be the same initial saving in costs as with the vee/vee connection.

(3) Third-harmonic voltages and currents are usually very small and of no practical importance.

(4) If desired, three-phase, two-phase, and single-phase loads may be supplied simultaneously.

(5) The neutral points are available for earthing or loading purposes. For the latter purpose it becomes more important that the halves of the main transformer windings should be interleaved.

(6) The iron loss of the group will be slightly less than with the vee/vee connection due to the lower voltage, $i.e.$, $\sqrt{3}/2$ across the teaser transformer.

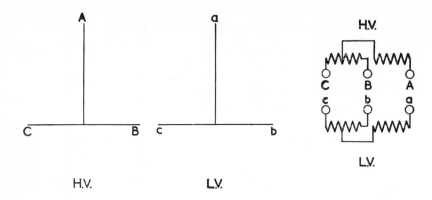

Fig. 7.6   Tee/tee

Disadvantages

(1) While only two single-phase transformers need to be installed, their total rated kVA must be $15\frac{1}{2}\%$ greater if the transformers are interchangeable, or $7\frac{3}{4}\%$ greater if non-interchangeable, than the actual load supplied.

(2) Corresponding halves of primary and secondary windings of the main transformer must be interleaved.

Applications

This connection, for use primarily to give a three-phase transformation, has found very little favour indeed. Theoretically its application would be most suitable for supplying three-, two-, and single-phase loads simultaneously, but as such loads are not usually found together in modern practice the connection has been very little adopted. It has the advantage over the vee/vee connection that primary and secondary neutral points are available.

Star/delta (Fig. 7.7)

Advantages

(1) Third-harmonic voltages are eliminated by the circulation of third-harmonic current in the secondary delta.

(2) The primary neutral may be earthed.

(3) The primary neutral is maintained stable by the secondary delta.

(4) Most desirable connection for large step-down transformers on account of the inherent characteristics of star windings for high voltages and delta windings for low voltages. See remarks under star/star and delta/delta connections.

(5) Any two or more transformers having connections listed in Class 2 can be phased-in whatever their internal meshing and direction of coil winding, simply by selecting the correct external connections to busbars.

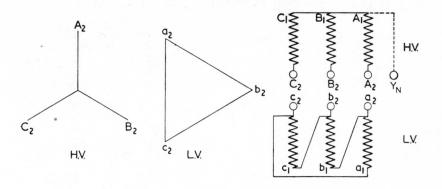

Fig. 7.7   Star/delta

143

Disadvantages

(1) No secondary neutral point is available for earthing or for giving a four-wire supply unless additional auxiliary apparatus be provided.

(2) A fault on one phase renders a three-phase unit or group inoperative until repaired.

(3) The delta winding may be mechanically weak in the case of a step-up transformer with a very high secondary voltage, or in the case of a transformer of small rated output.

Applications

The chief application of this connection is for stepping down from a high-voltage system employing large transformers.

Delta/star (Fig. 7.8)

Advantages

(1) Third-harmonic voltages are eliminated by the circulation of third-harmonic current in the primary delta.

(2) The secondary neutral may be earthed or it may be utilised for giving a four-wire supply.

(3) An *unbalanced* four-wire supply may be given and the resulting unbalanced voltages are relatively small, being proportional only to the internal impedances of the windings. Balanced and unbalanced loads may therefore be supplied simultaneously.

Disadvantages

(1) No primary neutral point is available for earthing. This is not necessarily a disadvantage, as the system on the primary side of the transformer is generally earthed at the generator or at the step-up transformer secondary.

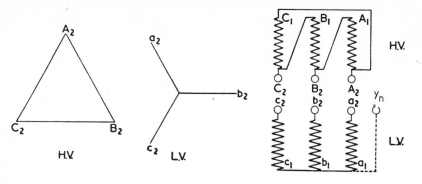

Fig. 7.8 Delta/star

(2) A fault on one phase renders a three-phase unit or group inoperative until repaired.

(3) The delta winding may be mechanically weak in the case of a step-down transformer with a very high primary voltage, or in the case of a transformer of small rated output.

Applications

(1) The chief application is for stepping down to supply a four-wire load which may be balanced or unbalanced. With this connection a combined load such as motors and lighting can be given.

(2) The connection is equally suited for stepping up to supply an h.v. distribution or transmission line, as third-harmonic voltages are eliminated the h.v. neutral is available for earthing, and both windings are employed under the best conditions.

Interconnected star/star (Fig. 7.9)

Advantages

(1) Neutral points are available for earthing on the primary side, and on the secondary side for supplying balanced or unbalanced four-wire loads.

(2) Third-harmonic voltages between lines and neutral on the primary side are eliminated by the opposition of such voltages in the halves of the windings which are connected in series to constitute one phase.

(3) Windings on both sides are mechanically robust. In this respect the connection is practically as good as the star/star connection.

Disadvantages

General (1) $15\frac{1}{2}\%$ additional copper with a corresponding increase in the total insulation is required for the primary winding. The frame size may therefore be larger, and the cost of the transformer is increased.

(2) On account of winding and assembly difficulties the interconnected star winding should be confined to the l.v. side of transformers having small ratings, and for this reason is an unsuitable primary for small step-down transformers.
Particular
Three-phase, core-type units (3) Third-harmonic voltages exist on the secondary side from each line to neutral, but with this type of transformer they do not as a rule exceed a small percentage of the fundamental. On the primary side third-harmonic voltages exist between the neutral point and the junction of the halves of the windings which are connected in series to form each phase, and also between each line terminal and the same junctions. As with the secondary side, they also do not exceed a small percentage of the fundamental.

Three-phase, shell-type units (3) On the secondary side third-harmonic voltages as high as 30 to 60 % of the fundamental may be present at the maximum flux densities which are quite satisfactory for three-phase, core-type transformers.

145

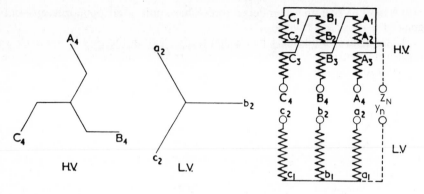

Fig. 7.9   Interconnected star/star

With an isolated secondary neutral the stress from the neutral point to earth is consequently increased as compared with the stress at the neutral of a three-phase, core-type transformer. If an earthing device is fitted to the neutral the insulating medium must be proportioned according to the estimated third-harmonic voltage at the neutral. On the primary side, third-harmonic voltages of the same order exist in the windings and the stress on the insulation of the coils is correspondingly increased, even though the harmonic voltages do not appear between lines or between lines and neutral.

Three single-phase, core- or shell-type units   (3) Same as (3) for three-phase, shell-type units.

Applications

(1) This connection was evolved as a substitute for either the star/delta or the delta/star. It was desired to obtain a connection which, while allowing unbalanced loads to be supplied and also eliminating third-harmonic voltages, gave a winding which possessed the mechanical rigidity of the star connection. The interconnected star/star combination was found to give the desired results with the exception that third-harmonic voltages were not eliminated on the star side.

The interconnected star/star connection may therefore be used for purposes that make the star/delta or the delta/star connection desirable, always bearing in mind, however, that with *certain types* of transformers it is not desirable to earth the neutral point of the star, while with all types the star winding should be placed on the h.v. side.

Star/interconnected star (Fig. 7.10)

Advantages, disadvantages and applications

As this combination is the exact reverse of the interconnected star/star and yet so closely similar, what has been said regarding the latter connection applies to this one also. It must be remembered, however, that when applying the remarks under the interconnected star/star combination to the star/interconnected star the words 'primary' and 'secondary' should be interchanged. In Great Britain at least, the star/interconnected star connection has been employed

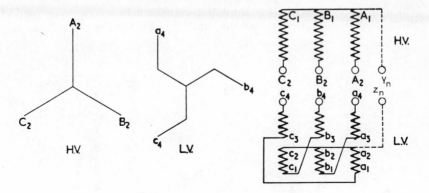

Fig. 7.10   Star/interconnected star

to replace the delta/star for step-down transformers of comparatively small outputs and high primary voltages with which a delta h.v. winding would not have been conducive to mechanical stability. The permissible unbalanced loading is greater with this connection than with the interconnected star/star connection.

Star/double star*† (Fig. 7.11)

Advantages

(1) The winding characteristics are similar to those for the star/star connection as regards (1), (2), and (5) of the latter.

(2) Both neutral points are available; the primary for earthing and the secondary for providing a d.c. or a.c. neutral.

(3) When supplying rotary converters the third-harmonic voltages are eliminated by the third-harmonic currents which circulate in the transformer and rotary converter windings. In split pole converters the third-harmonic voltage is utilised for converter voltage regulation.

(4)   †The double-star secondary only requires three coils as compared with six for the double delta.

---

*The remarks concerning three- to six-phase connections are to be taken as being a comparison between connections to give the same phase transformation only; that is, they are not compared with three- to three-phase connections.

†Strictly, Fig. 7.11 shows the diametrical connection; the true double-star winding consists of two separate windings of opposite polarity which give the same phasor voltages as shown in Fig. 7.11.

(5) Starting tappings for rotary converter transformers can most conveniently be taken from the double-star secondary.

Disadvantages

The only disadvantage of this connection is that a fault on one phase renders a three-phase unit or group inoperative until the defect has been repaired.

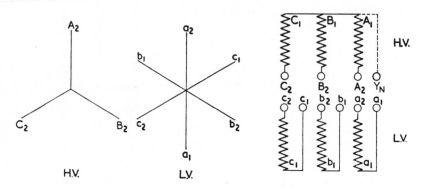

Fig. 7.11   Star/double star†

Applications

(1) For supplying six-phase rotary converters.
(2) For supplying three separate single-phase circuits for l.v. distribution purposes, in which case the mid-point is usually earthed on each secondary winding.

Delta/double delta (Fig. 7.12)

Advantages

(1) If one phase fails, a three-phase unit or group can be operated in vee/double vee to give a three- to six-phase output equal to $1/\sqrt{3}$ of its total normal output. The faulty phase should be treated in the same way as specified under the delta/delta connection, and the same limitations apply with the different types of transformers.
(2) The connection is used for low voltage, high current primaries and secondaries.
(3) Third-harmonic voltages are eliminated by circulation of third-harmonic currents in the primary and secondary deltas.

† See footnote on page 147.

Disadvantages

(1) No primary or secondary neutrals are available for earthing or for providing a d.c. or a.c. neutral.

(2) For high primary voltages the delta connection is not mechanically strong. Coil construction difficulties are greater and costs are higher with high line voltages.

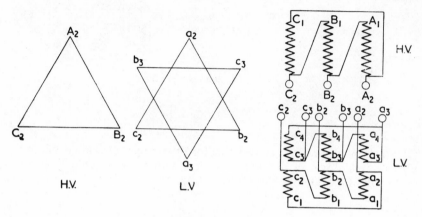

Fig. 7.12   Delta/double delta

(3) Six secondary coils are required which make the transformer somewhat larger and more expensive.

(4) Starting tappings cannot so easily be arranged as compared with the double star connection.

Application

For supplying six-phase rotary converters.

Star/double delta (Fig. 7.13)

Advantages

(1) The primary windings possess the mechanical rigidity and the minimum insulation stresses under both normal and transient conditions inherent in the star connection.

(2) The primary neutral may be earthed.

(3) The secondary windings possess the advantages inherent in the delta connection for low-voltage high currents.

(4) Third-harmonic voltages are eliminated by the circulation of third-harmonic currents in the delta-connected secondaries.

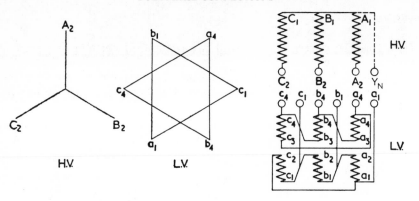

Fig. 7.13   Star/double delta

Disadvantages

(1) No secondary neutral is available for providing a d.c. or a.c. neutral.
(2) The breakdown of one phase renders a three-phase unit or group in-operative until the damage has been repaired.

Application

For supplying six-phase rotary converters.

Delta/double star* (Fig. 7.14)

Advantages

(1) Third-harmonic voltages are eliminated by the circulation of third-harmonic current in the primary delta.
(2) *The double-star secondary only requires three coils as compared with six for the double delta.
(3) Starting tappings for rotary converter transformers can most con-veniently be taken from the double-star secondary.
(4) A secondary neutral is available for providing a d.c. or a.c. neutral.

Disadvantages

(1) No primary neutral is available for earthing.
(2) For high primary voltages the delta connection is not mechanically strong. Coil construction difficulties are greater and costs are higher with high line voltages.
(3) If one phase fails, a three-phase unit or group becomes inoperative until the fault has been repaired.

*Strictly, Fig. 7.14 shows the diametrical connection; the true double-star winding consists of two separate windings of opposite polarity which give the same phasor voltages as shown in Fig. 7.14.

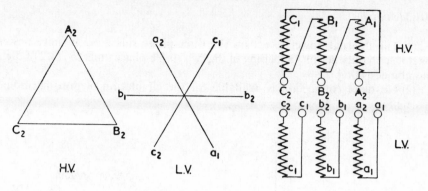

Fig. 7.14 Delta/double star

Applications

(1) For supplying six-phase rotary converters.
(2) For supplying three separate single-phase circuits for l.v. distribution purposes, in which case the mid point is usually earthed on each secondary winding.

Scott connection (Fig. 7.15)

The Scott connection is one employed for three- to two-phase transformation or vice versa, and generally its winding characteristics are similar, as regards mechanical rigidity, to the star/star connection. On the three-phase side a neutral point is available for earthing or for loading. On the two-phase side the windings may be connected to give a three-wire supply or four-wire supply as desired. The halves of the winding of the main transformer on the three-phase side must be interlaced, in order to avoid excessive reactance. Due to the internal power factor on the three-phase side even at unity power factor load, two single-phase units forming a Scott connection can only give an output of $\sqrt{3/2}$ of their combined single-phase kVA rating.

Small third-harmonic currents and voltages exist with this connection, but as a rule they are negligible.

Advantages

(1) Where a number of units are installed in a system the interchangeable unit can be employed to act as a spare for either the main or teaser in the event of a winding failure.
(2) Winding characteristics are similar to those of the star/star connection.
(3) On the three-phase side a neutral point is available for earthing or for loading.
(4) On the two-phase side the windings may be connected to give a two-, three- or four-wire supply.

Disadvantages

(1) The internal power factor on the three-phase side even at unity power factor load only permits a loading of the two single-phase units to $\sqrt{3/2}$ of their combined rating.

(2) No delta connection is available for the circulation of third-harmonic currents.

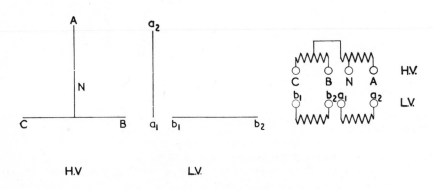

Fig. 7.15   *Scott connection*

(3) The separate halves of the main windings on the three-phase side must be interleaved to minimise leakage reactance effects, resulting in a degree of winding complication.

(4) The inherent single-phase construction and characteristics of this connection produce a comparatively bulky and a heavier transformer than compared with a normal three-phase transformer of the same rating.

Applications

(1) For supplying two-phase or single-phase loads from three-phase systems, or vice versa.

(2) For interconnecting three-phase and two-phase systems.

Le Blanc connection (Fig. 7.16)

For the conversion from three-phase to two-phase or vice versa, it has been British practice in the past to employ the Scott connection, whereas Continental practice has tended towards a more general use of the Le Blanc connection. A reason for this difference in general practice appears to be that the Scott connection became established in Great Britain a few years before the alternative connection was introduced and notwithstanding some advantages inherent in the Le Blanc connection, it did not enjoy the same popularity.

As far as parallel operation is concerned, a Scott connected transformer can be designed and constructed to operate in parallel with the appropriate Le Blanc

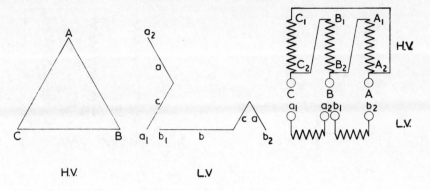

Fig. 7.16   Le Blanc connection

unit. The phasor diagram shown in Fig. 7.17 indicates that the Scott and Le Blanc connections are interchangeable and thus parallel operation presents no difficulties. In the case of three- to two-phase transformation, the Le Blanc connection can be provided with either a standard three-phase delta connected primary winding, or alternatively, the star connection can be employed. When the primary winding is connected in delta, the remarks made earlier in this chapter concerning third-harmonic voltages and fluxes equally apply to the Le Blanc connection.

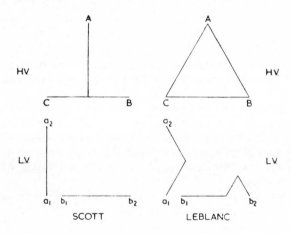

Fig. 7.17   Scott and Le Blanc connection

Scott-connected transformers must necessarily be built as two separate single-phase transformers mounted in a common tank or alternatively in separate tanks, or on a special two-phase, three-limb core having a cross-sectional area of the centre limb $\sqrt{2}$ times that of the outer limbs. An advantage of the Le Blanc connection is that a standard three-phase, three-limb core can be employed, and it follows that a transformer having this connection

153

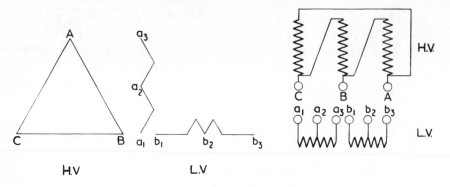

Fig. 7.18   Le Blanc connection

will occupy less space than the corresponding Scott group whether the latter is mounted in a common tank or in two separate tanks.

The Le Blanc connection can be arranged for either two-phase, three-wire, or two-phase, four-wire outputs as shown in Fig. 7.18, and the method used to give the mid points in each phase is clearly shown. When transforming from two-phase to three-phase for supplying three-phase loads, the transformer will operate under conditions similar to those stated earlier for a star/star transformer, and if a considerable unbalance between phases is anticipated on the three-phase side, a tertiary delta winding becomes necessary.

Advantages

(1) A star or delta-connected three-phase winding can be employed with the individual advantages applicable to each connection as detailed elsewhere in this chapter.

(2) The normal proportions and dimensions of a three-phase transformer are maintained with this connection, thus accommodation problems are eased.

(3) The ampere-turns of each phase of the primary windings are balanced by the phasor sum of the secondary ampere-turns on the same phase.

(4) A standard three-limb, three-phase core is employed which tends to simplify manufacture.

(5) This connection permits a more efficient use of the active materials and results in a lighter unit for a given rated kVA.

Disadvantages

(1) When employed for two- to three-phase transformation, the characteristics will be similar to those of a star/star transformer, and if it is required to supply unbalanced loads a tertiary delta winding is necessary.

(2) Referring to Fig. 7.16 it can be shown that on the two-phase side the sections of the windings labelled *a* must have a ratio of turns of $\sqrt{3}$; a similar condition applies to the sections labelled *c*. As only whole numbers of turns can be employed it follows that the choice of turns may be limited in certain cases.

154

Thus, the maximum kVA which can be transformed at any given voltage is also limited.

Applications

(1) For supplying two-phase or single-phase loads from three-phase systems, or vice versa.

(2) For interconnecting three-phase and two-phase systems.

Double Scott (Fig. 7.19)

The characteristics of this connection are very similar to those of the Scott connection. Third-harmonic voltages are practically eliminated by circulation

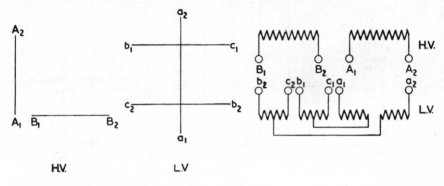

Fig. 7.19   Double Scott

of third-harmonic currents in the transformer and rotary converter windings. On the six-phase side the neutral point is available for earthing or loading.

Application

For supplying six-phase rotary converters from two-phase systems.

Auto transformers

On account of their relatively low initial cost, auto transformers appeal to many users of electrical plant, and while the question of cost is of prime importance, there are technical considerations which usually are the deciding factors in the adoption or otherwise of such transformers. As is generally known, auto transformers are constructed with a single winding only per phase, so that part of this is common to both primary and secondary sides.

The relationship between kVA of the transformer parts (or frame size) and the kVA transformed is given by the following expression:—

$$\frac{\text{frame kVA}}{\text{kVA transformed}} = 1 - \frac{1}{n}$$

where n = ratio of transformation, i.e., $\dfrac{E_1}{E_2}$

and Fig. 7.20 gives a graph showing the relative economy that can be effected at ratios up to 20.

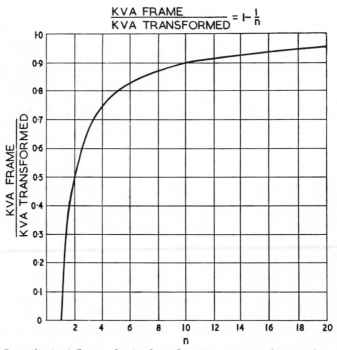

Fig. 7.20 *Curve showing influence of ratio of transformation upon equivalent transformer frame size*

From this curve it is clear that appreciable economy can only be effected when the ratio of transformation is low, especially bearing in mind that the cost does not vary to the same extent as the ratio of frame kVA to kVA transformed. In most cases an auto transformer is not a particularly favourable commercial proposition for voltage transformation ratios above 2.

The disadvantages from which auto transformers suffer are as follows:—

(1) The higher voltage is liable to be impressed upon the lower voltage circuit on account of the electrical continuity of the primary and secondary windings, and therefore the insulation of the entire low voltage circuit must, with certain connections, be designed to withstand the higher voltage, or breakdown in the low-voltage circuit is liable to ensue.

(2) On account of the electrical continuity of primary and secondary windings, and of the fact that part of the winding on each phase is common to both sides,

the leakage field between primary and secondary windings is small and the reactance correspondingly low. The auto transformer is, therefore, more likely to fail under external short circuit conditions than is the ordinary form of double-wound transformers, unless, of course, the auto transformer is protected by external reactors.

(3) Except when using the interconnected star connection, the connections on both primary and secondary side must be the same, *i.e.*, star/star or delta/delta, etc., and while this is in itself not necessarily a disadvantage, it may be so in practice on account of complications introduced due to changing primary and secondary phase angles.

(4) Again, on account of the single winding only per phase, it is not always permissible to earth the l.v. neutral point, if such is available, for the purpose of facilitating the operation of automatic protective gear or for reducing the electrical stress on the transformer windings. Earthing the l.v. neutral also earths the h.v. neutral of the auto transformers, as the neutral point is common to both windings, and as the neutral of the h.v. system will, in the majority of cases, be earthed elsewhere. Permission to earth the neutral should be sought from the supply authority.

(5) A further disadvantage of auto transformers is that it is more difficult to preserve the electromagnetic balance of the windings when voltage adjusting tappings are provided. In this connection it should also be remembered that the provision of adjusting tappings on an auto transformer may very considerably increase the equivalent kVA frame size, and if the range of the tappings is very large, the advantage gained in initial cost by using the auto connection may very largely be lost.

The following figures show the more common forms of connections for three-phase auto transformers, and these may briefly be referred to as follows:

Star (Fig. 7.21).

This connection for auto transformer service is perhaps the one most widely used, as it is the simplest and most robust, whilst a neutral point is available for earthing and for giving a four-wire supply. If the apparatus consists of a three-phase shell-type transformer or of a group of three single-phase transformers, third-harmonic trouble may be experienced when earthing the neutral point in just the same way as occurs with ordinary double-wound transformers. If, however, a three-phase core-type construction is adopted, no such trouble will result. The diagram applies equally well for step up and step down auto

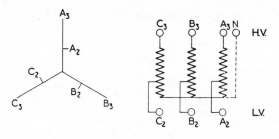

Fig. 7.21   Star

157

transformers, and it is only a question of supplying the terminals A_2, B_2, C_2 or A_3, B_3, C_3 respectively. Winding characteristics are similar to those of the star/star double wound transformers. The provision of a tertiary delta will, of course, eliminate any third-harmonic voltages which may otherwise be present.

Delta (Fig. 7.22).

Diagram 1 of the group of connections for delta-connected auto transformers shows the most symmetrical arrangement possible, but this can only be obtained with a ratio of two to one. This is obvious from the geometry of the figure.

Diagram 2 of the group shows the connections for step-up delta transformation in which the ratio is other than 2 to 1. The phase displacement between primary and secondary terminals is dependent upon the actual ratio of transformation, and this also will be understood from the geometry of the phasor diagrams.

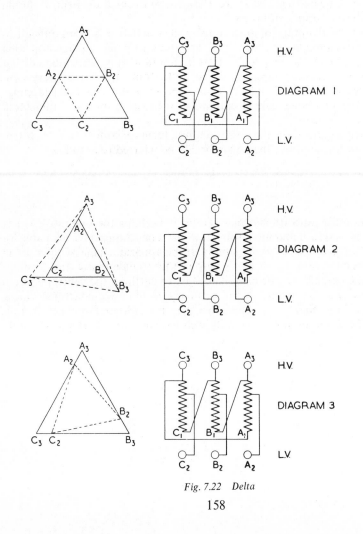

Fig. 7.22   Delta

158

Diagram 3 of the group shows similar connections, but for step-down transformation in which the ratio can have any desired value. In this case also the phase angle difference between primary and secondary terminals is dependent upon the actual ratio of transformation.

All these delta connections suffer from the disadvantage that no neutral point is available for earthing or for loading. Third-harmonic voltages will, however, be absent. Winding characteristics are similar to those of the delta double-wound transformer.

Vee (Fig. 7.23).

The connections for vee transformation are not very often used, and while the initial cost of the transformers is low, they suffer from the same disadvantages

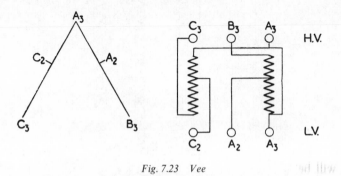

Fig. 7.23   Vee

as do the ordinary three-phase double-wound vee group. The connection is electrostatically unbalanced, and no neutral point is available. Winding characteristics are similar to those of the vee double-wound group.

Tee (Fig. 7.24).

This connection also is seldom used for three-phase auto transformers, but it has the advantage over the vee that a neutral point is available, and therefore the voltage balance of the connection can be maintained stable. Winding characteristics are similar to those of the tee double-wound group. The neutral points are not coincident and one only can be earthed.

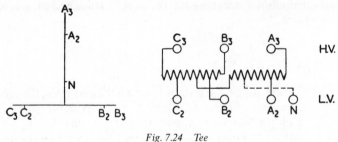

Fig. 7.24   Tee

159

Star/interconnected star, or interconnected star/star (Fig. 7.25).

This connection may at times be useful on account of the fact that such an auto transformer may be paralleled with a double-wound star/delta or delta/star transformer. Diagram 1 of the group shows the connections for a step-up star/interconnected star or a step-down interconnected star/star transformer. A neutral point is available for earthing or for loading, and third-harmonic

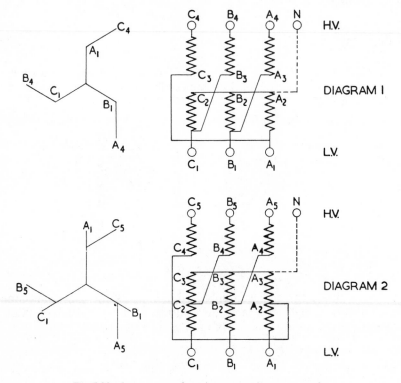

Fig. 7.25   *Interconnected star/star or star/interconnected star*

voltages will be absent on the interconnected star side. For the connections shown in this diagram the ratio of transformation must be 1 to $\sqrt{3}$.

Diagram 2 of the group shows the connections for star/interconnected star or interconnected star/star step-up or step-down transformers. Winding characteristics are similar to those of the star/interconnected star double-wound transformer.

Scott (Fig. 7.26).

In the case of true Scott-connected groups in which the end of the teaser winding is connected to the centre of the main winding, it is impossible to use the auto connection for giving a three-wire supply on the two-phase side. This is on account of the fact that in a three-wire two-phase system the ends of each of the

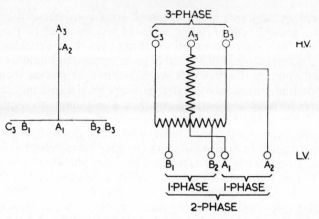

Fig. 7.26   Scott connection for three-phase three-wire to two-phase four-wire transformation

two phases are connected together, whereas in a true Scott-connected auto transformer one end of one phase must be connected to the centre of the other phase on the two-phase side. This is obvious from the Scott diagram, and an alternative arrangement is shown whereby it will be possible to give a four-wire supply on the two-phase side. With the Scott-connected group it is also impossible to obtain the same advantage in reduced frame size and initial cost, owing to the fact that in the main winding of the group the primary and secondary currents are not in exact opposition, and consequently the formula for obtaining the equivalent kVA frame size for an auto transformer will not apply in the case of the Scott connection, the actual equivalent kVA being appreciably higher. Apart from the question of actual auto windings, the advantages and disadvantages of the connection are comparable with those of the same connection for double-wound transformers.

There is one case which is very interesting regarding true Scott-connected auto transformers, and this is where the two-phase voltage is exactly $\sqrt{3}/2$ of the three-phase voltage. In this case the two-phase winding is exactly equal, both with regard to current and voltage, to the teaser winding on the three-phase side, and it is therefore possible to omit the teaser transformer altogether. The connections for this arrangement are shown in Fig. 7.27.

The applications of auto transformers, whenever used, are generally the same as those of double-wound transformers having similar winding connections.

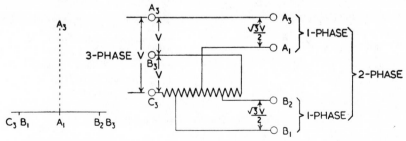

Fig. 7.27   Three-to two-phase transformation with one single-phase transformer only

The preceding diagrams in this chapter are simply basic diagrams only, showing the general phasor relationship of the voltages in the windings, together with the winding connections themselves. It is desirable, however, that more complete information should be given concerning line and transformer currents and voltages. Figures 7.28 to 7.32 show the phasor diagrams and the corresponding arrangement of the windings for the common connections used for three to three phase transformation. These diagrams apply to three-phase transformers and to three-phase groups of single-phase transformers, and they have been set out on the basis of balanced three-phase loads at 0·80 lagging power factor. The diagrams show the conditions corresponding to one set of windings only, which, of course, may be either primary or secondary, and the total effect on both primary and secondary sides of any given transformer may easily be obtained simply by setting out the two sets of phasor diagrams with the phase displacement between them corresponding to that given by the relative connections of the primary and secondary windings. Such a diagram is given in Fig. 7.33 which shows those corresponding to a star/delta transformer.

The diagrams have been drawn on the arbitrary assumption that the positive direction of the delta voltages and currents is clockwise round the circuit, and the positive direction of the star voltages and currents is radially outwards from the neutral point. This assumption is made purely for the purpose of facilitating the construction of the star phasor diagrams, as, of course, at any given instant the currents and voltages are not all radially outwards in a star winding, nor all acting in a clockwise direction in a delta-connected circuit. The Scott connection has not been included in these diagrams, as it forms the subject of a special chapter.*

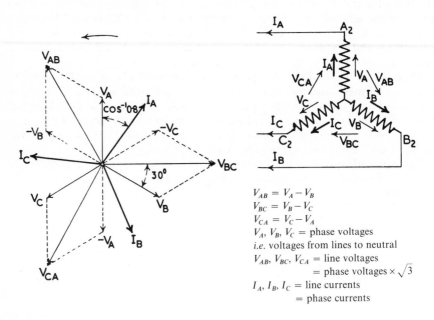

$V_{AB} = V_A - V_B$
$V_{BC} = V_B - V_C$
$V_{CA} = V_C - V_A$
V_A, V_B, V_C = phase voltages
i.e. voltages from lines to neutral
V_{AB}, V_{BC}, V_{CA} = line voltages
= phase voltages $\times \sqrt{3}$
I_A, I_B, I_C = line currents
= phase currents

Fig. 7.28   Star-connected windings

*See Chapter 8.

162

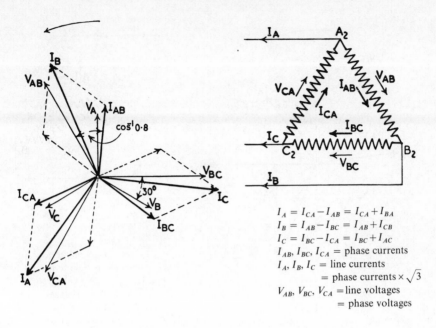

$I_A = I_{CA} - I_{AB} = I_{CA} + I_{BA}$
$I_B = I_{AB} - I_{BC} = I_{AB} + I_{CB}$
$I_C = I_{BC} - I_{CA} = I_{BC} + I_{AC}$
I_{AB}, I_{BC}, I_{CA} = phase currents
I_A, I_B, I_C = line currents
      = phase currents $\times \sqrt{3}$
V_{AB}, V_{BC}, V_{CA} = line voltages
      = phase voltages

Fig. 7.29   Delta-connected windings

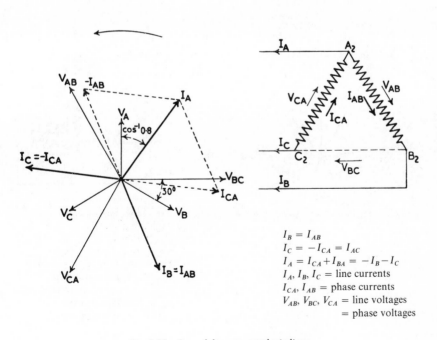

$I_B = I_{AB}$
$I_C = -I_{CA} = I_{AC}$
$I_A = I_{CA} + I_{BA} = -I_B - I_C$
I_A, I_B, I_C = line currents
I_{CA}, I_{AB} = phase currents
V_{AB}, V_{BC}, V_{CA} = line voltages
      = phase voltages

Fig. 7.30   Open delta-connected windings

163

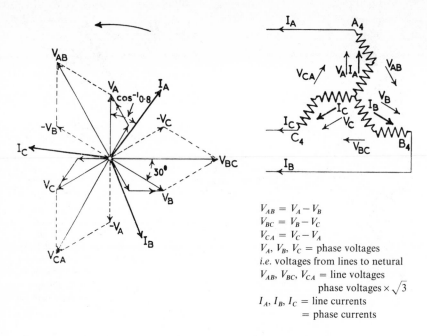

$V_{AB} = V_A - V_B$
$V_{BC} = V_B - V_C$
$V_{CA} = V_C - V_A$
V_A, V_B, V_C = phase voltages
i.e. voltages from lines to netural
V_{AB}, V_{BC}, V_{CA} = line voltages
phase voltages $\times \sqrt{3}$
I_A, I_B, I_C = line currents
= phase currents

Fig. 7.31   Interconnected-star connected windings

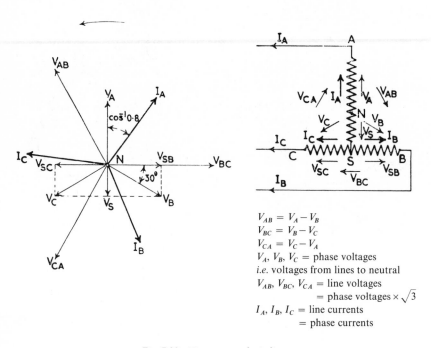

$V_{AB} = V_A - V_B$
$V_{BC} = V_B - V_C$
$V_{CA} = V_C - V_A$
V_A, V_B, V_C = phase voltages
i.e. voltages from lines to neutral
V_{AB}, V_{BC}, V_{CA} = line voltages
= phase voltages $\times \sqrt{3}$
I_A, I_B, I_C = line currents
= phase currents

Fig. 7.32   Tee-connected windings

164

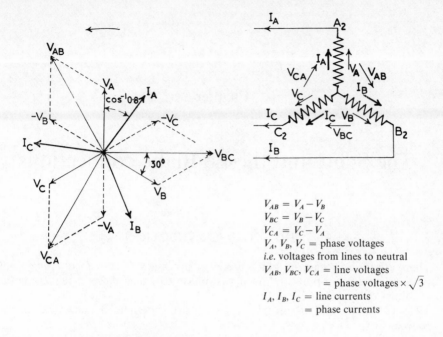

$V_{AB} = V_A - V_B$
$V_{BC} = V_B - V_C$
$V_{CA} = V_C - V_A$
V_A, V_B, V_C = phase voltages
i.e. voltages from lines to neutral
V_{AB}, V_{BC}, V_{CA} = line voltages
　　　　　　　　= phase voltages × $\sqrt{3}$
I_A, I_B, I_C = line currents
　　　　　　= phase currents

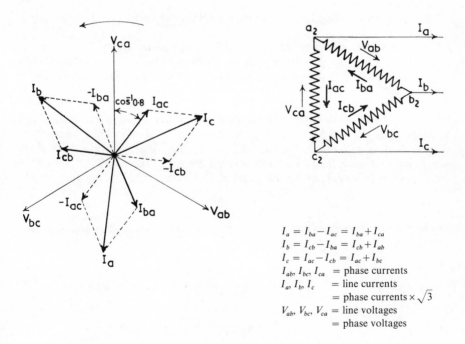

$I_a = I_{ba} - I_{ac} = I_{ba} + I_{ca}$
$I_b = I_{cb} - I_{ba} = I_{cb} + I_{ab}$
$I_c = I_{ac} - I_{cb} = I_{ac} + I_{bc}$
I_{ab}, I_{bc}, I_{ca} = phase currents
I_a, I_b, I_c = line currents
　　　　　　 = phase currents × $\sqrt{3}$
V_{ab}, V_{bc}, V_{ca} = line voltages
　　　　　　　 = phase voltages

Fig. 7.33　Star/delta connected windings

165

Chapter 8

The Scott and the Le Blanc connections

THE SCOTT CONNECTION

There are various reasons which render it necessary to transform either from three-phase to two-phase or from two-phase to three-phase, a few common examples of these being as follows:

(1) To give a supply to an existing two-phase system from a new three-phase source.

(2) To supply two-phase furnaces from a three-phase source.

(3) To supply three-phase motors from a two-phase source. This is sometimes desirable owing to the higher efficiency and more satisfactory operation of three-phase motors.

(4) To supply three-phase or six-phase rotary converters from a two-phase source.

(5) To interlink a two-phase system with a three-phase system.

(6) To supply a single-phase load which may or may not be divisible into separate loads, and at the same time maintain reasonable balance on the three-phase source.

The Scott connection is one method of making this phase transformation by means of two single-phase transformers connected to the system and to one another. In Fig. 8.1, if A, B and C represent the three terminals of a three-phase system and N represents the neutral point, the primary windings of three single-phase transformers forming a delta-connected three-phase bank may be represented by the lines AB, BC and CA. If it is desired to arrange the primary

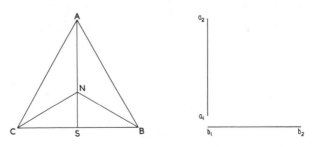

Fig. 8.1   Phasor derivation of the Scott connection

166

windings in star, the corresponding lines on the diagram are AN, BN and CN. If, in the diagram, AN is continued to the point S, the line AS is perpendicular to the line BC, and it is evident that it would be possible to form a three-phase bank consisting only of two single-phase transformers, their respective primary windings being represented in phasor form by the lines AS and BC. With this connection it is possible to form a three- to three-phase bank consisting of only two single-phase transformers. At the same time it is also possible, by giving each transformer a single secondary winding, to form a three- to two-phase bank. These secondary windings are represented in the diagram by the lines a_1a_2 and b_1b_2.

Non-interchangeable groups

The simplest possible form of Scott group is the so-called non-interchangeable group; this consists of two single-phase transformers, in one of which the ratio of the turns on the two windings is exactly equal to the line ratio of transformation. In this transformer it is necessary for the mid point of the winding on the three-phase side to be brought out in order that it may be connected to the other transformer. This transformer is known as the 'main' transformer. The other transformer is usually called the 'teaser', and the ratio of the windings is equal to to 0·866 of the ratio of the line voltages: the reason for this is apparent on reference to Fig. 8.1. ABC is evidently an equilateral triangle, and consequently the line AS must be equal to 0·866 times each of the sides. Each secondary winding is simply a single-phase winding, and the voltage across it and the current in it do not differ from what would be expected in an ordinary single-phase transformer. In the case of the three-phase side, however, it is of interest to consider the actual voltages and currents, which are as follows:

If V is the three-phase voltage, then:

voltage across main transformer $= V$

voltage across teaser transformer $= 0.866\ V$

$$\text{current in main transformer} \quad = \frac{1000 \times kVA}{\sqrt{3}\ V}$$

$$\text{current in teaser transformer} \quad = \frac{1000 \times kVA}{\sqrt{3}\ V}$$

By multiplying the voltage across each transformer by the current in it, the equivalent size of each transformer is obtained. In the case of the main transformer, this is equal to 0·577 times the group output; and in the case of the teaser transformer, 0·5 times the group output. Therefore, in a Scott-connected group, the two-phase windings are equivalent to the windings of two ordinary single-phase transformers of the same output, but on the three-phase side the winding of the main transformer is increased by 15·5% above what would be required in a single-phase transformer of the same output. Assuming that the primary and secondary winding of an ordinary single-phase transformer each occupies the same space, then, in a Scott-connected group it is necessary to have a transformer of 7·75% greater capacity for the main transformer, but the

167

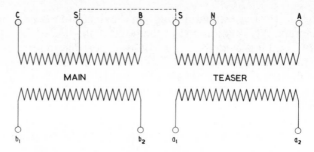

Fig. 8.2   Connections for non-interchangeable Scott group

capacity of the teaser transformer is not increased. A diagram showing the method of connecting a non-interchangeable Scott group is given in Fig. 8.2, in which the neutral point is indicated as being brought out on the three-phase side. It is not essential, however, to arrange a lead at the neutral point unless it is required for earthing or other purposes. If the neutral point is required, it is only necessary to bring out the lead at a point which divides the winding of the teaser transformer in the ratio of 2 to 1. That this statement is true geometrically is apparent on reference to Fig. 8.1.

Interchangeable groups

For convenience in operation, it is sometimes desirable that each transformer of a Scott group may be available for use either as a main transformer or as a teaser transformer. If such is the case, both transformers must have identical turns as for the main transformer, and must also be provided with tappings at the mid point, at a point from one end corresponding to 86·6% of the total winding, and also at the neutral point if it is desired that this should be brought out to a terminal. With regard to the equivalent capacity of the transformers, this will be the same as for non-interchangeable groups, except that each transformer will be equivalent to the main transformer of the non-interchangeable group. A diagram of connections for such a group is given in Fig. 8.3.

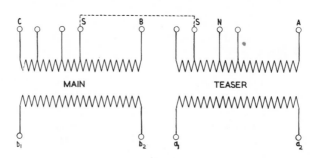

Fig. 8.3   Connections for interchangeable Scott group

Spares

There are two principal ways in which Scott groups can be arranged having a single transformer per group, as a spare:

(1) The commonest method is to provide three transformers all identical, such that any two may act as a Scott group, one transformer being the main transformer, and the other one the teaser. In this case all three transformers would be arranged, as described above, for interchangeable groups.

(2) This arrangement consists of a non-interchangeable group, together with a spare transformer, which is suitable for replacing either the main transformer or the teaser transformer. This spare transformer must be identical with one of the transformers forming an interchangeable group, while the other two must be exactly as described for a non-interchangeable group. Such an arrangement is slightly cheaper than three identical transformers, but in view of the convenience of the latter, it is perhaps hardly worth while making three transformers all different, as obviously they would be with this arrangement.

Two-phase to six-phase groups

In order to supply six-phase rotary converters from two-phase systems, it is necessary to provide transformer groups which shall make the necessary voltage and phase transformation. The phasor diagram is shown in Fig. 8.4, which shows that the arrangement consists of two-phase to three-phase Scott groups superimposed. It should be noted that the lines AS and A_1S_1 really lie in the same straight line, but they are shown slightly separated for convenience in indicating how the diagram is made up. It will be evident, on referring to the diagram, that this arrangement could be formed from four single-phase transformers, but since the lines AS and A_1S_1 and also B_1C_1 and BC are parallel to one another, it is possible to form such a group by means of two single-phase

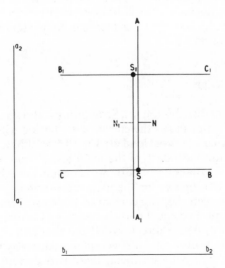

Fig. 8.4   Phasor diagram for two- to six-phase double Scott connection

transformers only. The windings represented by a_1a_2 and b_1b_2 are the respective primary windings, while AS and A_1S_1 and B_1C_1 and CB are the corresponding secondary windings. The various voltages and currents are as follows:

Main transformer: voltage across B_1C_1 and BC $= 0.866\ V$

Teaser transformer: voltage across AS and $A_1S_1 = 0.75\ V$

Main transformer: current in windings $= \dfrac{1000 \times kVA}{3\ V}$

Teaser transformer: current in windings $= \dfrac{1000 \times kVA}{3\ V}$

By multiplying corresponding voltages and currents as given above, it will be found that the equivalent sizes of the main and teaser transformers respectively are the same as for an ordinary three- to two-phase non-interchangeable group. Similarly, if such an arrangement is so designed that the two transformers are interchangeable, then each will have the same equivalent size as in the case of interchangeable groups transforming from three phase to two phase. If it is desired to provide a spare transformer per group, this may be done in the same two ways as described for three- to two-phase transformation.

If it is necessary to bring out the neutral point from such a group, it will depend upon the purpose for which the neutral is required as to whether it is desirable to bring it out from both the windings AS and A_1S_1 shown in Fig. 8.4, or from one of these windings only. In view of the fact that two- to six-phase transformer groups are almost always used for supplying power to rotary converters, it is probable that the neutral, if required at all, will be for the purpose of carrying the out of balance current from the mid wire of a three-wire d.c. supply. If this is the case, the question depends entirely upon the amount of out of balance current with which the transformer has to deal. If this is not more than approximately 25% of the full-load d.c. current, it will only be necessary to bring out the neutral point from one of the windings. If the six-phase supply is required for other purposes, it may be desirable to connect the neutral points on the two windings AS and A_1S_1 together in order to stabilise the two portions of the six-phase winding.

Features affecting design

In Fig. 8.5 the current distribution in a Scott group is shown under three different conditions. Figure 8.5(a) shows the current distribution when the secondary of the teaser transformer only is loaded; Fig. 8.5(b) shows the corresponding distribution when the secondary of the main transformer only is loaded; and Fig. 8.5(c) is a phasor diagram of currents showing a combination of the conditions in the first two figures for the main transformer only. On referring to Fig. 8.5(a) it will be seen that the current in the teaser windings on the three-phase side divides into two equal parts on reaching the main transformer, these two parts being in opposite directions. If the two halves of the three-phase winding on the main transformer are wound in such a way that there is a minimum magnetic leakage between them, these two currents will balance one another, and the main transformer will consequently offer very little impedance

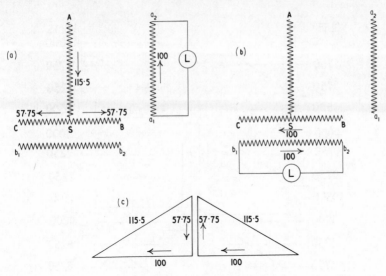

Fig. 8.5   *Loadings of Scott-connected groups*

to this current. If, however, the coupling between these two halves is loose, it is evident that the main transformer will act as a choke coil to the current from the teaser transformer. This may cause very serious unbalancing of the voltages. The effect is shown in phasor form in Fig. 8.6. AS and CB represent the voltages in the teaser and main transformers respectively, and SE represents the load current due to the load on the secondary of the teaser transformer. The line SS' at right angles to SE represents the voltage drop due to the reactive effect of the main transformer. Then AS' represents the effective voltage on the teaser transformer, and it is evident from the diagram that the fall in voltage will be very serious if the length of the line SS' is appreciable as compared with the length of AS and if the power factor is low. In order to minimise this effect it is necessary to arrange the coils on the main transformer in such a way that there is very little leakage between these two halves. One way of accomplishing this is shown in Fig. 8.7, which indicates a method of interleaving the coils on a core-type transformer. The method is also suitable for shell-type transformers, the

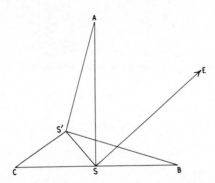

Fig. 8.6   *Phasor diagram showing voltage unbalancing due to loose coupling between the halves of the main unit on the three-phase side of Scott groups*

171

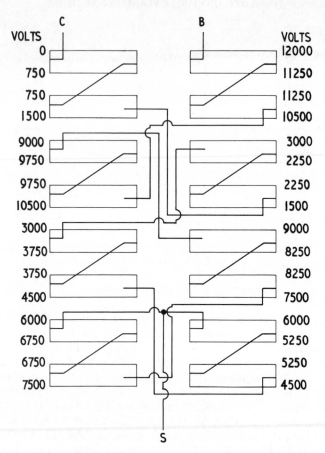

Fig. 8.7   Coil interconnection on the three-phase side of the main unit of Scott-connected core- and shell-type groups

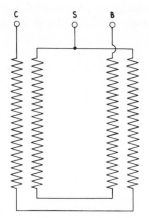

Fig. 8.8   Coil interconnection on the three-phase side of the main unit of Scott-connected core-type groups

only difference being that in the latter case there will be only one set of coils instead of two, as shown in the figure. The principal disadvantage of this method is that it introduces very high voltage differences between coils adjacent in the axial direction. The figures shown on either side of the coil diagram are the voltage differences between each face of each coil and the lead C for a line voltage of 12 000 V. It will be seen that in some cases the potential difference between adjacent faces of adjacent coils is very high. In order to insulate satisfactorily for these voltages, it is necessary either to increase the spacing between coils or to insulate the coils fairly heavily. The first of these methods has the disadvantage that it is costly in space, and also counteracts to a certain extent the advantage of interleaving, while the second method has the serious disadvantage that it lags the coils and so increases the tendency for hot spots to occur.

A much better method of reducing the leakage between the two halves of the main transformer winding is to use two concentric coils or coil groups on each limb of the transformer and cross-connect them as shown in Fig. 8.8. This has the advantages that the two coil groups can be insulated from one another in

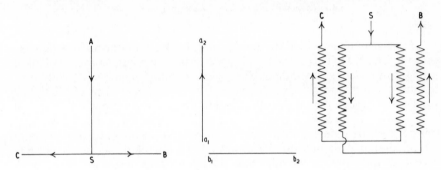

Fig. 8.9   Coil interconnection on the three-phase side of the main unit of Scott-connected core-type groups

the best possible manner, i.e., by means of an insulating cylinder extending the full length of the limb, and that the reactance between the two halves is very much lower than is generally possible with the other method of interleaving. If this method of winding is adopted, the voltage drop in the teaser transformer due to the impedance of the windings of the main transformer on the three-phase side is negligible. It is interesting to note that this way of interleaving the two halves of the main transformer windings is only applicable to core-type transformers. In the case of shell-type transformers the first method, as shown in Fig. 8.7, must be employed, and the attendant disadvantage must be overcome as far as possible by compromise.

The usual method of interleaving the windings of the main transformer on the three-phase side of a core-type group is as shown in Fig. 8.9, the arrows indicating the directions of the currents, assuming the teaser winding only to be loaded on the two-phase side.

For alternative series or parallel connection, the same coil grouping as given in Fig. 8.9 only is necessary, and the two connections are shown in Fig. 8.10, where (a) gives the connections and current distribution when the main windings are series connected, and (b) when they are connected in parallel.

173

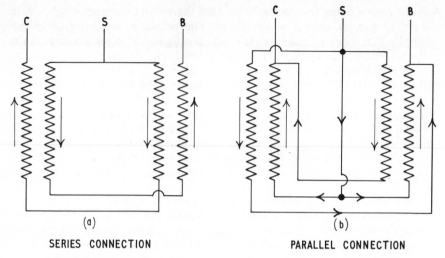

SERIES CONNECTION   PARALLEL CONNECTION

Fig. 8.10   Series and parallel connections of the coil grouping shown in Fig. 8.9

An alternative method of subdividing and interleaving the main windings is shown in Fig. 8.11. This consists of using a tertiary winding on each limb of the main transformer, these windings being connected up in opposition to each other but not connected to any external terminals. The ampere-turns of the tertiary windings are made the same as the ampere-turns of the main windings.

If a two- to three-phase, step-down, Scott-connected group of core-type transformers be required to supply a four-wire, three-phase load, it is almost certain that the loading on the group will be unbalanced, and it becomes necessary, therefore, to consider how the teaser transformer windings should be arranged on the three-phase side. The two extreme cases of loading are (a) single-phase load from the line terminal end of the teaser transformer winding to neutral, and (b) single-phase load from either line terminal end of the main

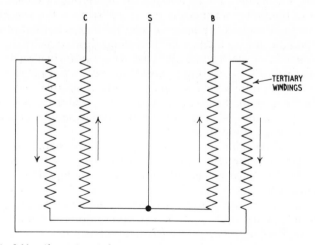

Fig. 8.11   Alternative winding arrangement employing tertiary windings

174

transformer winding to neutral. In all Scott-connected groups the main transformer winding on the three-phase side is split into four coils, which are wound upon the two limbs in pairs and interconnected, as shown in Fig. 8.9.

The result of this is that the load current on the three-phase side passing from the teaser winding, flows in opposite directions through each of the two concentric coils (on the same limb) of the main winding, and as the coils of this winding are wound in the same direction the flux leakages in each limb, due to the ampere-turns of the two coils, counteract each other, except for a small load leakage flux between the concentric coils on each limb, which produces a leakage reactance not exceeding a few per cent when based upon normal load rating of the group. So far as *load* fluxes in the three-phase side of the main transformer (due to current from the teaser transformer) are concerned, therefore, subdividing and interconnecting the coils of the main winding on the three-phase side produces the same self-eliminating effect as do the primary and secondary windings of an ordinary straight double-wound transformer. This is what is aimed at by subdividing and interconnecting the coils, but a small reactance between halves is unavoidable, due, of course, to the separation of the concentric coils on the limbs. If the main windings are not subdivided and interconnected, the currents from the teaser transformer, and therefore the corresponding fluxes in the two limbs of the main transformer, would flow in the

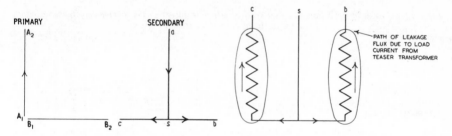

Fig. 8.12   Incorrect method of arranging main windings on three-phase side

same instantaneous direction up or down the limbs so that the fluxes would have to find return paths through the air or oil and the tank, so producing exceedingly high reactances; this is shown in Fig. 8.12. Tertiary windings can be provided, as shown in Fig. 8.11, as an alternative to subdividing and interconnecting the coils of the main windings on the three-phase side.

If, now, it is desired to give unbalanced four-wire supplies on the l.v. three-phase side, the current distribution under the conditions (a) and (b) already mentioned would be as shown qualitatively in Figs. 8.13 and 8.14 if the teaser winding on the three-phase side is not subdivided and interconnected, *i.e.*, if it is arranged as in a normal Scott group for three-phase, three-wire working. In both cases the active parts of the windings on the limbs of the teaser transformer would be badly unbalanced, as, in case (a), the whole of the teaser coil on one limb and one third only of the teaser coil on the other limb (on the three-phase side) would carry current, while in case (b), two thirds of the teaser coil on one limb only (on the three-phase side) would carry current.

Figures 8.15 and 8.16 show quantitatively typical relative magnitudes of load currents, number of turns and load ampere-turns in and of the windings of the teaser transformer under the two loading conditions (a) and (b) already referred to. The actual magnitudes of the currents and number of turns have been quite arbitrarily assumed and would vary with each individual design, but the *relative* ampere-turn values of the coils would always be the same. In case (*a*), Fig. 8.15, the whole of both the primary and secondary coils *z* and *y* on limb II carry currents which give unequal primary and secondary load ampere-turns on this limb, in the ratio of 2 to 3. For the secondary coil *y* on this limb the total load ampere-turns due to it can be considered to be split up into two parts, one part equalling and balancing the load ampere-turns of the concentric primary coil *z*, and the other part acting to supply the deficit load ampere-turns

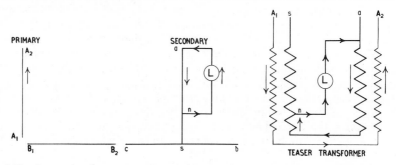

Fig. 8.13   *Current distribution in teaser transformer with incorrect arrangement of teaser winding, and with single-phase load from a to n*

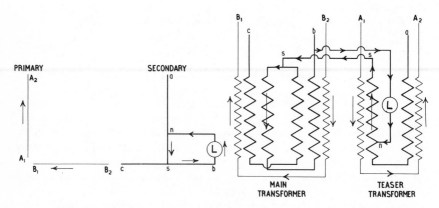

Fig. 8.14   *Current distribution in main teaser transformers with incorrect arrangement of teaser winding, and with single-phase load from b to n*

of the active part of the secondary coil *x* on limb I. From this viewpoint, therefore, the leakage fluxes due to the load ampere-turns of the coils *z* and *y* on limb II are as shown in Fig. 8.15, from which it can be seen the primary and secondary coils *z* and *y* can be considered to be electrically balanced from the point of view of leakage flux, *but* that the leakage flux between the coils *y* and *z* is augmented by the leakage flux set up between the coils *y* and *w* due to the excess load ampere-turns of coil *y*. It is interesting to note that the leakage flux between coils *y* and *w* only increases the leakage field between coils *y* and *z* but not between coils *w* and *x*, so that the voltage drop due to this leakage field is

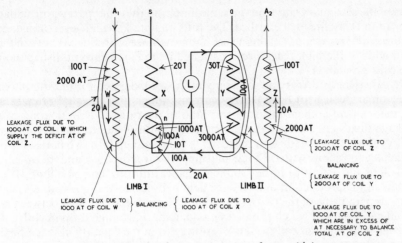

Fig. 8.15 *Current and leakage flux distribution in teaser transformer with incorrect arrangement of teaser winding, and with single-phase load from a to n*
Note—Currents and number of turns given above are arbitrarily assumed, but the ampere-turn relationships given always hold good

experienced by coil y and not by coil w. It will be appreciated, therefore, that a very considerable leakage field is set up in the space between the limbs by the interaction of coil y on limb II with coil w on limb I, and a very high reactance, of the order of 50 % of the limb voltage, results. On limb I, one third only of coil x, but the whole of coil w carry currents which also give unequal primary and secondary load ampere-turns on this limb in the ratio of 2 to 1. From the point of view of leakage flux between the concentric coils on this limb the load ampere-turns due to coil w can be considered to be split up into two parts, one part being equal to and acting with the load ampere-turns of the one third active

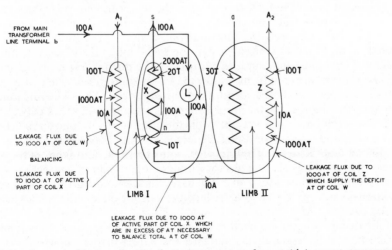

Fig. 8.16 *Current and leakage flux distribution in teaser transformer with incorrect arrangement of teaser winding, and with single-phase load from b to n*
Note—Currents and number of turns given above are arbitrarily assumed, but the ampere-turn relationships given always hold good

177

part of the secondary coil x on the same limb, and the other part acting with coil y on limb II, as already described. The coils on limb I are, however, unbalanced in space, and on account of this a very high reactance is produced between them. This adds to the high reactance occasioned by the load current distribution in the coils on limb II.

In case (b), Fig. 8.16, two thirds only of the secondary coil x, but the whole of the primary coil w of limb I carry currents which give unequal primary and secondary load ampere-turns on this limb in the ratio of 1 to 2. So far as the leakage fluxes are concerned, the load ampere-turns due to the active part of coil x can be considered, as in the case (a), to be split up into two parts, one part equalling and acting with the total load ampere-turns of coil w, and the other part equalling and acting with the total load ampere-turns of coil z on limb II. This results in the production of a leakage field between coils w and x due to equal load ampere-turns of the two coils and of an intense leakage field between coils x and z, the sum of which produces a high reactive voltage drop in coil x. The leakage field between coils w and x resulting from the equality of load ampere-turns of these two coils is augmented by the axial dissymmetry of the active parts of the two coils, so that the reactance on limb I, and consequently the voltage drop in coil x, becomes still further increased. There would thus exist a very high reactance in the teaser transformer, in case (a), between the one third of the secondary coil carrying current on the one limb and the whole of its corresponding primary coil on the same limb and between the same primary coil and the whole of the active secondary coil on the other limb, and in case (b), between the two thirds of the secondary coil carrying current on the one limb and the whole of the primary coils on *both* limbs.

In case (b) the currents in the secondary coils of the main transformer are balanced by currents in the corresponding primary coils on the same limbs, although there would be some slight unbalancing of reactance in the main transformer due to the secondary coils carrying current not being similarly placed with respect to the corresponding primary coils.

In order to avoid the high reactance in the teaser transformer, the winding on the three-phase side should be split up and interconnected in a manner similar to that employed for the three-phase side of the main transformer, except that the adjacent concentric coils on each limb of the teaser transformer would not have the same number of turns. In the main transformer such coils *have* the same number of turns, but in the teaser transformer one coil on each limb has double the number of turns of the other coil. The two coils (one on each limb) having the greater number of turns are joined in series between the teaser line terminal and the neutral, while the other two coils (one on each limb), having the lesser number of turns, are joined in series between the neutral and the other end of the teaser winding which joins up to the mid point of the main transformer windings. This is shown for an actual design in Fig. 8.17. By this means the limbs of the teaser transformer are balanced under all possible loading conditions, and Figs. 8.18 and 8.19 illustrate this for the conditions (a) and (b) already cited.

In the ordinary Scott-connected, two- to three-phase groups of single-phase transformers it is *essential* to interleave the halves of the main transformer windings on the three-phase side in order to minimise the reactive effects of the load currents from the teaser transformer winding. In such transformers the windings are usually connected, as shown in Fig. 8.20.

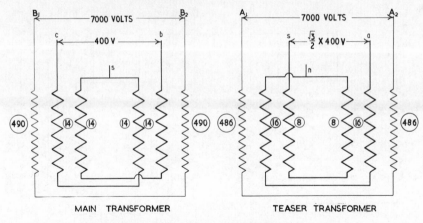

THE FIGURES ENCIRCLED ARE NUMBERS OF TURNS

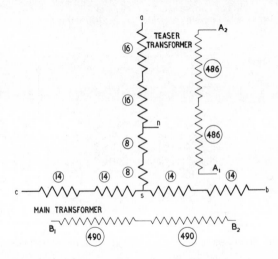

Fig. 8.17 Correct arrangement of windings and disposition of turns of main and teaser transformers for two/three-phase 4 line service.
500 kVA Scott group, two/three phase, 7000/400 volts
All the coils indicated above extend the full length of the core. Consequently, provided the ampere-turns on each limb are balanced, there is no voltage drop other than the normal regulation of the transformer group

In double Scott-connected, two- to six-phase transformer groups in which no provision has to be made for neutral d.c. out of balance current, there is no need to interleave the halves of the main transformer windings on the six-phase side, as those halves of these windings which are represented by phasors terminating at adjacent points on the circle embracing the six-phase phasors, carry currents from the teaser windings, which *naturally* neutralise each other.

Figure 8.21 shows the diagram of connections, the six-phase phasors, and the resulting current distribution, assuming current to be flowing in the teaser winding only on the two-phase side, and it will be seen that on the six-phase side the current flowing in the part cs neutralises that in the part bs, while the

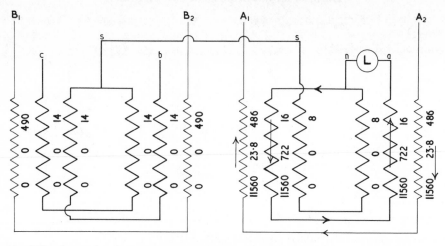

Fig. 8.18 *Current distribution in teaser transformer with correct arrangement of teaser winding, and with single-phase load from a to n*
500 kVA Scott group, two/three phase, 7000/400 volts

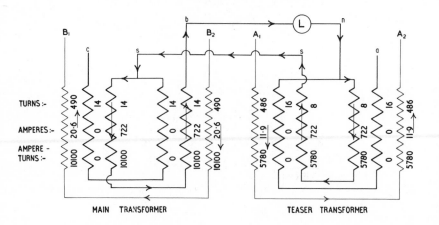

Fig. 8.19 *Current distribution in main and teaser transformers with correct arrangement of teaser winding, and with single-phase load from b to n*
500 kVA Scott group, two/three phase, 7000/400 volts

current in sc neutralises that in bs. There is thus no need to interlace the parts of the windings c_1s_1 with b_1s_1 or bs with sc.

In double Scott-connected transformer groups the neutral point on the six-phase side may be used for carrying out of balance d.c. current as already indicated. In transformers having a large number of turns on the six-phase side, the teaser windings on that side can easily be designed for one neutral lead only, if the out of balance d.c. current is not excessive. This arrangement of windings is shown in Fig. 8.22, the arrows indicating the directions of out of balance d.c. currents. This diagram shows that the out of balance ampere-turns counterbalance one another in the transformer windings so that the core is not magnetically affected thereby.

180

In transformer groups where the number of turns on the six-phase side is very small, corresponding to low voltages and heavy currents, it becomes impossible to arrange teaser windings as shown in Fig. 8.22. In such cases, therefore, the teaser windings on the six-phase side are arranged as shown in Fig. 8.23 in order to keep the d.c. out of balance current balanced on the two limbs of the teaser transformer. In this case it is necessary to bring a neutral point out from each secondary winding, as shown. The arrows in the diagram show the directions of out of balance d.c. currents flowing through the two neutral points, and it will be seen that a current balance is maintained on both limbs so that the transformer core is not affected magnetically by the d.c. current. Incidentally, the provision of two neutrals on the six-phase side permits a better distribution of out of balance d.c. current throughout the transformer windings.

In those cases where the provision of a double neutral point on the six-phase side may interfere with the starting arrangements of six-phase converters, the two neutral leads may be brought out separately and connected together by a suitable switch after the set has been run up.

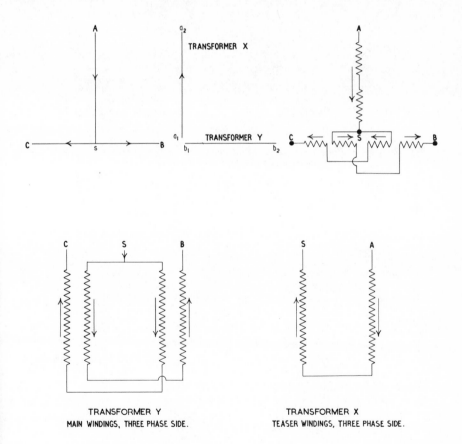

Fig. 8.20   Diagram showing arrangement of windings and current distribution in two/three-phase transformer groups with load on teaser windings only. All coils wound in the same direction

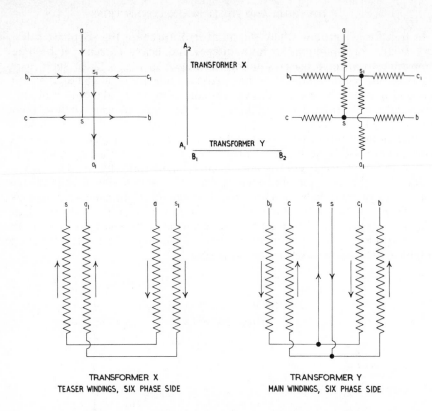

Fig. 8.21 Diagram showing arrangement of windings and current distribution in two/six-phase transformer group with load on teaser windings only

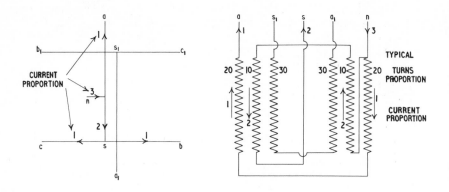

Fig. 8.22 Method of connecting teaser transformer windings on the six-phase side when only one neutral point is brought out

182

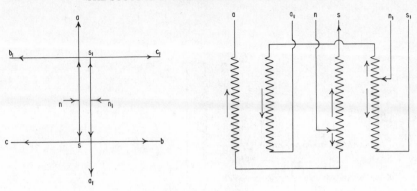

Fig. 8.23   Method of connecting teaser transformer windings on the six-phase side when two neutral points are brought out

THE REGULATION OF SCOTT-CONNECTED TRANSFORMERS

The calculation of the regulation of a Scott-connected group of single-phase transformers differs from that of normal single-phase or polyphase units or groups in that a number of resistances and reactances must be taken into account. The following study shows the data actually necessary, how they are obtained from tests on the units, and what relations should exist between the resistances and reactances of the units for satisfactory operation.

It will be as well to consider first the two types of connections in general use for core-type transformers with concentric windings; these are shown in Figs. 8.24 and 8.25, and, for simplicity, the units are non-interchangeable and without tappings for voltage variation. The teaser unit operates as a normal single-phase transformer, but when, in the case of two-phase to three-phase transformation, a line to neutral out of balance load has to be supplied, the windings on the three-phase side of this unit must be interconnected as, and in the proportions indicated, in Fig. 8.24, to prevent the reactance having an abnormal value when the parts of the winding on either side of the neutral point N are loaded separately.

The operation of the main unit is complicated by the fact that in addition to transforming half the group output it has to deal with the components of the teaser unit current. For this reason the windings must be arranged to have a low impedance between mid point and lines, when the teaser current divides equally into the two halves. The method most frequently used consists of interconnecting the windings on the three-phase side, as shown in Fig. 8.24(a). An alternative arrangement, shown in part (b), is to connect the windings on the two-phase side in parallel, but it should be noted that in this case the main unit actually transforms $1/\sqrt{3}$ of the group kVA, although, owing to the phase difference of the currents in the halves of the two-phase windings, one half only of the group output is available at the terminals. This latter type of main unit has certain advantages when the three-phase voltage is high relative to the kVA rating, and the saving in cost of insulation, apart from the fact that the windings will, in all probability, be more robust, may more than compensate for the cost of the additional copper required on the two-phase windings. There is a further alternative in which primary and secondary windings are connected as in a

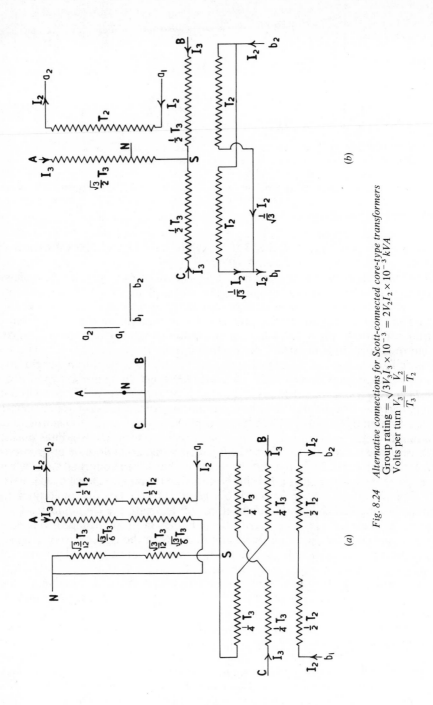

Fig. 8.24 Alternative connections for Scott-connected core-type transformers

Group rating $= \sqrt{3} V_3 I_3 \times 10^{-3} = 2V_2 I_2 \times 10^{-3} \, kVA$

Volts per turn $\dfrac{V_3}{T_3} = \dfrac{V_2}{T_2}$

normal single-phase transformer, and a tertiary winding is provided, the windings of which are connected in parallel, *i.e.*, as the two-phase windings of Fig. 8.24(*b*).

To calculate the regulation under all possible conditions of loading, six impedances, in general, must be known, and they are defined as follows, the symbols denoting percentage values.

(i) Z_M, the impedance of the main unit when transforming half the group output as a normal single-phase unit.

(ii) Z_M', the impedance of the main unit measured across mid point and lines when full-load, three-phase current divides equally into the two halves of the winding.

(iii) Z_M'', the impedance of the main unit measured across mid point and one line terminal when full-load, three-phase current flows in that half only of the three-phase winding.

(iv) Z_T, the impedance of the teaser unit when transforming its full output, *i.e.*, half the group kVA.

(v) Z_T', the impedance of the teaser unit when carrying full-load, three-phase current between line and neutral, *i.e.*, when transforming two thirds of its rated output.

(vi) Z_T'', the impedance of the teaser unit when carrying full-load, three-phase current between neutral and the terminal A (when the unit is non-interchangeable; see Fig. 8.24(*a*), *i.e.*, when transforming one third of its rated output.

The tests by which these impedances and the corresponding copper losses—the latter being denoted by C and the appropriate suffix—are measured are shown diagrammatically in Fig. 8.25, (i) to (vi). The measurements are indicated as being made on the three-phase sides of the units, but with the sole exception of test (ii), they may be made on the two-phase windings; reference to Table 8.1 will make this clear. In this table the conversion of the measured quantities to percentage values is shown; each impedance voltage is expressed as a percentage of the normal voltage of the portion of the winding across which it is measured, and, in order that the resistance voltages be similarly expressed, the copper losses should be divided by the appropriate multiple of the group output in kVA, the latter being denoted by the symbol K, shown in column 4, Table 8.1. The percentage reactances then follow from the usual formula,

$$X = \sqrt{Z^2 - R^2}$$

The tests indicated in Fig. 8.25 apply to the general case, but the actual number of tests necessary or possible depends on the type of connection used for the unit. Obviously, when the neutral point is not brought out Z_T' and Z_T'' cannot be measured. In the case of the main unit it can be demonstrated theoretically that for any type of connection,

$$R_M'' = R_M + \frac{1}{\sqrt{3}} R \qquad (8.1)$$

or,

$$C_M'' = C_M' + \tfrac{1}{3} C_M \qquad (8.2)$$

and,

$$X_M'' = X_M' + \frac{1}{\sqrt{3}} X_M \qquad (8.3)$$

185

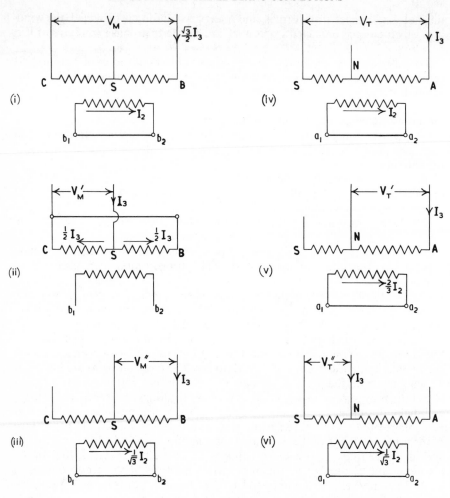

Fig. 8.25 Single-phase copper loss and impedance tests on Scott-connected transformers

In effect, two tests only are necessary on this unit. When the main transformer connection is of the type shown in Fig. 8.24(b), one test only suffices, for, as a little consideration will show, all the resistances and reactances are directly related; thus,

$$R_M = \sqrt{3}R_M' = \frac{\sqrt{3}}{2}R_M'' \tag{8·4}$$

or,

$$C_M = 3C_M' = \frac{3}{2}C_M'' \tag{8.5}$$

and,

$$X_M = \sqrt{3}X_M' = \frac{\sqrt{3}}{2}X_M'' \tag{8.6}$$

186

The relations given by eqns. 8.4, 8.5 and 8.6 are in accord with the general relations given in eqns. 8.1, 8.2 and 8.3.

In Table 8.2 are set out the various conditions, that are likely to occur in practice, and in each case the values of R and X shown are used with the usual regulation formula to calculate the regulation across the appropriate pair of terminals. The formula for the percentage regulation is, for full load,

$$\text{Percentage regulation} = V_R \cos \phi + V_X \sin \phi + \frac{(V_X \cos \phi - V_R \sin \phi)^2}{200}$$

where,
V_R = percentage resistance voltage or copper loss,
V_X = percentage reactance voltage,
$\cos \phi$ = load power factor.

The values of V_R and V_X in Table 8.2 show that a balanced regulation is obtained only when,

$$R_T = R_M - \frac{1}{\sqrt{3}} R_M', \text{ or } C_T = C_M - C_M'$$

and,

$$X_T = X_M - \frac{1}{\sqrt{3}} X_M'$$

and unless both conditions are satisfied, a group short-circuit test, usually made by supplying current to the three-phase terminals ABC, will not give the true group copper loss, for the current distribution will be as shown in Fig. 8.26. The complete mathematical analysis of the problem of finding the three-phase line currents, the group copper loss and impedance voltage on short-circuit has been omitted as being too involved and cumbrous for inclusion in a practical study, but the results have been set out in a form which will facilitate practical application. The data required for the calculations are obtained from the tests on the units, but obviously they can be obtained from the design details of the transformers, and the performance on-load and on-group short circuit thus predetermined; reference to Table 8.1 and Fig. 8.25 will show the conditions for which the various copper losses and reactances must be calculated.

The complicated expressions which result from the mathematical analysis are best dealt with in steps, and the following quantities A, B, C, D, p, q and r have been introduced for that purpose.

It is necessary to calculate,

$$A = X_T + \tfrac{1}{2}\left(R_M'' + \frac{1}{\sqrt{3}} X_M'' \right)$$

$$B = \frac{1}{\sqrt{3}} X_M' - \tfrac{1}{2}\left(R_M'' + \frac{1}{\sqrt{3}} X_M'' \right)$$

$$C = R_T - \tfrac{1}{2}\left(X_M'' - \frac{1}{\sqrt{3}} R_M'' \right)$$

$$D = \frac{1}{\sqrt{3}} R_M' + \tfrac{1}{2}\left(X_M'' - \frac{1}{\sqrt{3}} R_M'' \right)$$

187

Table 8.1

SINGLE-PHASE COPPER-LOSS AND IMPEDANCE TESTS ON SCOTT-CONNECTED TRANSFORMERS

Unit	Test	Copper loss Watts C	Percentage resistance	Impedance volts	Measurements made on three-phase side			Measurements made on two-phase side				Percentage reactance
					Short circuit to	Supply Current supplied	Percentage impedance	Short circuit to	Supply Current supplied		Percentage impedance	
Main	(i)	C_M	$R_M = C_M \div 5K$	V_M	$b_1 b_2$ BC	$\frac{\sqrt{3}}{2}I_3$	$Z_M = 100V_M \div V_3$	BC	$b_1 b_2$	I_2	$Z_M = 100V_M \div V_2$	X_M
	(ii)	C_M'	$R_M' = C_M' \div \dfrac{5}{\sqrt{3}}K$	V_M'	BC	BC, S I_3	$Z_M' = 100V_M' \div \frac{1}{2}V_3$	—	—	—	—	X_M'
	(iii)	C_M''	$R_M'' = C_M'' \div \dfrac{5}{\sqrt{3}}K$	V_M''	$b_1 b_2$ BS	BS I_3	$Z_M'' = 100V_M'' \div \frac{1}{2}V_3$	BS	$b_1 b_2$	$\frac{1}{\sqrt{3}}I_2$	$Z_M'' = 100V_M'' \div V_2$	X_M''
Teaser	(iv)	C_T	$R_T = C_T \div 5K$	V_T	$a_1 a_2$ AS	I_3	$Z_T = 100V_T \div \dfrac{\sqrt{3}}{2}V_3$	AS	$a_1 a_2$	I_2	$Z_T = 100V_T \div V_2$	X_T
	(v)	C_T'	$R_T' = C_T' \div \dfrac{10}{3}K$	V_T'	$a_1 a_2$ AN	I_3	$Z_T' = 100V_T' \div \dfrac{1}{\sqrt{3}}V_3$	AN	$a_1 a_2$	$\frac{2}{3}I_2$	$Z_T' = 100V_T' \div V_2$	X_T'
	(vi)	C_T''	$R_T'' = C_T'' \div \dfrac{5}{3}K$	V_T''	$a_1 a_2$ NS	I_3	$Z_T'' = 100V_T'' \div \dfrac{1}{2\sqrt{3}}V_3$	NS	$a_1 a_2$	$\frac{1}{3}I_2$	$Z_T'' = 100V_T'' \div V_2$	X_T''

In all cases $X = \sqrt{(Z^2 - R^2)}$

then,
$$p = \frac{I_B \cos \phi_B}{\frac{1}{2} I_A} = \frac{R_M A - X_M C}{X_M D - R_M B} \qquad (8.7)$$

$$q = \frac{I_B \sin \phi_B}{\frac{\sqrt{3}}{2} I_A} = \frac{AD - CB}{X_M D - R_M B} \qquad (8.8)$$

$$r = \frac{I_C \cos \phi_C - I_B \cos \phi_B}{I_A}$$

$$= \frac{X_M \left(R_T + \frac{1}{\sqrt{3}} R_M' \right) - R_M \left(X_T + \frac{1}{\sqrt{3}} X_M' \right)}{X_M D - R_M B} \qquad (8.9)$$

where, I_A, I_B, I_C, ϕ_B and ϕ_C are as defined in Fig. 8.26.

The group copper loss C_G' on short circuit, when the teaser current I_A is adjusted to normal full-load value, is,

$$C_G' = C_T + C_M q^2 + C_M'(1 + r^2)$$
$$= C_T + C_M(q^2 - \tfrac{1}{3} r^2) + C_M' + C_M'' r^2$$

and the percentage impedance voltage required is,

$$V_G' = (q R_M + \tfrac{1}{2} r X_M'') + j(q X_M - \tfrac{1}{2} R_M'') \qquad (8.10)$$

The value of I_B is, of course, obtained directly from p and q if required, and I_C is found from,

$$I_C \sin \phi_C = I_B \sin \phi_B$$
$$I_A = I_B \cos \phi_B + I_C \cos \phi_C$$

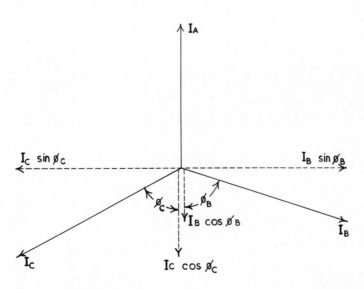

Fig. 8.26   *Phasor diagram, showing current distribution on group three-phase, short-circuit test*

189

Table 8.2

Transformer	Loading		Copper loss watts	Regulation across	Percentage R	Percentage X
Three phase to two phase	Balanced two phase		$C_T + C'_M + C_M$	$a_1 a_2$	$R_T + \dfrac{1}{\sqrt{3}}R'_M$	$X_T + \dfrac{1}{\sqrt{3}}X'_M$
				$b_1 b_2$	R_M	X_M
	Teaser		$C_T + C'_M$	$a_1 a_2$	$R_T + \dfrac{1}{\sqrt{3}}R'_M$	$X_T + \dfrac{1}{\sqrt{3}}X'_M$
	Main		C_M	$b_1 b_2$	R_M	X_M
	Balanced three phase	AB	$C_T + C'_M + C_M$	AB	$\dfrac{3}{4}\left(R_T + \dfrac{1}{\sqrt{3}}R'_M + \dfrac{1}{3}R_M\right) - \dfrac{\sqrt{3}}{4}\left(X_T + \dfrac{1}{\sqrt{3}}X'_M - X_M\right)$	$\dfrac{3}{4}\left(X_T + \dfrac{1}{\sqrt{3}}X'_M + \dfrac{1}{3}X_M\right) + \dfrac{\sqrt{3}}{4}\left(R_T + \dfrac{1}{\sqrt{3}}R'_M - R_M\right)$
		BC		BC	R_M	X_M
		CA		CA	$\dfrac{3}{4}\left(R_T + \dfrac{1}{\sqrt{3}}R'_M + \dfrac{1}{3}R_M\right) + \dfrac{\sqrt{3}}{4}\left(X_T + \dfrac{1}{\sqrt{3}}X'_M - X_M\right)$	$\dfrac{3}{4}\left(X_T + \dfrac{1}{\sqrt{3}}X'_M + \dfrac{1}{3}X_M\right) - \dfrac{\sqrt{3}}{4}\left(R_T + \dfrac{1}{\sqrt{3}}R'_M - R_M\right)$
Two phase to three phase	Single phase line to line		$C_T + C''_M$	AB	$\dfrac{\sqrt{3}}{2}\left(R_T + \dfrac{1}{\sqrt{3}}R'_M\right)$	$\dfrac{\sqrt{3}}{2}\left(X_T + \dfrac{1}{\sqrt{3}}X'_M\right)$
			C_M	BC	R_M	X_M
			$C_T + C''_M$	CA	$\dfrac{\sqrt{3}}{2}\left(R_T + \dfrac{1}{\sqrt{3}}R'_M\right)$	$\dfrac{\sqrt{3}}{2}\left(X_T + \dfrac{1}{\sqrt{3}}X'_M\right)$
	Single phase line to neutral		C'_T	AN	R'_T	X'_T
			$C''_T + C''_M$	BN	$\dfrac{1}{2}(R''_T + \sqrt{3}R''_M)$	$\dfrac{1}{2}(X''_T + \sqrt{3}X''_M)$
			$C''_T + C''_M$	CN	$\dfrac{1}{2}(R''_T + \sqrt{3}R''_M)$	$\dfrac{1}{2}(X''_T + \sqrt{3}X''_M)$

When the conditions for symmetry have been fulfilled, p and q are equal to unity and r is zero, and the group short-circuit loss equals the full-load copper loss, viz., $C_G = C_G' = C_T + C_M + C_M'$, while the percentage impedance is,

$$V_G = V_G' = R_M + jX_M.$$

It is unlikely that, in practice, the symmetrical conditions will ever be fulfilled completely; in general, it is fairly easy to design the units to meet the reactance requirement but quite another matter so to proportion the resistances, for, as will readily be appreciated, both units transforming the same kVA output will, with similar design, have copper losses of similar magnitude. The interconnected type of main unit shown in Fig. 8.24(a) usually gives values of R_M' and X_M' small compared with R_M and X_M, and consequently the correct values of R_T and X_T approach, in magnitude, the values of R_M and X_M. Thus, this type of unit is suitable for use in interchangeable groups, for, when operating as a teaser unit, both the resistance and the reactance are reduced in value and the symmetrical condition can be approached closely with suitable design.

With the type of main unit shown in Fig. 8.24(b), as there are but two windings per limb, the resistances in ohms corresponding to the percentage values R_M, R_M', and R_M'' are all equal. That is, putting r_M, r_M' and r_M'' as the ohmic values, we find,

$$r_M = r_M' = r_M''$$

Similarly, we find by putting x_M, x_M' and x_M'' as the ohmic values corresponding to X_M, X_M' and X_M'',

$$x_M = x_M' = x_M''$$

The symmetrical conditions now reduce to,

$$R_T = \tfrac{2}{3}R_M, \text{ or } C_T = \tfrac{2}{3}C_M = \tfrac{1}{2}(C_M - C_M'), \text{ and } X_T = \tfrac{2}{3}X_M$$

The reactance condition can always be designed for, but the resistance condition, requiring the teaser copper loss to be but half that of the main unit, presents considerable difficulty and would result in an uneconomical design. Were units of this type used to form an interchangeable group there would be a considerable difference in the regulation across the various pairs of terminals, depending, of course, on the direction of phase transformation, and while, in certain cases, this could be tolerated, it is important to note that a group short-circuit test virtually would be useless, for the group short-circuit loss and impedance voltage would be considerably in excess of the load values. On the other hand, for a non-interchangeable group, provided the reactances are correctly proportioned, it will be found that the regulation unbalance is small and a group short-circuit test will give reliable results. Fortunately with this type of main unit connection it is much easier to calculate the effect of asymmetry; the general formulae, eqns. 8.7 and 8.8, reduce to the following:

$$I_B \text{ or } I_C = I_A \sqrt{\left[\tfrac{1}{3} + \tfrac{3}{4}\left(\frac{Z_T}{Z_M}\right)^2 + \tfrac{1}{2}\frac{(R_M R_T + X_M X_T) \pm \sqrt{3}(R_M X_T - X_M R_T)}{Z_M^2}\right]}$$

the plus sign giving the current in line B, the minus sign that in line C; the main unit copper loss on group short circuit is,

$$\left[\tfrac{1}{3}+\tfrac{3}{4}\left(\frac{Z_T}{Z_M}\right)^2+\tfrac{1}{2}\frac{(R_M R_T+X_M X_T)}{Z^2{}_M}\right]$$

times its full-load value and the percentage impedance is,

$$V_G' = (R_T+\tfrac{1}{3}R_M)+j(X_T+\tfrac{1}{3}X_M)$$

It should be noted particularly that in the general case the percentage impedance on group short circuit can have three different values, the one selected in eqn. 8.10 corresponding to full-load current in the teaser unit. The point arises, however, as to what should be taken as the value of the impedance for group loading. Reference to Table 8.2, under 'Balanced three-phase loading', will show that it is really essential to know the values of R_T, R_M, R_M', X_T, X_M and X_M', for although the group copper loss can be stated definitely the value of the impedance—and regulation—depends upon the pair of terminals considered.

It has been observed that, in view of the relation between the resistances and reactances of the main unit, two tests on this unit give the data required; when the three-phase voltage is low, test (ii) may be inconvenient, and tests (i) and (iii), which differ only in the position of one supply lead, would be used. In this case the group copper loss can be evaluated directly as equal to,

$$C_G = C_T+\tfrac{2}{3}C_M+C_M''$$

As a matter of interest, when the main unit is interconnected on the three-phase side and voltage adjusting tappings are provided on this winding, it is obvious that equal numbers of turns in the two coil stacks of each limb at all tapping positions can be obtained only by providing tappings on each coil stack. This may present considerable difficulty, and if only one coil stack per limb is tapped there can be but one position when each coil stack has the same number of turns. In such a case, to prevent the reactance having abnormal values with the various tappings in use, the windings may be connected in parallel on the two-phase side. The two-phase winding current of the main unit in terms of the two-phase line current is,

$$I_W = \tfrac{1}{2}I_2\sqrt{\left[1+\tfrac{1}{3}\left(\frac{1}{100/\text{per cent tap}\pm 1}\right)^2\right]} \tag{8.11}$$

where the plus sign denotes an increase in three-phase turns over the number which makes the number of turns on each coil stack equal, and 'per cent tap' is the amount by which the tapped coil stack differs from the other as a percentage of the turns when the coil stacks have equal numbers of turns. The formula applies to 'constant kVA' tappings, and obviously minus tappings increase the two-phase winding current more than do plus tappings, although even a minus 20% tapping causes a current increase of 1% only. Equation 8.11 is applicable to the connection shown in Fig. 8.24(b), for in this case there is, in effect, a -50% tapping, giving an increase of $15\tfrac{1}{2}\%$ over $\tfrac{1}{2}I_2$.

THE LE BLANC CONNECTION

The alternative connection to the Scott for transforming from a three-phase to a two-phase supply is the Le Blanc connection. Although this latter connection has been accepted by engineers from the end of the nineteenth century it has not

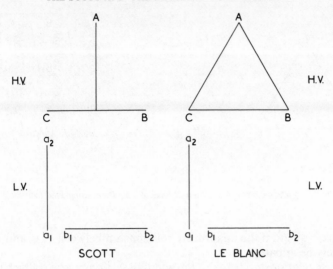

Fig. 8.27 *Phasor diagrams illustrating interchangeability of Scott and Le Blanc connections*

gained the same popularity as the Scott connection and is by no means so well known. Figure 8.27 shows the combined voltage phasor diagrams of the Scott and Le Blanc connections and from an inspection of these it will be seen that the phase displacement obtained by both methods is identical and that the connections are interchangeable. It follows therefore that transformers having these connections will operate satisfactorily in parallel with each other if the normal requirements of voltage ratio and impedance are met. The primary of the Le Blanc connected transformer shown in Fig. 8.27 is connected in three-phase delta which is the normal interphase connection in the case of a step-down unit supplied from an h.v. source. Where the primary three-phase winding is connected in delta the inherent advantage of this winding for the suppression of third-harmonic voltages will be apparent. For fuller details of this aspect reference should be made to Chapter 7. Where the three-phase side is the secondary, *i.e.*, when the transformer is operating two to three phase it would be more convenient to use a star connection on the three-phase side.

A core of the three-limb, three-phase design is employed for the construction of a Le Blanc-connected transformer compared with two single-phase cores for the Scott-connected transformer. In addition to a somewhat simpler standard core arrangement the Le Blanc transformer is less costly to manufacture due to the fact that for a given rating less active materials are required for its construction. The fact that a three-phase core, and hence a single tank can be employed to house the Le Blanc transformer means that the unit is more economical in floor space than the Scott transformer, particularly if compared with the arrangement of two separate single-phase cores each in its own tank. From the phasor and connection diagrams of Fig. 8.28, which is drawn to show the arrangement of windings for a three- to two-phase Le Blanc transformer, it will be seen that the h.v. primary is identical with that of any delta connected winding and is constructed as such. The voltage of the output winding is established across the four two-phase terminals a_1a_2 and b_1b_2 and the l.v. turns are so designed that the voltage phasor a_1a_2 is equal to b_1b_2. From the geometry

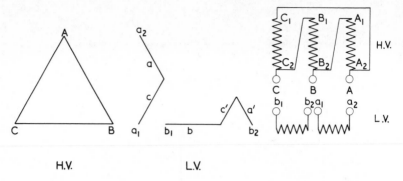

Fig. 8.28   Phasor and connection diagrams of a Le Blanc-connected transformer

of the phasor diagram the quadrature relationship between a_1a_2 and b_1b_2 will immediately be apparent.

The phase relationship between the winding sections a and c which comprise one phase of the two-phase output are 120° apart so that each section a and c must have 57·7% of the number of turns required to develop the specified phase voltage a_1a_2. Further, the winding sections a and c must have 1·73 times the number of turns of winding sections a′ and c′, resulting in winding sections a′ and c′ having 33·3% of the number of turns corresponding to the phase voltage b_1b_2. It follows that winding section b must have 66·6% of the number of turns corresponding to the phase voltage b_1b_2. These fixed relationships of number of turns between the winding sections a, a′, b, c and c′ follow from the basic voltage phasor diagram.

When transforming from a three-phase h.v. supply to an l.v. two-phase output quite definite limitations are therefore imposed upon the design of the secondary winding of a Le Blanc-connected transformer due to the fact that only whole numbers can be employed for the winding turns, whilst at the same time certain fixed ratios of turns must be maintained between sections of windings. These conditions are accentuated by an l.v. winding having comparatively few turns. In addition to these considerations of voltages of the various sections

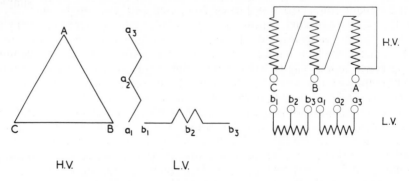

Fig. 8.29   Phasor and connection diagrams of a Le Blanc-connected transformer when mid points are required on the two-phase windings

of the two-phase side, the ampere-turns of each phase of the primary winding are balanced by the phasor sum of the ampere-turns of the components of the secondary windings of the two-phase winding on the same phase.

The Le Blanc connection can be arranged for either two-phase three-wire or four-wire output windings, and will transform from three- to two-phase or vice versa with the three-phase side connected in either star or delta. The former is invariably employed for three-phase l.v. secondary windings and the latter for h.v. three-phase primary windings.

When supplying a balanced three-phase load from a star-connected secondary the regulation of the Le Blanc transformer will be comparable with that of a three-phase star/star connected transformer and if it is required to load the transformer windings between line and neutral, and so cause appreciable unbalanced loading, a tertiary delta-connected winding should be provided.

The phasor and winding diagrams shown in Fig. 8.29 illustrate the modification necessary to the two-phase side of a Le Blanc transformer when the mid points are required to be available on the two-phase winding. Compared with the arrangement of the windings of Fig. 8.28 it will be seen that each winding section a, a′, b, c and c′ of that diagram are each subdivided into halves and interconnected to provide the mid points at a_2 and b_2 of Fig. 8.29.

Chapter 9

The interconnected star and the open delta connections*

THE INTERCONNECTED STAR CONNECTION

In practice the interconnected star or zigzag connection of three-phase transformer windings finds its greatest application when adopted as the method for
connecting the secondary windings of three-phase transformers or transformer
groups in conjunction with a star connected primary, and the combined connection is frequently used in place of the delta/star connection. It would be as
well, therefore, to compare the star/interconnected star connection with the
delta/star, particularly as the latter connection is perhaps the one most widely
employed, and certainly so in Great Britain. The chief features involved are as
follows:

(1) The effect of the connection chosen on the economical limits of voltage
and rating
(2) parallel operation of star/interconnected star with delta/star connected
transformers
(3) unbalanced loading
(4) harmonics
(5) availability of the neutral point.

Effect of the connection chosen on the economical limits of voltage and rating

As a high-voltage distribution system extends into less dense areas, it is often
found that more or less scattered loads of a few hundred kilowatts require
supplying, the only voltage immediately available being that of primary
distribution, which may be, say, 11 000 or 22 000 V. Now it is at once apparent
that there must be some economic relation between the output and voltage of
a transformer, other things being equal; in other words, for a given kVA rating
there is a limiting voltage per phase beyond which a transformer could not
economically be designed, particularly having in mind the needs of standardisation. Conversely, for a given phase voltage there is a minimum kVA rating below

*See Chapter 7.

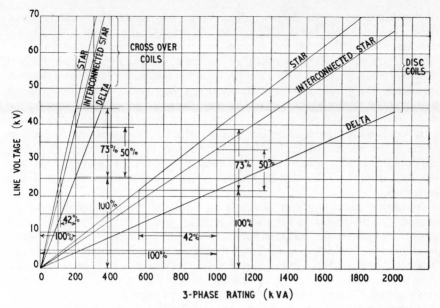

Fig. 9.1   Average relations between economic voltage and three-phase ratings of transformers

which it would not be economical to go. The economical upper and lower limits of voltage and rating are dictated by considerations of mechanical strength and rigidity of the windings, and not by electrical or magnetic features of the design, so that outside these limits, while the proposed coil conductors might be suitable from the standpoint of current carrying capacity, losses and voltage drop, they would not be mechanically strong enough to withstand the short circuit conditions occurring on the system, nor would they lend themselves to the best methods of modern winding and insulating processes. Such coils could be inherently weak, due to the low copper space factor.

Figure 9.1 shows an average relationship between the economical voltage and the three-phase rating.

Broadly speaking, we are not so much concerned with the voltage *per phase* for which the transformer is designed, as with the line voltage of distribution, so that if the same or closely similar operation conditions can be attained with two different types of connections, obviously the one to select is that having the greater mechanical strength of the transformer windings.

Consider first the delta/star connection for a step-down transformer of given line voltage ratio.

On the h.v. side the number of turns per phase is a maximum as the phase voltage is equal to the line voltage. The space taken up by the insulation is also a maximum due to maximum number of turns. The cross section of the winding conductor need only be large enough to carry 58 % of the line current, so that in the region of voltage and rating limits we have all the characteristics of a mechanically weak winding, viz., 100 % turns, 100 % insulation, and 58 % copper section.

If, however, a star connection is used on the primary side, the number of turns per phase is 58 % of the foregoing, the space taken up by insulation is 58 %

197

of that taken up by the delta winding (as the best modern practice is *not* to reduce the insulation between turns and layers on account of the star voltage), and the cross-sectional area of the conductor is a maximum as the phase and line currents are equal. Hence a much more robust winding is produced which is inherently better able to withstand the arduous conditions imposed by service on a large system. The total space taken up by the copper and insulation of the high-voltage winding is therefore less with the star connection than with the delta, as the total copper section in the coil is the same, but the star connection only requires the same thickness of insulation for 58 % of the maximum number of turns. So far as the high-voltage winding alone is concerned, therefore, a smaller magnetic circuit can theoretically be used by adopting the star winding, though due to standardisation of parts it may not always be found convenient.

In certain cases it is not desirable to retain the star connection on the secondary side with a star-connected primary, particularly if loads are to be connected between line terminals and neutral. The reason for this is referred to later. Neither is a delta connection suitable for somewhat similar reasons. If, however, the secondary windings are designed for and connected interconnected star, the objectionable operating features of the star or delta secondary with star primary for three-phase four-wire service are overcome.

With such a winding, due to the phase difference of the two halves of the windings on different limbs, which are connected in series and comprise each phase, the number of turns per phase is 67% of the maximum, *i.e.*, of those obtaining with a delta connection. The space taken up by the insulation is also 67% of the maximum, while the cross-sectional area of the individual conductors is a maximum as the phase and line currents are equal. Consequently a winding is produced which mechanically is almost as good as a star winding, and which, in conjunction with a star-connected primary, is electrically as good as the delta/star combination.

So far as the l.v. winding alone is concerned, a slightly larger magnetic circuit is theoretically required to accommodate the interconnected star winding as compared with the straight star, as there is $15\frac{1}{2}\%$ more copper and insulation per phase. This may be offset to some extent by the gain obtained in using a star-connected primary, and often the delta/star standard magnetic circuit will accommodate the interconnected star secondary. It may happen, of course, that the nearest standard frame size suitable for the particular output and voltages will not take the star/interconnected star winding, so that the net result is a somewhat larger magnetic circuit, and consequently increased cost of the transformer.

With, at the most, a small increase in frame size and in copper weight we have therefore obtained a design and connection of windings electrically equivalent to the delta/star, and mechanically much stronger, and in addition, one in which the h.v. phase voltage is only 58 % of the maximum, *i.e.*, of the line voltage.

In effect, therefore, the upper limit of line voltage for a given rated kVA has been increased by 73% or the lower limit of rated kVA has been reduced by 42% for a given line voltage by changing from the delta/star to the star/interconnected star connection.

Owing to the higher copper space factor the secondary l.v. winding takes much less room in proportion to the h.v. winding, and for the purpose of this argument can be ignored.

Given, therefore, equal possible technical performance between two combinations, this is the main reason for using the star/interconnected star connection for a step-down transformer supplying a consumer.

Parallel operation of star/interconnected star with delta/star connected transformers

With regard to the question of parallel operation of star/interconnected star transformers with delta/star transformers, it is not necessary to consider this aspect in great detail, and it is sufficient to state that delta/star connected transformers or transformer groups will operate satisfactorily in parallel with any star/interconnected star transformer or transformer group. This holds good whatever the meshing of the phases may be or whatever may be the direction in which the primary and secondary coils are wound, and it is only a question of selecting a correct set of external connections from the transformer terminals to the busbars in each case, as explained in Chapter 18.

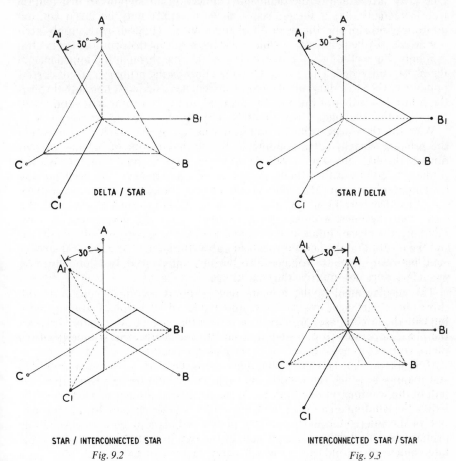

Fig. 9.2

Fig. 9.3

Phasor diagrams showing the comparison between connections

It has already been stated that electrically a star/interconnected star transformer is equivalent to a delta/star connected one, though no proof has been given for this. The fact can best be illustrated by means of a phasor diagram, and Fig. 9.2 shows a comparison between the two connections. From this it will be seen that there is the same angular phase displacement in both cases between primary and secondary phasors, viz., 30°, and therefore both primary and both secondary line currents and voltages are in phase.

Of course, cases may also occur in practice where the interconnected star winding is required on the primary side to be used in conjunction with a star-connected secondary, and in this case the connection would be electrically equivalent to the star/delta. Figure 9.3 shows a comparison between these two combined connections, from which it will be seen that there is again the same angular displacement in both cases between primary and secondary phasors.

Unbalanced loading

One great advantage of the delta/star connection for step-down distribution transformers is that a three-phase four-wire supply may be given without appreciably distorting the position of the neutral. The four-wire supply need not necessarily be balanced, in which case the resulting distortion of the neutral is simply due to the out of balance current flowing through the impedance of the phase concerned. There is no choking effect as the primary current corresponding to the secondary out of balance current has a perfectly free path through the primary winding of the phase concerned and the two line wires connected to it; the other two phases are not affected by the out of balance current.

With the three-wire primary and four-wire secondary star/star connection the primary current corresponding to the secondary out of balance current flowing through any one phase of a four-wire system must flow through the primary windings of the remaining two phases, and as there are no balancing load ampere-turns on the secondary windings of these two phases the primary out of balance current acts wholly as no-load current to such phases, with the result that the phase voltages of the two lightly loaded phases increase considerably, the phase voltage of the heavier loaded phase correspondingly drops, and the position of the neutral is considerably displaced. The angular displacement between the phase voltages also become unbalanced. A delta connected secondary does not improve the conditions.

This displacement of the primary neutral point becomes acute only with three-phase shell-type transformers and banks of single-phase transformers, but with the three-phase core-type transformer the magnetic interaction between the phases operates to preserve, to a far greater degree, the symmetry of the phase voltages to neutral, and so to stabilise the neutral.

If, however, the secondary winding is connected up in the interconnected star manner in which the winding on each limb is split up into two halves, the half of the winding on one limb being electrically connected in series with the half of the winding on another limb, any out of balance current flowing through one of the interconnected secondary phases is balanced by a corresponding primary current flowing through each of the two primary phase windings, the interconnected secondaries of which carry the out of balance current. As a consequence the primary out of balance current has a perfectly free path through the two windings affected and the two line wires, and the neutral in all cases is only slightly disturbed on account of the voltage drop due to the out of balance current. As with the delta/star arrangement there is no choking effect.

These remarks apply equally well to three-phase transformers and three-phase banks of single-phase transformers.

Harmonics

It is generally known that due to the variations in permeability with changes in flux density in the transformer core, also on account of the shape of the hysteresis loop, the m.m.f., and consequently the magnetising current wave form required to produce a sine wave of flux in the magnetic circuit will contain a third harmonic; if the third harmonic is absent from the wave of magnetising current it will be present in the flux wave, and therefore in the induced voltage waves.

As with a symmetrical star-connected generator the third harmonic and its multiples are eliminated from the terminal voltages, it follows that they also will not appear in the applied line (or phase) voltages of the delta-connected primary of a transformer, but third harmonics will appear in the wave of the magnetising current. Such a delta winding provides a closed path for the circulation of the triple frequency component of the magnetising current, and there exists no third-harmonic voltage to neutral on the secondary star-connected windings. Looking at it from another point of view, the triple frequency current circulating round the primary delta supplies the third-harmonic component of the magnetising current so that the flux and consequently the induced voltage waves are sinusoidal, and no third-harmonic voltages exist.

Third-harmonic voltages of any appreciable magnitude can only occur in single-phase transformers (core and shell type), and in three-phase shell-type transformers, as in each of these types the magnetic circuit presents a completely closed magnetic path to the third harmonic flux. They cannot be of any serious value in the usual construction of three-phase core-type transformers having

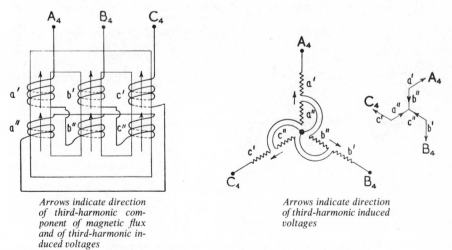

Arrows indicate direction of third-harmonic component of magnetic flux and of third-harmonic induced voltages

Arrows indicate direction of third-harmonic induced voltages

Fig. 9.4 *Neutralising effect of third-harmonic voltages*

201

three limbs, whatever the connection, as the resulting triple frequency instantaneous flux flows in the same direction in each limb, and so must take the return path through the air (or oil) and tank, this path being of very high reluctance.

When changing to a star-connected primary it is not permissible to retain the star connection on the secondary side, nor is a delta secondary desirable, as a neutral point is usually required for giving a four-wire supply or for earthing purposes. Both of these requirements, of course, can be obtained by artificial means, but it is better to incorporate them with the transformer secondary winding if possible.

By using an interconnected star winding on the secondary side of the transformer, the third-harmonic voltages which are induced in the same direction in each half of the windings on each limb are eliminated from lines to neutral by opposition, as shown in Fig. 9.4.

This, therefore, becomes an important feature *per se*, when dealing with single-phase shell- and core-type transformers and with three-phase shell-type transformers, but not so to any practical extent in the case of three-phase core-type transformers. It does, however, give an advantage even with the latter type, but is not a decisive factor in the selection of the interconnected star secondary.

Availability of neutral point

As stated earlier, one of the chief features of the delta/star connection is that the neutral point on the secondary star-connected side is available either for supplying as three-phase four-wire load or for earthing purposes. On the primary side, if a system requires earthing, this is usually done at the generator or by the creation of an artificial neutral point. With the star/interconnected star connection a neutral point is available on both primary and secondary sides of the transformer, so that the same facilities are still available on the secondary side, with additional provision on the primary side for earthing if desired. This may at times be a very useful feature, but it should be borne in

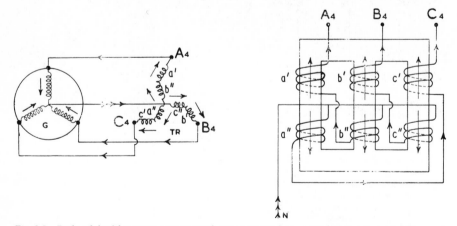

Fig. 9.5   Paths of third-harmonic currents with interconnected star transformer primary winding and fourth wire between generator and transformer neutrals

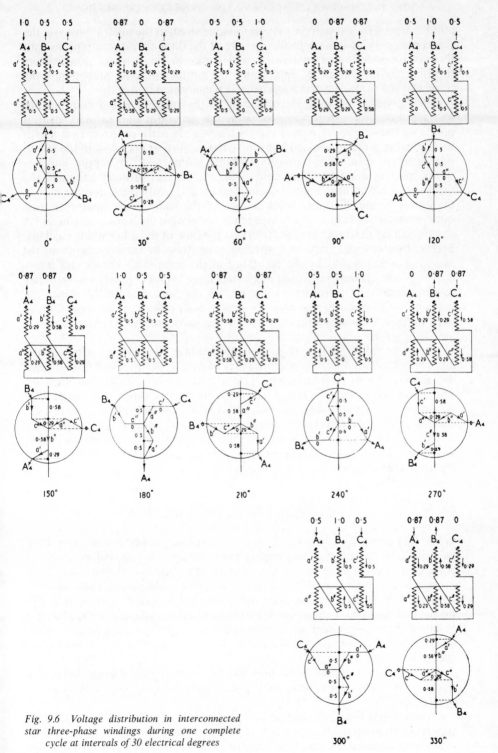

Fig. 9.6 Voltage distribution in interconnected star three-phase windings during one complete cycle at intervals of 30 electrical degrees

203

mind that generally it will only be permissible to earth the neutral point on the star connected side if the transformers are of the three-phase core-type, as with three-phase shell-type transformers and banks of single-phase transformers third-harmonic voltages of appreciable magnitudes may be present just the same as if the secondary windings were also connected in star.

If the primary winding is connected in the interconnected star manner and supplied direct from the generator, the generator and primary neutral points should not be connected together, either via earth or direct by a fourth wire. If the neutrals are joined a third-harmonic current may flow in the fourth wire resultant from third-harmonic voltages in the source of supply, and due to the splitting and interconnection of the transformer primary windings, the third harmonic exciting ampere-turns, and consequently the induced e.m.fs in the halves of the windings on each limb of the transformer neutralise each other, thereby providing a low-impedance path (virtually a short circuit so far as transformer reactance is concerned) for the flow of third-harmonic currents. Such currents would only be limited in magnitude by the impedance of the generator winding and by the resistance of the connecting cables and transformer windings, and they would reach such large values in practice as seriously to overload the transformer and generator windings. Figure 9.5 shows the paths of the third-harmonic currents, and in both diagrams the opposition of third-harmonic exciting ampere-turns in the halves of the windings on each limb will be seen.

It may be interesting to show how the voltage distribution varies in the interconnected windings at different parts of each cycle. Figure 9.6 shows this distribution during one complete cycle at intervals of 30 electrical degrees. The figure shows the winding diagrams with relative directions and instantaneous values of the voltages in each half winding and in the lines, and also the corresponding positions of the phasors. The phasor diagrams also show the distribution of voltage in the windings, together with their instantaneous values, which are obtained by projection on to the verical line passing through the centre of each diagram.

THE OPEN DELTA CONNECTION

While it is generally appreciated that a three-phase supply can be given from two single-phase transformers having both primary and secondary windings connected in open delta or vee, the absence of a third transformer frequently arouses a certain amount of mystification in so far as the current distribution is concerned. References to the open delta connection usually state that except for the voltage unbalancing caused by the unequal impedances of the phases, a balanced three-phase load can be supplied with a corresponding balance of the line currents and voltages, though the loading of the transformers must receive special attention.

The following remarks show how the line voltage and current balance is maintained when using two single-phase transformers only, connected in open delta.

Consider first Fig. 9.7, which shows a star connected three-phase alternator supplying an open delta connected transformer bank, the secondary windings being open circuited. Except for the small no-load currents, which will be ignored,

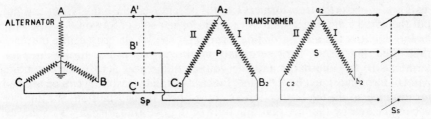

Fig. 9.7   Open-circuit conditions with an open delta transformer bank

there are no currents in the lines or in the transformer windings. Therefore, the alternator phase and line voltages are the same as with no transformer bank connected, and the line voltages, as shown by the phasor diagram of Fig. 9.8, appear across the primary switch terminals A′, B′, C′. That is, voltage differences can be measured by suitable instruments, across the switch terminals A′ to B′, B′ to C′, C′ to A′, and the magnitude and phase relationship of these are the same as the voltage differences generated across the alternator terminals and indicated in phasor form in Fig. 9.8. Now, if the transformer primary switch S$_p$ be closed, obviously the voltage differences across A′ to B′, B′ to C′, and C′ to A′, which are identical in all respects with the voltages across the alternator terminals AB, BC, and CA, are impressed across the transformer bank primary terminals A$_2$ to B$_2$, B$_2$ to C$_2$, C$_2$ to A$_2$, that is across transformer I, the open phase, and transformer II. Now, bearing in mind that at the moment voltage differences in an open-circuited bank of transformers only are being discussed, it will be clear that the omission of a third transformer in no way affects the distribution of the voltages, as this is absolutely fixed by the alternator. For

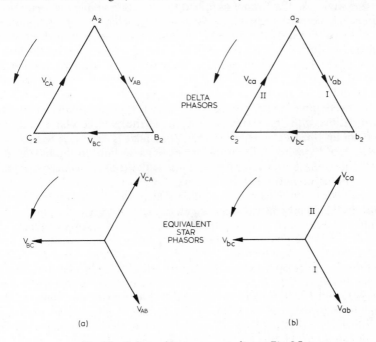

Fig. 9.8   Voltage phasors corresponding to Fig. 9.7

205

instance, suppose the transformers I and II be removed, the bank is non-existent, but the voltage differences across the transformer side of S_p would be the same as those across the alternator line terminals. The whole crux of the matter is that three transformer bank terminals are directly connected to the three line terminals of the alternator and the voltage differences across corresponding pairs of terminals must be the same. Precisely similar reasoning can be applied, for instance, to the tee connection for three-phase transformation, but this case is complicated by the fact that one only of the two transformers is connected directly across one pair of supply terminals.

Now, in the open delta bank two single-phase transformers are each connected across the lines of a three-phase source, and so far as all normal induced effects are concerned, the transformers act as individual single-phase units. That is, whatever happens under normal operating conditions on the primary side of each transformer (apart from magnetisation effects), its exact counterpart is produced on the secondary side, modified in magnitude, of course, according to the voltage ratio; the transformers *are* single phase and *act* exactly as such. On the secondary sides of the two transformers forming the group there is induced across the windings of each a voltage in opposition to that applied to its primary, and when the two secondaries are connected in open delta the voltages across the three secondary terminals bear exactly the same relationship to one another as do the corresponding primary terminal voltages. This is shown in Fig. 9.8. These voltages appear at the secondary switch terminals and, when the switch S_s is closed upon a load, at the load terminals also, and at every part of each cycle the voltage balance, as shown, is maintained.

Assume now that the load, consisting of, say, three-phase synchronous motors drawing current at unity power factor, is connected to the transformer bank secondary. As the voltages applied to the motors have the formation of a balanced symmetrical three-phase system, the currents in the secondary line wires will have precisely the same balanced three-phase relationship, being 30° out of phase, however, with the applied voltages. This is so as the star load currents in the lines depend entirely upon the delta voltages applied to the motors and upon the impedance of the entire secondary circuit. The conditions indicated in Fig. 9.8 are now arrived at which must, by commonsense reasoning, infallibly occur, and the next thing to consider is how the results so far reached are compatible with the statement that a balanced three-phase load can be given from an open delta transformer bank. That is to say, it is necessary to consider what happens in the transformer bank to link up the balanced symmetrical three-phase alternator line voltages with the balanced line voltages and currents applied to and taken by the load.

In a balanced symmetrical three-phase system the line currents are of equal magnitude and equal phase displacement, *i.e.*, 120° between each, and therefore their phasor sum is zero. Now consider Fig. 9.9 which shows in phasor form, in their true relative positions, the secondary line voltages V_{ab}, V_{bc}, V_{ca}, and the line currents I_a, I_b, I_c, taken by the load. At the instant when I_a reaches its maximum, I_b and I_c each have half their maximum amplitudes and flow in an opposite direction to I_a. If I_a is assumed to flow from the transformer bank to the load, I_b and I_c will flow from the load to the transformer bank, I_b flowing through transformer I, and I_c through transformer II. Due to the absence of the third transformer across phase BC the currents in the two transformer secondaries are equal to the line currents and lead and lag respectively by 30°. That

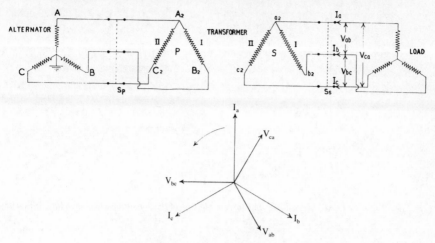

Fig. 9.9 *Open delta transformer bank supplying a balanced three-phase load*

is, even at unity power factor load the transformers have internal power factors of 86·6%, one leading and the other lagging.

At this instant both transformers obviously are transforming the same amount of power which is given for each transformer by the expression:—

Transformer I

Power transformed $= \dfrac{V_{ab}I_b}{1000}$ where the *instantaneous* values for V_{ab} and I_b are inserted.

For
$$V_{ab} = 3000 \text{ V as the } 100\% \text{ value}$$
$$I_b = 1000 \text{ A as the } 100\% \text{ value}$$

the instantaneous power transformed $= \dfrac{2600 \times 500}{1000} = 1300 \text{ kW}.$

Transformer II

Power transformed $= \dfrac{V_{ca}I_c}{1000}$ where the *instantaneous* values for V_{ca} and I_c are inserted.

For
$$V_{ca} = 3000 \text{ V as the } 100\% \text{ value}$$
$$I_c = 1000 \text{ A as the } 100\% \text{ value}$$

the instantaneous power transformed $= \dfrac{2600 \times 500}{1000} = 1300 \text{ kW}.$

At this instant the line voltages V_{ab} and V_{ca} have 86·6% of their full values, which accounts for the figures of 2600 in the above equations. The total instantaneous power transformed by the bank is therefore 2600 kW.

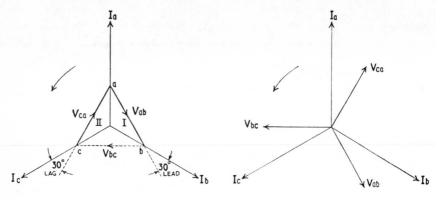

Fig. 9.10   *Impressed and load voltages and current phasors of the conditions corresponding to Fig. 9.9*

Take next the instant when I_b reaches its maximum, that is 120° later in time according to the direction of phase rotation shown in Fig. 9.10. I_b will now be flowing from the transformer bank to the load, and I_a and I_c in the reverse direction, each at half their maximum amplitudes. I_a—at half its maximum amplitude—will flow through transformer I, while I_c—also at half its maximum amplitude—will flow through transformers II and I in series. Therefore, at this instant, transformer II carries half the maximum line current and transformer I the full maximum line current. Although the total power transformed by the bank at this instant is the same at the instant when I_a reached its maximum, the transformers do not share the power equally; in fact all the power transformation is done by transformer I, and this will be clear from the following expressions:

Transformer I

Power transformed $= \dfrac{V_{ab} I_b}{1000}$ where the *instantaneous* values for V_{ab} and I_b are inserted.

For $V_{ab} = 3000$ V as the 100% value
$I_b = 1000$ A as the 100% value

the instantaneous power transformed $= \dfrac{2600 \times 1000}{1000} = 2600\,\text{kW}.$

Transformer II

Power transformed $= \dfrac{V_{ca} I_c}{1000}$ where the *instantaneous* values for V_{ca} and I_c are inserted.

For $V_{ca} = 3000$ V as the 100% value
$I_c = 1000$ A as the 100% value

the instantaneous power transformed $= \dfrac{0 \times 500}{1000} = 0 \, \text{kW}.$

At this instant the line voltages V_{ab} and V_{ca} have 86·6% *and* 0% of their full values respectively, which accounts for the figures of 2600 and 0 in the above equations.

The total instantaneous power transformed by the bank is therefore still 2600 kW.

At a period of time corresponding to a further 120° phasor rotation, I_c reaches its maximum, flowing from the transformer bank to the load. I_a and I_b at this instant each have half their maximum value and flow in an opposite direction to I_c. I_a—at half its maximum amplitude—will flow through transformer II, while I_b—also at half its maximum amplitude—will flow through transformers I and II in series. Therefore, at this instant, transformer I carries half the maximum line current and transformer II the full maximum line current. At this instant, also, the transformers are not sharing the power transformed, although the total bank output still remains the same. The following expressions show how the power load is distributed:

Transformer I

Power transformed $= \dfrac{V_{ab} I_b}{1000}$ where the *instantaneous* values for V_{ab} and I_b are inserted.

For $\qquad V_{ab} = 3000$ V as the 100% value
$\qquad\qquad I_b = 1000$ A as the 100% value

the instantaneous power transformed $= \dfrac{0 \times 500}{1000} = 0 \, \text{kW}.$

Transformer II

Power transformed $= \dfrac{V_{ca} I_c}{1000}$ where the *instantaneous* values for V_{ca} and I_c are inserted.

For $\qquad V_{ca} = 3000$ V as the 100% value
$\qquad\qquad I_a = 1000$ A as the 100% value

the instantaneous power transformed $= \dfrac{2600 \times 1000}{1000} = 2600 \, \text{kW}.$

At this instant the line voltages V_{ab} and V_{ca} are zero and 86·6% respectively of their full values, which accounts for the figures of 0 and 2600 in the above equations.

The total instantaneous power transformed by the bank is therefore again 2600 kW.

The current distribution has now been determined in the entire secondary circuit at three separate instants, namely, when I_a, I_b, and I_c reach their respective maxima, and in order to get a perfectly clear conception of what happens during each complete cycle the current distribution will now be determined at intervals of 30 electrical degrees throughout the cycle. Figure 9.11 shows, at 30° intervals, the equivalent star phasors for the secondary line currents and voltages. Commencing with the one shown at the instant marked 0°, that is, when I_a reaches its maximum (this being arbitrarily chosen as a starting point), it will be seen that the diagram is an exact replica of the star phasor diagram of Fig. 9.10 and as such is self-explanatory.

Now, in setting out to decide the current flow at particular instants, instantaneous currents in the three secondary lines must be derived from the star phasor diagrams which represent the maximum values only that are reached

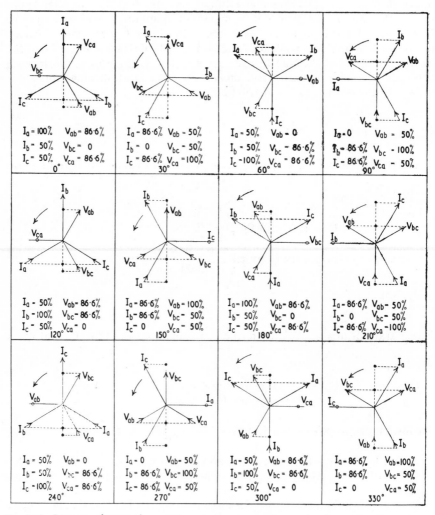

Fig. 9.11   Diagrams showing the variations, during one complete cycle, of secondary line currents and voltages

Note. The disposition of the phasors at 360° is the same as at 0°. Dotted lines show construction for determining instantaneous values.

210

at intervals of 120 electrical degrees or one third of a cycle. This can most easily be done by drawing any line through the point from which the star vectors radiate, this line serving as a fixed reference axis for determining instantaneous values from the star phasors. For the purpose a vertical axis is selected as shown on the diagrams of Fig. 9.11. If, now, the extremity of any one or more of the star phasors be projected horizontally on to the reference axis, the distances between the points so obtained and the origin of the system of phasors gives, to the same scale as that to which the diagram is drawn, the instantaneous values of the phasors concerned, and the lines connecting the derived points and the origin indicate the instantaneous phase relationship and direction of the star phasors. Referring again to Fig. 9.11, instant $0°$, the extremities of I_a, I_b and I_c projected on to the vertical axis of reference give, as instantaneous values, I_a equal to 100%, assumed to be flowing outwards, *i.e.*, from the transformer

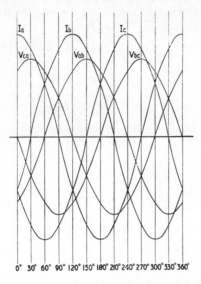

Fig. 9.12   Line voltage and current sine waves, corresponding to phasors of Fig. 9.11

bank to the load, I_b and I_c each equal to 50%, and flowing in the opposite direction to I_a, *i.e.*, from the load to the transformer bank. Corresponding instantaneous values must, of course, be in phase with one another, as they are taken at the same instant of time. From the diagram at $0°$ it will be observed that instantaneous values balance out, that is, the same total current flows instantaneously towards the origin as flows away from it. Now the subsequent diagrams of Fig. 9.11, which show the positions of the phasors at intervals of $30°$, are identical with the first one so far as sequence of the phasors is concerned, and the entire system is simply rotated through $30°$ at each step in a counter-clockwise direction, the position of the vertical reference axis, however, remaining unaltered. In every case horizontal projections of the extremities of the phasors representing the maximum values of I_a, I_b and I_c on to the vertical reference axis give the values of the currents in the lines a, b and c, at the instants concerned.

211

The inclusion of the line voltage phasors in the diagrams shows how the instantaneous values of these, too, vary throughout the cycle. Figure 9.12 shows the wave formation of the secondary line currents and voltages to which the phasor diagrams of Fig. 9.11 correspond, and the instantaneous values derived from Fig. 9.11 can easily be checked from Fig. 9.12 by taking the height of the ordinate, above or below the zero axis, measured from the zero axis to the point of intersection of the 'instant' ordinate with the wave concerned.

The instantaneous current distribution in the transformers with respect to both magnitudes and phase angles can now be determined throughout the complete cycle.

Figure 9.13 shows this distribution, at 30° intervals, throughout the cycle these corresponding to the diagrams of Fig. 9.11. The construction of the diagrams is as follows, considering, for example, the diagram at the instant 0°. Lay out the equilateral triangle a, b, c to a convenient scale, so that each side

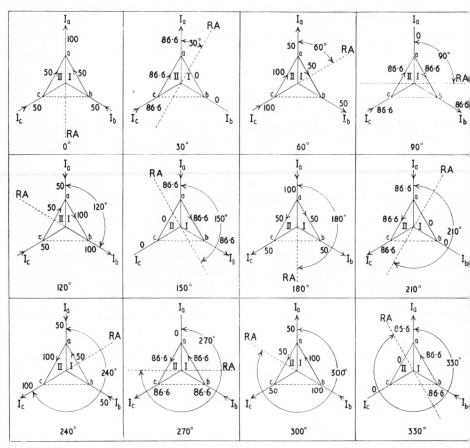

Fig. 9.13   *Diagrams showing percentage instantaneous current magnitudes in transformer secondary windings and in lines*

Note. The conditions at 360° are the same as those at 0°.

212

represents the secondary line voltage. The two sides drawn in full represent the two transformer secondary windings, and the dotted base the open phase. The letters a, b and c designate the lines connected to the points of the triangle similarly marked. From the centroid of the triangle lay out the star phasors I_a, I_b and I_c to represent to scale the full values of the secondary line currents. Draw in the vertical dotted reference axis RA, through the medium of which instantaneous current and voltage values are derived. Now, diagram 0° of Fig. 9.13 shows that the instantaneous current in line a is the full 100% value, flowing from the transformer bank, while in lines b and c the current is 50% of the full value, but flowing towards the transformer bank. The phasor I_a, of diagram 0°, Fig. 9.13 is therefore arrowed to indicate current flowing away from the transformer bank, and is also marked 100. I_b and I_c are arrowed to indicate current flowing into the bank and each marked 50. The diagram is now complete, and it only remains to determine the current in the transformers. At this instant the value of the transformer currents is obvious from an inspection of the diagram. In transformer I the current is the same as in line b, i.e., 50% of the full maximum line current, flowing from b to a. In transformer II the current is the same as in line c, i.e., also 50% of the maximum line current, flowing from c to a. As all instantaneous currents are in phase, the transformer currents combine arithmetically to give the full 100% current flowing out over line a.

Projections from phasor extremities on to and normal to the reference axis give instantaneous current and voltage values as already described.

Now consider the second instant of Fig. 9.13, i.e., at 30°. From the corresponding diagram of Fig. 9.11 it will be seen that the instantaneous line currents are 86·6% of the full maximum value and flowing away from the transformer bank in line a, 86·6% of the full maximum value and flowing into the transformer bank in line c, and zero in line b. The diagram is the same as before, except that the reference axis is rotated 30° in a clockwise direction. In this set of diagrams the reference axis is rotated and the current and voltage phasors held stationary; the reference axis must, of course, be rotated in an opposite direction to that given to the system of phasors in the diagrams of Fig. 9.11. The phasor I_a of diagram 30°, Fig. 9.13, is arrowed to indicate current flowing away from the transformer bank, and is marked 86·6, while I_c is arrowed to indicate current flowing into the bank and is also marked 86·6. The phasor I_b is not arrowed, but is marked 0 as there is no current in line b. Again, from an inspection of the diagram it is clear that there is no current in transformer I as there is none in line b, while transformer II carries all the current from line c, passing it out at the same value to line a. The current in transformer II is therefore 86·6% of the full maximum line current, and flows from c to a.

The third diagram of Fig. 9.13 shows the current distribution 30° later in phase, that is, when the current in line c is the full 100% value flowing into the transformer bank, and 50% of the full maximum value in lines a and b, flowing away from the bank. The diagram is the same as before, except that the reference axis is rotated a further 30°. The phasor I_c is arrowed to indicate current flowing into the transformer bank and marked 100 as given by the corresponding diagram of Fig. 9.11. I_a and I_b are similarly arrowed and marked to indicate 50% current flowing away from the bank. Now, current enters transformer II from line c, and as this line current has no alternative path through which to flow, the full value of the current in line c must flow through transformer II, which, therefore, carries 100% line current from c to a. Upon reaching the junction a between the two transformers, 50% of the full line current must flow through line a to the load, and hence the remaining 50% from line c flows through transformer I from a to b. The 50% current in transformer I flows out through line b, meeting the current requirements of that line.

Finally, consider the fourth diagram of Fig. 9.13 at the instant 90°. The corresponding diagram of Fig. 9.11 shows that the line currents are zero in line a and 86·6% of the full maximum value in lines b and c, and flowing into the transformer bank through line c and away from it through line b. Again, the general diagram is as before, except that the reference axis is rotated a further 30°. The phasor I_c is arrowed to indicate current flowing into the transformer bank, and is marked 86·6. I_b is arrowed to indicate current flowing away from the bank, and is also marked 86·6. I_a is not arrowed, but is marked 0 as line a carries no current at this instant. From a simple inspection of the diagram it is easily seen that, as there is no current outlet at a, the 86·6% current value from line c flows through transformer II from c to a at the same value, and through transformer I from a to b also at the same value, finally passing out from the bank through line b, which meets the current requirements of that line.

The remaining diagrams of Fig. 9.13 are similar to one or another of the first four described above, and the explanatory remarks already made can easily be applied to obtain an understanding of them.

A summary of the line and transformer current distribution at 30° intervals through a complete cycle is given in Table 9.1.

Although, as shown in Fig. 9.13 and Table 9.1, the two transformers are not always equally loaded, the total power transformed by the bank at all instants is the same. Table 9.2 shows that this is so, it being assumed for this illustration that the line current is 1000 A and the line voltage 3000 V. The table shows that whatever deficit there is in the power transformed by one unit as compared with the average of the bank, the other unit transforms a corresponding increase, so maintaining the level of total power transformed.

The current distribution in the transformer secondary windings has now been determined, and it has been shown that currents can flow therein at suitable values, so as to meet the current requirements in the secondary lines supplying a balanced symmetrical three-phase load. It has already been stated that the transformers act individually as single-phase units, and that whatever happens on one side, produces similar results on the other, the relative current and voltage magnitudes depending, of course, upon the ratio of transformation. The voltages have already been dealt with. The transformer primary currents will be exact replicas of those in the secondary windings, that is, at all instants currents will flow in the transformer primary windings having the same power factors (apart from the effect of no-load currents) and the same percentage values—taking the full maximum primary line current as the 100% standard— as the transformer secondary currents. If, as has been shown, the transformer secondary currents satisfy the requirements of the balanced secondary line currents drawn by a balanced three-phase load, it is fairly obvious that replica currents flowing in the transformer primary windings will, in turn, draw similarly balanced three-phase currents through the three primary line wires from the alternator. At any given instant, therefore, primary and secondary line currents will also be replicas of one another, and the complete distribution of current from the alternator to the load has therefore been established, while it has been found that, although the transformer load currents are not evenly distributed,

Table 9.1

SUMMARY OF INSTANTANEOUS CURRENT FLOW IN TRANSFORMERS AND LINES, AS SHOWN IN FIGS. 9.11 AND 9.13

Instant corresponding to Figs. 9-11, 9-13	In line a		In line b		In line c		In transformer I			In transformer II		
	Percentage amplitude	Direction	Percentage amplitude	Direction	Percentage amplitude	Direction	Percentage amplitude	Direction	Flowing from	Percentage amplitude	Direction	Flowing from
0°	100	T-L	50	L-T	50	L-T	50	b-a	Line b	50	c-a	Line c
30°	86·6	T-L	0	—	86·6	L-T	0	—	—	86·6	c-a	Line c
60°	50	T-L	50	T-L	100	L-T	50	a-b	Line c and Tr. II	100	c-a	Line c
90°	0	—	86·6	T-L	86·6	L-T	86·6	a-b	Line c and Tr. II	86·6	c-a	Line c
120°	50	L-T	100	T-L	50	L-T	50 50	a-b a-b	Line a Line c. Tr. II	50	c-a	Line c
150°	86·6	L-T	86·6	T-L	0	—	86·6	a-b	Line a	0	—	—
180°	100	L-T	50	T-L	50	T-L	50	a-b	Line a	50	a-c	Line a
210°	86·6	L-T	0	—	86·6	T-L	0	—	—	86·6	a-c	Line a
240°	50	L-T	50	L-T	100	T-L	50	b-a	Line b	50 50	a-c a-c	Line a Line b and Tr. I
270°	0	—	86·6	L-T	86·6	T-L	86·6	b-a	Line b	86·6	a-c	Line b and Tr. I
300°	50	T-L	100	L-T	50	T-L	100	b-a	Line b	50	a-c	Line b and Tr. I
330°	86·6	T-L	86·6	L-T	0	—	86·6	b-a	Line b	0	—	—
360°	100	T-L	50	L-T	50	L-T	50	b-a	Line b	50	c-a	Line c

T-L means in the direction from the transformers to the load.
L-T means in the direction from the load to the transformers.
Percentage amplitudes throughout are based upon the full line currents equalling 100%.

215

Table 9.2

CALCULATIONS SHOWING POWER TRANSFORMED DURING ONE COMPLETE
CYCLE AT INTERVALS OF 30 ELECTRICAL DEGREES

Instant corresponding to	Transformer I		Transformer II		Total power kW
	$(V_{ab}I_{ab}) \times 10^{-3}*$	kW	$(V_{ca}I_{ca}) \times 10^{-3}*$	kW	
0°	$2600 \times 500 \times 10^{-3}$	1300	$2600 \times 500 \times 10^{-3}$	1300	2600
30°	$1500 \times 0 \times 10^{-3}$	0	$3000 \times 866 \times 10^{-3}$	2600	2600
60°	$0 \times 500 \times 10^{-3}$	0	$2600 \times 1000 \times 10^{-3}$	2600	2600
90°	$1500 \times 866 \times 10^{-3}$	1300	$1500 \times 866 \times 10^{-3}$	1300	2600
120°	$2600 \times 1000 \times 10^{-3}$	2600	$0 \times 500 \times 10^{-3}$	0	2600
150°	$3000 \times 866 \times 10^{-3}$	2600	$1500 \times 0 \times 10^{-3}$	0	2600
180°	$2600 \times 500 \times 10^{-3}$	1300	$2600 \times 500 \times 10^{-3}$	1300	2600
210°	$1500 \times 0 \times 10^{-3}$	0	$3000 \times 866 \times 10^{-3}$	2600	2600
240°	$0 \times 500 \times 10^{-3}$	0	$2600 \times 1000 \times 10^{-3}$	2600	2600
270°	$1500 \times 866 \times 10^{-3}$	1300	$1500 \times 866 \times 10^{-3}$	1300	2600
300°	$2600 \times 1000 \times 10^{-3}$	2600	$0 \times 500 \times 10^{-3}$	0	2600
330°	$3000 \times 866 \times 10^{-3}$	2600	$1500 \times 0 \times 10^{-3}$	0	2600
360°	$2600 \times 500 \times 10^{-3}$	1300	$2600 \times 500 \times 10^{-3}$	1300	2600

$V = 3000$ V. $I = 1000$ A.   u.p.f. load.

*Instantaneous values of V and I are taken so that the power factor does not appear in these expressions.

the line currents on both sides are symmetrical, and are such as would occur with a symmetrical three-phase transformer or bank.

In the preceding remarks no account has been taken of the effect upon the line voltages of the load currents flowing through the unequal impedances of the phases, as it is not essential to this particular treatment.

For loads other than at unity power factor, all the current phasors would be rotated through the same angle, i.e., through an angle ϕ, where the power factor $= \cos \phi$, and the general conclusions arrived at would not be affected. The line voltages would become appreciably unbalanced, however, due to the unequal impedances of the phases.

It is sometimes asked, 'how is the third phase derived?' and this study shows that the currents which would normally flow in the windings of the third transformer are shunted into the windings of the two transformers in open delta, returning, however, to the normal balanced paths in the lines. The line voltages, it has been shown, are transferred by ordinary magnetic induction.

Chapter 10

Multiwinding transformers: including tertiary windings

MULTIWINDINGS

Transformers having more than two windings are now frequently employed in power and distribution systems. They are generally designed and constructed with concentric windings, the relative disposition of the various windings depending upon the operating conditions of the transformer. The arrangement of the windings of a multiwinding transformer can be varied to distribute the leakage reactances in such a way as to obtain the optimum value of voltage regulation. On a multiwinding transformer, the open circuit no-load voltage of a winding will change with variation of loading on another separate winding, or windings, although it may be unloaded itself.

Applications of multiwinding transformers are as follows:

(1) the interconnection of several power systems operating at different supply voltages
(2) the regulation of the system voltage and of reactive power by means of a synchronous capacitor connected to the terminals of one winding
(3) to enable the short circuit power to be controlled by the use of subdivided windings
(4) to provide a delta-connected stabilising winding, which may also be designed to provide an external load.

Multiwinding transformers raise a number of problems which cannot be solved as easily as for two-winding transformers and probably the most important of these is that of voltage regulation and leakage reactance. The three windings of a multiwinding transformer are each interlinked with the magnetic leakage fields of the other windings. Load currents in one circuit affect voltages in another and a lagging current in one winding can, under certain circumstances, cause a voltage rise in one or more of the other windings. It is therefore essential to understand the leakage impedance behaviour of a multiwinding transformer and to be in a position to calculate the voltage regulation of such transformers. Numerous methods have been evolved for the calculation of voltage regulation, each claiming various degrees of accuracy and practicability. The following

217

text* indicates a method which is applicable to two-winding transformers and can be extended to three-winding transformers enabling the voltage regulation to be calculated with an accuracy comparable with that of the test measurements or the data available. It assumes that the currents in the windings remain constant both in magnitude and phase angle even though the output terminal voltages change, due to the voltage regulation, from the no-load values. As has been stated earlier in this chapter, the open circuit no-load voltage of a winding of a multiwinding transformer will change when current flows in the other windings even though the winding itself remains unloaded.

In Chapter 1 it has been shown that for a two-winding transformer the percentage voltage regulation at rated load and at unity power factor is approximately equal to

$$V_R + \frac{V_X^2}{200} \tag{10.1}$$

This value is always positive and indicates a voltage drop.

At any other loading the percentage voltage regulation for a times the rated load at a power factor of $\cos \phi_2$ is equal to:

(a) For transformers having an impedance voltage up to and including 4%.

$$a(V_R \cos \phi_2 + V_X \sin \phi_2) \tag{10.2}$$

(b) For transformers having an impedance voltage up to and including 20%.

$$a(V_R \cos \phi_2 + V_X \sin \phi_2) + \frac{a^2}{200}(V_X \cos \phi_2 - V_R \sin \phi_2)^2 \tag{10.3}$$

where, V_R = percentage resistance voltage at full load.

V_X = percentage leakage reactance voltage at full load.

$\cos \phi_2$ = power factor of the load.

In all cases $\cos \phi_2$ is taken as positive for a lagging power factor and negative for a leading power factor. When the resultant value of voltage regulation obtained is positive, then a voltage drop occurs between no-load and full load; a negative value indicates a voltage rise.

It should be noted that the voltage regulation of a winding on a three-winding transformer is expressed with reference to its no-load open circuit terminal voltage when only one of the other windings is excited and the third winding is on no-load, i.e., the basic voltage for each winding and any combination of loading is the no-load voltage obtained from its turns ratio.

For the common case of two output windings W_2 and W_3, and one input winding W_1, shown diagrammatically in Fig. 10.1, the voltage regulation is usually required for three loading conditions:

W_2 only loaded
W_3 only loaded
W_2 and W_3 both loaded.

For each condition two separate values would be calculated, namely, the

---

*Based on Appendix G of B.S.171: 1959, 'Power transformers' and subsequent issues by permission of the British Standards Institution, Newton House, 101–113 Pentonville Road, London WIX 4AA, from whom copies of the complete standard may be obtained.

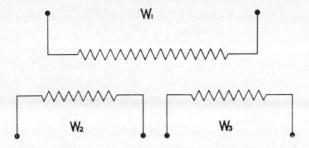

Fig. 10.1   Diagram of a three-winding transformer

regulation of each output winding W_2 and W_3 (whether carrying current or not) for constant voltage supplied to the winding W_1.

The voltage regulation between W_2 and W_3 relative to each other, for this simple and frequent case is implicit in the values (W_1 to W_2) and (W_1 to W_3) and nothing is gained by expressing it separately.

The data required to obtain the voltage regulation are the impedance voltage and load losses derived by testing the three windings in pairs and expressing the results on a basic kVA which can conveniently be the rated kVA of the smallest winding. Such data should if possible be obtained from test results on the transformer.

From these data an equivalent circuit is derived, as shown in Fig. 10.2. It should be noted that this circuit is a mathematical conception and is not an indication of the winding arrangements or connections. It should, if possible, be determined from the transformer as built.

The equivalent circuit is derived as follows:

Let, a_{12} and b_{12} be respectively the percentage resistance and reactance voltage referred to the basic kVA and obtained from test, short circuiting either winding W_1 or W_2 and supplying the other, with the third winding W_3 on open circuit,

a_{23} and b_{23} similarly apply to a test on the windings W_2 and W_3 with W_1 on open circuit,

a_{31} and b_{31} similarly apply to a test on the windings W_3 and W_1 with W_2 on open circuit,

d = the sum $(a_{12}+a_{23}+a_{31})$, and
g = the sum $(b_{12}+b_{23}+b_{31})$

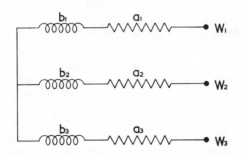

Fig. 10.2   Equivalent circuit of a three-winding transformer

219

Then the mathematical values to be inserted in the equivalent circuit are:

Arm W_1: $\qquad a_1 = d/2 - a_{23}$ $\qquad\qquad b_1 = g/2 - b_{23}$

Arm W_2: $\qquad a_2 = d/2 - a_{31}$ $\qquad\qquad b_2 = g/2 - b_{31}$

Arm W_3: $\qquad a_3 = d/2 - a_{12}$ $\qquad\qquad b_3 = g/2 - b_{12}$.

It should be noted that some of these mathematical quantities will be negative or may even be zero, depending on the actual physical relative arrangement of the windings on the core.

For the desired loading conditions the kVA operative in each arm of the network is determined and the regulation of each arm is calculated separately. The regulation with respect to the terminals of any pair of windings is the algebraic sum of the regulations of the corresponding two arms of the equivalent circuit.

The detailed procedure to be followed subsequently for the case of two output windings and one supply winding is as follows:

(1) Determine the load kVA in each winding corresponding to the loading being considered.
(2) For the output windings, W_2 and W_3, this is the specified loading under consideration; evaluate n_2 and n_3 for windings W_2 and W_3, being the ratio of the actual loading to the basic kVA used in the equivalent circuit.
(3) The loading of the input winding W_1 in kVA should be taken as the phasor sum of the outputs from the W_2 and W_3 windings, and the corresponding power factor $\cos \phi$ and quadrature factor $\sin \phi$ deduced from the in-phase and quadrature components.

Where greater accuracy is required, an addition should be made to the phasor sum of the outputs and they should be added to the quadrature component to obtain the effective input kVA to winding W_1,

$$(\text{the output kVA from winding } W_2)\frac{b_2 n_2}{100} +$$

$$(\text{the output kVA from winding } W_2)\frac{b_3 n_3}{100},$$

n for each arm being the ratio of the magnitude of the actual kVA loading of the winding to the basic kVA employed in determining the equivalent circuit.

A more rigorous solution is obtained by adding the corresponding quantities (a, n, output kVA) to the in-phase component of the phasor sums of the outputs, but this has rarely an appreciable effect on the voltage regulation.

Equations 10.2 and 10.3 are now applied separately to each arm of the equivalent circuit, taking separate values of n for each arm as defined earlier.

To obtain the voltage regulation between the supply winding and either of the loaded windings, add algebraically the separate voltage regulations determined for the corresponding two arms, noting that one of these may be negative. A positive value for the sum determined indicates a voltage drop from no-load to the loading considered while a negative value for the sum indicates a voltage rise.

Repeat the calculation described in the preceding paragraph for the other loaded winding. This procedure is applicable to auto transformers if the equivalent circuit is based on the effective impedances measured at the terminals of the auto transformers.

220

In the case of a supply to two windings and output from one winding, this method can be applied if the division of loading between the two supplies is known.

An example of the calculation of voltage regulation of a three-winding transformer is given in the following.

Assume that:

W_1 is a 66 000 V primary winding

W_2 is a 33 000 V output winding loaded at 2000 kVA and having a power factor $\cos \phi_2 = 0.8$ lagging.

W_3 is an 11 000 V output winding loaded at 1000 kVA and having a power factor $\cos \phi_3 = 0.6$ lagging.

The following information is available, having been calculated from test data, and is all related to a basic loading of 1000 kVA.

$$a_{12} = 0.26 \qquad b_{12} = 3.12$$
$$a_{23} = 0.33 \qquad b_{23} = 1.59$$
$$a_{31} = 0.32 \qquad b_{31} = 5.08$$

whence, $\qquad d = 0.91 \qquad$ and $\qquad g = 9.79$

Then for $\qquad W_1, a_1 = 0.125$ and $b_1 = $ plus 3.305

$\qquad\qquad W_2, a_2 = 0.135$ and $b_2 = $ minus 0.185

$\qquad\qquad W_3, a_3 = 0.195$ and $b_3 = $ plus 1.775.

The effective full-load kVA input to winding W_1 is,

(i) With only the output winding W_2 loaded $= 2000$ kVA at a power factor of 0.8 lagging

(ii) With only the output winding W_3 loaded $= 1000$ kVA at a power factor of 0.6 lagging

(iii) With both the output windings W_2 and W_3 loaded $= 2980$ kVA at a power factor of 0.74 lagging.

Applying expressions 10.2 or 10.3 separately to each arm of the equivalent circuit, the individual regulations are, in

W_1 under condition (i) where $n_1 = 2.0$, the value of 4.23%
W_1 under condition (ii) where $n_1 = 1.0$, the value of 2.72%
W_1 under condition (iii) where $n_1 = 2.98$, the value of 7.15%
W_2 $\qquad\qquad$ where $n_2 = 2.0$, the value of -0.02%
W_3 $\qquad\qquad$ where $n_3 = 1.0$, the value of 1.53%.

Summarising these calculations therefore the total transformer voltage regulation is:

(i) With output winding W_2 fully loaded and W_3 unloaded,
 at the terminals of winding W_2, the value of $4.23 - 0.02 = 4.21\%$
 at the terminals of winding W_3, the value of $4.23 + 0 \quad = 4.23\%$

(ii) With output winding W_2 unloaded and W_3 fully loaded,
 at the terminals of winding W_2, the value of $2.27 + 0 \quad = 2.72\%$
 at the terminals of winding W_3, the value of $2.72 + 1.53 = 4.25\%$

(iii) With both output windings W_2 and W_3 fully loaded,
 at the terminals of winding W_2, the value of $7.15 - 0.02 = 7.13\%$
 at the terminals of winding W_3, the value of $7.15 + 1.53 = 8.68\%$

221

TERTIARY WINDINGS

Tertiary windings, in the form of a delta connection, have been used on star/star connected three-phase transformers and groups for many years, but at the same time it cannot be said that in Great Britain their use has become common. The probable reason for this is that the delta/star connection for both step-up and step-down supplies has for many years so satisfactorily filled all the requirements (particularly when bearing in mind the considerable interconnection of networks and the general restrictions regarding multiple earthing) that star/star-connected units or groups would not, as a rule, now fit very well into the general scheme of things. For supplies given by individual radial feeders their use may be suitable, but it is not always known at the time of installation whether what is initially a single radial feeder will remain so or eventually form part of a general inter-connected network. For these probable reasons the star/star connection has not been adopted widely for main transmission and distribution in Great Britain, although this connection is now being installed in greater numbers by the supply authorities.

As its name implies, a tertiary winding is simply a third winding of a trans-former unit or group and the general, though not essential, form is the closed delta for three-phase working. The star/star connection by itself has been regarded hitherto with some disfavour on account of its third-harmonic phenomena and its behaviour when transforming very seriously unbalanced loads, so that the desire for the elimination of these two defects is the principal reason for incorporating tertiary delta windings. The other advantages are of the kind that become virtues by necessity, as removal of the two objections cited above secures, without any further effort, the remaining advantages.

The uses of the tertiary delta windings for star/star connected transformers and groups are thus as follows:

(1) to reduce third-harmonic voltage components
(2) to permit the transformation of unbalanced three-phase loads
(3) to supply an auxiliary load in addition to the main load.

(1) It is well known that star/star connected three-phase transformers and groups operating on three-wire circuits on both sides have induced in their phase windings third-harmonic voltage components. The third-harmonic voltage component operates to increase the peak value of the resultant com-posite phase voltage wave. The chief disadvantage, however, of the third-harmonic voltages is that the neutral point oscillates above and below ground at a voltage equal to the magnitude of the third-harmonic component of each phase. So long as the neutral point is not earthed this oscillation is generally of no great importance apart from the extra stress thrown on to the transformer insulation. If, however, the neutral point is earthed, the third-harmonic voltage components are transferred to each line terminal, and electrostatic induction may be set up in nearby communication circuits should these be present. In addition to this, triple frequency currents flow through the ground, completing their circuits through the capacitances formed between each line wire and ground. In general, the heating effect in the transformer windings from these currents is of little importance,* but electromagnetic interference with telephone circuits

---

*Unless amplified by line capacitance; see Chapter 26.

employing earth returns may take place if these are adjacent to the power line. It is generally known that if one of the main windings is delta connected, third-harmonic voltages are substantially eliminated by the circulation of third-harmonic currents in the closed delta, and as such currents do not circulate in the lines it is obvious that any closed delta winding, whether a part of one of the main windings or separate therefrom, will provide the means for counteracting third-harmonic voltages. To meet these requirements the tertiary delta winding in conjunction with the star/star connections came into use, particularly for those cases where it was thought desirable to have star points on both sides for earthing purposes without the disadvantages attendant upon the use of straight star/star connections. In such cases the tertiary windings are isolated from all outside sources, and they simply provide the necessary magnetising ampere-turns to eliminate the third-harmonic voltages which would otherwise be inherent in the transformer design.

(2) The second chief disadvantage of the straight star/star connection employing a three-wire primary is that unbalanced loads on the secondary four-wire side may produce voltage unbalancing, particularly with three-phase shell-type transformers and three-phase groups of single-phase transformers. This is due to the fact that on the primary side two of the phases carry return load currents from the third loaded phase, and as they have no corresponding secondary load ampere-turns they act as choke coils and produce severe distortion of the phase voltages. It is well known that a delta/star connected transformer or group eliminates this phenomenon and permits the supply of greatly unbalanced loads without producing abnormal voltage drops. If, however, the star/star with tertiary delta connection be used, the load currents in the primary phases corresponding to the unloaded secondaries are balanced by the flow of load currents in the tertiary delta, as shown typically in Fig. 10.3. In this case the tertiary delta must be designed to be sufficiently large to provide the necessary triple-frequency magnetising ampere-turns, and at the same time to handle the maximum load current that may flow in it due to the most onerous conditions of load unbalancing.

The tertiary winding is normally placed between the core and the l.v. winding, and if a one-to-one turns ratio of all windings be assumed, the relative current distribution is as shown in Fig. 10.3. The current value $3I_s$ is, of course, the value of the total current taken by the load.

Allied with the question of the connection to handle satisfactorily unbalanced loads is that of the flow of short-circuit fault currents. This, of course, has special

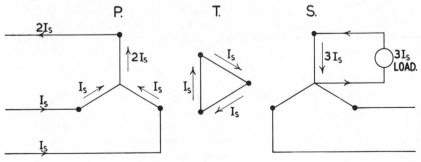

Fig. 10.3   Single-phase load to neutral

reference to the short circuits caused by line earth faults with the neutral points earthed. In such cases an earth fault on one line is equivalent to a heavy single-phase load on the phase concerned, and the general current distribution for different conditions is shown in Fig. 10.4. For case (a) of Fig. 10.4 the magnitude of the fault current is,

$$I_S = \frac{100I}{IZ_{PT}} \qquad (10.4)$$

for case (b) the fault current is

$$I_S = \frac{100\,I}{2\,IZ_{PS} + IZ_{TS}} \qquad (10.5)$$

for case (c) the fault current is

$$I_S = \frac{100\,I}{IZ_{PT}} \qquad (10.6)$$

and for case (d) the fault current is

$$I_S = \frac{100\,I}{2\,IZ_{PS} + IZ_{TS}} \qquad (10.7)$$

where I_S = the fault current shown in Fig. 10.4(a), (b) and (c)

       I_{SP} = the fault current due to the primary supply in Fig. 10.4(d)

       I_{SS} = the fault current due to the secondary supply in Fig. 10.4(d)

       I = normal full-load current of the transformer

       IZ_{PS} = the percentage normal full-load impedance per phase between primary and secondary windings

       IZ_{PT} = the percentage normal load impedance per phase between primary and tertiary windings

       IZ_{TS} = the percentage normal full-load impedance per phase between tertiary and secondary windings.

Expressions 10.4 to 10.7 apply strictly to a one-to-one turns ratio of all windings, and the true currents in each can easily be found by taking due account of the respective turns ratios.

It will be appreciated that from the point of view of continuous and short time loads the impedances between the tertiary windings and the two main windings are of considerable importance. The tertiary winding must be designed to be strong enough mechanically, to have the requisite thermal capacity, and to have sufficient impedance with respect to the two main windings to be able to withstand the effects of short circuits across the phases of the main windings and so as not to produce abnormal voltage drops when supplying unbalanced loads continuously.

It is incumbent upon the purchaser to provide sufficient information to enable the transformer designer to determine the worst possible external fault currents that may flow in service. This information (which should include the system characteristics and details of the earthing arrangements) together with a knowledge of the impedance values between the various windings, will permit an accurate assessment to be made of the fault currents and of the magnitude of the currents that will flow in the tertiary winding. It is not sufficient for a purchaser to state that a tertiary winding should have a rating of, say, 25 or 33·3% of that of the main windings. A satisfactory value of the rating of the tertiary winding will be dependent upon the impedances between windings of the transformers and upon other factors as stated earlier, and in this regard the particular impedance values to be employed should be left to the designer who will consider all relevant

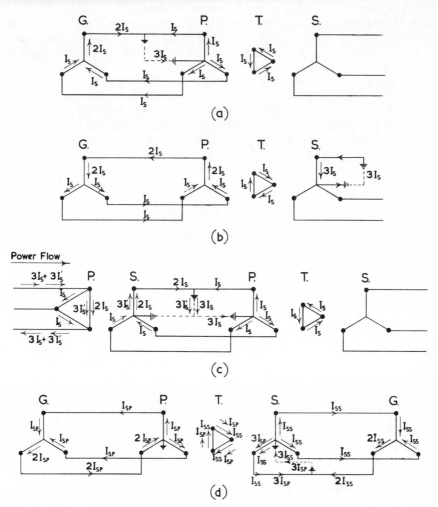

Fig. 10.4   Fault currents due to short circuits to neutral

facts and produce the most economical and reliable design. Reference to Fig. 10.3 will immediately show, from the 3-to-1 ratio between the currents in the secondary and tertiary windings, how the practice of specifying a 33·3% rating of the tertiary winding has arisen. Considering the currents to be earth fault currents and not load currents (as is qualitatively quite permissible) it will be seen that the tertiary winding current is one third of that of the faulted phase.

With a straight star/star connected transformer or group operating on a three-wire primary circuit, and with an earthed neutral on the secondary side, a secondary line earth fault may not cause sufficient fault current to flow to operate the automatic tripping gear, due to the choking effect of the two sound phases. The tertiary delta remedies the defect by permitting currents to flow therein to balance the return currents in the two sound phases so that the transformer would be tripped out of service irrespective of whether the main secondary windings were connected to a load or not at the time of the failure.

(3) Having available a tertiary winding, the main requirements of which are to provide certain magnetising ampere-turns, it needed little imagination to suggest its utility for supplying auxiliary loads in addition to the loads handled by the main windings. In this case, therefore, the tertiary windings have to carry the additional current of the auxiliary load, while an extra provision is necessary for the main primary windings. The tertiary windings can, of course, easily be designed for whatever voltages may be required for the auxiliary loads. In main power station transformers particularly, the tertiary delta is useful for supplying the station auxiliaries, while for transmission it has found considerable use abroad for supplying synchronous condensers for power-factor improvement on long transmission lines.

Chapter 11

Transformer tappings

One of the simplest and most inexpensive methods of providing for adjustments in supply voltages is to arrange tappings upon transformer windings. Although, in the majority of cases, there is no objection to this procedure, there are certain cases in which the provision of tappings introduces difficulties in design or manufacture, or unsatisfactory operating conditions. In order to appreciate such difficulties it is advisable first to review the various arrangements of tappings and their several uses.

Tappings are usually provided for one of the following purposes:

(1) For maintaining the secondary voltage constant with a varying primary voltage.

(2) For varying the secondary voltage.

(3) For providing an auxiliary secondary voltage for a special purpose, such as lighting.

(4) For providing a reduced voltage for starting rotating machinery.

(5) For providing a neutral point either for earthing, or for dealing with out of balance current in single-phase, three-wire circuits, in three-phase, four-wire circuits, etc.

There are three common types of tapping, and the choice usually accords with the frequency with which changes are likely to be required:

Tappings to oil level

This is the cheapest type of tapping arrangement, and is usually adopted when tapping changes are likely to be infrequent, as, for instance, when tappings are provided to allow for voltage differences at various parts of a distribution system, although the voltage at any one point does not vary appreciably. To change this type of tapping the tank cover must be removed, and when an oil conservator is provided it is also necessary to draw off part of the oil.

Tappings outside the tank

This type of tapping is very convenient although it is liable to be expensive if the winding to which the tappings are connected is for a very high voltage. The expense may be reduced to a certain extent in cases where there are several tappings, by bringing out leads, which are at approximately the same potential, through a single terminal bushing. Apart from the general convenience of tappings brought to the outside of the tank, there are occasions when this type of tapping is essential, as when two or more voltages are to be used simultaneously.

Tappings operated by selectors

This arrangment probably represents the most convenient method of dealing with tapping changes. There are three main types:

(1) Tappings outside the tank and operated by an external selector. This type has the advantage that it is possible to operate the selector on load, provided a suitable circuit is incorporated in the switching arrangements to avoid opening the external load circuit. Such an on-load selector is regarded as an integral part of the transformer.

(2) Tappings inside the tank and operated by an internal handle. This type has little advantage over the ordinary method of bringing tappings to oil level and can be ignored.

(3) Tappings inside the tank and connected to an internal selector which is operated by an external handle. This is the commonest type of tapping selector and is usually designed for off-circuit operation.

The various limitations to the usefulness of transformer tappings, and the difficulties which occasionally arise, may best be considered by subdividing them according to certain aspects of design.

VOLTS PER TURN

When the voltage of the winding from which tappings are to be taken is high compared with the output of the transformer, it is possible to arrange the tappings, within very fine limits, to give any desired voltage. If, however, adjusting tappings are required on a comparatively low voltage winding, there are usually very severe limitations to the voltages obtainable by means of tappings. Figure 11.1 shows average values of volts per turn against output in kVA for three-phase, 50 Hz transformers designed for voltages up to, say, 33 kV. It is the number of turns in the winding under consideration which limits the fineness of adjustment obtainable by tappings, and it is evident that this number of turns is dependent upon the volts per turn and the voltage of the winding. Suppose, for example, it is desired to fit adjusting tappings to a 433 V star connected winding on a 600 kVA three-phase transformer. From Fig. 11.1 it is seen that the volts per turn will be approximately 10, and therefore the number of turns per phase is

$$N = \frac{433}{10\sqrt{3}} = 25$$

Since tappings can only be provided at integral number of turns, the adjustment obtainable is evidently in steps of 4%. Thus, if adjusting tappings are required to give voltages of 5 and 10% below normal, the nearest that can be obtained are 4 and either 8 or 12%. It is for this reason that tappings should

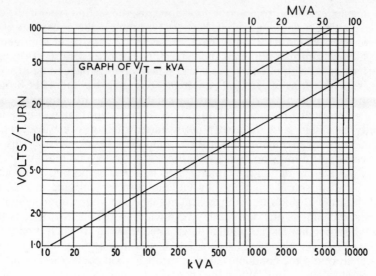

Fig. 11.1   Relation between kVA and volts per turn for three-phase, 50 Hz transformers

generally be fitted on the higher-voltage winding. When two or more l.v. voltages are required simultaneously the tappings must be provided on the l.v. winding, and consequently the exact voltages desired may not be attainable. In some cases the difficulty may be overcome by means of a special design, utilising either a non-standard core diameter or a value for the induction density other than that normally employed. Such methods are, however, usually expensive and are not therefore to be recommended.

CURRENT

The magnitude of the current in windings to which tappings are to be fitted has a pronounced bearing upon the question as to whether the presence of such tappings will have an adverse effect. This is largely due to the fact that the magnitude of the current usually determines the type of coil to be used, and the type of coil employed is an important factor in determining whether or not tappings are desirable.

For small currents, up to about 20 A, the 'crossover' type of coil is employed. It consists of a number of turns of round wire wound in layers. The bringing out of tappings from such coils is a process demanding great care, especially if they have to be brought out from inner layers of the winding. In order to avoid risk of damage to adjacent turns, the lead from the tapping point, consisting generally of thin copper strip, must be well insulated, and the insulation employed should be chosen principally for its mechanical properties, owing to the fact that damage due to tappings is usually caused by the extra mechanical pressure applied to the adjacent turns. When the current, and consequently the wire, are very small, the difficulties and attendant risks are greatly accentuated, and for this reason it is advisable to avoid the use of tappings on windings for small outputs at relatively high voltages.

229

For heavier currents, from about 20 A upwards, the section type or the continuous disc type of winding is used. In either type the conductor consists of one or more strips wound flat, and in parallel. Each coil consists of a number of discs insulated from one another by shaped sectors. Each disc consists of a number of turns arranged with one turn per layer. When tappings have to be provided on windings consisting of such coils, it is highly desirable they should be brought out from the outside of the discs where the conductor runs from one disc to the other. This is usually possible, although in certain cases it may require a little ingenuity in design, and frequently necessitates winding a few discs with a different shape of conductor. When the tappings can be arranged in this way their presence introduces no difficulty or extra expense beyond that involved in bringing out and insulating the tapping leads and their respective terminals.

The so-called 'spiral' winding may be employed for currents in excess of 100 A. These coils, which are really helical in shape, are wound upon cylindrical formers and comprise one or more layers of winding, the conductor consisting of a number of square or rectangular wires taped together. When it is necessary to fit tappings to such coils, the process consists in making a stirrup of thin copper strip round the conductor, and ensuring a good contact by means of a well-poured soldered joint. When the current is large the stirrup may consist of several copper strips in parallel, the individual thicknesses of the strips being minimised for the sake of flexibility. Such tappings occupy an appreciable amount of space, both axially and radially, and this must be allowed for when designing. At the same time the extra radial depth at the point where the tapping is brought out must be allowed for when determining the necessary width of oil duct, if the coil is an inner one, as is frequently the case with this type of coil. In cases where a number of tappings are brought out from an inner coil, the problem of bringing out the tapping leads at the ends of the winding may introduce considerable difficulty, and if the current is at all heavy, extra end clearance may have to be allowed. Another difficulty, introduced by the provision of tappings on spiral coils, is due to the necessity for interchanging the relative positions of the individual wires of the conductor, in such a manner that each wire, in its passage from terminal to terminal, occupies each different position relative to the path of the leakage flux for an equal distance. The points where the interchange of wires is effected are known as 'crossovers', and it is evident that it is futile to introduce them if their effect is to be nullified by making electrical contact between the individual wires at a number of tapping points. In cases where the tappings are intended to carry only a small proportion of the normal full-load current, or to carry current only for short periods, the difficulty may be overcome by taking the tapping from the top layer of wires only. This method may be adopted generally with absolute safety when the tappings are for one of the following purposes:

> to supply a small auxiliary load such as for lighting
> to provide a return path for out-of-balance current
> for starting rotating machinery.

When the current required from a tapping is too heavy to justify the bringing out of the tapping from the top layer of wires only, each section of winding between adjacent tapping points must be regarded as a separate coil for the purpose of arranging the crossovers if the risk of excessive eddy-current loss is to be avoided. If, for instance, a certain spiral coil has four wires in parallel radially, and the effect of crossing over the wires is completely neutralised owing to the provision of a number of tappings, then the eddy-current loss in the coil will be 16 times as great as it would have been in the absence of such tappings. If, therefore, the eddy-current loss under normal conditions is appreciable, as may be the case in transformers of large currents or high reactance, the provision of tappings without suitable precautions will cause the eddy-current loss to be so high as to render the total copper losses excessive.

VOLTAGE

The effect of the voltage, for which a winding is designed, upon the problem of providing tappings is dependent upon two factors, firstly the current which is, of course, inversely proportional to the voltage for a given output; and secondly, the amount of insulation it is necessary to provide on the tapping leads to guard against dielectric failure. The first of these factors has already been dealt with under the heading 'current', and it is only necessary to add that when the voltage is high, the difficulties introduced in fitting tappings to small-current windings are greatly accentuated. The adequate insulation of tappings connected to high-voltage windings is a matter of expense rather than difficulty in design or manufacture. The provision of suitable insulation at points where the tapping leads pass through supports is a problem which entails careful consideration, and while there is no real difficulty, the cost of such insulation is considerable, and, in the case of small transformers for very high voltages, may represent an appreciable proportion of the total cost.

Table 11.1

MULTIPLYING FACTORS APPLICABLE WHEN PRIMARY OR SECONDARY TURNS ARE REDUCED IN THE RATIO 1:r BY MEANS OF A TAPPING

Tappings on primary or secondary	Variations of primary or secondary	Constant output or current	Multiplying factor for							
			Current		Turns		Amp. turns	Volts per turn	React-ance	Out-put
			Prim.	Sec.	Prim.	Sec.				
primary	primary	output	$\frac{1}{r}$	1	r	1	1	1	1	1
primary	primary	current	1	r	r	1	r	1	r	r
primary	secondary	output	1	r	r	1	r	$\frac{1}{r}$	r^2	1
primary	secondary	current	1	r	r	1	r	$\frac{1}{r}$	r^2	1
secondary	secondary	output	1	$\frac{1}{r}$	1	r	1	1	1	1
secondary	secondary	current	r	1	1	r	r	1	r	r
secondary	primary	output	r	1	1	r	r	$\frac{1}{r}$	r^2	1
secondary	primary	current	r	1	1	r	r	$\frac{1}{r}$	r^2	1

REACTANCE

The factors in the formula for reactance which may be varied by the presence of tappings are current, turns, volts per turn, and axial length of winding. The percentage reactance varies directly as the first two of these terms, and inversely as the remaining two. In addition the cutting out of appreciable portions of the winding by means of tappings tends to distort the leakage field, thereby affecting the reactance. Any alteration in the length of a winding due to tappings usually affects one winding only, and this also causes a distortion of the leakage field. It is therefore convenient to consider this term in the reactance formula in conjunction with the consideration of the effect of distortion. The variations in reactance due to changes in current, number of turns, and volts per turn, caused by the presence of tappings, are illustrated in Table 11.1. For the purpose of this table it has been assumed that the normal condition is when all turns are in circuit, and that under this condition the two windings are equal in length. The table shows the factors by which the various items are multiplied when the turns in one winding are reduced to *r* times their original value by means of a tapping. The first two columns indicate the winding from which the turns are cut out, and the object to be achieved. The third column is introduced to provide for the fact that sometimes full output is required under all conditions of operation, while occasionally it is permissible to design the windings to give full output only under the most favourable condition, and to reduce the output in proportion to any reduction in voltage which may occur when tappings are in use. This latter case has been described as 'constant current'; it will, however, be appreciated that the current is not necessarily constant in both windings, but that no increase of current is permitted in either winding when the tappings are employed. As a matter of interest, the output obtainable subject to this limitation is shown in the table.

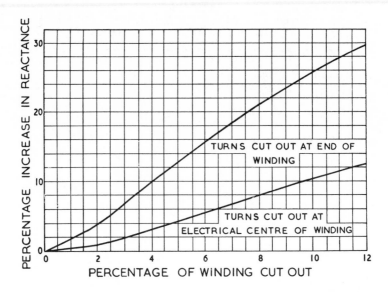

Fig. 11.2   Increase in reactance due to leakage field distortion caused by cutting out a portion of one winding by means of a tapping

The effect upon reactance, of leakage field distortion due to the cutting out of portions of the winding by means of tappings, is illustrated in Fig. 11.2. Two curves are given, one showing the effect of tappings at the end of a winding, and the other the effect of tappings at the electrical centre of a winding. These curves are simply representative, as the percentage variation in reactance, due to a given percentage reduction in the number of turns in circuit, varies with different designs. The curves may, however, be taken as a fair approximation for all transformers of normal output and voltage. It will be noted that the variation in reactance due to this cause is far greater when the tappings are at the ends of windings than when they are approximately at the electrical centre. This is a further reason for arranging the tappings in this latter manner. The changes in reactance indicated by Table 11.1 and Fig. 11.2 are cumulative.

If, for example, the use of a certain tapping involves a decrease of 10% in the reactance according to Table 11.1 and an increase of 5% according to Fig. 11.2, the net decrease will be,

$$100 - 100(0 \cdot 9 \times 1 \cdot 05) = 5 \cdot 5\%$$

LOSSES

Variations in copper loss due to the use of tappings are mainly due to changes in current and in the total lengths of conductors in circuit. Table 11.2, which is prepared on similar lines to Table 11.1, indicates the extent of the variations for different conditions. In compiling this table only variations in current and

Table 11.2

MULTIPLYING FACTORS FOR LOSSES WHEN PRIMARY OR SECONDARY TURNS
ARE REDUCED IN THE RATIO $1:r$ BY MEANS OF A TAPPING

Tappings on primary or secondary	Variations of primary or secondary	Constant output or current	Multiplying factor for			
			Copper loss		Iron loss	
			Watts	Per cent	Watts	Per cent
primary	primary	output	$\dfrac{1+r}{2r}$	$\dfrac{1+r}{2r}$	1	1
primary	primary	current	$\dfrac{r(1+r)}{2}$	$\dfrac{1+r}{2}$	1	$\dfrac{1}{r}$
primary	secondary	output	$\dfrac{r(1+r)}{2}$	$\dfrac{r(1+r)}{2}$	$\dfrac{1}{r^2}$	$\dfrac{1}{r^2}$
primary	secondary	current	$\dfrac{2(1+r)}{2}$	$\dfrac{r(1+r)}{2}$	$\dfrac{1}{r^2}$	$\dfrac{1}{r^2}$
secondary	secondary	output	$\dfrac{1+r}{2r}$	$\dfrac{1+r}{2r}$	1	1
secondary	secondary	current	$\dfrac{r(1+r)}{2}$	$\dfrac{1+r}{2}$	1	$\dfrac{1}{r}$
secondary	primary	output	$\dfrac{r(1+r)}{2}$	$\dfrac{r(1+r)}{2}$	$\dfrac{1}{r^2}$	$\dfrac{1}{r^2}$
secondary	primary	current	$\dfrac{r(1+r)}{2}$	$\dfrac{r(1+r)}{2}$	$\dfrac{1}{r^2}$	$\dfrac{1}{r^2}$

resistance have been taken into account, the losses due to eddy currents having been assumed to be a constant percentage of the total loss. For the sake of simplicity it has also been assumed that the initial losses are equally divided between the high-voltage and low-voltage windings. When the use of tappings involves an appreciable distortion of the leakage field, the former assumption is not strictly true, but the error introduced would not be appreciable in the majority of cases. It may be advisable to make some allowances for increased eddy-current losses, when appreciable leakage field distortion takes place, in transformers having heavy current windings and high reactance values.

Variations in iron loss are due to changes in magnetic flux density, which are proportional to the changes in volts per turn indicated in Table 11.1. The extent of these variations is shown in Table 11.2, the calculations being based on the assumption that the loss varies as the square of the flux density. This assumption is fairly accurate for small changes in the flux density.

PARALLEL OPERATION

Most of the difficulties introduced into the design of transformers for parallel operation with others of different output have already been discussed in part, under the headings 'Volts per turn' and 'Reactance'. When two or more transformers have to operate in parallel it is essential that they should have identical values of turns ratio under all conditions of operation, and that their respective impedance values should be approximately equal. It has already been pointed out that it is often a matter of great difficulty to obtain specified ratios with any degree of accuracy if tappings are to be provided on a low-voltage winding. The difficulty is accentuated if both primary and secondary windings are for comparatively low voltages, as in this case it may not be possible to adjust the turns on the winding having no tappings, in such a way as to reduce the mean error in the various tappings ratios to a minimum. The ratio error in transformers for parallel operation should not exceed $\frac{1}{2}$%, and even this error may in some cases be excessive. From this it follows that difficulties will probably arise in fitting tappings to windings having less than one hundred turns, and by utilising the curve given in Fig. 11.1 it is possible to determine the lowest-voltage winding to which it is convenient to fit tappings for any size of transformer. There may be, and frequently are, cases in which it is possible to obtain the desired ratios sufficiently accurately, even though there are far fewer than one hundred turns in the winding to be tapped, but in all such cases it is necessary to consult the designer. In this connection there is one very important aspect to remember; even though no difficulty may be experienced in obtaining the desired ratios by means of tappings on the lower-voltage winding, it is undesirable to do so, because in the event of further transformers of different output being required to operate in parallel at a later date, the same ratios may be unobtainable, especially if the new output is appreciably greater than that initially employed.

Reactance, the remaining factor affecting parallel operation, has already been discussed in certain of its aspects under the heading 'Reactance'. Transformers which are required to operate in parallel are required to do so at all available ratios, and from a consideration of the remarks already made concerning the effects of tappings upon reactance, it will be appreciated that trouble may arise due to the existence of tappings, if care is not taken to ensure that

Table 11.3

DISPOSITION OF VOLTAGE-ADJUSTING TAPPINGS

Tappings used for obtaining voltage variations		Variations of voltages from the normal service voltages		Outputs and currents capable of being dealt with when using primary tappings or secondary tappings*	Percentage of rated full-load current in windings to give full-rated output at plus and minus 5% voltage variations from normal service voltages	
On primary windings	On secondary windings	Primary voltage	Secondary voltage		Primary windings	Secondary windings
minus	plus	constant service voltage	higher than service voltage	full-rated output / reduced-rated secondary current† / full-rated primary current	100	95
plus	minus	constant service voltage	lower than service voltage	reduced-rated output† / full-rated secondary current / reduced-rated primary current†	100	105
plus	minus	higher than service voltage	constant service voltage	full-rated output / full-rated secondary current / reduced-rated primary current†	95	100
minus	plus	lower than service voltage	constant service voltage	reduced-rated output† / reduced-rated secondary current† / full-load primary current	105	100

*This column applies only to transformers, the windings of which are fitted with tappings to give a total range of voltage variation exceeding 10% of the service or mean voltages.
†All output and current reductions are proportional to the percentage increases or decreases represented by the tappings in use.

transformers for parallel operation have approximately equal reactance values at all ratios. If, for instance, an existing transformer is provided with tappings at the end of a winding and a new transformer is designed to operate in parallel with it, and is provided with tappings at the electrical centre of the corresponding winding, it does not follow that because the two reactance values are equal for a given ratio that they will necessarily be equal at all the ratios obtainable.

Another difficulty which occasionally arises is due to an existing transformer having tappings on a low-voltage winding. In order to obtain the correct ratios on a new transformer, the volts per turn must be equal to, or an exact sub-multiple of, the voltage between adjacent tappings on the existing transformer. When the voltage of the winding under consideration is very low, this involves a close limitation upon the value which may be selected for the volts per turn, and as the reactance of a transformer, other factors being constant, varies inversely as the square of the volts per turn, it is evident that there may be considerable difficulty in obtaining the desired reactance value. The difficulty is greatly accentuated if the output be large, or if the adjusting tappings in question are arranged for small variations of voltage.

These various difficulties generally can be overcome by ingenuity in design, but it is highly desirable that they should be avoided, whenever possible, by a careful consideration of the question of tappings when the first transformers are to be installed.

SPECIAL CASES

There are certain special cases, either in the type of transformer, or in the duty to be performed, in which the provision of tappings necessitates special consideration. A few examples of these are as follows:

Single-phase, core-type transformers

All transformers, to which this description strictly applies, have two limbs, the windings being equally divided between the two and being generally connected in series. When adjusting tappings are provided they should be arranged in such a manner that the number of turns of each winding, in circuit at any ratio, is always equally divided between the two limbs. Non-compliance with this rule may give rise to excessive values of reactance and regulation, under those conditions which involve an unbalanced arrangement of primary and secondary windings. In small transformers a certain degree of unbalancing is permissible, although it should be reduced to a minimum; in large transformers, however, it is advisable to keep the windings completely balanced with all possible tapping arrangements.

Interconnected star winding

When tappings have to be provided on windings which are connected in this manner, it is necessary to duplicate the tappings on the two portions comprising any given phase, as otherwise there will be a phase displacement, the extent of which will depend upon the extent to which the two portions are unbalanced. There are, of course, certain cases in which a slight phase displacement may be permissible. Generally, however, it should be avoided, as difficulties may subsequently arise when extra transformers are introduced. This would certainly be the case if transformers having different connections were required to operate in parallel at a later date, as, for instance, new delta/star transformers with existing star/interconnected star transformers. Both methods of arranging tappings on interconnected star windings are illustrated in Fig. 11.3.

Scott-connected groups

The Scott connection is commonly used for converting from three phase to two phase and vice versa. In addition to such adjusting tappings as may be required, the main transformer must be provided with a midpoint tapping on the three-phase side, and in the case of interchangeable groups, each transformer, in addition to the midpoint tapping, must have a tapping or tappings to give $\sqrt{3}/2\%$ of the total winding. When adjusting tappings are required in addition to those necessitated by the Scott connection, the same general considerations apply as have already been discussed with one additional feature. If the group is of the

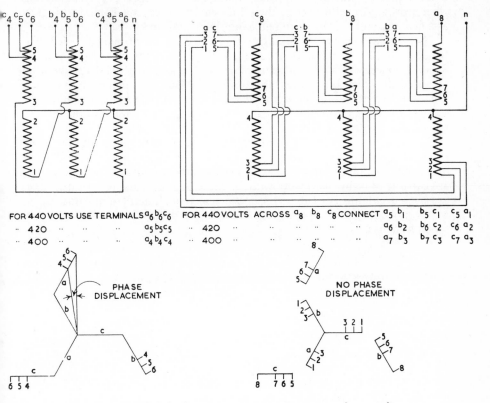

Fig. 11.3   Two methods of arranging tappings on interconnected star windings

interchangeable type, adjusting tappings should preferably be on the two-phase side, subject to there being no serious objection to this arrangement on the score of low voltage. The reason for this is that the main and teaser transformers have unequal amounts of winding in circuit on the three-phase side, and consequently similar adjusting tappings will represent unequal percentages of the total windings in circuit in the two units, if these tappings are on the three-phase side.

Auto transformers

The foregoing remarks concerning tappings on double-wound transformers apply generally to auto transformers, provided references to current and voltage are understood to apply to actual windings and not line values. There is, however, one very important additional aspect. The size of an auto transformer is dependent not only upon its output, but also upon its ratio. Consequently the provision of tappings, by allowing for changes in ratio, may also materially affect the size and, therefore, the cost of such transformers. For example, an auto transformer for stepping up from 400 to 440 volts would be increased in weight and cost by 30 to 40% by the provision of a plus 5% tapping on the secondary side. It is important, therefore, that the worst condition should be considered when determining the equivalent size of auto transformers having tappings.

It is important that proper precautions be taken in respect of current loadings when working transformers on their different tappings, and Table 11.3 is given as a general guide to the permissible limiting conditions.

The standard range of tappings where fitted to the primary or secondary windings is 10% of the normal service voltage, or of the mean voltage of the winding concerned. Any of these tappings may be used to give full rated output, provided they be used on their appropriate voltages. The percentages of rated full-load current in the windings to give full rated output at plus and minus 5% voltage variations from the normal service voltages are shown in the last two columns of Table 11.3.

For those cases where the total maximum range of tappings fitted to either primary or secondary winding exceeds 10% of the normal service voltage, or of the mean voltage of the winding concerned, the outputs and currents which are capable of being dealt with, without exceeding the usual temperature rises, are given in column 5 of Table 11.3, in which all reductions of output and current are proportional to the percentage increases or decreases represented by the tappings in use.

Chapter 12

Voltage variation by tap changing

The voltage of a system may be varied through the medium of adjusting tappings on the power transformers or by some form of auxiliary voltage regulator which also may be a tapped transformer or other type of variable voltage apparatus. The voltage may be varied, on the one hand in steps or, on the other hand, by stepless control. In most cases it is found that voltage variation of transmission and distribution systems can be carried out quite effectively in steps without creating objectionable disturbances on the system. This variation is generally

Fig. 12.1   300 kVA, three-phase 50 Hz, 3300/565 V, dry-type transformer with h.v. link type off-circuit tappings. (Bonar Long & Co. Ltd.)

achieved by means of tappings on the power transformer which, because of the smaller currents to be dealt with, are normally located on the higher-voltage winding.

Transformer tap-changing equipments fall naturally into two categories, (a) those by which the changes are made when the transformer is isolated from the supply on both sides and (b) those by which changes are carried out without interrupting the load current.

OFF-CIRCUIT TAP CHANGING

Dealing first with off-circuit tap changing, the simplest arrangement is that in which the power transformer tappings are terminated just below oil level and there changed manually by means of swinging links or plugs mounted on a suitable terminal board. The drawback to this arrangement is that it necessitates removing the transformer tank cover or handhole cover. It is, however, extremely simple and is the cheapest tap-changing device. This arrangement is shown in Fig. 12.1.

It is important to design the tap-changing link device with captive parts as otherwise there is always the danger that loose nuts, washers, etc. may fall into the tank whilst the position of the links is being altered.

The next stage in the evolution of tap-changing technique is typically illustrated by Fig. 12.2 which shows a three-phase, core-type transformer fitted on the high-voltage side with an off-circuit tapping selector which can be operated from outside the tank. With the particular unit shown, tappings are changed by a rotary movement of the selector handwheel. An arrangement of this kind is particularly valuable in the case of outdoor transformers where it is impracticable to remove the tank cover, while in the case of conservator type transformers an externally operated tapping selector is invaluable. Figure 12.3 shows a design of high-voltage off-circuit tapping selector suitable for use up to voltages of 150 kV.

ON-LOAD TAP CHANGING

In order to control large high-voltage distribution networks and to maintain correct system voltages on industrial and domestic supplies, it is now common practice to provide means of on-load voltage variation on the majority of main transmission and base load substation transformers. Before the inception of extensive transmission systems, it was the general practice to obtain reasonable voltage regulation on small individual systems by adjustment of generated voltage, but this method has been almost entirely discontinued and it is now usual to provide on-load tap-changing equipment on transformers for the same purpose. The main transformers step up the constant generated voltage at the power station terminals to the variable-voltage high-tension networks, step-down transformers then supply the medium voltage systems and finally, the last transformation is arranged to feed the smaller distribution systems; all these transformers are now frequently provided with on-load tap-changing gear.

Wherever voltage adjustment under varying load conditions is required the on-load tap-changing transformer usually has an application. There are many industrial applications where variable voltage is required for manufacturing processes and typical instances where on-load tap-changing transformers are used, are arc furnaces, electrolytic plants, chemical manufacturing processes, etc.

In the following sections the principal circuits used for on-load tap-changing equipments are described and examples of construction are given together with information regarding the methods of control employed.

TAP CHANGER CIRCUITS

All forms of on-load tap-changing circuits possess two fundamental features, (a) some form of impedance which is introduced to prevent short circuiting of the tapping section, and (b) a duplicate circuit which is provided so that the

Fig. 12.2   750 kVA three-phase 11 000/433 V, 50 Hz transformer with h.v. off-circuit tap selector.
(Bonar Long & Co. Ltd.)

241

load current can be carried by one circuit whilst switching is being undertaken in the other.

The impedance can take the form of either a resistor or a centre-tapped reactor and on-load tap changers can, in general, be classified as resistor or reactor types.

In early designs of on-load tap changers, although it was recognised that resistor transition had considerable advantages in longer contact life, due to the relatively short arcing times associated with unity power factor switching, the use of centre-tapped reactors, as the tap-changing impedance, was in general more popular in spite of the inevitable shorter contact life.

This tendency arose because reactors could be continuously rated, whereas transition resistors had a definite time rating due to the high power dissipated when in service. This would have been of no consequence if positive mechanical tap-changer operation could have been assured, but, although various attempts at achieving this were generally successful, there was a risk of damage if a tap changer failed to complete its operation. Thermal protection arrangements were usually introduced, necessitating tripping and the isolation of the transformer concerned during such an emergency.

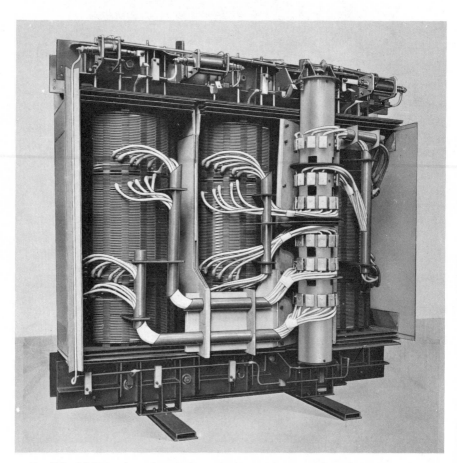

Fig. 12.3   7·5 MVA, three-phase, 50 Hz, 97/23 kV transformer with h.v. off-circuit tap selector

These early tap changers operated at low speeds, and contact separation was slow enough for arcing to persist for several half-cycles, arc extinction finally taking place at a current zero when the contact gap was wide enough to prevent a re-strike. In general contacts were of plain copper.

The mechanical drive to these earlier tap changers, both resistor or reactor transition type, was either direct drive or of stored energy type, using a flywheel or springs, but such drives were associated with complicated gearing and shafting and the risk of failure had to be taken into account.

To a very large extent these older designs have now been superseded since the introduction of high-speed resistor transition tap changing. Reliability of operation has been vastly improved, very largely by the practice of building the stored energy drive into the switching device itself, thus eliminating many of the hazards of the earlier drives. Furthermore, the introduction of contacts using copper/tungsten alloy arcing tips has brought about a substantial improvement in contact life and a complete change in switching philosophy. It is now recognised that long contact life is associated with short arcing-time and breaking at the first current zero is the general rule.

The bridging resistors are still, of necessity, short-time rated but, with improving mechanical methods of switch operation there is only a negligible risk of resistor damage as the resistor is only in circuit for a few milliseconds.

A further advantage of high-speed resistor transition is that improved oil life has been achieved. The transformer oil surrounding the making and breaking contacts becomes contaminated with carbon formed in the immediate vicinity of the switching arc. This carbon formation is directly proportional to the load current and arcing-time and whereas with earlier, slow speed designs oil had to be treated or replaced after a few thousand operations, a life some ten times greater is now generally the rule.

One of the principal advantages of mid-point reactor transition was that twice as many active working positions as there were transformer tappings could be obtained. This proved of considerable advantage where a large number of tapping positions was required and the arrangement is still commonly used by North American manufacturers. A number of special switching arrangements including shunting resistors have been introduced to improve contact life when reactors are employed, but there are definite limits to the safe working voltage appearing across the ends of the reactor and to the recovery voltage when interrupting circulating current. As has already been said, reactor transition has now been almost entirely superseded in Great Britain and Western Europe.

It is of significance to record that it is only in recent years that international recommendations with regard to on-load tap changing have been formulated. These are covered by Publication 214 of the International Electrotechnical Commission and are primarily written to set performance standards for high-speed resistor type equipment.

In some of the earliest designs of tap changers the transformer winding was divided into two circuits which normally operated in parallel. Duplicate switching of tappings was used and the load was transferred to one winding while the tappings were being changed on the other. The operation was completed after both windings were finally connected in circuit on the same tapping position. This scheme had the objection that the halves of the winding were overloaded in turn and obviously a failure in the switching mechanism could be disastrous. This method consequently had a short life and is no longer employed.

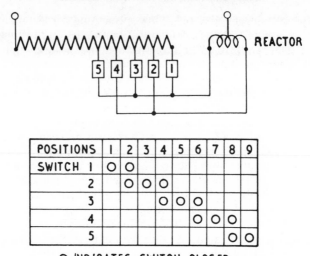

POSITIONS	1	2	3	4	5	6	7	8	9
SWITCH 1	O	O							
2		O	O	O					
3				O	O	O			
4						O	O	O	
5								O	O

O INDICATES SWITCH CLOSED

Fig. 12.4   On-load tap changing by reactor transition

It is useful to explain the methods of tap changing which have been used in the past and those which are in current use today and the various arrangements follow.

The simplest form of reactor switching is that shown in Fig. 12.4. There is only a single winding on the transformer and a current breaking switch is connected to each tapping. Alternate switches are connected together to form two separate groups which are connected to the outer terminals of a separate mid-point reactor, the windings of which are continuously rated.

The sequence of changing tappings is shown in the table on the diagram. In the first position, switch no. 1 is closed and the circuit is completed through half the reactor winding. To change taps by one position, no. 2 switch is closed in addition to switch no. 1. The reactor then bridges a winding section between two tappings and gives a mid-voltage position. For the next tap change, no. 1 switch is opened and switch no. 2 left closed so that the circuit is then completed via the second tap on the transformer winding.

The necessity for a relatively large number of current breaking switches involved large dimensions and oil quantities and frequently a modification of the circuit described, in which two off-load tapping selectors and two current breaking or diverter switches were used. This arrangement is shown in Fig. 12.5.

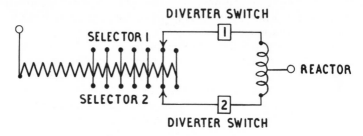

Fig. 12.5   On-load tap changing by reactor transition using diverter switches

The selector and diverter switches are interlocked by mechanical gearing so that (a) no. 1 diverter switch is opened, (b) no. 1 tap selector is moved, (c) no. 1 diverter switch is re-closed, (d) no. 2 diverter switch is opened, (e) no. 2 tap selector is moved, and (f) no. 2 diverter switch is re-closed. Fundamentally this circuit is similar to that of Fig. 12.4 except that the impedance voltage drop across the reactor is eliminated in the operating positions, and this is the usual circuit employed for reactor switching.

Fig. 12.6 5 MVA, three-phase, 50 Hz, 33/11·5 kV transformer core and winding designed for reactor transition tap changing. These have been almost entirely superseded in Great Britain and Western Europe

The inductance of the reactor, which usually had air gaps in its magnetic circuit, was a compromise between conflicting ideals. In order to avoid too wide a voltage variation during a tap change a low value of reactance is desirable, but too low a value would result in an excessive circulating current when the ends of the reactor are connected to adjacent tappings and it was usual to design the reactor core and windings for about 50 to 60 % of the full-load current of the transformer under this condition.

Fig. 12.6 shows a transformer with three single-phase reactors designed for

245

this type of tap changing. It will be seen that the accommodation of relatively small reactors in the same tank as the main transformer is difficult to achieve with economy of space, and this is one factor which has led to the use of high-speed resistor type transition equipments. Further, on transformers operating at very high voltages when the insulation to earth of a small reactor presents insulation difficulties, it is easier and more economical to use resistors instead, which can be mounted in the tap changer, and the transformer tank need only be designed to accommodate the transformer core and windings.

As already mentioned, high-speed resistor tap changing has now almost completely superseded the mid-point reactor method and, in general, there are two basic circuit arrangements in use. These can be divided into those types which carry out selection and switching on the same contacts and use one transfer resistor, and those types which have tapping selectors and separate diverter switches which, in general, use two resistors.

With a single resistor, load current and resistor circulating current have to be arranged to be subtractive, which dictates use with unidirectional power flow or reduced rating with reverse power flow. When two resistors are employed the duty imposed on the diverter switch is unchanged by a change in the direction of power flow.

The two types fall into two classes, single and double compartment tap changers. Most designs of the single compartment type employ a rotary form of selector switch and Fig. 12.7 shows diagrammatically the various switching arrangements for resistor type changers. Figure 12.7(a) illustrates the method employed for the single compartment tap changer and is known as the pennant cycle, whilst Fig. 12.7(b) to (d) shows the connections when two resistors and separate diverter switches are employed and is known as the flag cycle.

Figure 12.8(a) illustrates a typical selector switch assembly of a three-phase, high-speed, on-load, single compartment tap changer. In Fig. 12.8(b) the switching sequence in moving from one tap to the next is described.

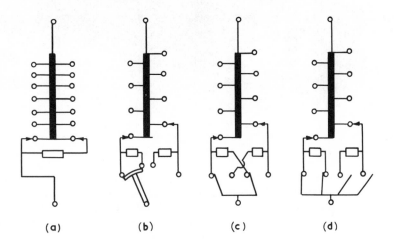

(a)   (b)   (c)   (d)

Fig. 12.7   Types of resistor transition tap changing
(a) Pennant cycle
(b), (c) and (d) Flag cycle

246

Fig. 12.8 (a) *Selector assembly of a three-phase, high-speed on-load compartmented tap changer.*
(Associated Tapchangers Ltd.)

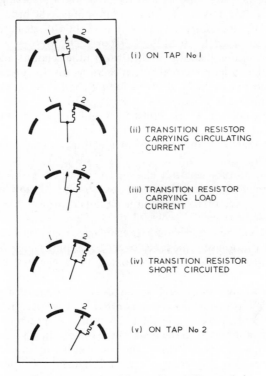

(i) ON TAP No 1

(ii) TRANSITION RESISTOR
    CARRYING CIRCULATING
    CURRENT

(iii) TRANSITION RESISTOR
    CARRYING LOAD
    CURRENT

(iv) TRANSITION RESISTOR
    SHORT CIRCUITED

(v) ON TAP No 2

Fig. 12.8 (b) *Switching of the tap changer illustrated above*

247

On larger transformers the on-load tap-changing equipment is more usually arranged with separate tap selectors and diverter switches. The tap selectors are generally arranged in circular form but 'in-line' contact and crescent arrangements are also used and in fact provide a greater voltage withstand over the tapping range. The diverter switches have contacts operating in rapid sequence with usually four separate make and break units. Figure 12.9 shows typical selector and switching arrangements; diagram A shows a tap changer which has a linear selector arrangement for 14 steps, 15 positions.

In the process of changing from one tapping to the next, the main contact M1 and the auxiliary contact A1 are closed so that connection is made to tapping 10. To change to tapping 9, M1 is opened first, and this transfers the load current via A1 with one resistor R1 in series. A2 then closes. In this position the two resistors are in series across tappings 9 and 10 setting up a circulating current. The load current is divided passing through each of the resistors to each of the tappings. A1 then opens and interrupts the circulating current and the load is now transferred to tapping 9 passing through the resistor $R2$. Finally M2 closes completing the tap change. In the majority of designs this sequence is completed in between 50 and 80 ms. For a tap change in the opposite direction the sequence is reversed, but it will be noted that for the first reversal the tap selectors do not have to be moved. This feature is obtained by using a lost motion coupling in the drive to the tap selectors.

When the tapping range is large or the system voltage very high, thus producing a considerable voltage between extreme tappings, it is an advantage to halve the length of the tapping winding and to introduce a reversing or changeover selector. This not only halves the number of tappings to be brought out from the main winding of the transformer, but also halves the voltage between the ends of the tapping selector switch—see Fig. 12.9B or 12.9C. In diagram B, the tapped portion of the winding is shown divided into nine sections and a further untapped portion has a length equal to ten sections. In the alternative diagrams C and D a section of the transformer winding itself is reversed. The choice of tap changer employed, whether as in Fig. 12.9B or C, will depend on the design of the transformer. In diagrams A, B and C, the tappings are shown at the neutral end of a star connected winding, and in diagram D the tap changer is shown connected to an auto transformer with reversing tappings at the line end of the winding.

In the three examples where a changeover selector is shown, the tapping selectors are turned through two revolutions—one revolution for each position of the changeover selectors; thus with the circuits shown, eighteen voltage steps would be provided.

An important factor in designing transformers using on-load tap-changing gear is that of variation of impedance over the tapping range. Wide variations can often be avoided by the use of the coarse/fine range switching circuits just described.

An example of a single-phase high-speed resistor type tap changer for a double-wound 22/400 kV transformer, as illustrated in Figs. 12.9B or C and with tappings at the neutral end of the 400 kV winding, is shown in Fig. 12.10. The tapping leads are connected to the back of the tap selector housing, on the right of the illustration, and diverter switch and resistors are mounted in the chamber on the left. Three such tap changers are mechanically coupled together and driven by a common driving gear.

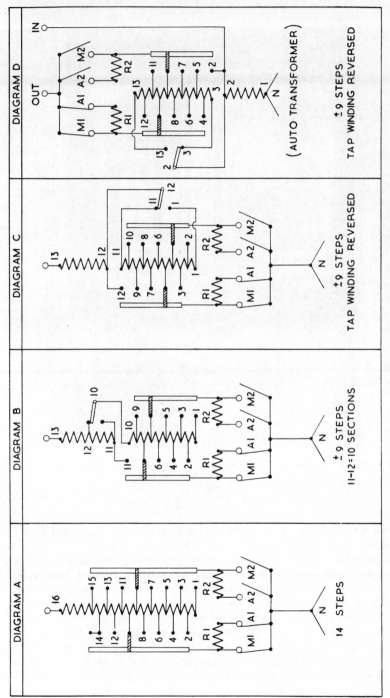

Fig. 12.9 On-load tap-changing circuits for resistor transition using diverter switches

249

Fig. 12.10   Single-phase tap changer for connection to tappings at the neutral side of a 400 kV winding.
(GEC Transformers Ltd.)

On the British grid system there are now many 275/132 kV and 400/132 kV auto-transformers where the on-load tap changers are at the 132 kV point on the auto winding. Earlier designs employed a reversing arrangement as shown in Fig. 12.11(a), utilising a separate reversible regulating winding. More recently a linear arrangement has been used with the tapping section of the winding forming part of the main winding as shown in Fig. 12.11(b). In either case the tapping winding is usually a separate concentric winding.

Earthed neutral end tappings have also been used more particularly on 400/132 kV auto transformers. This introduces simultaneous changes in the effective number of turns in both primary and secondary and also results in a variation in the core flux density and associated complications of variable tertiary voltage. The latter can be corrected by introduction of a tertiary booster fed from the tapping winding.

250

On-load tap changers, particularly when associated with high-voltage systems, have to be designed to meet the surge voltages arising under impulse conditions. In earlier high-voltage tap changers, it was quite a common practice to fit nonlinear resistors across individual tappings or across a tapping range. These nonlinear resistors had an inherent characteristic whereby the resistance decreased rapidly as the surge voltage increased. In modern tap changers these have, in general, been eliminated by improvements in design and positioning of contacts such that appropriate clearances are provided where required. There is now much better understanding of basic transformer design features and many of the unpredictable high-voltage surge conditions arising in the past have been eliminated.

Figures 12.12 and 12.13 show high-speed resistor tap changers of the double compartment type.

The diverter switches are often housed in an oil-filled drop-down tank to permit easy maintenance, but increased reliability and the long life of modern diverter switches often allows a bolted cover to be used instead. A tap changer of this construction, mounted in its tank, is shown in Fig. 12.13. The selectors are mounted on the cast-resin barrier boards in the upper compartment. Their operating spindles are coupled by shafts and gearing, first to the motor driving gear in the lower right-hand compartment and then to the diverter switches mounted in the lower left-hand compartment. The initial setting of the shafts and gearing ensures that the movements of the selectors and diverter switches take place in the required sequence. This type of tap changer is suitable for current ratings up to 800A for tappings at the earthed neutral end of windings for system voltages up to 145 kV.

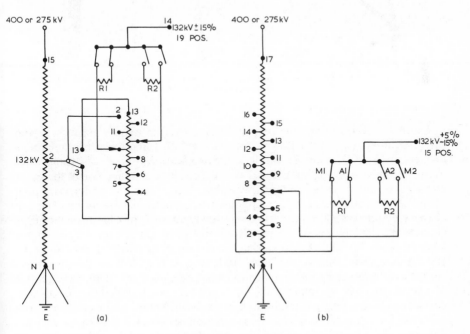

Fig. 12.11   Diagrams of three-phase 400/132 kV and 275/132 kV auto-transformer with 132 kV high-speed resistor type tap changer

251

Fig. 12.12   Three-phase high-speed resistor type 33 kV, 400 A, on-load tap changer. (Associated Tapchangers Ltd.)

Another method of voltage regulation employed in main transmission and distribution systems is one in which shunt regulating and series booster transformers are used. The former unit is connected between phases whilst the latter is connected in series with the line. Tappings on the secondary side of the shunt transformer are arranged to feed a variable voltage into the primary winding of the series transformer, these tappings being controlled by an on-load tap changer equipment. The frame size or equivalent kVA of each transformer is equal to the 'throughput' of the regulator multiplied by the required percentage buck or boost. It should be noted that the voltage of the switching circuit of the regulating transformer to which the on-load gear is connected can be chosen to suit the design and rating of the tap changing equipment. This arrangement of transformers is normally used for 'in-phase' regulation, but can also be employed for 'quadrature' regulation, or for both. Figure 12.14 shows the connections for a typical 'in-phase' and 'quadrature' booster employing two tap changers. Such a unit can be used for the interconnection of two systems with small variations of phase angle.

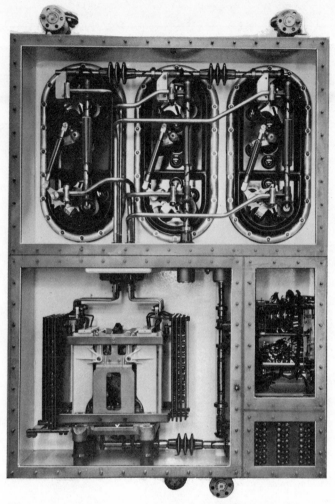

Fig. 12.13 Three-phase high-speed neutral end tap changer. (GEC Transformers Ltd.)

CONSTRUCTION OF TAP CHANGERS

It is a fundamental requirement of all tap changers that the tap selector and diverter switch combination shall operate in the correct sequence. Many methods have been used to achieve the desired alternating movement, the more usual employing the 'Geneva' wheel. Figure 12.15 shows one on each spindle of a tapping selector with a common operating wheel in the centre which operates each selector shaft in turn. A similar sequence of movements uses an elliptical gear train and Fig. 12.16 shows a gear box of this type partly dismantled. The centre spindle is driven at constant speed and the outer spindles operate alternately, completing about 330° operation in turn for each 180° movement of the centre spindle; thus alternate motion of the two selectors is obtained. An extension to this gear train is also available for the movement of a changeover

253

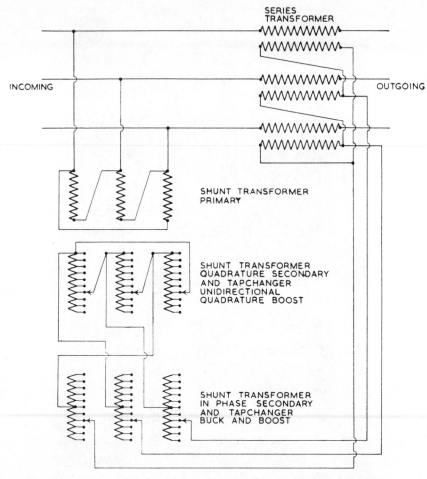

Fig. 12.14   Diagram of connections for a three-phase 'in-phase' and 'quadrature' booster

selector. These alternating drives are coupled to the tap selectors only, the latter usually mounted in a separate compartment from the diverter switches.

The selectors can be maintained under a head of oil provided by the conservator and thus brought within the zone of the main transformer gas relay protection. The diverter switches are normally housed below the selectors in a free oil level compartment with inspection covers for maintenance purposes. A tap changer of this type is shown in Fig. 12.12 with these covers removed. The selectors are mounted in the upper compartment and operated via an elliptical gear box from vertical shafting and worm reduction gear. The same shafting, directly driven from the motor operated driving mechanism, is connected through a worm reduction gear to the diverter switch drive. In some cases a drop down tank construction is used for the diverter switch.

Another arrangement is to mount the tap selectors in compartments at the phase centres along the transformer tank, with diverter switches separately housed at the end. This arrangement has the advantage of short tapping leads

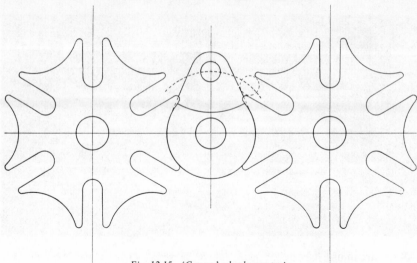

Fig. 12.15 'Geneva' wheel gear train

within the transformer tank, and this feature can be seen in Fig. 12.17, which illustrates one tap selector compartment on the transformer tank. Figure 12.18 shows a 60 MVA transformer fitted with this type of tap changer. The three tap selection compartments and the diverter switches can be seen mounted at the end of the transformer tank.

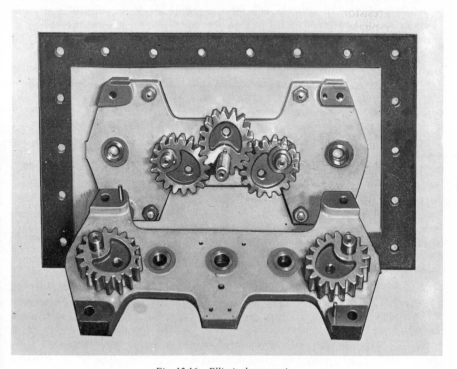

Fig. 12.16 Elliptical gear train

255

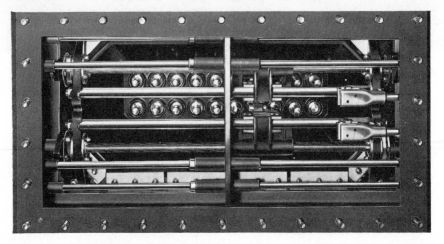

Fig. 12.17   Tap selector compartment of a three-phase neutral end tap changer. (Ferranti Ltd.)

Several forms of high-speed diverter switches have been devised and a large number are in use. In general, these operate from 'over' toggle link mechanisms, a spring being charged across the toggle prior to this being tripped mechanically. Reliable operation has now reached an exceedingly high standard and in all cases very considerable contact life can be guaranteed. In fact diverter switch contacts will now, in general, last for the useful life of the transformer itself. Two types of diverter switch for high-speed resistor transition are shown in Figs. 12.19 and 12.20. The switching circuit is similar to that shown in Fig. 12.9, using two resistors.

A three-phase single compartment tap changer suitable for 33 kV system transformers is illustrated in Fig. 12.21. The rotary diverter switches have duplicate moving contacts arranged so that one makes before the other breaks, with the transition resistor between them. The selectors are mounted directly on a cast-resin barrier board and are driven through a train of insulated gear wheels by the spring operated quick-motion mechanism on the right. This tap changer is suitable for currents up to 450 A.

An external view of a 33 kV tap changer, also of the single compartment type, being fitted to an 11·5/23 MVA, continuous emergency rated system transformer, is shown in Fig. 12.22.

At the other end of the tap-changer range, Figs. 12.23 and 12.24 show two examples of single-phase fully insulated 132 kV tap changers, and in each case they are fitted to 240 MVA, 400/132 kV auto transformers.

The units shown in Fig. 12.23 are arranged for mounting on the tank side, with the tap selectors mounted within the individual phase tanks bolted to the side of the transformer. The tap changers in Fig. 12.24, however, have their tap selectors mounted on turrets supported from the top cover of the transformer. In each case accommodation is provided for current transformers, within the turrets.

The high-speed resistor diverter switches of both designs are mounted on the top of the 132 kV porcelain insulators which take the place of the normal transformer bushings, the 132 kV line terminal lead being taken from the diverter switch housing. Both of these tap changers are designed for a power frequency test to earth of 300 kV.

CONTROL OF ON-LOAD TAP CHANGING GEAR

Many advances have been made in the design of control circuits associated with on-load tap changing. Mention has been made of driving mechanisms and the fundamental circuits associated with the starting up of the motor for carrying out a tap change, whilst varying from one maker to another, are comparatively

Fig. 12.18   60 MVA transformer with a neutral end tap changer. (Ferranti Ltd.)

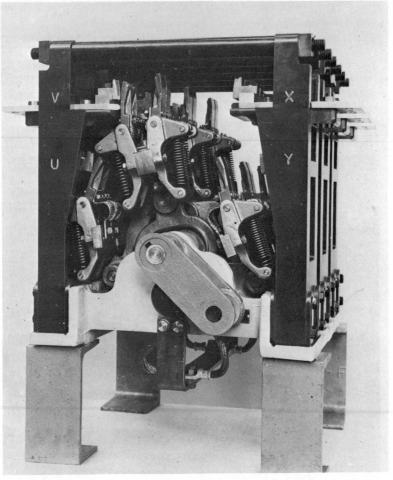

Fig. 12.19   Three-phase 600 A diverter switch. (Associated Tapchangers Ltd.)

simple; in general, the motor is run up in one direction for a 'raise' tap change and in the reverse direction for a 'lower' tap change. In some cases a brake is employed to bring the motor to rest while in others clutching and de-clutching is carried out electrically or mechanically and arranged so that only one tap change at a time is performed for each initiation. It is, however, in the bringing in of the initiation of the tap change that the main interest lies.

Mechanical hand operation is always made available for emergency use and in some cases tap changers are arranged for hand operation only.

Simple pushbutton control is still used, but there has been a tendency towards unattended automatic voltage control at substations arranged so that a pre-determined constant or compensated busbar voltage can be maintained. In general, a tap changer is provided on a transformer for maintaining a pre-determined outgoing voltage where the incoming voltage is subject to variation due to voltage drops and other system variations. In order to initiate the voltage

Fig. 12.20 Diverter switches for a three-phase tap changer; tank removed. (GEC Transformers Ltd.)

Fig. 12.21 33 kV, 450 A, three-phase, single compartment tap changer. (Associated Tapchangers Ltd.)

Fig. 12.22   33 kV, 450 A, three-phase, single-compartment tap changer being fitted. (Ferranti Ltd.)

controlling apparatus a voltage transformer is necessary, energised from the controlled voltage side of the transformer. Its output voltage is used to energise a voltage sensitive relay with contacts to start a tap change in the required direction as the voltage to be controlled varies. It is usual to introduce a time delay element either separately or within the voltage relay itself to prevent unnecessary operation or 'hunting' of the tap changer during transient voltage changes. The 'balance' voltage of the relay, namely the value at which it remains inoperative, can be pre-set using an adjustable series resistor in the voltage coil circuit of the relay so that any predetermined voltage within the available range can be maintained.

As an additional feature, load compensation is frequently added. It is often an advantage to arrange for an increase in the voltage to be maintained with increasing load and this is achieved by means of a line drop compensator. This device comprises a combination of a tapped resistor and a tapped reactor, fed from the secondary of a current transformer whose primary winding carries the load current and is arranged to subtract a voltage drop proportional to the load on the main transformer from the voltage applied to the voltage relay. By suitable adjustment of the resistance and reactance components, which will depend upon the outgoing line characteristics, it is possible to obtain constant voltage at some distant point on a system irrespective of the load or power factor. In practice, the selected settings for line drop compensators are usually compromise values and they are arranged to obtain the most acceptable voltage variation under varying loads on a distribution network. Figure 12.25 shows the principle of the compensator, which for simplicity is here applied to a single-phase circuit; it consists of a potentiometer rheostat and a tapped reactor connected in series

260

Fig. 12.23   Three single-phase, fully insulated 132 kV tap changers fitted to a 240 MVA 400/132 kV auto transformer. (Hawker Siddeley Power Transformers Ltd.)

Fig. 12.24 Another example showing single-phase fully insulated 132 kV tap changers fitted to a 240 MVA/132 kV auto transformer, (GEC Transformers Ltd.)

with the main voltage coil of the regulating relay. The tappings from the reactor are brought to a multicontact switch, and the connections from the moving contact of this switch and of the potentiometer rheostat are connected to the main current transformer secondary circuit.

A voltage equal to the phasor sum of the voltage drops across the potentiometer and the reactor is thus injected into the main regulating voltage coil circuit, that is, into the voltage transformer secondary circuit. By means of the potentiometer and tapping switch these voltages may be adjusted to be proportional to the line resistance and line reactance voltage drops, that is, to the line impedance drop. These voltages are also proportional to the line current at any moment.

The voltage across the compensator is arranged to 'buck' the voltage transformer secondary voltage and thus, as the former increases with increased load

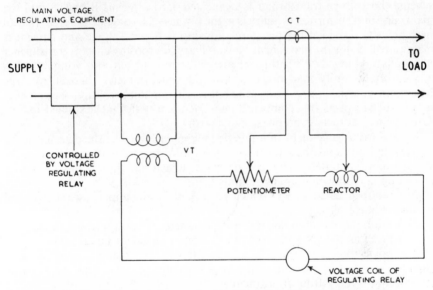

Fig. 12.25   *Single-phase diagram showing the principle of line drop compensation*

current, the voltage regulating relay becomes unbalanced and operates the main regulating device to raise the line voltage at the sending end by an amount equal to the line impedance drop and so restores the relay to balance. The reverse action takes place when the load current decreases.

The regulating relay and compensator are, of course, usually employed on three-phase circuits, but since in any case the relay voltage coil circuit is single phase (usually taken across two phases) the only fundamental difference between the arrangement used and that shown in Fig. 12.25 is in the arrangement of the main voltage transformer and current transformer primary connections which must be such as to provide the proper phase relationship between the voltage and the current.

The compensator is designed for operation at 0·5A and either one or two auxiliary current transformers are required according to which one of the following alternative methods of connection is adopted.

In the first method the voltage transformer is connected across the A and C phases and the current transformer is in the A phase. Different phases may be used provided the phase relationship is maintained. The compensation afforded by this method is not strictly correct since there is a 30° phase angle between the voltage and the load current at unity load power factor. Since line drop compensation is usually a compromise this method will suit certain cases.

In Fig. 12.25, a single current transformer is shown in the line connection for the current supply to the compensator. It is usual practice to have an interposing current transformer in order to obtain the correct full-load secondary current but, at the same time, provide protection against damage due to overloads or fault current in the line. The interposing current transformer is specifically designed to saturate under such conditions, thus avoiding the introduction of high overload current to the compensator circuit.

If greater accuracy is desired an alternative scheme may be used. With this scheme the voltage transformer is connected across A and B phases and the main current transformer primary is in the C phase. Here again, different phases may be used provided the phase relationship between voltage and current is maintained. As for the first scheme, an auxiliary interposing current transformer is usually employed. With this scheme since the current and voltage are in quadrature, at unity load power factor, the potentiometer rheostat and the reactor provide the line reactance and line resistance compensation respectively; in all other respects this compensator is identical with that described for the first scheme but there is no phase angle error.

A standard arrangement of a line drop compensator has the external series resistor and mean setting adjustment rheostat for the regulating element of the voltage relay mounted in the compensator, which has three adjusting knobs providing the following ranges of adjustment:

(a) A knob marked ZERO which gives $\pm 10\%$ adjustment of the nominal no-load voltage setting.

(b) A knob marked R giving 0 to 15% compensation for line resistance.

(c) A knob marked X giving 0 to 15% compensation for line reactance.

If compensation is required for line resistance only, a simple potentiometer rheostat is supplied instead of the complete compensator and the external resistor and mean setting adjustment rheostat are supplied separately.

When the compensator has been installed and all transformer polarities correctly checked, the regulating relay may be set to balance at the desired no-load voltage value by means of the knob marked ZERO. The resistance and reactance drops calculated from the line characteristics may then be applied by means of the control knobs R and X and their associated scales.

A pair of voltage control cubicles with voltage regulating relay and line drop compensation is shown in Fig. 12.26. The voltage regulating relay is of the balanced plunger electro-mechanical type and its arrangement is shown diagrammatically in Fig. 12.27. It is clear that the timing element is of straightforward eddy disc type, and whereas the time delay period can be adjusted, it is in no way related to the degree of voltage unbalance. There is an obvious advantage in providing some means by which a sudden wide change in voltage can be more quickly corrected, and a number of solid-state voltage relays are now available having this characteristic. These relays have a solid-state voltage sensing circuit and also an inverse characteristic timing circuit so that the delay is inversely proportional to the voltage change.

Fig. 12.26   Control cubicles with voltage regulating relay and line drop compensator. (Hawker Siddeley Power Transformers Ltd.)

Where two or more transformers with automatically controlled on-load tap changing gear are operating in parallel, it is normally necessary to keep them on the same tapping position or, as a minimum, not more than one step apart. Obviously, if transformers are operated in parallel on different tappings circulating currents will be set up. In general, one step is the most that can be tolerated.

Many different schemes of parallel control have been devised and several are in regular use. If it is considered necessary to operate always on the same tapping this can be achieved either by a master-trailer system or by a simultaneous operation method.

With the master trailer system, the first tap changer of a group is initiated from the automatic relay and on completing its tap change, passes on an impulse to the second tap changer which in turn passes an impulse to the next unit as required. Lockout circuits are provided so that the first unit cannot operate again until all units in the group have come into line. This system involves auxiliary step by step master and follower switches and somewhat

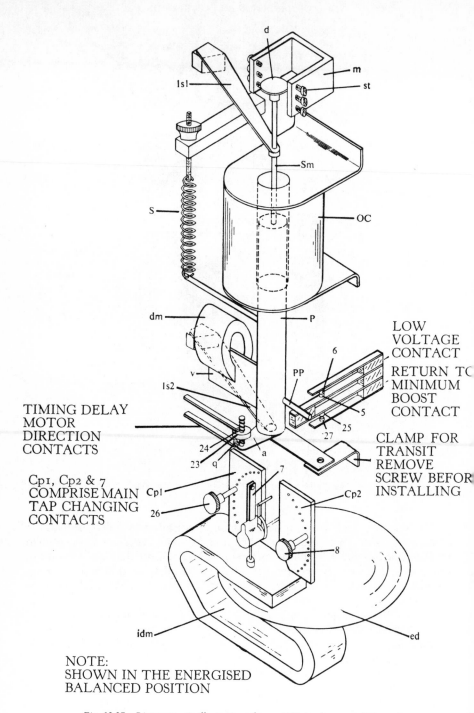

Fig. 12.27   *Diagrammatic illustration of type AVE 4 voltage regulating relay*

complicated interconnections. However, it has the advantage that the overall voltage change of a group is made in small steps, as each unit makes its individual change. It is usual to have selection arrangements such that any unit of a group can operate as the master.

With simultaneous operation, all tap changes of a group are arranged to start their operation at the same time. This is a simpler arrangement although it is still necessary to provide lockout arrangements to take care of any individual failure.

Where tap-changer operation one step apart can be tolerated, random operation from individual voltage relays can be permitted. Here again it is necessary to provide adequate lockout protection arrangements to prevent units becoming widely out of step due to some control fault. These can be operated from a monitor of actual out-of-balance circulating currents or from step-by-step switches. Another method is to use a component of the out-of-balance to bias the line drop compensators, thus automatically bringing units into step.

There has been much development in the supervisory control of system voltages, and on some systems centralised control has been achieved by the operations of tap changers by remote supervisory methods. This is usually confined to supervisory remote pushbutton control with supervisory indication of tap-changer position, but more complicated schemes have been installed and are being satisfactorily operated where tap changers are controlled from automatic relays on their respective control panels, with supervisory adjustment of their pre-set voltage and supervisory selection of groups operating in parallel, with all necessary indications reported back by supervisory means to the central control room.

Symbol Key to Fig. 12.27

d	Soft iron position locating disc
m	Position location permanent magnet
st	Position location insensitivity adjustment screws.
ls1 and ls2	Movement supporting leaf springs
Sm	Movement plunger stem
S	Movement balancing spring
OC	Operating coil
dm	Movement damping magnet
P	Movement plunger
v	Movement damping vane
PP	Auxiliary contact operating pin
5 and 6	Moving and fixed low voltage contacts
25 and 27	Moving and fixed return to minimum boost contacts
a	Timing motor contact carrying arm.
q	Timing motor high voltage contact
24	Timing motor high voltage contact
23	Timing motor contact
Cp1 and Cp2	Raise and lower tap changing contact plates
7	Tap changing common contact
26 and 8	Raise and lower tap changing contact screws adjustable for timing
ed	Timer eddy disc
idm	Induction disc motor

Chapter 13

Performance and type tests

The establishment and maintenance of a high standard of both materials and workmanship can only be achieved by continuous inspection during the manufacturing stages and by subsequent testing of the components and finished product.

Throughout the operations which the preservation of such a standard entails, no influence is greater than that of the system of tests to which the product is subjected, and the final tests, with which this chapter deals, are checks upon all preliminary inspection made throughout the period of manufacture.

Therefore the stringency and thoroughness with which these are carried out is a matter of vital importance, and in order to show to transformer users how these facts are realised this chapter gives a detailed description of the various methods employed.

In order to obtain accurate results it is essential that low power factor wattmeters, precision grade ammeters, voltmeters, and class 'AL'* current and voltage transformers are employed. These instruments should be checked at intervals not exceeding six months to ensure that the requisite accuracy is maintained.

In accordance with the requirements of B.S.171, power transformers are subjected to the following tests in order to ensure their being electrically and mechanically sound in every respect, and also to make certain that they meet the guaranteed performance:

(1) Ratio and polarity
(2) Load losses
(3) Impedance
(4) Insulation resistance
(5) Resistance of windings
(6) No-load losses
(7) No-load current
(8) Voltage tests—windings
   (*a*) separate source
   (*b*) induced overvoltage
(9) Core insulation voltage test

*See B.S.3938 and 3941.

(10) Temperature rise test
(11) Impulse voltage test
(12) Noise level test.

Tests 1 to 9 are known as routine tests to which all transformers are subjected. Tests 10–12 are type tests and are not made unless included in the particular contract. If tests are required to be conducted in the presence of the purchaser or his representative this must be specified when inviting tenders.

These tests are briefly described for three-phase transformers in the following text. The procedure is generally similar for single-phase units.

Ratio and polarity test

Ratio and polarity tests are carried out on every transformer to ensure that the turns ratio of the windings is correct and also that the tappings on any of the windings have been made at the correct position. The B.S. tolerance for such ratios is plus or minus 0.5% of the declared no-load ratio or a percentage equal to 1/10th of the percentage impedance voltage at the rated load, whichever is the smaller. This (combined) test is initially carried out on the core and windings before assembly in the tank in order to facilitate access to the various tappings. The ratio is again checked when the transformer is in its tank and finally when assembled with terminals and any tap changing mechanism.

The ratio of the transformer can be checked by the voltmeter method, that is by measuring the voltage on both h.v. and l.v. sides of the transformer when energised from a low-voltage supply, but this is not considered to be sufficiently accurate, particularly in the case of high turns ratios. The ratiometer method is far more satisfactory.

Ratiometer method (i)

The diagram of connections for this test is as shown in Fig. 13.1 which illustrates a star/star transformer connected for a ratio test. The ratiometer itself consists of a single-phase, double-wound transformer, having a constant primary winding and a variable secondary winding, the voltage variation being effected through the intermediary of adjusting tappings and two dial switches. Various secondary tappings are arranged to provide fine and coarse regulation. In carrying out the test, the h.v. winding of the transformer being tested is connected in opposition to the primary winding of the ratiometer, while the l.v. winding of the transformer is connected to the terminals of the dial switches through an ammeter. A single-phase voltage, not exceeding 2000 V*, is applied across the primary winding of the ratiometer, and the two dial switches are adjusted until the ammeter registers zero. Under this condition the voltage across the l.v. winding of the transformer under test is equal to the voltage across the effective part of the secondary winding of the ratiometer, *i.e.*, that part of the winding between the two dial switch terminals. As the voltage across the h.v. winding of the transformer and across the primary winding of the ratiometer is the same,

---

*This may be considerably less than 2000 V when testing transformers having h.v. voltages less than 2000 V.

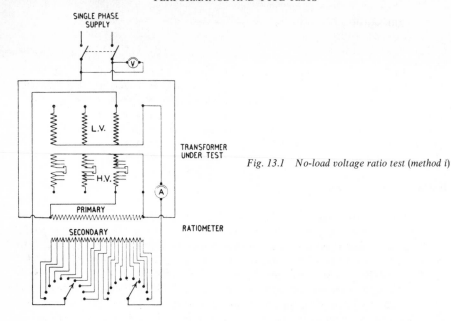

SINGLE PHASE
SUPPLY

L.V.

H.V.

PRIMARY

SECONDARY

TRANSFORMER
UNDER TEST

RATIOMETER

Fig. 13.1   No-load voltage ratio test (method i)

the ratio of the windings of the transformer must equal the ratio of the ratio-meter, which, of course, is observed for any given setting of the dial switches. The actual voltages are determined by multiplying the different test ratios by the specified transformer l.v. phase voltage. With three-phase transformers of any interphase connection each limb is tested separately, while if tappings are provided, the test is repeated for each tapping ratio.

When using the ratiometer, care must be taken as it is possible to damage the ammeter. This will occur if the two transformers (*i.e.*, the transformer under test and the ratiometer transformer) are not connected up in opposition, and in order to safeguard the ammeter it is safer, after making the necessary connections, initially to apply only a low voltage to the circuit. If this precaution is not taken, and the ratio of the ratiometer differs very greatly from that of the transformer under test, a very heavy circulating current may flow. If, however, as is normally the case on test, the ratio of the transformer is known, the dial switches should be adjusted so that the two ratios are equal. If a large current then flows with a small applied voltage, the two connections on the dial switch terminals should be reversed.

The applied voltage may then be increased and the dial switches adjusted until balance is obtained. If it is found impossible to obtain an absolute balance, the dial switch giving fine regulation should first be put on the stud before, and then on the stud after, the one which gives the least current as indicated on the ammeter.

In order to calculate the correct l.v. turns the following procedure should be adopted:

Let a = ammeter reading on the stud before the one on which the best balance is obtained

270

b = ammeter reading on the stud after the one on which the best balance is obtained

x = dial switch reading on coarse regulation switch

y = dial switch reading on fine regulation switch for ammeter reading a above.

$$\text{then l.v. turns} = x+y+\left(\frac{2a}{a+b}\right).$$

In order to protect the ammeter, it should be provided with shunts so that the total scale reading may be reduced, thus providing greater sensitivity for the final readings.

The winding polarity of the transformer is automatically checked by this ratio test, as if the windings were connected up in such a manner as to give a polarity the reverse of that required, it would not be possible to obtain a zero reading of the ammeter connected between the transformer under test and the ratiometer. In those cases where a separate polarity test is required, the method adopted for phasing in any two transformers for parallel operation* may be applied, it being borne in mind that the voltmeter will, however, never register zero as the transformer ratio is not unity.

Ratiometer method (ii)

The diagram of connections for this test is shown in Fig. 13.2.

This method has the advantage that the high voltages likely to be obtained in method (i) are unnecessary. In fact the highest voltage employed is that of the mains supply. Furthermore, the ratio is given directly by the instrument dial settings for the condition of zero balance when the ratiometer is connected across the h.v. and l.v. windings of the transformer being tested.

The ratiometer is designed to give accurate measurement to within 0.1% over a range of ratios of approximately 1110 to 1. The principle of this ratiometer is the 'bridge' circuit, in which the voltages of the windings are balanced against the voltage drop across a variable and a fixed resistor. The variable resistor which is calibrated, is normally adjusted during the test until a balance is obtained and the galvanometer has zero deflection. At balance the ratio of the windings under test can be read directly from the dial settings, *i.e.*, dial readings indicate the ratio of the transformer to unity. The ratiometer indicates the correctness of polarity, since should the polarity of one of the windings be incorrect with correct test connections a null reading of the bridge cannot be obtained.

Polarity of windings and phasor group connections

Polarity and interphase connections are checked whilst measuring the ratio by either of the ratiometer methods but care must be taken to study the diagram of connections and the phasor diagram for the transformer before connecting up for test. A ratiometer may not always be available and this is usually the case on site so that the polarity must be checked by voltmeter. The primary and

*See Chapter 18.

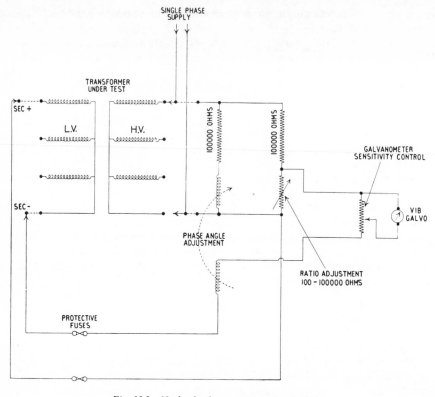

Fig. 13.2   No-load voltage ratio test (method ii)

secondary windings are connected together at one point as indicated in Fig. 13.3. A low-voltage three-phase supply is then applied to the h.v. terminals. Voltage measurements are then taken between various pairs of terminals as indicated in the diagram and the readings obtained should be the phasor sum of the separate voltages of each winding under consideration.

Load loss test and impedance test

These two tests are carried out simultaneously, and the connections are shown in Fig. 13.4. The two-wattmeter method is employed for measuring the load (copper) loss of a three-phase transformer, one instrument normally being used, the connections from which are changed over from any one phase of the transformer to any other by means of a double pole switch. Closing the double pole switch on phase A places the ammeter and the current coil of the wattmeter in series with that phase. The wattmeter voltage coil and the voltmeter are connected across phases A and B, both leads from the wattmeter voltage coil being taken direct to the transformer terminals. When the double pole switch is subsequently closed on phase C, the ammeter and the wattmeter current coil are in series with that phase, and the wattmeter voltage coil and the voltmeter will be connected across phases B and C, the one voltage coil lead being changed from phase A to phase C. Voltage is applied to the h.v. winding, the l.v. being short circuited. The links in A and C phases being closed, a low voltage is first applied to the h.v. windings, the initial value being a fraction of the calculated impedance voltage. The double pole switch is then closed and the link on phase A opened. The applied voltage is gradually increased until the ammeter in the h.v. circuit indicates the normal full-load current when wattmeter, ammeter and voltmeter readings are noted. The link in phase A is then closed, the double pole switch changed over, the link in C opened, and the wattmeter voltage coil

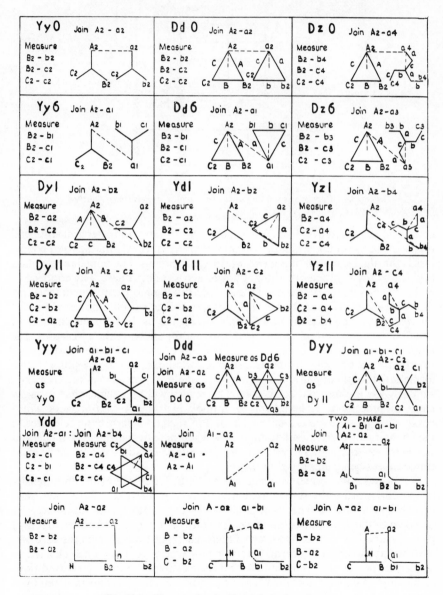

Fig. 13.3   Diagrams for checking polarity by voltmeter

273

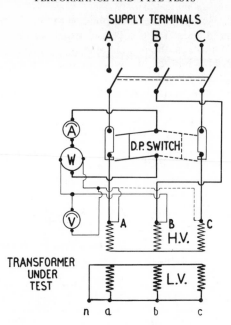

Fig. 13.4   Copper loss and impedance voltage test; two-wattmeter method

connection changed over from A to C phase. Wattmeter, ammeter and voltmeter readings are again taken. These readings complete the test and the total copper loss is the algebraic sum of the two wattmeter readings. The impedance voltage is given by the voltmeter reading obtained across either phase. The copper loss would be the same if measured on the l.v. side, but it is more convenient to supply the h.v. winding. It is important that a copper-loss test should be carried out at the frequency for which the transformer is designed, as the frequency affects the eddy-current copper-loss component, though not affecting the I^2R losses. The connections given and procedure outlined are exactly the same whatever the interphase connections of the transformer windings. For single-phase transformers the total copper loss is given by a single wattmeter reading only, and similarly for the impedance voltage. The reader will appreciate that in many cases in practice it is necessary to employ instrument transformers whilst conducting the tests described earlier, and in such an instance the reference to the changing of current and voltage coils when making wattmeter connections would of course equally refer to the secondary circuits of any instrument transformers being employed in the test.

For power transformers having normal impedance values the flux density in the core during the short-circuit test is very small, and the iron loss may therefore be neglected. The losses as shown by the wattmeter readings may thus be taken as the true copper loss, subject to any instrument corrections that may be necessary. In the case, however, of high-reactance transformers the core loss may be appreciable. In order to determine the true copper loss on such a transformer the power input should first be measured under short-circuit conditions as already explained, and then with the short circuiting connection removed (*i.e.*, under open-circuit conditions) the core loss should be measured with an applied voltage equal to the measured impedance voltage. This second test will give the iron loss at the impedance voltage, and the true copper loss will be obtained by the difference between these two loss measurements.

When making the copper-loss test it must be remembered that the ohmic resistance of the l.v. winding may be very small, and therefore the resistance of the short circuiting links may considerably affect the loss. Care must be taken to see that the cross-sectional area of the short circuiting links is adequate to carry the test current, and that good contact is obtained at all joints.

To obtain a true measurement it is essential that the voltage coil of the wattmeter be connected directly across the h.v. windings, and the necessary correction made to the instrument reading.

The temperature of the windings at which the test is carried out must be measured accurately and also the test must be completed as quickly as possible so as to ensure that the winding temperature does not change during the test. Should several copper loss and impedance tests be required on a transformer (*i.e.*, on various tappings) then it is advisable to carry out these tests at reduced currents, though in no case less than one half of the rated current, and correct the results to rated values of current.

The three-wattmeter method can also be adopted for copper-loss measurement with advantage where the test supply is unbalanced. This test is essentially the same as the two-wattmeter method where one winding is short-circuited and a three-phase supply is applied to the other winding, but in this case the wattmeter current coil is connected to carry the current in each phase whilst the voltage coil is connected across the terminals of that phase and neutral. The sum of the three readings taken on each phase successively is then the total copper loss of the transformer. During this test the current in each phase can be corrected to the required value before noting wattmeter readings. On large transformers where the impedance of the transformer causes a low power factor it is essential that wattmeters designed for such duty are employed.

The copper loss and impedance are normally guaranteed at 75°C but in fact both are normally measured at test room temperature and the results obtained corrected to 75°C on the assumption that the direct load loss (I^2R) varies with temperature as the variation in resistance and the stray load loss varies with the temperature inversely as the variation in resistance.

The tolerance allowed by B.S.171 on impedance is $\pm\frac{1}{10}$ for a two-winding transformer and $\pm\frac{1}{10}$ to $\frac{1}{7}$ for a multiwinding transformer, both on the principal tapping. The copper loss at 75°C is subject to a tolerance of $\frac{1}{7}$ but iron plus copper losses in total must not exceed $\frac{1}{10}$ of the guaranteed value.*

The test connections for a three-phase, interconnected star earthing transformer are shown in Fig. 13.5. The single-phase current I in the supply lines is equal to the earth fault current and the current in each phase winding is one third of the line current. Under these loading conditions the wattmeter indicates the total copper loss in the earthing transformer windings at this particular current while the voltmeter gives the impedance voltage from line to neutral. The copper loss measured in this test occurs, of course, only under system earth fault conditions. Normally earthing transformers have a short time rating (*i.e.*, for 30 s) and it may be necessary to conduct the test at a reduced value of current, and to omit the measurement of the copper loss, thus testing impedance only.

---

* Refer to Chapter 4 for details of B.S. 171 tolerances.

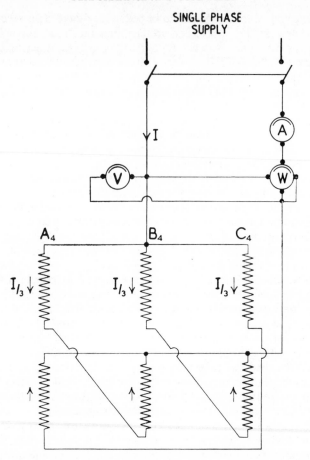

Fig. 13.5 Copper loss and impedance voltage test for a three-phase interconnected star neutral earthing transformer

At the same time as the copper loss is being measured on the three-phase interconnected star earthing transformer, the zero phase sequence impedance Z_0 and resistance R_0 can be obtained as follows:

Z_0 per phase (ohms) $= \dfrac{3V}{I}$ where I = current in the neutral during test and

R_0 per phase (ohms) $= \dfrac{3 \times \text{power (watts)}}{I^2}$.

All other tests on earthing transformers are carried out in the same way as for power transformers.

Insulation resistance test

Insulation resistance tests are carried out on all windings, core, and core clamping bolts. The standard Megger testing equipment is used, the 'line' terminal of which is connected to the winding or core bolt under test. When making the test on the windings, so long as the phases are connected together, either by the neutral lead in the case of the star connection or the interphase connections in the case of the delta, it is only necessary to make one connection between the Megger and the windings. The h.v. and l.v. windings are, of course, tested separately, and in either case the procedure is identical. In the case of the core bolts, each bolt is tested separately.

Should it be required to determine exactly the insulation of each separate winding to earth or between each separate winding, then the guard of the Megger should be used. For example, to measure the insulation resistance of the h.v. winding to earth the line terminal of the Megger is connected to one of the h.v. terminals, the earth terminal to the transformer tank, and the guard terminal to the l.v. winding. By connecting the windings and the instrument in this way any leakage current from the h.v. winding to the l.v. winding is not included in the instrument reading and thus a true measurement of the h.v. insulation to earth is obtained.

Resistance of windings

The d.c. resistances of both h.v. and l.v. windings can be measured simply by the voltmeter/ammeter method, and this information provides the data necessary to permit the separation of I^2R and eddy-current losses in the windings. This, of course, is necessary in order that transformer performances may be calculated at any specified temperature.

The voltmeter/ammeter method is not entirely satisfactory and a more accurate method such as measurement with the Wheatstone or Kelvin double bridge should be employed. It is essential that the temperature of the windings is accurately measured, remembering that at test room ambient temperature the temperature at the top of the winding can differ from the temperature at the bottom of the winding. Care also must be taken to ensure that the direct current circulating in the windings has settled down before measurements are made. In some cases this may take several minutes depending upon the winding inductance unless series swamping resistors are employed. If resistance of the winding is required ultimately for temperature rise purposes then the 'settling down' time when measuring the cold winding resistance should be noted and again employed when measuring hot resistances taken at the end of the load test.

Iron-loss test and no-load current test

These two tests are also carried out simultaneously and the connections are shown in Fig. 13.6. This diagram is similar to Fig. 13.4, except that in Fig. 13.6 voltage is applied to the l.v. windings with the h.v. open circuited, and one watt-meter voltage coil lead is connected to the transformer side of the current coil. The two-wattmeter method is adopted in precisely the same way as described for the copper-loss test, the double pole switch being first closed on phase A. The rated l.v. voltage at the specified frequency (both of which have previously been adjusted to the correct values) is first applied to the l.v. windings, and then readjusted if necessary, the links being closed in phases A and C. The double

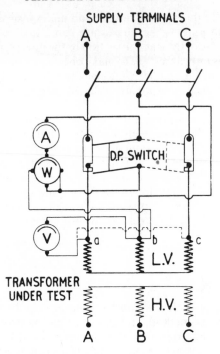

Fig. 13.6   Iron loss and no-load current test; two-wattmeter method

pole switch is then closed, the link opened in phase A, and wattmeter, ammeter and voltmeter readings noted. The wattmeter is then changed over to phase C, and one voltmeter connection changed from phase A to phase C. Wattmeter, ammeter and voltmeter readings are again noted. These readings complete the test, and the total iron loss is the algebraic sum of the two wattmeter readings. The no-load current is given by the ammeter reading obtained in each phase. The iron loss would be the same if measured on the h.v. side, but obviously the application of voltage to the l.v. winding is more convenient. The no-load current would, however, be different, and when checking a test certificate, note should be taken of the winding on which this test has been carried out.

The connections given and procedure outlined are exactly the same whatever the interphase connections of the transformer windings. For single-phase transformers the iron loss is obtained simply by one wattmeter reading.

For all transformers except those having low-voltage primary and secondary windings this test is conducted with the transformer in its tank immersed in oil.

If the l.v. voltage is in excess of 1000 V, instrument transformers will be required and the remark made earlier equally applies.

In making this test it is generally advisable to supply to the l.v. winding for two reasons; firstly, the l.v. voltage is more easily obtained, and secondly, the no-load current is sufficiently large for convenient reading.

The supply voltage can be varied either by varying the excitation of the alternator or by using an induction regulator. A variable resistor in series with the transformer winding should not be used for voltage adjustment because of the effect upon the voltage wave shape and the transformer iron loss.

The iron loss will, of course, be the same if measured on either winding, but the value of the no-load current will be in inverse proportion to the ratio of the turns. This no-load loss actually comprises the iron loss including stray losses due to the exciting current, the dielectric loss and the I^2R loss due to the exciting current.

In practice the loss due to the resistance of the windings may be neglected.

It is sometimes more convenient to measure the iron loss by the three-wattmeter method, particularly when the l.v. voltages are of a high order. In all cases low power factor wattmeters must be used.

If three separate wattmeters are available the method of connection is, of course, obvious, but if only one is available a possible method of connection is shown in Fig. 13.7.

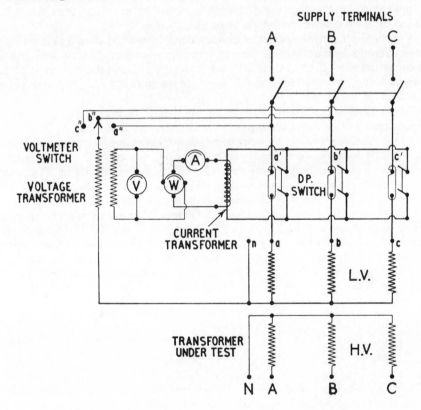

Fig. 13.7   *Iron-loss and no-load current test; three-wattmeter method*

The test is conducted as follows: the double pole switch a′ is closed and the link opened, switches b′ and c′ being open with their corresponding links closed. The voltmeter switch is put on to the contact a″. The supply voltage is adjusted until the voltmeter reads the correct phase voltage. The frequency being adjusted to the correct value, ammeter, voltmeter and wattmeter readings are taken.

The link on the double pole switch a′ is then closed and the switch opened. Switch b′ is closed and the corresponding link opened, while the voltmeter switch is moved to contact b″. Any slight adjustment of the voltage that may be necessary should be made and the meter readings again noted. This operation is again repeated for phase C.

The algebraic sum of the three wattmeter readings will then give the total iron loss.

The B.S. tolerance on iron loss is $\frac{1}{7}$ but iron losses plus copper losses in total must not exceed $\frac{1}{10}$ of the guaranteed value.*

It is essential that the supply voltage waveform is approximately sinusoidal and that the test is carried out at the rated frequency of the transformer under test.

For normal transformers, except three-phase transformers without a delta-connected winding, the voltage should be set by an instrument actuated by the mean value of the voltage wave between lines but scaled to read the r.m.s. value of the sinusoidal wave having the same mean value.

For three-phase transformers without a delta-connected winding the no-load losses should be measured at the r.m.s. voltage indicated by a normal instrument actuated by the r.m.s. value of the voltage wave, and the waveform of the supply voltage between lines should not contain more than 5 % as a sum of the 5th and 7th harmonics.

In all cases when testing iron losses, the rating of the alternator must be considerably in excess of the input to the transformer under test.

In the *routine* testing of transformers it is not necessary to separate the components of hysteresis and eddy-current loss of the magnetic circuit, but for research or investigational purposes or for any iron-loss correction, which may be necessary on account of non-sinusoidal applied voltage, such procedure may be required. The losses may be separated graphically or by calculation, making use of test results at various frequencies. Generally, loss tests at a minimum of three frequencies, say, at 25, 50 and 60 Hz, are sufficient for the purpose but additional tests provide confirmatory data. All the tests are carried out in the standard manner already indicated, but at the required frequencies and at a constant *flux density*, the value of the latter usually being that corresponding to the normal excitation condition of the transformer. The two methods are then as follows:

(a) Graphical

This method is illustrated by Fig. 13.8. The measured losses are converted into total energy loss per cycle by dividing the total power by the frequency, and the results are then plotted against the respective frequencies. The resulting graph should be a straight line, intercepting the vertical axis as shown. The ratio of the ordinate value at the vertical axis (*i.e.*, at zero frequency) to the ordinate value at any other frequency gives the ratio of hysteresis loss per cycle to total measured iron loss per cycle at that frequency, and the hysteresis loss per cycle in watts can then be determined. At power frequencies and normal flux densities cold rolled steel has a ratio of hysteresis to eddy-current loss of unity.

---

* Refer to Chapter 4 for details of B.S.171 tolerances.

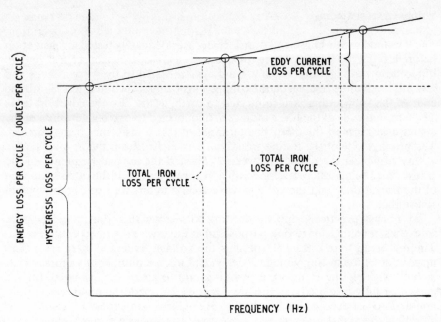

Fig. 13.8   *Graphical method of separating hysteresis and eddy-current losses*

(b) *By calculation*

It is well known that while the hysteresis loss is a constant loss per cycle, and that, therefore, the hysteresis loss at any given frequency varies directly as the frequency, the eddy-current loss varies as the square of the frequency. If the total iron loss be measured at two different frequencies and at such applied voltages that the flux density is constant, it is a comparatively simple matter to separate the total iron loss into its constituent parts. The relationship between total iron loss P_f and frequency f may be written as

$$P_f = fP_h + f^2 P_e \qquad (13.1)$$

where P_h and P_e are the hysteresis loss and eddy-current loss respectively. Now, suppose iron-loss tests have been made at 25 Hz and 50 Hz and that the total iron losses are P_{f25} and P_{f50} respectively. Then, from eqn. 13.1

$$P_{f50} = 50P_h + 2500P_e \qquad (13.2)$$
$$P_{f25} = 25P_h + 625P_e \qquad (13.3)$$

Multiplying eqn. 13.3 by 2 and subtracting the result from eqn. 13.2 eliminates P_h and so enables P_e to be determined, as then

$$P_e = \frac{P_{f50} - 2P_{f25}}{1250} \qquad (13.4)$$

Substitution of the value of P_e in eqns. 13.2 or 13.3 then enables P_h to be determined at the relevant frequency.

281

Voltage tests—windings

The h.v. and l.v. windings of all transformers are adequately tested for insulation before leaving the factory. These tests consist of (a) separate source test, and (b) induced overvoltage test. The h.v. windings are tested to the l.v. windings and any other windings and earth, the connections being shown in Fig. 13.9. All the ends of the l.v. windings are connected to earth, as are also the core and tank. The terminal ends of the h.v. windings are connected to one h.v. terminal of the testing transformer, the other h.v. terminal of this transformer being earthed. The primary side of the testing transformer is excited from the normal supply at any convenient frequency between 25 Hz and the normal frequency of the transformer being tested. It is preferable to apply this test at the rated frequency of the transformer and the voltage wave should in any case be approximately sinusoidal.

It is normal practice to apply to the transformer windings during the separate source test a voltage for testing purposes of at least twice the rated voltage of the winding being tested. B.S.171 specifies the voltage levels for these tests. The magnitude of the testing voltage is dependent upon a number of factors, which include whether the transformer windings are (i) air or oil insulated, (ii) are graded or fully insulated and (iii) are designed for operation on an electrically non-exposed or exposed system. In the case of (iii) a non-exposed system is one in which the transformer is not subject to atmospheric surge voltages and generally transformers connected to cable networks fall in this category. Those, however, which are connected to overhead transmission lines are considered to be electrically exposed installations. For a more detailed explanation of the

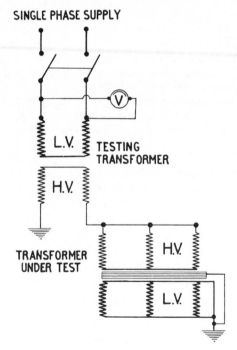

SINGLE PHASE SUPPLY

L.V.

H.V.

TESTING TRANSFORMER

TRANSFORMER UNDER TEST

H.V.

L.V.

Fig. 13.9   Voltage tests: (a) *Separate source test*

considerations which affect the selection of the transformer insulation level, and hence the test voltage, reference should be made to the appropriate Appendix of B.S.171. The following tables indicate the power-frequency test voltages based upon these considerations.

Table 13.1
ONE MINUTE TEST: DRY-TYPE
TRANSFORMERS

System highest voltage, kV, r.m.s.	Power frequency test voltage, kV, r.m.s.
Less than 1·1	2·5
1·1	3
3·6	8
7·2	15
12·0	25
17·5	36

The standard test levels given in Tables 13.1 to 13.3 apply to installations at altitudes up to and including 1000 metres above sea level. For operation at altitudes between 1000 metres and 3000 metres above sea level, the separate source test voltage should be increased by 1·25% for each 100 metres in excess of 1000 metres above sea level, above that for tests made at an altitude up to 1000 metres for dry type AN and AF transformers.

This test has the effect of raising the whole of the h.v. windings of the transformer under test to the predetermined test voltage above earth. If V equals the voltage applied across the primary of the testing transformer and n equals the testing transformer ratio, then the test voltage is given by nV. The l.v. windings are tested to earth by connecting the terminal ends of the windings to one of the h.v. terminals of the testing transformer in a similar manner to that outlined for the h.v. windings. In this case the core and one end of the h.v. windings are earthed. The magnitude of the test voltage is obtained in the same way as in the case of the h.v. windings. In the case of transformers having large electrostatic capacitance, the peak value of the test voltage is determined either by means of

Table 13.2
ONE MINUTE TEST: OIL IMMERSED
TRANSFORMERS ELECTRICALLY
NON-EXPOSED INSTALLATIONS*

System highest voltage, kV, r.m.s.	Power frequency test voltage, kV, r.m.s.
Less than 1·1	2·5
1·1	3
3·6	8
7·2	15
12·0	25
17·5	36

*Non impulse tested transformers.

Table 13.3
ONE MINUTE TEST: OIL IMMERSED TRANSFORMERS
ELECTRICALLY EXPOSED INSTALLATIONS*

System highest voltage, kV, r.m.s.	Power frequency test voltage kV, r.m.s.	Insulation to earth
3·6	16	
7·2	22	
12·0	28	
17·5	38	
24	50	uniform
36	70	
52	95	
72·5	140	
100	150	
123	185	
145	230	
170	275	
245	395	graded
300	460	
362	510	
420	630	

*See Chapter 15, Impulse testing of transformers.

a carefully calibrated sphere gap mounted on the testing transformer or by an electrostatic voltmeter.

If for special reasons the full voltage cannot be applied then a reduced voltage may be applied for longer time periods as follows:

Table 13.4
VALUE OF INSULATION TEST VOLTAGE WITH REFERENCE TO
DURATION OF TEST

Duration of test in multiples of standard period	Per cent of standard test voltage	
	Test at works	Test on site
1	100	75
2	83	70
3	75	66
4	70	62
5	66	60
10	60	54
15	57	50

In the case of transformers which have been in service, the transformer should not be recommissioned after repair until it has passed separate source and induced high-voltage tests equivalent to 75 % of the original test levels. If direct current is used for h.v. testing, the crest value of the rectified alternating voltage employed should not exceed the crest value specified for an alternating current test voltage.

Figure 13.10 shows the diagram of connections for carrying out the induced overvoltage test, which is made by supplying the specified test voltage to the l.v. windings through an h.v. testing transformer and at a frequency higher than the normal value for which the transformer under test is designed. Power transformers are subjected to an induced overvoltage test for one minute equal in value to twice the normal rated voltage of the h.v. windings. During this test the supply frequency should be increased by at least 100% above the normal value in order to avoid an excessive exciting current. The h.v. windings are left open-circuited, the test voltage being applied to the l.v. windings. The voltage is measured on the l.v. side of the transformer under test, either directly or by

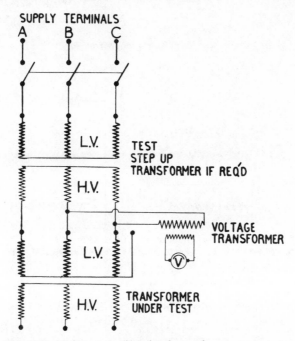

Fig. 13.10   Voltage tests: (b) Induced overvoltage test

means of a voltage transformer. If V equals the voltage measured across the secondary of the voltage transformer, and n_1 the ratio of the voltage transformer, then the value of the induced voltage test on the l.v. windings equals $n_1 V$, and the value on the h.v. side of the transformer under test equals $n_1 n_2 V$ where n_2 equals the ratio of the transformer being tested.

The connections given and procedure outlined for all the voltage tests are exactly the same for single-phase and three-phase transformers whatever the interphase connections of the windings. These tests are, of course, carried out with the transformer in its tank and immersed in oil, and complete with all fittings.

During this test, for which the supply should be obtained from a voltage source of approximately sinusoidal waveform, care must be taken to avoid

SINGLE PHASE TESTS ON THREE PHASE TRANSFORMERS WITH FULLY GRADED INSULATION DESIGNED FOR OPERATION WITH SOLIDLY EARTHED NEUTRAL POINTS

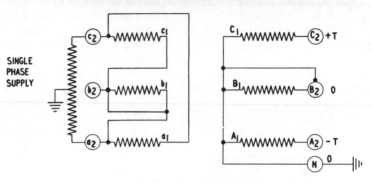

(a) TEST ON THE TWO OUTER LIMBS

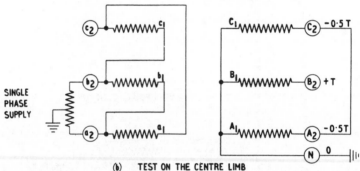

(b) TEST ON THE CENTRE LIMB

TRANSFORMERS WITH PARTIALLY GRADED INSULATION PERMITTING NEUTRAL POINT EARTH CONNEXION TO BE REMOVED DURING TEST

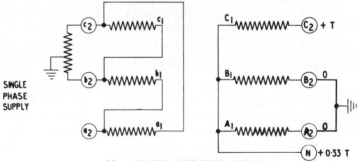

(c) CONNEXIONS FOR TEST ON ONE LIMB C
(TEST TO BE REPEATED ON THE OTHER LIMBS A & B
USING THE APPROPRIATE CONNEXIONS)

NOTE:- "T" DENOTES THE TEST VOLTAGE BETWEEN LINE TERMINAL & EARTH

Fig. 13.11   Induced overvoltage tests

286

abnormal exciting currents and also to ensure that excessive voltages do not occur across the windings. Any winding not having graded insulation may be earthed at any convenient point during the test. Any winding having graded insulation should be earthed at such a point as will ensure the required test voltage arising between each line terminal and earth, the test being repeated under other earthing conditions when this is necessary, to ensure the application of the specified test voltage to every relevant terminal.

The r.m.s. value of the test voltage can be measured directly by a suitable voltmeter; alternatively the actual crest value of the voltage induced in the higher-voltage winding can be determined by a sphere gap or by an electrostatic voltmeter.

The test should be commenced at a voltage of approximately one third of the test voltage and increased to the specified value as rapidly as is consistent with the magnitude being indicated by the measuring instrument. At the end of the test the test voltage should be reduced rapidly to approximately one third of the full value before switching off.

The standard duration of the induced voltage test is 60 s for any test frequency up to and including twice the rated frequency. When the test frequency exceeds twice the rated frequency the duration of the test should be equal to

$$60 \left(\frac{2 \times \text{rated frequency}}{\text{test frequency}} \right), \text{seconds}$$

but in no case should the duration of the test be less than 15 s.

In the case of polyphase transformers, especially three-phase high-voltage units, it is permissible to apply the test voltage to individual phases in succession. Figure 13.11 shows the connections for this test.

The induced-voltage test on series parallel windings should be made with the windings connected in series and repeated in parallel.

Core insulation voltage test

The insulation on all limb and yoke bolts and between clamping plates and core laminations undergoes a routine separate source test of 2000 V.

Temperature rise test

When a transformer is provided with a standard tank it is not necessary to make a test for temperature rise, as in such cases the tank dissipation constant is known beforehand and consequently it is only necessary to measure the transformer losses and to calculate the temperature rise of the oil and the windings on continuous full load. If however, the tank is non-standard, it may be necessary to carry out a temperature rise test on the transformer, and the various methods of conducting this test are as follows:

(a) short-circuit equivalent test
(b) back to back test
(c) delta/delta test
(d) open circuit test.

Method (a)

The general procedure under this method is as follows: One winding of the transformer is short circuited and a voltage applied to the other winding of such a value that the power input is equal to the total normal full-load losses of the transformer at the temperature corresponding to continuous full load. Hence it is necessary first of all to measure the iron and copper losses as described earlier in this chapter. As these measurements are generally taken with the transformer at ambient temperature, the next step is to calculate the value of the copper loss at the temperature corresponding to continuous full load. Assuming the copper loss has been measured at 15°C, the copper loss at the continuous full-load temperature will be equal to the measured copper loss increased by a *percentage* equal to 0·4 times the anticipated temperature rise. This calculation assumes the copper loss varies directly as the resistance of the windings. This is not quite true, however, since a portion of the copper loss consists of eddy-current loss, and this portion will decrease as the resistance of the windings increases. The inaccuracy is slight, however, and has the advantage that it tends to increase the power supplied and consequently to shorten the test. Before commencing the test it is desirable to calculate also the approximate current required in order to avoid an excessive current density. At the commencement of the test this will be given by

$$\text{normal current} \times \sqrt{\left(\frac{\text{iron loss} + \text{hot copper loss}}{\text{cold copper loss}}\right)}$$

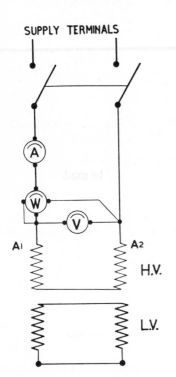

SUPPLY TERMINALS

Fig. 13.12   Single-phase short-circuit equivalent

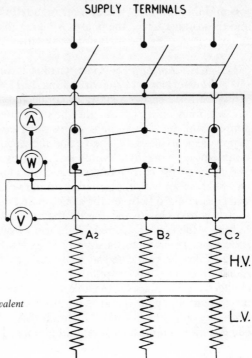

SUPPLY TERMINALS

A₂ B₂ C₂

H.V.

L.V.

Fig. 13.13   Three-phase short-circuit equivalent

and at the end of the test by

$$\text{normal current} \times \sqrt{\left(1 + \frac{\text{iron loss}}{\text{hot copper loss}}\right)}$$

However, to ensure greater accuracy, the test is made by measuring the power input, which is finally increased to include the hot copper loss, though the current obtained by the above calculation indicates how much the winding will be over-loaded from the current density point of view. In general it will be seen that this test is most suitable when the copper loss is high compared with the iron loss, and conversely discretion is needed when dealing with transformers having relatively high iron losses.

When the normal temperature rise is approached the copper loss should be measured and any necessary current adjustment should then be made in order to correct the power input to obtain the true losses under normal full-load conditions, *i.e.*, as regards current and temperature rise.

The short-circuit equivalent test should not be adopted when the ratio of copper loss to iron loss is less than two to one; for loss ratios below the figure mentioned the open-circuit test is preferable.

Single-phase transformers The l.v. winding is short-circuited and the h.v. winding connected to a single-phase supply with an ammeter, voltmeter and wattmeter in circuit, as shown in Fig. 13.12. The current in the h.v. winding is adjusted until the power input is equal to the sum of the calculated hot copper loss and the iron loss. The current required is in excess of the full-load current, and the voltage across the phases is higher than the impedance voltage in order to compensate for the inclusion of the iron loss with the copper loss.

Three-phase transformers   The various means of utilising this test for three-phase transformers are shown in Figs. 13.13, 13.14 and 13.15.

Figure 13.13 shows a star/star connected transformer ready for the test, the h.v. windings of the transformer being connected to a low-voltage three-phase supply, and the l.v. windings being short-circuited. Links are provided in the supply leads to phases A and C, and the various instruments are connected to a double-pole changeover switch such that by closing the switch in either phase and opening the corresponding link, the ammeter and wattmeter current coil will be in series with that phase, and the voltmeter and wattmeter voltage coil will be connected between the same phase and phase B. The three-phase supply switch is first closed and the double-pole switch then closed in phase A, the link in A then being opened. The supply voltage is increased until the current shown by the ammeter is slightly in excess of the full-load h.v. current. This current may be calculated as previously explained. The wattmeter reading is then noted. The link in phase A is next closed and the double-pole switch changed over to phase C, the link in this phase being then opened, and the wattmeter reading again noted. This process is repeated until, after making the necessary adjustments, the algebraic sum of the two wattmeter readings is equal to the sum of the iron and hot copper losses.

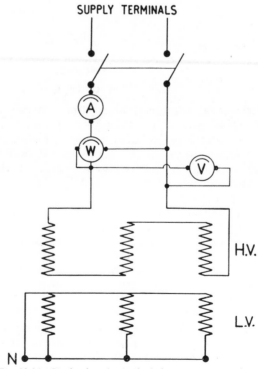

Fig. 13.14   Single-phase 'series h.v.' short-circuit equivalent

SUPPLY TERMINALS

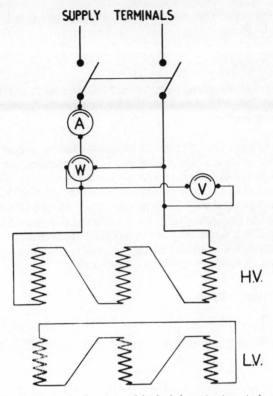

Fig. 13.15 Single-phase 'open delta h.v.' short-circuit equivalent

Figure 13.14 shows an alternative method of connecting up a star/star transformer for test. The l.v. windings in this case are short-circuited through the
neutral, the h.v. being temporarily 'series' connected. The two open ends of the
h.v. windings are then connected to a single-phase supply through a wattmeter
and ammeter. The current is adjusted until the power input is equal to the sum
of the iron and hot copper losses. This current is somewhat higher than the
normal full-load line current if the transformer is normally star connected, and
somewhat higher than the normal full-load line current divided by $\sqrt{3}$ if the
transformer is normally delta connected. The corresponding value of the applied
single-phase voltage required will be somewhat higher than three times the
transformer impedance voltage per phase.

Figure 13.15 shows a further method in which the l.v. windings are connected
in closed delta, and the h.v. in 'open delta'.* This method is applicable to any
three-phase transformer whatever the normal interphase connections, and
temporary connections are made as necessary. The test should be confined to
transformers of low and medium impedances, however, and it should not be
used for transformers of high impedance. For the latter the short-circuit
equivalent test illustrated by Fig. 13.14 is recommended. The h.v. windings are
connected to a single-phase supply, and the same procedure as described for

*This must not be confused with the so-called open delta or vee connection for giving a three-phase
supply from two single-phase transformers.

Fig. 13.14 is followed. The current and voltage required will be the same as given for Fig. 13.14.

Method (b)

In this method, known as the back to back (or Sumpner) test, the transformer is excited at normal voltage and the full-load current is circulated by means of an auxiliary transformer.

Single-phase transformers   Figure 13.16 shows the method of connection for single-phase transformers. The transformers (two identical units are required) are placed not less than 1 m apart with the h.v. sides adjacent. The h.v. windings are then connected in opposition through an ammeter. The l.v. winding of one transformer is connected to a single-phase supply, and the other is connected in parallel with it, but the l.v. winding of a suitable auxiliary transformer is included in this circuit. The h.v. winding of the auxiliary transformer is either supplied from a separate source as shown in Fig. 13.16 or is placed in parallel across the other mains with a variable resistor in series with it.

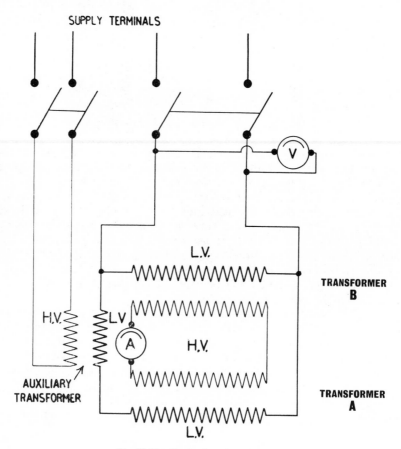

Fig. 13.16   Single-phase back to back

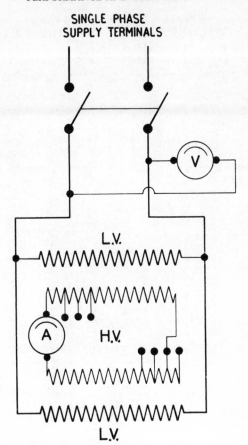

Fig. 13.17 *Single-phase back to back*

Normal l.v. voltage at the correct frequency is then applied to the l.v. windings in parallel, and the supply voltage to the h.v. winding of the auxiliary transformer is adjusted at correct frequency until the ammeter in the h.v. circuit of the transformer under test reads the normal full-load current. If the variable resistor connection is used for the auxiliary transformer, its resistance is adjusted until the ammeter in the h.v. circuit of the transformer under test indicates the normal h.v. full-load current.

It should be noted that in this method no wattmeter is used, as the actual full-load conditions, *i.e.*, normal excitation and full-load current, are reproduced. The copper and iron losses must therefore be those which would normally occur, and there is consequently no need to measure them during this test.

The machine supplying the l.v. windings in parallel must be capable of giving the *normal* l.v. voltage of the transformer under test and twice the no-load current, and it is this circuit that supplies the iron losses.

The l.v. winding of the auxiliary transformer must supply twice the impedance voltage of the transformer under test at the normal l.v. full-load current, and when the method shown in Fig. 13.16 is used, the machine supplying the auxiliary

293

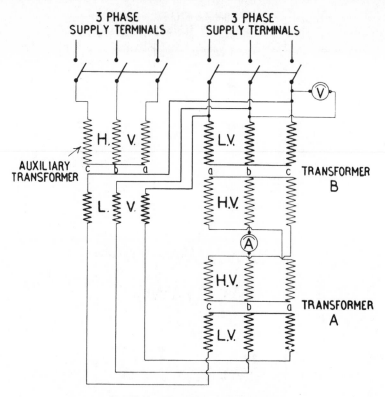

Fig. 13.18   Three-phase back to back

transformer must be capable of giving a voltage equal to the ratio of transformation of the auxiliary transformer multiplied by twice the impedance voltage of the transformer under test, and a current equal to the l.v. current of the transformer under test divided by the ratio of transformation of the auxiliary transformer. This circuit supplies the copper losses to the transformers under test.

There is a further method of making a back to back test on two similar single-phase transformers which is possible when the transformers are provided with suitable tappings. The transformers are connected as shown in Fig. 13.17 which is similar to the previous method except that the auxiliary transformer is omitted and the current circulation is obtained by cutting out a portion of the h.v. winding of one of the transformers. It will be evident that the percentage difference between the numbers of turns in the two h.v. windings should be approximately equal to the sum of the percentage impedances. For example, if the transformers are provided with plus and minus 2·5 and 5% tappings and the impedance of each is 3·75%, this test could be made by using the plus 5% tapping on one transformer and the minus 2·5% tapping on the other transformer. An ammeter is connected in the h.v. side, as in the previous test, and the supply to the l.v. windings in parallel is given at the normal voltage and frequency. If it is found that with the best available tappings the ammeter does not indicate exactly the correct full-load h.v. current, the supply voltage may be varied slightly up or down and the power input adjusted as already described for method (a), i.e., the short-circuit equivalent test. When it is necessary to raise the supply voltage above normal in order to obtain the correct power input, it is evident that the transformers have a greater iron loss and lower copper loss than would be the case under normal full loading and excitation. The converse, of course, holds true when it is necessary to lower the supply voltage below normal in order to obtain the correct power input.

It should be noted that the tappings are assumed to be on the h.v. winding as this arrangement is more common, but the test may be made equally well if the tappings are on the l.v. winding.

Three-phase transformers   The diagram of connections for the test on three-phase transformers is shown in Fig. 13.18 which corresponds to Fig. 13.16 for single-phase transformers. The diagram shows two star/star connected trans-formers, but the external connections are the same for any other combination of interphase connections. The ammeter on the h.v. side of the transformers under test is, for the sake of simplicity, shown permanently connected in the middle phase, but it would actually be arranged for connecting in any phase by means of changeover switches. The same remark applies also to the voltmeter across the supply. The method of procedure is the same as described for single-phase transformers connected as in Fig. 13.16.

Figure 13.19 indicates the connections for two star/star transformers, using the voltage adjusting tapping method, though these would be the same irrespec-tive of the normal interphase connections, temporary connections being made

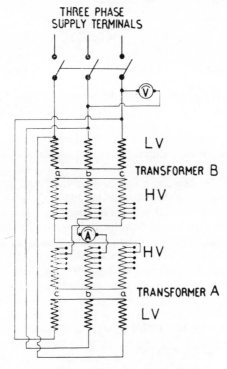

Fig. 13.19   Three-phase back to back

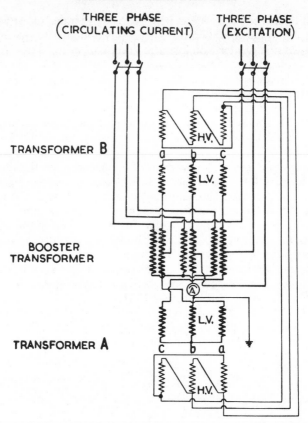

THREE PHASE
(CIRCULATING CURRENT)

THREE PHASE
(EXCITATION)

TRANSFORMER B

H.V.

a   b   c

L.V.

BOOSTER
TRANSFORMER

LV.

TRANSFORMER A

c   b   a

H.V.

Fig. 13.20   Three-phase back to back, employing three-phase excitation and current circulation

as desired. The general procedure is identical with that outlined for the single-phase transformers shown in Fig. 13.17. The l.v. windings of the two transformers are connected in parallel and excited at the normal voltage while the h.v. windings are connected in opposition, but at the same time suitable tappings are selected to give the voltage difference necessary to provide the circulating full-load current.

When these methods of testing are used it will be found that one transformer has a temperature rise higher than that of the other. This is due to the fact that the copper loss is supplied by means of a common circulating current, whereas the iron loss is supplied to the two transformers in parallel. The no-load current is out of phase with the circulating current, but not actually in quadrature with it, and consequently the phasor sum of the no-load and circulating currents in one l.v. winding is greater than the corresponding sum in the other l.v. winding. The difference is not great but may be difficult to understand without this explanation.

The back to back tests illustrated by Figs. 13.16 to 13.19 inclusive may, of course, be applied to delta/star and to star/interconnected star transformers.

Two alternative forms of three-phase, back to back temperature rise tests are illustrated in Figs. 13.20 and 13.21. The arrangement shown in Fig. 13.20 may be applied to three-phase transformers of any type, of any combination of primary and secondary connections, and of any impedance, it only being necessary that the two transformers under test are identical. As shown in the diagram, an auxiliary booster transformer is used for providing the circulating current passing through the windings of the transformers under test, and the normal excitation supply is applied to the centre points of the secondary winding of the booster transformer. In the event of no centre points being accessible on the booster windings, the normal excitation supply may be applied to the terminals of either transformer, in which case one transformer would have a slightly lower voltage across its terminals than the other, due to the impedance drop in the secondary windings of the booster transformer. Where the normal excitation is applied to the centre points of the booster transformer, the supply voltage should be slightly higher than the rated voltage of the transformers under test in order to compensate for the impedance drop in the secondary winding of the booster transformer.

The copper losses are supplied from the three-phase source which provides the necessary circulating currents via the primary windings of the booster

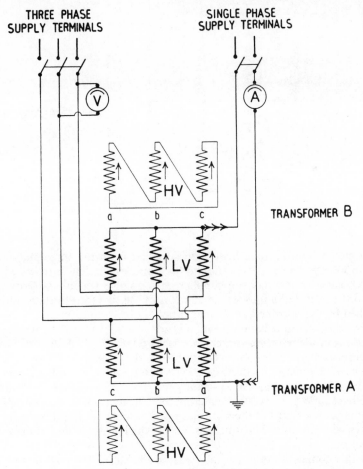

Fig. 13.21(a)   *Three-phase back to back on two delta/star connected transformers*

THREE PHASE
SUPPLY TERMINALS

SINGLE PHASE
SUPPLY TERMINALS

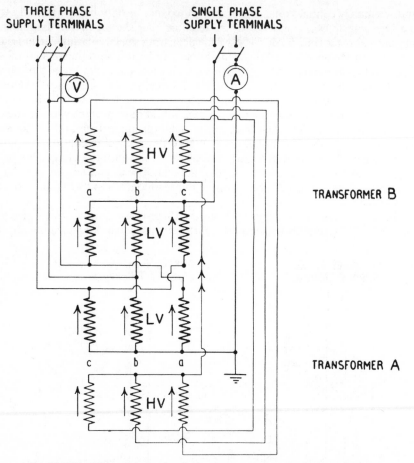

Fig. 13.21(b)   *Three-phase back to back on two star/star connected transformers*

transformer, while the iron losses are supplied from the three-phase source which supplies the normal excitation to the transformers. The primary windings of the booster transformer are supplied at a voltage which is approximately equal to the sum of the impedance voltages of the two transformers under test multiplied by the booster transformer ratio.

This method has an advantage that it is not necessary to make any temporary connections inside the transformers, nor is it necessary to reinforce any connections temporarily to carry any special heavy test currents.

Figures 13.21(*a*) and (*b*) illustrate a type of test which is applicable to three-phase delta/star and star/star transformers. The l.v. windings are connected back to back, and current is circulated in them from a single-phase supply. The l.v. windings are excited at their normal rated voltage, so that this method also simulates very closely the heating conditions which arise in the ordinary course of operation.

With this connection the neutral leads on the star sides must be reinforced to carry three times the normal full-load current. Circulating current is supplied

298

from a single-phase source, so that the currents in all three limbs are equal and in phase. The leakage flux between windings returns partly through the tank walls, and for this reason the method indicated by Fig. 13.21 should not be used for transformers where the impedance exceeds 5%. Otherwise it is quite a satisfactory method of conducting a load test and one which is frequently used.

Method (c)

This method, known as the delta/delta test, is applicable to single- as well as three-phase transformers where the single-phase transformers can be connected up as a three-phase group.

Figure 13.22 shows the diagram of connections often employed. The l.v. windings are connected in closed delta, and supplied from a three-phase source. The h.v. windings are connected in open delta* and include an ammeter. Voltmeters are connected between phases in the l.v. circuit. Three-phase voltage at the correct frequency is applied to the l.v. windings and is adjusted until it equals the normal l.v. voltage. Single-phase current is supplied separately to the h.v. windings and is adjusted to the normal h.v. full-load current.

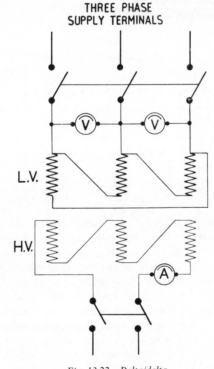

Fig. 13.22   *Delta/delta*

*This must not be confused with the so-called open delta or vee connection for giving a three-phase supply from two single-phase transformers.

299

This method may be used whatever the normal internal connections of the transformer, temporary connections being made if necessary. The voltages and currents required under this test for various normal interphase connections are given in Tables 13.5 and 13.6.

Table 13.5

Application of voltage or current	L.V. connection	
	Delta	Star
Voltage applied to l.v.	V	$V/\sqrt{3}$
Current applied to l.v.	I_0	$I_0 \times \sqrt{3}$

V = normal line voltage
I_0 = normal no-load current.

Table 13.6

Application of voltage or current	H.V. connection	
	Delta	Star
Voltage applied to h.v.	$V_2 \times 3$	$V_2 \times \sqrt{3}$
Current applied to h.v.	$I/\sqrt{3}$	I

V_2 = h.v. impedance voltage
I = normal line current.

If the normal h.v. voltage is of the order of 11 000 V and above, the method shown in Fig. 13.23 is safest. In this method the h.v. winding is simply closed delta connected, the l.v. being connected in 'open delta'*. A three-phase voltage equal to the normal l.v. *phase* voltage is applied to the l.v. winding at the correct frequency, and the l.v. copper loss current is supplied single-phase.

Method (d)

Occasionally it may happen that a transformer possesses a high iron loss as compared with its copper loss, and when the ratio of copper loss to iron loss is low it is generally impossible to conduct a temperature rise test by the short circuit method. This is on account of the fact that the required power input necessitates an excessive current in the windings on the supply side of the transformer so that a prohibitively high current density would be reached. In such cases it may be possible to test the transformer on open circuit, the normal losses being dissipated in the iron circuit. If a supply at a frequency considerably below the normal rated frequency of the transformer is available, a condition may be obtained whereby the total losses are dissipated at a test voltage and current in the neighbourhood of the normal rated voltage and current of the

---

*This must not be confused with the so-called open delta or vee connection for giving a three-phase supply from two single-phase transformers.

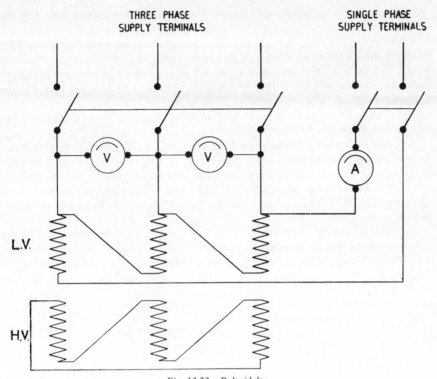

Fig. 13.23   Delta/delta

transformer. If, however, a lower frequency supply is not available, the transformer may be run at the normal rated frequency with a supply voltage greater than the normal rated voltage, and of such a value that the total losses are dissipated in the iron circuit. Assuming that the iron loss varies as the square of the voltage, the required voltage under these conditions is given by the formula:

$$\text{normal voltage} \sqrt{\left(1 + \frac{1 \cdot 2 \times \text{cold copper loss}}{\text{normal iron loss}}\right)}$$

Either side of the transformer may be supplied according to which is the more convenient. The method can be applied to both single-phase and polyphase transformers.

It is, of course, important that instruments connected in h.v. circuits should be earthed; alternatively voltmeters and ammeters should be operated through voltage and current transformers respectively.

Temperature readings

The oil temperature of the transformer under test is measured by means of a thermometer so placed that its bulb is immersed just below the upper surface of the oil in the transformer tank.

When bulb thermometers are employed in places where there is a varying magnetic field, those containing alcohol should be employed in preference to the mercury type, in which eddy currents may produce sufficient heat to yield misleading results.

When measuring the temperature of a surface such as a core or a winding, the bulb should be surrounded by a single wrapping of tin foil having a thickness of not less than 0·025 mm. The foil should be turned up at one end to form a complete covering for the bulb, which should then be secured in contact with the surface under test. The exposed part of the wrapped bulb should be completely covered with a pad of heat insulating material without unduly shielding the test surface from normal cooling.

The cooling air temperature should be measured by means of several thermometers placed at different points around and at a level approximately midway up the transformers and at a distance away of 1 to 2 metres, protected from draughts and abnormal heat radiation. The value to be employed for the cooling air temperature during the test should be the mean of three thermometers placed as described earlier.

In order to avoid errors due to the time lag between the variations in the temperature of the transformer and that of the cooling air, precautions must be taken to reduce these variations. Thus the thermometers for determining the cooling air temperature may be immersed in a suitable liquid, such as oil in a heavy metal cup which should be a cylinder approximately 75 mm in diameter and approximately 140 mm long. The cup should be full of oil which should completely cover the thermometer bulb.

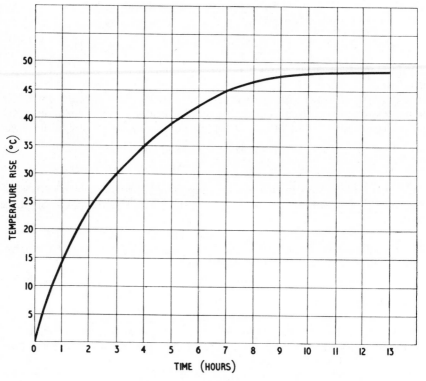

Fig. 13.24   Typical time/temperature rise curve used in conjunction with Fig. 13.25

The temperature test of a transformer should be of such duration that sufficient evidence is available to show that the temperature rise would not exceed the guaranteed limits if the test were prolonged until steady final temperature conditions were reached. One way of determining this is by taking readings of the top oil temperature at regular intervals and plotting a curve on linear co-ordinate paper between the temperature rise and the rate of change during the subsequent interval. Extrapolating to zero rate of change by the best straight line through the plotted points will give the ultimate temperature rise. Figure 13.24 illustrates a typical time/temperature rise curve obtained from the test certificate record of Fig. 13.25. Alternatively, the temperature test may be continued until the temperature rise does not exceed 1°C per hour.

In addition to ascertaining the temperature rise of the oil, it is usual to calculate the temperature rise of the windings from measurements of the increase of resistance. To do this, it is necessary to measure the resistance of the windings before the test (R_c) carefully noting the temperature of the windings at the time of taking the reading, and also to measure the resistance (R_h) at the close of the test. Since, over the normal working temperature range, the resistance of copper is directly proportional to its temperature above minus 235°C, the temperature T_h corresponding to the resistance R_h may be obtained from the relationship

$$\frac{R_h}{R_c} = \frac{T_h + 235}{T_c + 235}$$

whence, if the ambient air temperature is T_a, the winding temperature rise

$$= T_h - T_a$$
$$= \frac{R_h}{R_c}(T_c + 235) - 235 - T_a$$

To obtain an accurate value of the temperature rise of the windings the temperature T_c must be the temperature at which the resistance of the windings is R_c. Due care must be taken in the measurement of T_c, particularly in the case of large transformers because even if a transformer is left unenergised for days, the oil temperature usually varies from the top to the bottom of the tank, so that the top oil temperature may differ from the mean temperature of the windings by some degrees.

At the end of the temperature rise test, when the power supply to the transformer is shut off, the temperature of the windings T_h is appreciably higher than the mean temperature of the cooling medium, which is the oil around the coils in the case of oil immersed transformers or the surrounding air in the case of air cooled transformers. Consequently, the windings cool in an exponential manner towards the cooling medium temperature, the thermal time constant of this phase of the cooling being that of the windings only, and of short duration, e.g., 5 to 20 minutes.

Some time interval must elapse between shut down and the measurement of winding resistance, so that the power supply connections may be removed, the Kelvin bridge connected, and the measuring current allowed to become steady. To obtain the winding resistance R_h at shut down as accurately as possible, allowance must be made for the drop in winding resistance during this time.

TRANSFORMER TEST CERTIFICATE

Customer		Date of test	
Customer's Order No.		Our Order No.	
Inspection by		Specification B.S.171	
Rated kVA 500 3-phase 50 Hz		Rating: Continuous. Type: Indoor	

Winding	h.v.	l.v.
Rated line voltage	6435	400
Rated line current	44·9	722
Connection	Delta	Star

B.S.171   Phasor group reference   Dy 11

Diagram of connections No.   Outline drawing No.

Tappings on h.v. wdg. for full capacity, controlled by off-circuit selector

Cooling ONAN   Oil quantity 640 litres   Cold oil depth 1500 mm

No-load voltage ratio 6756-6596-6435-6274-6112/400

Transformer serial No.		54119C		
Winding referred to, at 75°C		6435/400		
No-load loss at normal voltage ratio	watts	996		
Load loss at rated current	watts	7140		
Impedance voltage at rated current	%	4·49		
No-load average line current at normal voltage ratio	%	0·75		
Average resistance per phase: ohms	h.v.	1·605		
	l.v.	0·00204		
Excitation current with	A.phase	—		
ref. to l.v. winding	B.phase	—		
	C.phase	—		
Percentage efficiency	100 % rated current	98·4		
at unity power factor	75 % ,, ,,	98·68		
	50 % ,, ,,	98·90		
	25 % ,, ,,	98·86		
Percentage efficiency	100 % rated current	98·01		
at 0·8 lagging power	75 % ,, ,,	98·35		
factor	50 % ,, ,,	98·63		
	25 % ,, ,,	98·58		
Percentage regulation	1·0 unity p.f.	1·52		
at full load	0·8 lagging p.f.	3·72		
Temperature rise,	°C by thermometer in oil	45		
from type test	by resistance h.v. wdg.	54		
	l.v. ,,	58		
Insulation resistance, megohms to other windings, core, frame and earth at 23°C (2500 volts d.c.)	h.v.	2000		
	l.v.	2000		
Separate-source voltage withstand test from indicated winding to other windings, core, frame and earth for 60 s at 23°C	h.v.	22 kV		
		2,5 kV		
Induced overvoltage withstand test at 100 Hz for 60 s at 23°C		100 %		

CHIEF TESTING ENGINEER   CHIEF DESIGNER

Date of issue

Fig. 13.25

The winding resistance at the instant of shut down R_h may be obtained in different ways as follows:

(a) To the winding temperature obtained from the resistance measurement made after shut down can be added an arbitrary correction such as 1°C per minute of time elapsing between shut down and resistance measurement. For this method to be accurate, this arbitrary correction must be correct, and the single resistance measurement taken on the winding must be free from any inductive effects.

(b) A resistance/time graph may be plotted for the cooling of the winding after shut down and extrapolated back to the instant of shut down. If plotted with linear scales this graph has a marked curvature and the accuracy of the extrapolation is open to question.

(c) The resistance/time graph mentioned in (b) may be plotted so that it forms a straight line which can then be extrapolated accurately. This may be achieved in different ways as follows:

    (i) Resistance is plotted on a linear scale against time from shut down (t) expressed as a fraction of the winding thermal time constant (τ) on an exponential scale.

    Typical results are shown in Fig. 13.26.

    The value of τ may be determined by trial.

    As shown in Fig. 13.26, if the correct value of τ is used, the graph is a straight line as (B).

    If the value of τ used is too low, the graph curves downwards, (A); and similarly if the value chosen is too high the graph curves the other way (C).

    (ii) The measured resistance of the winding minus the resistance at the mean oil temperature is plotted on a logarithmic scale against time on a linear scale.

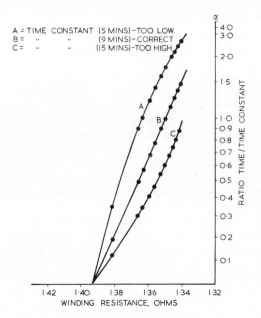

Fig. 13.26 Cooling curve for a 750 kVA transformer winding plotted on linear/exponential scales

305

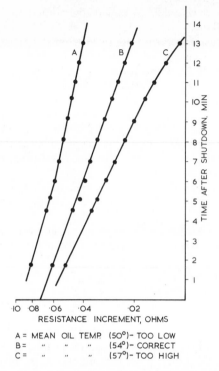

A = MEAN OIL TEMP. (50°)- TOO LOW
B =   „   „   „   (54°)- CORRECT
C =   „   „   „   (57°)- TOO HIGH

Fig. 13.27   Cooling curve for a 750 kVA transformer winding, plotted on log/linear scales

Typical results are shown in Fig. 13.27.

The mean oil temperature is estimated from thermometer readings taken in the top layer of the oil and from measurements of temperature taken at the surface of cooling tubes.

If the estimated value of mean oil temperature be too low the graph curves upwards as (A), if it is too high the curve drops as (C). If estimated correctly the graph is a straight line as (B).

This method is also a means of obtaining the value of mean coolant temperature around the coils and hence of the temperature difference between windings and coolant oil.

The measurement of temperature rise by increase in resistance necessitates the ability to measure small increments of resistance. A Kelvin double bridge has been used and found to be satisfactory, providing the winding resistance is not lower than 0·005 Ω.

In all resistance measurements, care has to be taken that winding inductive effects are reduced to an absolute minimum. The time taken for the measuring current to become constant should be noted during the cold resistance measurement. Later, when taking the hot resistance values, which may be changing quite rapidly due to cooling effects, it is imperative that a similar time is allowed to elapse so that inductive effects are avoided. The inductive effect may last for some minutes and if present will give a higher resistance value than in fact is true. Thus, if a cooling curve is plotted on suitable scales as described earlier in this chapter and is found to be a straight line except for the initial one or two points which lie above the line, these values should be ignored.

Duration of temperature-rise tests

In general, temperature-rise tests last from six to fifteen hours. They may be shortened, if necessary, by overloading the transformer at the commencement of the test and then reverting to full-load losses as the final temperature is approached, but this method should only be adopted in special cases for if insufficient time is allowed for the windings to attain their correct steady temperature, errors will be introduced. As an alternative method it is possible, in the case of separate radiators or coolers, to restrict the normal oil flow and so accelerate the temperature rise of the oil in the early stages of the test. Further information is given in B.S.171 regarding the measurement of oil and winding temperatures at the end of a temperature-rise test.

Test certificate

At each stage in the testing of a transformer the results are recorded on the testing department's records and subsequently these are transferred to an official test certificate for transmission to the customer. Typical test certificates are shown in Figs. 13.25 and 13.28.

Impulse voltage test

The impulse testing of transformers has been standardised as a type test, as a means of ensuring a satisfactory level of insulation and the ability of the transformer to withstand, within certain limits, such overvoltage surges as may occur in service.

It is considered that the technique involved is sufficiently specialised to justify a detailed treatment, and reference should be made to Chapter 15.

Noise level tests

Reference should be made to Chapter 28, for details of transformer noise measurement.

TRANSFORMER TEMPERATURE RISE TEST CERTIFICATE

Customer   Date of test
Customer's Order No.   Our Order No. N.5829
Inspection by   Specification B.S.171
Rated kVA 500   3-phase 50 Hz   Rating: Continuous. Type: Outdoor
Winding   h.v.   l.v.
Rated line voltage   11000   364
Rated line current   26·2   794
Connection   Delta   Star (N. out)
B.S.171   Phasor group reference   Dy11
Diagram of connections No.   Outline drawing No.
Tappings on h.v. wdg. for full capacity, controlled by off-circuit tapping selector.
Cooling ONAN   Oil quantity 800 litres   Cold oil depth 1500 mm
Guaranteed temperature rise: Oil 50°C   Winding 60°C
No-load voltage ratio 11550-11275-11000-10725-10450/364

Transformer Serial No. 54359C.   All temperatures in degrees Celsius.
Type of test: Short circuit in accordance with B.S.171
Supplying total hot losses to h.v. with short circuit on l.v.

No-load loss —	993 watts
Load loss at 75°C —	7076 watts
Total input	8069 watts

Time from start hours	Actual temperatures					Ave. amb-ient	Top oil rise	Line current A	Line voltage V	k watts input
	Top oil			Headers or tubes						
				Top	Bottom					
0	18·5			18·5	18·5	18·5	0	29·2	566	8·07
1	32·5			30·0	21	18·5	14	28·8	554	8·07
2	42			39	26	19	23	28·5	549	8·07
3	49			44·5	28	19·5	29·5	28·2	545	8·07
4	54			49	30	20	34	27·8	541	8·07
5	57			51·5	32·5	20	37	27·8	541	8·07
6	59			54	34	20·5	38·5	27·8	541	8·07
7	60·5			55	35	20·5	40	27·8	541	8·07
8	62			56	36	21	41	27·8	541	8·07
9	63			57	36·5	21·5	41·5	27·8	541	8·07
10	63·5			58	37	21·5	42	27·8	541	8·07
11	64			59	38	21·5	42·5	27·8	541	8·07
12	64			59	38	21·5	42·5	27·8	541	8·07
13	64			59	38	21·5	42·5	27·8	541	8·07
14	64			59	38	21·5	42·5	27·8	541	8·07
Input reduced to rated current of 26·2A for 1 h										
15	63·5			58	37	21·5	42	26·2	500	7·1
Shut down										

Fig. 13.28   Typical transformer temperature rise test certificate

TRANSFORMER TEMPERATURE RISE TEST CERTIFICATE

Temperatures after shut down   *continuation sheet*

Actual temperatures in degrees Celsius

Time min. from shut down	Ave. amb- ient	Top oil	Headers or tubes		
			Top	Bottom	
0	21·5	63·5	58	37	
15	21·5	63·0	57·5	37	

RESISTANCES

H.V. between terminals A8-B8	L.V. between terminals a2-b2
Cold 2·768 Ω at 22°C	Cold 0·002726 Ω at 22°C
Hot 3·303 Ω after 0·95 min	Hot 0·003227 Ω after 1·96 min
Hot 3·273 Ω after 3·05 min	Hot 0·003194 Ω after 4·08 min
Hot 3·248 Ω after 5·14 min	Hot 0·003165 Ω after 6·65 min
Hot 3·222 Ω after 8·03 min	Hot 0·003143 Ω after 9·26 min
Hot 3·205 Ω after 10·5 min	Hot 0·003128 Ω after 11·7 min
Hot 3·190 Ω after 13·18 min	Hot 0·003115 Ω after 14·56 min
Hot Ω after min	Hot Ω after min
Hot Ω after min	Hot Ω after min
Hot Ω after min	Hot Ω after min
Hot Ω after min	Hot Ω after min
Hot Ω after min	Hot Ω after min
Hot Ω after min	Hot Ω after min

Remarks:

RESULTS

Final temperature rises; degrees Celsius

TOP OIL	42·5
H.V. WINDING	53·5
L.V. WINDING	52
HOT SPOT	62·2 (Calculated from test data)

CHIEF TESTING ENGINEER   CHIEF DESIGNER

Chapter 14

Phasor representation of transformer test conditions

While there is nothing intrinsically difficult to understand in the electric and magnetic conditions of a transformer during the various factory tests usually applied, phasor diagrams generally help considerably in visualising what actually is taking place, and those now given show clearly the currents, voltages, and magnetic fluxes present in the transformer under the different test conditions.

The tests illustrated are,
(1) ratio and polarity
(2) load losses and impedance
(3) no-load loss and no-load current
(4) voltage tests: (a) separate source, (b) induced overvoltage
(5) temperature rise tests: (a) short circuit, (b) open circuit, (c) back to back.

In all cases tests on single-phase transformers are dealt with as affording the simplest illustrations of the testing conditions; for polyphase transformers the conditions are generally similar to those for single-phase, and the complete phasor representation of the former can easily be derived from the latter.

The phasor nomenclature used throughout this study is as follows:

$$V_1 \quad \text{primary terminal voltage}$$
$$I_m \quad \text{primary magnetising current}$$
$$I_o \quad \text{primary no-load current}$$
$$I_c \quad \text{primary core loss current}$$
$$\phi_m \quad \text{maximum (peak) value of magnetic flux}$$
$$I_1 X_1 \quad \text{primary reactance voltage drop}$$
$$I_1 R_1 \quad \text{primary resistance voltage drop}$$
$$I_1 Z_1 \quad \text{primary impedance voltage drop}$$
$$E_1 \quad \text{primary induced voltage}$$
$$E_2 \quad \text{secondary induced voltage}$$
$$V_2 \quad \text{secondary terminal voltage}$$
$$I_2 X_2 \quad \text{secondary reactance voltage drop}$$
$$I_2 R_2 \quad \text{secondary resistance voltage drop}$$
$$I_2 Z_2 \quad \text{secondary impedance voltage drop}$$
$$I_2 \quad \text{secondary load current}$$
$$I_2' \quad \text{load component of total primary current}$$

I_1 total primary current (including I_0 and I_2')
$\cos \phi_2$ secondary load power factor
$\cos \phi_1$ primary total load power factor
$I_o R_1$ resistance voltage drop due to I_0
$I_2 Z_e''$ total impedance voltage drop
$I_2 R_e''$ total resistance voltage drop
$I_2 X_e''$ total reactance voltage drop.

It should be noted in the phasor representation of the transformer test conditions, that certain phasors do not represent full-load values and they should not be confused with similar phasors referred to in earlier chapters. For the sake of clarity the same symbols have been employed in this chapter.

RATIO AND POLARITY TEST

The diagram of connections for this test is shown in Fig. 14.1 (*a*), and the phasor relationships of currents, voltages and magnetic flux are shown in Fig. 14.1 (*b*). The ratiometer consists of a single-phase double-wound transformer having a constant primary winding and a variable secondary, the voltage variation being effected through the intermediary of adjusting tappings and two or more dial switches. In carrying out the test the h.v. winding of the transformer under test is connected in parallel with the primary winding of the ratiometer, while the l.v. winding of the transformer is connected to the two terminals of the dial

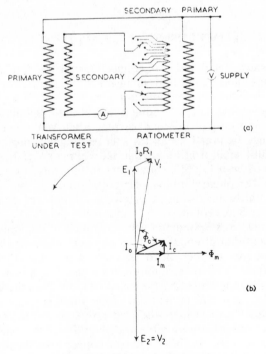

Fig. 14.1 Ratio and polarity test

311

switches through an ammeter, care being taken to ensure that the respective e.m.f.s are in opposition. A single-phase voltage is applied across the primary winding of the ratiometer, and the two dial switches shown in Fig. 14.1 (a) are adjusted until the ammeter registers zero. The voltage across the l.v. winding of the transformer under test is then equal to the voltage across the effective part of the secondary winding of the ratiometer, that is, across that part of the winding between the two dial switch terminals. As the voltage across the h.v. winding of the transformer and across the primary winding of the ratiometer is the same, the ratio of the transformer windings must equal the ratio of the ratiometer, which is known for any setting of the dial switches. The actual voltages are determined by multiplying the different test ratios by the specified transformer l.v. voltage. If tappings are provided on the transformer under test, the test is repeated for each tapping ratio. It will be seen from Fig. 14.1 (b) that all we have present in the transformer under test is the applied primary voltage V_1, the induced primary and secondary e.m.f.s E_1 and E_2, the magnetising flux Φ_m, and the no-load current I_o, giving its magnetising and core-loss components I_m and I_c corresponding to the particular excitation. It should be noted that as this is simply a ratio test the transformer is not necessarily excited at its full voltage, but only at some convenient voltage which it is safe to apply, both to the primary winding of the ratiometer and to the h.v. winding of the transformer under test. The maximum value is usually below 2000 V, but it should not be decreased unnecessarily owing to the reduced sensitivity thereby entailed. The resistance voltage drop in the primary winding is ignored in practice as the error involved is negligible. Similar phasor relationships hold for the alternative ratio tests given in Chapter 13.

LOAD LOSSES AND IMPEDANCE

Figure 14.2 (a) shows the diagram of connections for this test, and Fig. 14.2 (b) the corresponding phasor conditions in the transformer. In this test one winding, usually the l.v., is short circuited and a low voltage first is applied to the h.v. winding. The applied voltage is gradually raised until the ammeter in the h.v. circuit registers the normal full-load current when wattmeter, ammeter and voltmeter readings are noted. The wattmeter reading indicates the total copper loss plus a certain small iron loss, while the voltmeter reading measures the impedance volts. Figure 14.2 (b) shows that in the transformer under test we have primary and secondary full-load currents I_1 and I_2 in opposition, primary and secondary impedance voltage drops, I_1Z_1 and I_2Z_2, each with their resistance and reactance voltage drop components I_1R_1, I_1X_1, I_2R_2 and I_2X_2, and a small flux Φ_m corresponding to the excitation at the impedance voltage. It should be noted that although the phasor diagram shows the primary and secondary impedance voltage drop triangles separately, these are not so measured but are derived by calculation from the design. The total effective impedance drop is represented by the phasor V_1, which is the applied voltage circulating full-load currents in the transformer windings.

In high reactance transformers the iron loss, due to the excitation at impedance voltage, may be considerable and should be deducted from the total short-circuit loss in order to obtain the true copper loss. The diagram of connections and the phasor diagram corresponding to the test for this iron loss

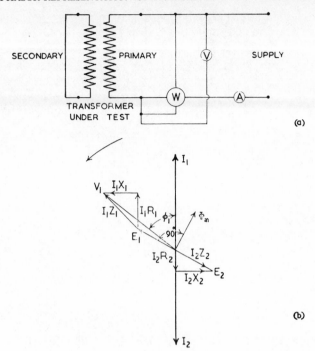

Fig. 14.2   *Load losses and impedance test*

measurement are the same as shown in Figs. 14.3 (*a*) and (*b*). It will be appreciated that the magnitude of the applied voltage to be employed for this supplementary test is the impedance voltage obtained from the load loss and impedance test.

NO-LOAD LOSS AND CURRENT

The diagram of connections for this test is shown in Fig. 14.3 (*a*) and the corresponding phasor conditions in Fig. 14.3 (*b*). The phasor diagram is exactly the same as that for the ratio test, the only difference being that in the iron loss test the phasors correspond to the full excitation, whereas in the ratio test they frequently correspond to a reduced excitation. The resistance voltage drop of the primary winding is generally ignored as it is of negligible magnitude. The ammeter indicates the total no-load current I_o, from which the iron loss current I_c and magnetising current I_m can be calculated by the following formulæ:

$$I_c = \frac{\text{iron loss (watts)}}{V_1}$$

$$I_m = \sqrt{(I_o^2 - I_c^2)}$$

313

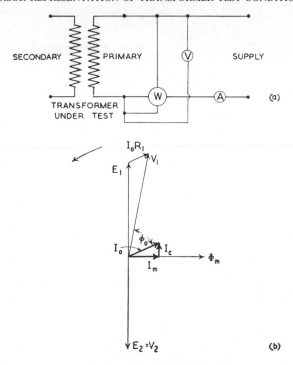

Fig. 14.3   *No-load loss and no-load current test*

VOLTAGE TESTS

(*a*) Separate source

Figure 14.4 (*a*) shows the diagram of connections for the test and Fig. 14.4 (*b*) the corresponding phasor relationships. Each winding is tested to earth and the other winding in turn. Both terminals of the winding under test are connected to one terminal of the high-voltage testing transformer, so that the whole winding is raised to the full test voltage above earth, there being no voltage difference across the terminals of the winding. The transformer core is, of course, not magnetised, and the only currents flowing in the transformer under test are small capacitance currents between the windings and from the winding under test to earth. Due to the electrostatic capacitance between windings and from each winding to earth, a voltage may be induced electrostatically on the secondary winding if the latter be unearthed, as shown by Fig. 14.4 (*c*). The value of this static voltage may be derived from the expression

$$V_i = \frac{V_t}{\dfrac{C_1}{C_2} + I}$$

Where V_t = testing voltage in volts
V_i = static induced voltage in volts

314

C_1 = capacitance between h.v. and l.v. windings
C_2 = capacitance between l.v. winding and earth.

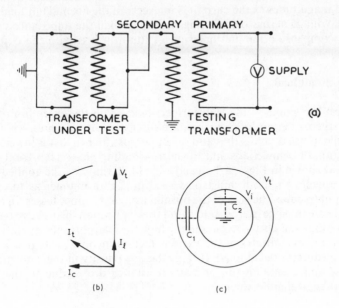

Fig. 14.4   Voltage test. Separate source

(b) Induced overvoltage

The diagram of connections and phasor representations for this test are the same as Figs. 14.3 (a) and 14.3 (b) for the no-load loss test, except that the applied and induced voltage phasors correspond to the value of the test instead of to the normal excitation. The magnetic flux Φ_m and no-load current I_o may vary from normal excitation conditions, although not in the same proportion as the voltage increases, on account of the fact that induced overvoltage tests are conducted at a frequency higher than the normal frequency for which the transformer is designed, in order to keep the no-load current down to a reasonable value such that it can be handled safely by both the testing transformer and the transformer under test. The wattmeter is, of course, omitted in this test.

TEMPERATURE RISE TESTS

(a) Short circuit

The diagram of connections for this test is the same as shown in Fig. 14.2 (a), while the phasor conditions are the same as shown in Fig. 14.2 (b). In this test the whole of the transformer losses, with the exception of the relatively small iron loss at the impedance voltage, is dissipated in the windings. Consequently it should not generally be adopted for transformers where the ratio of copper loss to iron loss is less than about 1·5 owing to the possibility of developing

315

excessive temperatures in the copper. The current input to the transformer is such as to give a wattmeter reading equal to the sum of the iron and copper losses. Due to this the phasors shown in Fig. 14.2 (b) represent quantities of increased magnitudes as the current is in excess of the normal full-load current, while the voltage across the windings is higher than the impedance voltage in order to compensate for the inclusion of the iron loss with the copper loss.

(b) Open circuit load

In this test the whole of the transformer losses, with the exception of a small I^2R loss due to the no-load current, are dissipated in the core. It is therefore inadvisable to use it unless the ratio of copper loss to iron loss is lower than 1·0. The diagram of connections and the corresponding phasor representation are the same as shown in Figs. 14.3 (a) and 14.3 (b). In this case the applied voltage, or the frequency, or both, must be varied in such a manner as to cause the power input to equal the sum of the normal iron and copper losses. In attaining this result care must be exercised to avoid the application of an excessive voltage to the windings and also to avoid unduly high no-load currents. A consideration of the relative magnitudes of the various factors involved will give some idea of the procedure to be adopted. If B_m is the flux density in the core and f is the frequency, and assuming the primary resistance drops, due to the no-load current, to be negligible, then,

$$V_1 \propto B_m f$$

Also, if P_f is the iron loss, $\qquad P_f \propto B_m^2 f$

then $\qquad\qquad\qquad B_m \propto \dfrac{V_1}{f}$

and $\qquad\qquad\qquad P_f \propto \dfrac{V_1^2}{f}$

If, therefore, P_f is to be increased to a value P_{ft} representing the sum of the iron and copper losses, and the new values of V_1 and f are V_{1t} and f_t respectively, then,

$$\frac{P_{ft}}{P_f} = \frac{V_{1t}^2}{f_t}\frac{f}{V_1^2}$$

The applied voltage must not be increased unduly. Assume that an increase of 20% is permissible, then the limiting condition is given by the equation,

$$\frac{P_{ft}}{P_f} = \frac{1·44V_1^2}{f_t}\frac{f}{V_1^2}, \text{ i.e., } f_t = 1·44f\frac{P_f}{P_{ft}}$$

The new value B_m, that is B_{mt}, can be found from the following,

$$\frac{B_{mt}}{B_m} = \frac{V_{1t}}{f_t}\frac{f}{V_1}$$

$$B_{mt} = B_m\frac{1·2V_1}{1·44f}\frac{P_{ft}}{P_f}\frac{f}{V_1} = 0·833B_m\frac{P_{ft}}{P_f}$$

If this value of B_m is such as to give an excessive no-load current, then the method is not applicable. A special case in which this method is very useful is in transformers for use at very high frequencies. These transformers usually have very low values of flux density and low ratios of copper loss to iron loss. With such transformers it is usually possible to conduct an open circuit run without exceeding the normal voltage.

It should be noted that equation $P_f \propto B_m^2 f$ given on page 316, and consequently the equations derived from it, are only approximate and therefore the actual input to the transformer must always be determined by wattmeter.

The phasors in Fig. 14.3 (b) represent quantities of different magnitudes as compared with the normal excitation due to the changed excitation and frequency required for the inclusion of the copper loss with the iron loss. The voltages, the flux Φ_m, the no-load current I_o, the magnetising current I_m, and the iron loss current I_c are all generally greater than those corresponding to normal excitation.

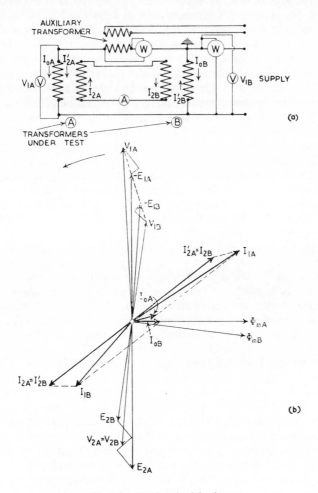

Fig. 14.5   Back to back load test

(c) Back to back load

In this test the h.v. windings are connected in opposition through an ammeter, while the l.v. windings are in parallel and connected to a suitable source of supply. The l.v. winding of a suitable auxiliary transformer is included in the l.v. supply circuit for the purpose of circulating load current therein. The h.v. winding of the auxiliary transformer is supplied from a separate source, or it may be connected in parallel across the main supply terminals in series with a variable resistance. Normal rated voltage at the correct frequency is applied to the l.v. windings in parallel, and the supply voltage to the h.v. windings of the auxiliary transformer is adjusted at correct frequency until the ammeter in the h.v. circuit of the transformer under test reads the normal full-load current. No wattmeter is necessary with this test as the actual full-load conditions, namely, normal excitation and full-load current, are reproduced. Wattmeters, however, sometimes may be inserted with advantage as a check upon the loading conditions. The l.v. winding of the auxiliary transformer must supply the impedance voltages of the two transformers under test at the normal l.v. full-load current. The diagram of connections is as shown in Fig. 14.5 (a), while the phasor conditions are shown in Fig. 14.5 (b). The latter figure shows how the auxiliary transformer supplies a voltage in the l.v. circuit $V_{1A} - V_{1B}$ equal to the sum of the impedance voltages of the two transformers under test. It will also be clear from the phasor diagram why, under this test, one transformer attains a slightly higher temperature than the other, this being due to the relative phase relationships of the no-load currents I_{OA}, I_{OB}, and of the load currents I_{1A}, I_{1B}. In transformer A the load and no-load currents are in the same quadrant, and therefore give a sum greater than the load current alone. In transformer B the load and no-load currents are in different quadrants and give a sum less than the full-load current alone. The fluxes and no-load characteristics are slightly different in the two transformers, due to the influence of the voltage injected into the l.v. circuit by the auxiliary transformer. They are, however, substantially the same as under normal excitation conditions. In the foregoing description of this test it has been assumed that the supply is to the l.v. windings, as this is usually the more convenient arrangement. When, however, the h.v. voltage is not very high, or when the l.v. voltage is extremely low, and the current correspondingly high, it may be more convenient to supply to the h.v. winding. The terms h.v. and l.v. should, therefore, be regarded as interchangeable. Also, in cases where the supply is to the h.v. windings the auxiliary transformer may be used for stepping up. It will be noticed that in all cases the wattmeter potential coils and the voltmeters are always connected to the load side of the wattmeter current coils as the magnitudes and phase of the currents taken by them are easily predetermined and thus can be allowed for. The voltage drops in the wattmeter current coils and in the ammeters are not easily predetermined either in magnitude or phase. In 14.5 (b) the phasors are drawn as though the two supplies were from the same source or were correctly synchronised. Actually the phasors for the normal voltage supply and the circulating current supply are not necessarily correlated and may even be at different frequencies. In the case of transformers having high internal reactance it is frequently an advantage to supply the circulating current at a frequency lower than the normal value in order to reduce the kVA required. In such cases it is customary to place wattmeters in both supply circuits and to adjust the circulating current so that the total power input is equal to the total losses under normal operating conditions.

Chapter 15

Impulse testing of transformers

Voltage surges have always occurred on transmission systems and these surges, whether arising from lightning or switching, are liable to be propagated along the transmission line and into the windings of a transformer. To prevent breakdown of the transformer and interruption of the supply, the equipment must be designed to withstand surges, the peaks of which are many times the normal working voltage of the system. These surge voltage withstand characteristics apply not only to the windings of the transformer but to all the associated equipment—switchgear, bushings, tap changers, etc.

During the last twenty to twenty-five years, voltage impulse testing has played an important part in the development of the modern power transformer with its high level of reliability in service. Voltage impulses produced in the laboratory to simulate lightning or switching surges were, and still are, used to obtain the surge characteristics of insulation materials for use in transformer construction. Standardised tests with voltage impulses on windings and on the completed transformer are now common practice and are a check on material uniformity, workmanship and design.

IMPULSE TEST LEVELS

Impulse test voltage levels have been chosen after many years' study of surges on supply systems. These levels are based on fully insulated systems but as the trend at the present time is to earth effectively systems of over 100 kV a reduced insulation level has been recommended for these higher system voltages.

The International Electrotechnical Commission (I.E.C.) has issued a report on standard insulation levels entitled 'Recommendations for Insulation Co-ordination' (Publication No. 71.). In addition, standard impulse voltage test levels for oil immersed transformers have been standardised in B.S.171 and values of these test levels appropriate to the system voltage are given in Table 15.1.

The system highest voltage, upon which these levels are based, is the highest r.m.s. line-to-line voltage permitted under normal operating conditions at any time and at any place on the system. It excludes temporary voltage variations due to fault conditions or the sudden disconnection of large loads, and is generally about 10% above the nominal system voltage.

319

Transformers to be impulse tested are completely erected with all fittings in position, including the bushings, so that in addition to applying the surge voltage to the windings, the test is applied simultaneously to all ancillary equipment such as tap changers, etc., together with a test on physical clearances between bushings and to earth. An impulse test upon a transformer is a type test and is undertaken only when specified on a particular contract.

Impulse voltage wave shapes

It has been generally agreed that a double exponential wave of the form $v = V(e^{-\alpha t} - e^{-\beta t})$ be used for laboratory impulse tests. This wave shape is further defined by the nominal duration of the wave front and the total time to half value of the tail, both times being given in microseconds and measured from the start of the wave. B.S.923, the British Standard Specification for impulse testing, defines the standard wave shape as being 1·2/50 microseconds and gives the methods by which the duration of the front and tail can be obtained. The nominal wave front is 1·25 times the time interval between points on the wave front at 10% and 90% of the peak voltage; a straight line drawn through the same two points cuts the time axis ($v = 0$) at O_1 the nominal start of the wave. The time to half value of the wave tail is the total time taken for the impulse voltage to rise to peak value and fall to half peak value, measured from the start as previously defined. The tolerances allowed on these values are plus and minus 30% on the wave front, and plus and minus 20% on the wave tail.

Table 15.1
STANDARD IMPULSE VOLTAGE TEST LEVELS FOR
OIL IMMERSED POWER TRANSFORMERS

System highest voltage kV, rms	Impulse voltage level		Insulation to earth
	1·2/50 full wave test kV, peak	1·2/50 chopped wave test kV, peak	
3·6	45		
7·2	60		
12·0	75	The peak value of the voltage	
17·5	95	applied for the chopped waves	
24	125	shall at least be equal to that	uniform
36	170	of the specified full wave	
52	250		
72·5	325		
100	450		
100	380		
123	450		
145	550	The peak value of the voltage	
170	650	applied for the chopped waves	
A 245	900	shall at least be equal to that	graded
300	1050	of the specified full wave	
362	1175		
420	1425		

A typical wave shape, the method of measuring it and the tolerance allowed are shown in Fig. 15.1. The standard American waveshape is not more than $2.5\mu s$ to crest and not more than $40\mu s$ to half value on tail.

Another waveform used in transformer impulse testing is the 'chopped wave' which simulates an incoming surge chopped by flashover of the co-ordination gaps close to the transformer. During this test a rod gap or equivalent is used to produce a sudden collapse (chop) of the voltage at between 2 and 6 microseconds from the nominal start of the wave. To ensure that flashover of the rod gap occurs as desired, the gap spacing is set for 50-50 flashover at the standard test level, and the actual applied voltage during a chopped wave is increased to approximately 115% of the standard full-wave level.

A rod gap consists of two electrodes cut off at right angles and left with the sharp corners untrimmed, and mounted horizontally in supports so that a length of electrode equal to or greater than half the gap overhangs the inner edge of the supports. The clearances of the electrodes from floor, walls and earthed metal in all directions must be adequate. A chopped wave shape is also shown in Fig. 15.1 and can be compared with the $1.2/50$ microsecond wave shape.

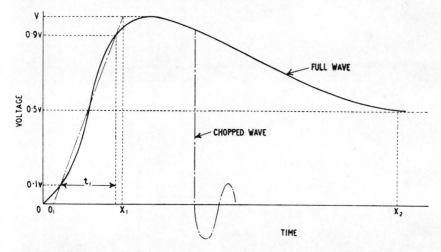

Fig. 15.1   Standard impulse voltage wave shape: $1.2/50$ microseconds. Nominal wave front $O_1X_1 = 1.2 \mu s$, tolerance $\pm 30\%$. Nominal wave tail $O_1X_2 = 50 \mu s$, tolerance $\pm 20\%$

IMPULSE GENERATORS

The production of voltage impulses is usually achieved by the discharge of a capacitor or number of capacitors into a wave forming circuit and the voltage impulse so produced is applied to the object under test.

For conducting high-voltage impulse tests the multistage generator (as in Fig. 15.2), a modified version of Marx's original circuit, is now generally used.

Fig. 15.2   Impulse generator, having a maximum voltage output of 1700 kV, and a stored energy of 30 kW seconds. The capacitors, each having a capacitance of 0·25 μF, may be arranged in parallel banks to give a complete impulse generator circuit of 1, 2, 3, 4, 6 or 12 stages. The d.c. charging voltage used is 142 kV

322

This consists of a number of capacitors initially charged in parallel and discharged in series by the sequential firing of the interstage spark gaps. A simple single stage impulse generator is shown in Fig. 15.3. The generator consists of a capacitor C which is charged by direct current and discharged through a sphere gap G. A resistor R_c limits the charging current whilst the resistors R_t and R_f control the wave shape of the surge voltage produced by the generator. The output voltage of the generator can be increased by adding more stages and frequently up to twenty stages are employed for this purpose. Additional stages are shown in Fig. 15.4 and as will be seen from this diagram all stages are so arranged that the capacitors C_1^*, C_2, C_3, etc., are charged in parallel. When the stage voltage reaches the required level V the first gap G_1 discharges and the voltage V is momentarily applied to one electrode of the capacitor C_2. The other electrode of C_2 is immediately raised to $2V$ and the second gap G_2 discharges. This process is repeated throughout all stages of the generator and if there are n stages the resultant voltage appearing at the output terminal is nV. This output is the surge voltage which is applied to the test object.

Impulse voltage measurement

There are a number of devices available for the measurement of impulse voltage, the three most common basic methods being as follows:

The first and most widely accepted is the sphere gap. Details of this method and the required gap settings are given in B.S.358.

This method has the disadvantages of requiring a large number of voltage applications to obtain the 50% flashover value and of giving no indication of the shape of the voltage wave.

The second and third methods require a voltage divider and their accuracy depends chiefly on the accuracy with which the rapid changes in the applied wave can be reproduced by the divider with a constant reduction ratio. To satisfy this requirement dividers are constructed of resistors or capacitors or a combination of both.

The second method of measurement requires a voltage divider and a high-speed cathode-ray oscilloscope. These oscilloscopes use sealed off tubes with accelerating voltages of 10 to 25 kV or continuously evacuated tubes with accelerating voltages of up to 60 kV. Besides giving the amplitude of the voltage

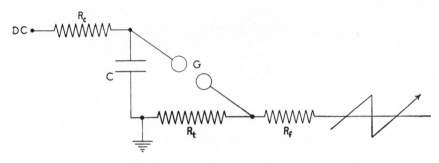

Fig. 15.3   Single-stage impulse generator

323

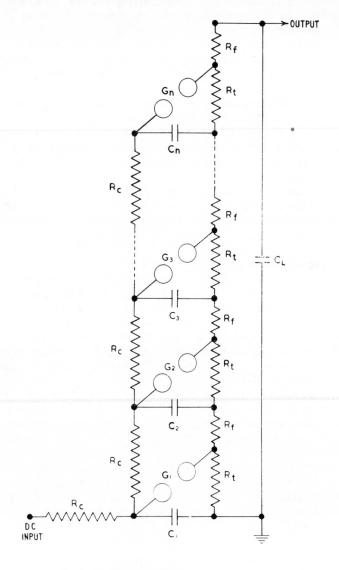

C_1, C_2, C_3 *Stage capacitors*
C_L *Divider capacitors*
R_C *Charging resistors*
R_f *Series resistors controlling wave front*
R_t *Discharge resistors controlling wave tail*
G_1, G_2, G_3 *Stage gaps*

Fig. 15.4   Impulse generator having n stages

wave, the cathode-ray oscilloscope can also be used to provide a photographic record from which the wave front time and the time to half value on the tail of the wave can be determined.

The full peak voltage cannot be applied directly to the deflecting plates of the cathode-ray oscilloscope as the input voltage to these instruments is usually limited to 1 or 2 kV. The necessary reduction in voltage is obtained by means of the voltage divider. The ratio of the divider can be determined accurately and hence by suitable calibration and measurement at the low-voltage tapping point, the amplitude of the impulse voltage can be ascertained.

The third method employs a voltage divider and a crest voltmeter. This method gives a direct and accurate measurement of the peak voltage but no indication of the actual wave shape.

IMPULSE TESTS ON TRANSFORMERS

The test voltage impulses to be applied to a transformer under test are laid down in B.S.171 as has been stated earlier in this chapter, and the test voltages are required to be applied in the following order:

(1) adjustment and calibration shot at between 50 % and 75 % of the standard insulation level

(2) two full-wave shots at the standard insulation level.

The application of voltages (1) and (2) comprises a standard impulse type test and they are applied successively to each line terminal of the transformer. If during any application, flashover of a bushing gap occurs, that particular application shall be discounted and repeated.

Where chopped waves are specified, the test sequence shall be as follows:

(a) one reduced-level full-wave, at 50–75 % of the test level

(b) one full-wave at the test level

(c) two chopped waves

(d) one full-wave at the test level.

If during the application of a chopped wave the chopping time is greater or less than the permitted 2–6μs, the particular application shall be discounted and repeated.

The time interval between successive applications of voltage should be as short as possible.

These tests employ the 1·2/50 microsecond wave shape and the chopped waves can be obtained by setting a standard rod gap in parallel with the transformer under test. Values of rod gap setting are given in Table 15.2.

The rod gap spacings given in Table 15.2 are for standard atmospheric conditions, i.e.,

$$
\begin{array}{ll}
\text{barometric pressure } (p) & 760 \text{ millimetres} \\
\text{temperature } (t) & 20\,°\text{C} \\
\text{humidity} & 11 \text{ grams of water} \\
& \text{vapour per cubic metre.}
\end{array}
$$

(11 g/m³ = 64 % relative humidity at 20 °C).

For other atmospheric conditions a correction should be made to the rod gap spacing as follows. The spacing should be corrected in an *inverse* proportion to the relative air density d, at the test room where,

Table 15.2
STANDARD ROD GAP SPACING FOR CRITICAL
FLASHOVER ON 1·2/50 MICROSECOND WAVE

Impulse test level, full-wave, 1·2/50 microseconds kV peak	Spacing of standard rod gap	
	Positive polarity cm	Negative polarity cm
45	4·5	4·0
60	6·5	5·5
75	8·9	7·0
95	11·5	9·0
125	16·5	13·5
170	23·5	19·5
250	38	29
325	51	40·0
380	60	48·5
450	71	58
550	88	72
650	105	89
900	149	127
1050	175	152
1175	198	172
1425	240	212

$$d = 0.386 \frac{p}{273+t}$$

The gap spacing should be increased by 1·0% for each 1 g/m^3 that the humidity is *below* the standard value and vice versa.

In some cases when testing large transformers, particularly those having comparatively few winding turns, the inductance may be so low that the standard wave shape of 1·2/50 microseconds cannot be obtained from the impulse generator even with a number of stages connected in parallel. It is permissible in such cases for a shorter wave shape than the standard to be agreed between the purchaser and the transformer manufacturer.

Voltage oscillograms are recorded for all shots and, in addition, as part of the fault detection technique, oscillographic records can be taken of one more of the following:

(a) the current flowing in the earthed end of the winding under test
(b) the total current flowing to earth through a shunt connected between the tank insulated from earth and the earthing system
(c) the transformed voltage appearing across another winding.

These records are additional to those obtained of the applied surge voltage and the method adopted from either (a), (b) or (c) is chosen by the transformer manufacturer according to which is the most appropriate and effective for the particular transformer under test. During an impulse test the transformer tank is earthed, either directly or through a shunt which may be used for current measurement. The winding under test has one terminal connected to the impulse generator whilst the other end is connected to earth. In the case of star connected windings having no neutral point brought out to a separate terminal, or in the case of delta connected windings, it is usual to connect the two remaining terminals together and earth via a measuring shunt unless otherwise

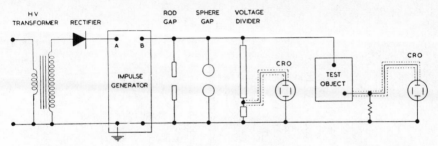

Fig.15.5   General arrangement of equipment for an impulse test (Diagrammatic only)

agreed between the manufacturer and the purchaser, and shall be unchanged throughout. It is essential that all line terminals and windings not being tested shall also be earthed directly or through a suitable resistance in order to limit the voltage to a level appropriate to the particular winding.

Where arcing gaps are fitted to transformer bushings they should be set to the maximum permissible gap in order to prevent flashover during testing. The transformer under test, the impulse generator, sphere gap, rod gap, voltage divider and the measuring and recording equipment should be grouped together as closely as possible consistent with the practical requirements in the testing laboratory. Negative polarity shall be used for the impulse tests.

The general arrangement of the various pieces of equipment employed for an impulse test on a transformer is shown diagrammatically in Fig. 15.5.

Fault detection during impulse tests

Detection of a breakdown in the major insulation of a transformer usually presents no problem as comparison of the voltage oscillograms with that obtained during the calibration shot at reduced voltage level gives clear indication of this type of breakdown. The principal indications are as follows:

(1) Any change of wave shape, apart from differing amplitudes, as shown by comparison with the full-wave voltage oscillograms recorded during the full-wave shots taken before and after the chopped-wave shots.

(2) Any difference in the chopped-wave voltage oscillograms, up to the time of chopping, by comparison with the full-wave oscillograms.

(3) The presence of a chopped wave in the oscillogram of any application of voltage for which no external flashover was observed.

A breakdown between turns or between sections of a coil is, however, not always readily detected by examination of the voltage oscillograms and it is to facilitate the detection of this type of fault that current or other oscillograms are recorded. A comparison can then be made of the current oscillograms obtained from the full-wave shots and the calibrating oscillograms obtained at reduced voltage.

The differential method of recording neutral current is occasionally used and may be sensitive to single turn faults. All neutral current detection methods lose sensitivity when short-circuited windings are magnetically coupled to the winding being tested. Connections for this and other typical methods of fault detection are shown in Fig. 15.6 (*a*) to (*e*).

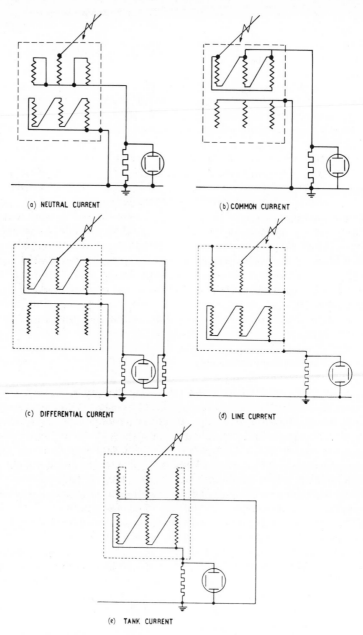

(a) NEUTRAL CURRENT

(b) COMMON CURRENT

(c) DIFFERENTIAL CURRENT

(d) LINE CURRENT

(e) TANK CURRENT

Fig. 15.6 Connections used for fault detection when impulse testing transformers

Note: Terminals of windings not under test shall be earthed, either directly or through resistors. Each phase should be damped by a suitable resistor.

328

In all cases the current is taken to earth through a non-inductive shunt resistor or resistor/capacitor combination and the voltage appearing across this impedance is applied to the deflection plates of an oscillograph.

Another indication is the detection of any audible noise within the transformer tank at the instant of applying an impulse voltage. This has given rise to a completely different method of fault detection known as the electro-acoustic probe, which records pressure vibrations caused by discharges (sparks) in the oil when a fault occurs. The mechanical vibration set up in the oil is detected by a microphone suspended below the oil surface. The electrical oscillation produced by the microphone is amplified and applied to an oscilloscope, from which a photographic record is obtained. In addition, audio indication by a loudspeaker or headphones, and visual indication on a meter are provided.

Fault location

The location of the fault after an indication of breakdown is often a long and tedious procedure which may involve the complete dismantling of the transformer and even then an interturn or interlayer fault may escape detection. Any indication of the approximate position in the winding of the breakdown will help to reduce the time spent in locating the fault.

Current oscillograms may give an indication of this position by a burst of high-frequency oscillations or a divergence from the 'no-fault' wave shape.

Since the speed of propagation of the wave through a winding is about 150 m/μs, the time interval between the entry of the wave into the winding and the fault indication can be used to obtain the approximate position of the fault, provided the breakdown has occurred before a reflection from the end of the winding has taken place. The location of faults by examination of current oscillograms is much facilitated by recording the traces against a number of different time bases. Distortion of the voltage oscillogram may also help in the location of a fault but it generally requires a large fault current to distort the voltage wave and the breakdown is then usually obvious.

Figure 15.7 illustrates a typical set of voltage and neutral current oscillograms associated with a normal impulse withstand test, and Fig. 15.8 those obtained with increasing impulse voltage levels up to breakdown, which is clearly shown in Fig. 15.8 (f). It should be noted that these tests were conducted in accordance with an earlier issue of B.S.171.

A wave of negative polarity and having a wave shape of 1·06/48 microseconds was employed for all tests. The voltage calibration line corresponds to 107·4 kV and the time calibration corresponds to 10 μs on oscillograms (a), (b), (d) and (e), and to 1 μs on (c).

LOW-VOLTAGE SURGE TESTS

The insulation of a transformer must be proportioned to the surge voltages which will appear at the various points throughout the windings. High-voltage surge tests on a completed transformer are costly and take a great deal of time. In order to obtain the maximum possible amount of information it is desirable to have electrical contact with the maximum number of points on the winding.

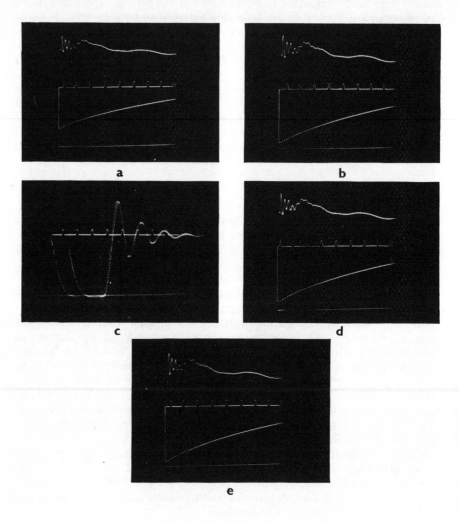

Test sequence	Test
a	75 % calibration voltage and neutral current
b	100 % full-wave voltage and neutral current
c	115 % chopped-wave voltage (two tests)
d	100 % full-wave voltage and neutral current
e	100 % full-wave voltage and neutral current

Fig. 15.7   Oscillograms of 'no-fault' wave shapes

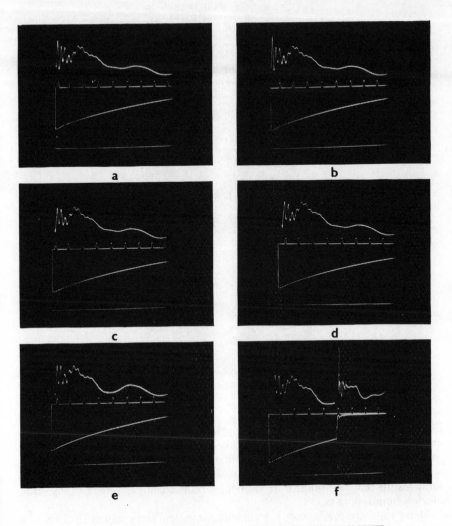

Test sequence	Test
a	145% full-wave voltage and neutral current
b	150% full-wave voltage and neutral current
c	155% full-wave voltage and neutral current
d	160% full-wave voltage and neutral current
e	165% full-wave voltage and neutral current
f	170% full-wave voltage and neutral current

Fig. 15.8   Oscillograms of 'no-fault' and fault wave shapes

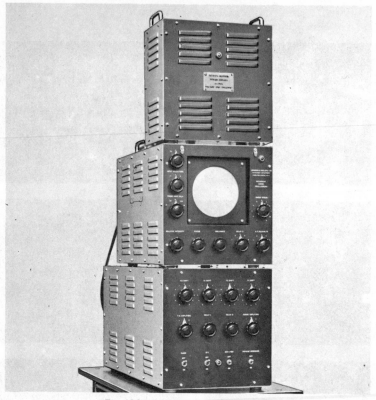

Fig. 15.9   Recurrent surge oscilloscope

Furthermore, for high-voltage transformers the core and windings must be immersed in oil and mounted in the tank. This condition does not facilitate the collection of data. Tests have shown that the surge voltage distribution in a winding is independent of the magnitude of the applied voltage and that the same results may be obtained by applying a reduced surge voltage, of the order of a few hundred volts.

These tests are made with a recurrent surge generator which consists of a capacitor charged to a suitable voltage and discharged by means of a thyratron into a circuit which is designed to generate the required low-voltage surge of the standard wave shape. The charge and discharge sequence is repeated fifty times per second. The output voltage from the recurrent surge generator is applied to the terminal of the transformer winding under investigation, in a similar manner to that in which a high-voltage surge test would be conducted, whilst the surge voltage appearing at any point of the winding can be measured and displayed on the screen of the cathode-ray oscilloscope. The time base is arranged so that it is synchronised with the recurrent discharge of the capacitor. By this means it is possible to obtain a standing picture on the screen of the applied voltage and of the voltage appearing at points along the winding, together with a time calibration wave which can be viewed directly by the operator or photographed for permanent record and later analysis.

In order to increase the usefulness of the recurrent surge oscilloscope for development and research investigations, facilities to vary the wave front and wave tail, to produce chopped waves, and to give variable time sweeps and timing waves, are incorporated in the equipment. Figure 15.9 shows a recurrent surge oscilloscope employed for low-voltage surge tests. The equipment is normally brought alongside the transformer which it is desired to test, the connections between the instrument and the transformer being made by means of low capacitance co-axial lead in order that the resulting oscillograms obtained shall be unaffected by the capacitance of the connection. For obtaining permanent records, a camera (which is not shown in Fig. 15.9) would be attached to the front panel in order to photograph the trace appearing on the screen of the cathode-ray tube.

Initial voltage distribution

The presence of capacitance between coils and to earth within the winding of a transformer causes the winding to behave as a capacitance and not as an inductance when surge impulses of rectangular wave shape are applied to the winding. Figure 15.10 shows the simplified equivalent circuit of a uniform single

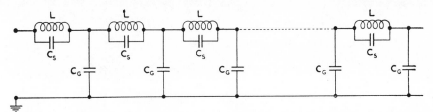

Fig. 15.10   Simplified equivalent circuit of a uniform single winding

winding. The initial voltage distribution of approximately $1\ \mu s$ duration is governed by the series and shunt capacitive elements within the windings and depends upon the factor α where

$$\alpha = \sqrt{\frac{C_g}{C_s}}$$

in which C_g is here the total capacitance of the winding to earth and C_s is the total series coil capacitance from one end of the coil stack to the other.

Great variation in the voltage gradient occurs throughout a transformer winding and the gradient at the line end may be many times the mean value. The greater the value of α the greater the concentration of voltage at the line end and the smaller the voltage in the interior of the winding. The maximum voltage per turn at the line end is equal to α times the value of the voltage corresponding to a uniform voltage distribution. High-voltage transformers having disc type h.v. windings and spiral l.v. windings arranged concentrically have values of α varying from five to thirty.

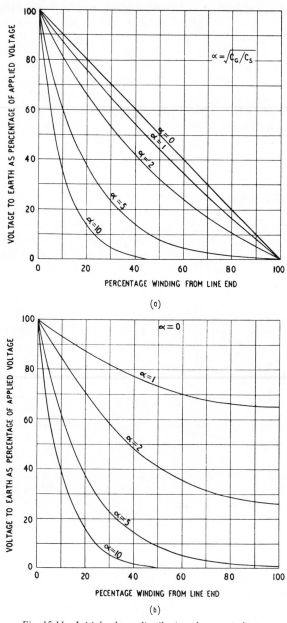

Fig. 15.11   Initial voltage distribution along a winding.
(a) With neutral earthed
(b) With neutral isolated

334

The initial voltage distribution throughout a transformer winding when the capacitive elements are charged can be calculated from the formula

$$v = V_0 \frac{\sinh \alpha x/\ell}{\sinh \alpha}$$

for the case of the earthed condition, and

$$v = V_0 \frac{\cosh \alpha x/\ell}{\cosh \alpha}$$

for the case where the neutral is isolated,

where, V_0 is the value of the voltage to earth of the transformer line terminal

x is the distance along the winding

and ℓ is the length of the winding

i.e. x/ℓ is the fractional distance along the winding.

The value of α is that given earlier, namely the square root of the ground capacitance to the series capacitance $\sqrt{C_g/C_s}$. Plotting the ratio v/V_0 against x/ℓ gives the family of curves shown in Figs. 15.11(a) and (b). For $\alpha = 0$, that is for $C_g = 0$, the result is a straight line, i.e., a uniform voltage distribution. As stated, the greater the value of α, the greater is the concentration of voltage at the line end of the winding.

If, when making measurements with the recurrent surge generator, the maximum voltage to earth is noted for various points along the winding these values

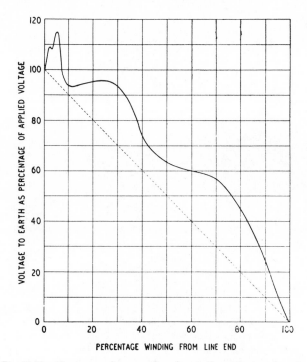

Fig. 15.12   Maximum voltage envelope for winding with neutral earthed

335

can be plotted and a curve drawn through them gives a maximum voltage envelope. Such a curve is shown in Fig. 15.12. This envelope is not the voltage distribution at any given time, but the maximum value at each point irrespective of time. It gives an indication of the voltage stress upon the major insulation to earth which the particular winding must withstand. The maximum voltage appearing between tappings, sections, etc., can also be assessed by the same method.

Chapter 16

Dispatch, installation and maintenance of oil immersed transformers, including drying out on site

DISPATCH

According to circumstances, transformers are dispatched from the factory for both the home and overseas markets either:

(*a*) In their tanks with sufficient oil to cover the windings and insulation, the remaining oil being dispatched separately in sealed steel drums, or in bulk by means of a road tanker. Alternatively the full complement of oil may be contained in the transformer tanks.

(*b*) In their tanks without oil, all the oil being provided separately. In this case the tank is filled either with dry air or inert gas to a slight pressure above atmosphere.

When a transformer is dispatched in its tank either with or without oil, special care is exercised when finally fitting it into the tank, in order to prevent movement of the core and windings during transit. The tank cover is always bolted into position preparatory to final packing.

A transformer adequately secured in its tank may be packed in a strong wooden crate or case for dispatch, or may be sent with lesser protection, according to circumstances. Terminals projecting outside the tank which would otherwise be exposed, are protected. In the case of very high voltage transformers the h.v. terminals are packed separately, but temporary electrical connections are brought out from the h.v. and l.v. windings and the core so that measurements of insulation resistance can be made.

Transformers for the home market in which the tank is of the tubular type need no external protection apart from the hoods temporarily fitted over any projecting terminals. For export however, transformers in tubular tanks may be protected by fitting a number of specially shaped substantial wooden fenders, which are designed to protect the cooling tubes and prevent any damage which might otherwise be incurred during transit, or they are packed in wooden crates.

337

Examination upon arrival

Immediately a transformer is received it should be unpacked and thoroughly examined externally for possible damage which may have occurred during transit. The complete unit should be signed for according to the result of this examination or as 'not examined', and any damage found should be reported to the carriers and to the manufacturer within three days, or otherwise a claim cannot be lodged with the responsible party.

When receiving oil which has been dispatched separately in drums, each drum should first be examined to ascertain if the seal is intact, and if any is found broken the supplier should be advised. If for any reason oil sent in steel drums cannot be used immediately it arrives at its destination, the drums should be placed in a dry and, if possible, warm place, with the seals downwards. Drums should not be stored on end, as moisture would be liable to collect in the dished ends. The seal should not be broken until the oil is required for filling the transformer tank, immediately prior to placing the transformer into commission. When cold drums of oil are received into a warm atmosphere, condensation of moisture takes place on the outside of the drums. The seals on the bungs of the drums should not be broken until the oil has attained the temperature of the room in which it is placed, and any condensed moisture has dried from the surface of the drums.

INSTALLATION

If the transformer has been dispatched in oil, the oil level should be checked and if, when any allowance for a change of temperature has been made, it is low, an investigation must be made to ascertain the cause of the deficiency.

In the case of transformers dispatched under a slight pressure of dry air or inert gas, it is particularly undesirable that the core and windings should become exposed to the ambient atmosphere. This is best achieved by filling the tanks with oil under vacuum, and the method adopted depends upon the plant available. A check should first be made that there is still a positive pressure inside the tank.

An insulation resistance test should then be carried out between windings, and between windings and earth (core and tank) as a measure of the dampness: if the readings obtained compare favourably with those recorded at the works immediately prior to dispatch, the tank may be filled with oil. However, if the values obtained are low it may be necessary first to dry out the transformer.

If a power oil pump is available the oil can be pumped into the tank (at a rate not exceeding 2200 litres/h) after a vacuum of 350 mm. of mercury has been applied to the tank for 6-8 h. The oil should, of course, be tested in accordance with B.S.148 immediately before use. If the oil is de-aerated it should be pumped in from the bottom of the tank, but if not, from the top of the tank.

If a power pump is not available, oil filling must be effected by using a hand pump and an intermediate oil vessel. If the transformer is designed for a conservator, this should be fitted and filled with oil. The transformer tank is evacuated, as before, for 6-8 h and the conservator for 30 min to 350 mm of vacuum. The interconnecting pipeline is then opened and the oil allowed to enter the main tank by gravity feed. This process should then be repeated until the tank is filled with oil to a level sufficient to cover the core, windings and major insulation.

In either case a vacuum of 350 mm of mercury should be maintained for a further 6–8 h and then broken slowly to admit dry air e.g. through a silica-gel breather.

An intermediate dryness test should now be made, after which the final erection of pipework, coolers, etc. is completed. Final oil filling is carried out as above and the whole system then maintained under a vacuum of 350 mm of mercury for a further 6–8 h.

Finally a check must be made on the insulation resistance and the values obtained should compare favourably with those taken at the works.

Owing to the hygroscopic nature of the oil and the various types of insulating materials employed, some moisture may be absorbed by both when in contact with the atmosphere. The oil is less susceptible than the dry fabricated insulation to the influence of atmospheric moisture, and it is good practice to dispatch transformers, whenever possible, in their tanks with at least sufficient oil to cover the windings.

It is essential that all transformers should be free from moisture before being commissioned, as the presence of moisture results in a lowering of the insulation resistance. Consequently if the values obtained for the insulation resistance between windings and between each winding and earth do not compare favourably with those obtained at the works previous to dispatch, the core and windings must be dried out, either alone or under oil. In the case of transformers dispatched in oil, a sample of the oil should be drawn from the bottom of the tank and tested for moisture content immediately prior to placing the transformer into commission. If only a small amount of moisture is shown to be present the oil alone should be dried out, but if a large proportion of moisture is indicated it is advisable to dry out the transformer itself in addition.

DRYING OUT ON SITE

The following brief notes describe some of the methods used for drying out on site, but the ultimate choice is largely influenced by the facilities available. These methods are by: (1) oil immersed resistor heating; (2) short circuit; (3) hot air circulation.

In addition, some transformer manufacturers, oil companies and supply authorities have mobile filter plants and test equipment available to undertake the filling of transformers with oil and for any subsequent treatment. Where the equipment is available for site use, modern practice tends to employ the method in which oil is circulated under vacuum in the oil treatment plant. The principles and application of some equipments are dealt with later in the chapter.

(1) Oil immersed resistor heating method

This method consists of drying out the transformer and oil simultaneously in the transformer tank by means of specially constructed resistor units, such as are illustrated by Fig. 16.1, which are lowered between the windings and tank to the bottom of the latter.

Fig. 16.1   Typical protected resistance unit used for drying out transformers in oil, and for oil alone

The tank should be filled with oil up to the working level in the case of plain tanks, or up to a point just below the top of the cooling tubes in the case of tubular tanks. The oil should be allowed to stand for, say, an hour or so. The tank cover should be raised at least 300 mm, or, better still, removed altogether in order to allow perfectly free egress of the moisture vaporised during the drying out process. In order to conserve the heat generated in the resistors and also to cut down the energy consumption, the four sides of the transformer tank should preferably be well lagged all over, using, say, wagon sheets, sackings, or any similar coverings which may be available. The resistors should be spaced as symmetrically as possible round the inside of the tank in order to distribute the heat. The resistors themselves should be banked in series or parallel, according to the supply voltage available, and they should be connected to the low-voltage supply by means of suitable insulated leads. This supply may be either single-phase a.c. or d.c. which may be controlled by an external adjustable resistor if necessary. The drying out resistance values should be chosen according to the voltage available so that during the steady temperature period of the drying out process the top oil temperature can be maintained at a value not exceeding 85°C. It should be borne in mind that in the immediate vicinity of the resistor units the oil will be at a higher temperature than is indicated at the top of the tank, and consequently the temperature near the resistors is the limiting factor. If this temperature is too high the oil will become overheated locally and will deposit sludge. During the heating up period, larger currents may be employed in order to speed up this part of the operation, or alternatively the tank may be more efficiently lagged, always bearing in mind, however, that the oil must not reach too high a temperature in the immediate vicinity of the resistors. The temperature may be measured by a thermometer, the bulb being immersed in the top layers of the oil. If, either by direct application or through control, the voltage available is such as will not permit a steady drying out temperature, the resistors should be arranged so as to pass a higher current than that required continuously for holding the specified temperature steady. In this case the resistor should be loaded intermittently, being switched off when the oil temperature is within a few degrees of the maximum specified and switched on again when it has dropped a few degrees below that figure. In this way the average temperature may be kept approximately at the steady temperature recommended.

During the drying out process the following readings should be taken at frequent regular intervals: (a) insulation resistance between h.v. and l.v. windings and between each winding and earthed metal; (b) temperature; (c) power input. There are three stages in the complete process. First, the heating up stage, which is of relatively short duration, when the temperature is increasing from the ambient to the recommended maximum for drying out and the insulation resistance of the windings is falling. Second, the longest and real drying period, when the temperature is maintained constant and the insulation resistance becomes approximately so but commences to rise at a certain point towards the end of this period. Third, which is again of short duration, when the supply

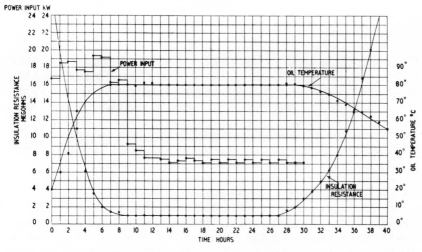

Fig. 16.2   Drying out curves of a 500 kVA three-phase 3300/440 V, 50 Hz transformer. Transformer and oil dried out together by the resistance grid method

to the resistors is cut off, the temperature falling, and the insulation resistance increasing. Usually, curves are plotted of insulation resistance and also temperature as ordinates, against time as abscissae, and these curves give an indication of the progress of the drying out operation. It is when the curve of insulation resistance against time begins to rise that the supply to the resistor is cut off, as the rising of the curve indicates that nearly all moisture has been removed. Typical drying out curves are given in Fig. 16.2, which shows the actual drying out of a 500 kVA three-phase transformer. The windings' and any terminals normally under oil must be covered with oil, and the tank should preferably be lagged. On no account must normal excitation be applied until the windings and oil are thoroughly dried out.

This method has particular application for sites at which mobile plants are not available and, provided reasonable care is exercised in avoiding excessively high local oil temperature, is safe and perfectly satisfactory. Unless a transformer has had a thorough soaking, any moisture present will be partly in the oil and partly deposited on the outside layers of the insulation, and will not have penetrated much to the inside layers. With this method, therefore, the highest temperature is applied where it is most required, namely, to the oil and to the surface of the winding.

The advantages of the method are as follows:

(a) The transformer core, windings, oil and tank are dried out simultaneously, thus economising in labour and expense.

(b) Either a.c. or d.c. voltage can be used.

(c) Heat is generated at the bottom of the tank where most of the moisture is likely to have settled, and the maximum temperature is applied where it is most required.

(d) The insulation of both the bottom and top coils is thoroughly dried.

(e) There is no fear of local heating taking place in the interior of the windings, and consequently the insulation at these points cannot become damaged.

Important notice. On no account should a transformer be left unattended during any part of the drying out process.

(2) Short-circuit method

This method is also frequently used (a), for drying out the transformer and oil simultaneously in the transformer tank; and (b), for drying the transformer only, out of its tank. Dealing first with (a), the same initial precautions are taken as with method (1), the tank cover being raised at least 300 mm, or removed altogether. The tank sides should also preferably be lagged. The l.v. winding, preferably, is short circuited, a low single-phase or three-phase voltage being applied to the h.v. windings, and of a value approaching the full-load impedance voltage of the transformer. If a suitable single-phase voltage only is available, the h.v. windings should temporarily be connected in series, as shown in Fig. 16.3. A voltmeter, ammeter and fuses should be connected in circuit on the h.v. side. If the voltage available is not suitable for supplying the h.v. winding, but could suitably be applied to the l.v., this may be done and the h.v. winding instead short circuited. In this case special care must be taken to avoid breaking the short circuiting connection as, if this is broken a high voltage will be induced in the h.v. winding which will be dangerous to the operator. The temperature

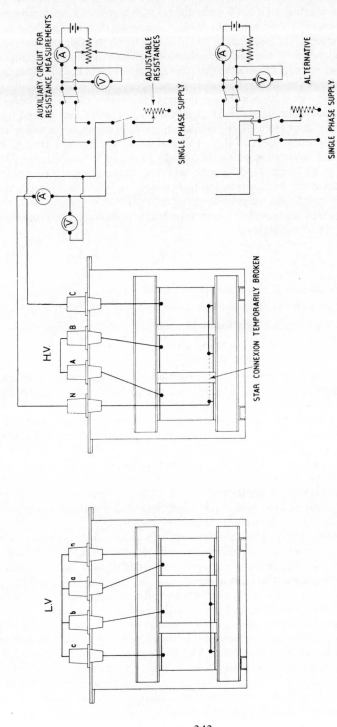

Fig. 16.3 Connections for drying out a three-phase transformer by the single-phase short-circuit method

343

should be measured both by a thermometer in the oil and, if possible, by the resistance of the windings. In the former case it is better to use spirit thermometers, but if mercury thermometers only are available, they should be placed outside the influence of leakage magnetic fields, as otherwise eddy currents may be induced in the mercury, and the thermometers will give a reading higher than the true oil temperature. If resistance readings cannot be taken, it is an advantage to place thermometers in the vertical oil ducts between the h.v. and l.v. windings, and in this case it is essential that the spirit type of thermometer should be used. The resistance measurements are taken periodically, say, every half-hour during the actual drying out period. These measurements are made by utilising any suitable d.c. supply available, and Figs. 16.3 and 16.4 indicate the connections. If tappings are fitted to either winding the tapping switch or link device should be positioned so that the total winding turns are in circuit during the drying out process. The a.c. supply for heating the transformer is, of course, temporarily interrupted when taking d.c. resistance measurements.

The temperature in degrees Celsius corresponding to any measured resistance is given by the following formula:

$$T_2 = \frac{R_2}{R_1}(235 + T_1) - 235$$

where T_2 = temperature of the windings when hot

T_1 = temperature of the windings when cold

R_2 = resistance of the windings when hot

R_1 = resistance of the windings when cold

Temperature rise of the windings is $T_2 - T_1$

The maximum average temperature of each winding measured by resistance should not be allowed to exceed 95°C. If it is not possible to take the resistance of the windings, the top oil temperature should not exceed 85°C.

Dealing next with (b), the transformer being dried out separately and out of its tank, the method is electrically the same as for (a), but the applied voltage must be lower. The transformer should be placed in a shielded position to exclude draughts, and the steady drying-out temperature measured by resistance must not exceed 95°C.

The value of drying out currents will, of course, be less than when drying out the transformer in oil, but the attainment of the specified maximum permissible temperature is the true indication of the current required.

Apart from the question of oil, all other remarks made under (a) apply to (b). The oil, after being dried out separately, may be poured direct into the transformer tank after placing the transformer in position. The tank must, of course, previously be thoroughly dried out also.

The only advantage of this method of drying out is its simplicity, while it has several disadvantages. The first of these is the possible occurrence of local heating of the inner windings, though, to a certain extent, a check can be kept on this if it is possible to make the resistance measurements previously referred to. In this case it will be seen that the maximum quantity of heat is obtained just where it is not wanted, namely, within the windings, for in the large majority of cases any moisture that may be present will be on the outside of the coils and in the oil. Insulation resistance measurements must, of course, be taken and interpreted as outlined in the previous method.

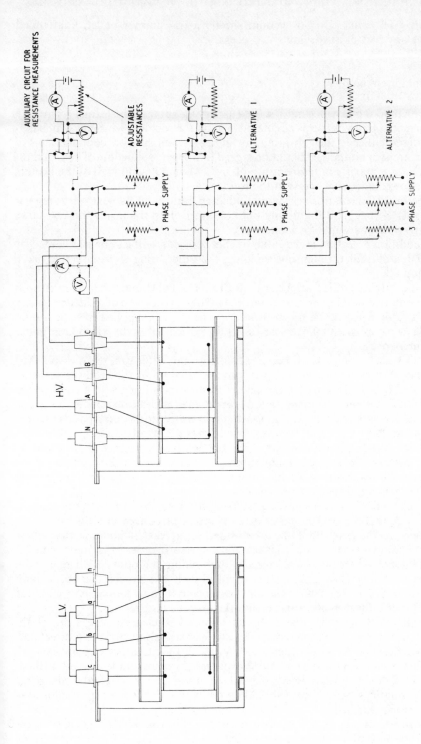

Fig. 16.4 Connections for drying out a three-phase transformer by the three-phase short-circuit method

345

Important notice. On no account should a transformer be left unattended during any part of the drying out process.

(3) Hot air circulation

This method is one which may be used for drying out the transformer separately. A current of clean hot air is circulated around the core and windings which may be housed either in their own tank or in some other suitable container.

The tank or container should be lagged to prevent undue loss of heat and the hot air, at a maximum temperature of $95°C$, should be admitted at the bottom and allowed adequate egress at the top.

Readings of insulation resistance and temperature should be taken at regular intervals of time, and the drying out continued until the insulation resistance begins to rise, as stated for method (1).

If additional heating is required, the hot air ciculation can be augmented by passing a current through one winding, the other being short circuited, as in method (2).

If the oil is supplied separately, in drums, samples must be taken from a representative number of drums and tested for the presence of moisture. In any case a sample should be taken from every drum. Oil supplied in bulk should similarly be tested and if necessary should be dried out as described later under 'Maintenance'.

Both transformer and oil being satisfactory the latter can, when dispatched separately, now be run or pumped into the tank up to the requisite level, as indicated by the oil gauge. In the case of transformers fitted with a conservator the oil filling should be made to the correct level as indicated by the oil gauge and a check should be made that the ambient temperature corresponds to, or is not greatly different from, the transformer maker's standard filling calibration level. It will be appreciated that, if the temperature difference is great due to either high or low ambient temperature, the conservator may overflow at full load on the transformer, or the minimum level may not be obtained at low ambient temperatures.

Care should be taken when transferring oil from the drums to the tank to make sure that no foreign matter or moisture is introduced into the oil.

The oil in the tank should be left to stand for at least 24 hours so that all air bubbles may rise to the top of the tank and disperse. In order to facilitate removal of air bubbles which may have lodged in the winding, it is desirable to agitate the oil periodically. When a breather is supplied it should be seen that this is fully charged with silica-gel before placing the transformer into commission, and that the oil seal is filled in accordance with the maker's instructions.

Insulators projecting outside the tank should be cleaned with a dry cloth.

The transformer tank and cover should be effectively earthed in a direct and positive manner, while in order to comply with the Statutory Regulations the l.v. neutral point of substation and similar transformers should also be earthed. Earthing of the neutral prevents the accumulation of an electrostatic charge on the l.v. windings. Special care should be taken to see that the core itself is also satisfactorily earthed.

In order to obtain the full rated output and any overloads, indoor transformers should be accommodated in a well-ventilated chamber, which, at the same time, affords the protection required against rain and dripping water. Too great a stress cannot be laid upon the necessity for providing adequate ventilation, for it is principally the thermal conditions which decide the life of a transformer. Badly ventilated and inadequately sized chambers undoubtedly shorten the useful life of a transformer, and hence should be avoided. Where several transformers are placed in one building or chamber there should be a clear space between any two of approximately 1 metre, and the same clearance should be allowed between any one transformer and an adjacent wall. If this distance is decreased appreciably, the transformer tanks will have the effect of lagging each other and the temperature rise will increase, possibly to such an extent at full load as, in time, to cause insulation deterioration.

In unattended substations it is an advantage to fit each transformer with a maximum indicating thermometer, so that a check can be kept upon the temperature rise. In large attended substations and in power stations, transformers, particularly the large ones, are often fitted with alarm thermometers or winding temperature indicators having two alarms, one to ring a bell as a warning of rising temperature and the other to trip the transformer out of service when the temperature attains a certain predetermined maximum.

The setting of these alarms is dependent on local ambient and loading conditions, but is based on the B.S. maximum oil temperature of 90°C. Alarm thermometers, which depend upon oil temperature, might be set at 85°C and 90°C respectively to take account of the inherent time lag between maximum and top oil temperatures. Winding temperature indicators, which more closely follow variations of winding temperature, are used for all large transformers and might have a warning alarm set at 105°C and a trip at 110°C: these values are similarly subject to local ambient and loading conditions.

It must be borne in mind that there will be a temperature gradient between the actual maximum temperature of the copper conductors and that registered in the top of the oil, the former, of course, being the higher. This accounts for the differences suggested between the permissible continuous temperature and the alarm temperatures.

If the transformer is not required to operate in parallel with other transformers, the voltage may now be applied. It is desirable to leave the transformer on no-load for as long a period as possible preceding its actual use, so that it may be warmed by the heat from the iron loss, as this minimises the possible absorption of moisture and enables any trapped air to be dispelled by the convection currents set up in the heated oil. The same objective would be achieved by switching in directly on load, but for transformers fitted with gas actuated relay protection the supply may be interrupted by the dispelled gas from the oil actuating the relays, which would then trip the supply breaker.

If, however, the transformer has to operate in parallel with another unit, it should be correctly phased in, as described in the chapter dealing with parallel operation, before switching on the primary voltage. It is essential that the secondary terminal voltages should be identical, otherwise circulating currents will be produced in the transformer windings even at no-load. Transformers of which the output ratio is greater than three to one should not be operated in parallel. A watch should also be kept on the ratio of the resistance to reactance voltage drop for different transformers when operating in parallel as, if this ratio is different with such transformers, there is the danger that some of the transformers may be overloaded when the bank is supplying its total rated load.

This possibility is due to the phase difference between the load currents in the individual transformers.

Switching in or out should be kept to an absolute minimum. In the case of switching in, the transformer is always subject to the application of steep fronted travelling voltage waves and current inrushes, both of which tend to stress the insulation of the windings, electrically and mechanically, so increasing the possibility of ultimate breakdown and short circuit between turns. From the point of view of voltage concentration it is an advantage, wherever possible, to excite the transformers from the l.v. side, although, on the other hand, the heaviest current inrushes are experienced when switching in on the l.v. side. The procedure adopted will therefore be one of expediency, as determined from a consideration of voltage surges and heavy current inrushes. The destructive effects on transformers of faults on the system are, of course, nowadays largely minimised by the inclusion of suitable discriminative automatic protective gear, and some of the well known arrangements have been proved in practice to be very effective in disconnecting a transformer in which a short circuit between turns has occurred. In a large number of cases the operation of the automatic oil circuit breaker has been so rapid as to localise the short circuit to the coil in which it has originated, and often this lowers the cost of repairs very considerably.

MAINTENANCE

It can generally be said that the day is now past when a transformer is looked upon as a piece of apparatus which requires no attention whatever because its parts are stationary. It is now recognised by experienced operating engineers that it pays to inspect transformers regularly in service, and the benefits resulting from such inspection have been proved over and over again. A general examination and overhaul once a year cannot be anything but beneficial, providing the work is done with care; it should be remembered that as a transformer ages the insulation becomes more brittle, and it is necessary to expend more care upon the handling of the apparatus. At the same time the older a transformer becomes, the less likely is it to require frequent adjustment, and this is justification for relaxing the inspections. Like so many other engineering features, this question of inspection on site often resolves itself into one of expediency, as the periodic inspection of a large number of transformers on a system may be a costly procedure. In such cases it is usual to find that an inspection is not made unless the temperature rise of the oil, which is taken periodically, indicates the presence of a fault.

As it has been found in practice that explosive mixtures may be generated in the enclosed space in the top of the tanks or conservator vessels, no exposed flame should be allowed in the area when removing the cover for inspection purposes, or when emptying or filling the tanks.

Periodical inspection should include tightening up the coils either by means of the adjustable clamping screws, which are supplied on the larger transformers, or by inserting insulating packing blocks between the top of the high-voltage coils and yokes. When tightening the coils it is very important that undue pressure is not applied, as if they are compressed too much there is the danger that individual turns of the high voltage windings will become displaced, so increasing the likelihood of a short circuit between turns. All nuts should be examined and thoroughly tightened where necessary, particularly those for current-carrying terminals and for the bolts clamping the core and yoke plates together. Insulated bolts passing through core laminations might advantageously be Meggered (at a maximum of 2000 V) to afford a qualitative idea of the condition of the bolt insulation. The windings too, can similarly be Meggered if desired. Insulators should be cleaned periodically with a dry cloth.

If a breather is supplied, the air-drying agent, which is usually silica-gel, should be checked regularly and replaced or dried as is found necessary.

If a conservator vessel is fitted to the transformer, any moisture which may have become trapped at the bottom of the vessel should be drawn off periodically by means of the drain provided for the purpose. Naturally such periods vary with the climatic conditions, but this operation must be undertaken as circumstances prescribe. Clean, dry oil should be introduced to make up for any wet oil removed.

Oil maintenance should be conducted in accordance with the recommendations given in Chapter 17, and with those of the British Standards Institution Code of Practice, C.P.1009, entitled 'Maintenance of Insulating Oil'. This code refers to the deterioration or contamination of insulating oil whether in storage, handling or service, and describes routine site methods of sampling and testing in order to determine whether the oil is suitable for further service. A section is included on the treatment of oil on site.

Contamination due to moisture or solids can be dealt with satisfactorily on site. Centrifugal separators are effective in removing free water and, generally, finely divided solid impurities, whereas filters, although generally capable of removing small quantities of water, deal more effectively with solid impurities.

The choice of oil temperature for purification depends upon the circumstances. Sludge and free water are more soluble in hot oil than in cold and are therefore most effectively removed by cold treatment. Also the tendency of the oil to oxidation is minimised if the temperature at atmospheric pressure is limited to 50°C. Dissolved water and small amounts of free water are effectively removed by oil circulation under vacuum, in which case temperatures up to 70°C are advantageous. Figure 16.5 illustrates in graphical form the water content of transformer oil at saturation related to the oil temperature.

The method selected is largely dependent upon the equipment available. Various designs of equipment are on the market and transformer manufacturers, oil companies and supply authorities all have mobile plant available both to undertake the filling of transformer tanks with oil on site, and for any subsequent treatment.

Some brief notes on such equipment are given below, but it should first be mentioned that whilst a filter or a centrifuge can improve the physical condition of an oil by the removal of water and foreign matter there is really only one satisfactory way of dealing with oxidised or partly oxidised oil in an old transformer, that is, when the acidity of the oil has risen to a value approaching 0·5 mg KOH/g. In such cases the transformer core and windings should be thoroughly drained and carefully washed down with a spray or jet of clean oil before the introduction of a fresh charge. The degraded oil can be returned for reconditioning to an oil supplier who has the necessary plant to carry out such work. Careful and thorough washing down of the core and windings of the unit is imperative if contamination of the fresh charge of oil is to be reduced to a minimum and its life accordingly prolonged.

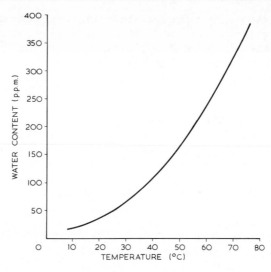

Fig. 16.5   Water content of transformer oil at saturation, related to the oil temperature

Oil filter press

A typical oil filter press is shown in Fig. 16.6.

The apparatus consists essentially of a series of cast iron frames and plates illustrated in Fig. 16.7. The frames and plates are assembled alternately, each plate supporting the filter paper, which is generally used in five ply thickness. The whole surface of the plates, with the exception of a 12 mm plain rim round the edge, is grooved horizontally and vertically in such a way as to give a honeycombed plate allowing passage for the oil, and at the same time providing ample area of flat surface to support the filter paper. Plates, frames and filter paper are pierced with two holes, one at each of the two lower corners. When the filter is assembled, these holes register, forming the inlet and outlet channels for the oil, which enters at the left-hand corner at the head of the filter press, passes along the channel formed by the holes and enters on one side of all the plates in parallel, passes through the filter paper and honeycombed plates, and thence to the outlet channel formed by the holes at the right-hand corner of the plates.

The outlet from the press should be fitted with a suitable strainer in case the operator should not take sufficient care to change the filter paper before it becomes saturated with moisture, as if this is allowed to happen the strength of the papers would be so reduced that one of them might burst and so enter the clean oil pipeline.

A pump mounted in the base of the apparatus produces a working pressure up to $4 \cdot 0 \text{ kg/cm}^2$ for forcing the oil through the filter. It is usually found that the best results are obtained when working at a pressure of about $2 \cdot 0 \text{ kg/cm}^2$.

As the oil passes through the several frames in parallel, the area of filtering surface may be increased by increasing the number of frames as well as by increasing the size of frame used.

350

Fig. 16.6   Typical motor driven filter press (J. O. Buchanan (Engineering) Ltd.)

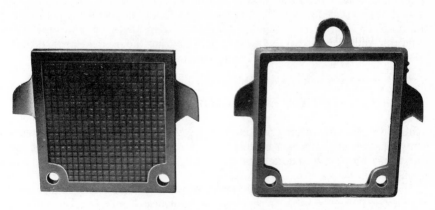

Fig. 16.7   Oil filter press frame and plate

As commercially made, the presses have filtering surfaces varying from about $0.3\ m^2$ up to about $3.0\ m^2$ in area. The capacity of the filters dealing with warm oil in average condition is approximately 5 litres per minute for each square metre of filtering area. The capacity thus varies from about 15 litres/min for the smaller sizes up to about 150 litres/min for the larger unit. It will be appreciated that outputs based upon such considerations must be regarded as only approximate. Some time must obviously be allowed for the removal of paper from the filter press and for its replacement. Further, as has already been stated, the initial condition of the oil will have a bearing upon the throughput and in the case of badly contaminated oil it may be necessary to arrange for the oil to be passed through the filter press more than once.

The motors for driving the oil pumps can be arranged for any available l.v. power supply. The power required varies from about 750 watts for the smaller filters up to about $4.5\ kW$ for the larger units.

Before the filter paper is used it must be dried thoroughly. This is normally accomplished in drying ovens made of double lined sheet steel filled with non-conducting lagging material.

The successful operation of the filter press in removing moisture from the oil depends entirely upon the filter paper being thoroughly dry. It is now possible to purchase packs of pre-dried paper and, provided that they have been stored under the conditions specified by the supplier, further drying is not necessary.

When the oil is in poor condition, containing either moisture or suspended dirt and fibres, the filter paper should be changed frequently, about once every half hour. With oil in better condition the filter paper requires changing much less frequently.

In practice the presses usually run for thirty to sixty minutes, depending on the condition of the oil, and then stopped. Of the sheets of filter paper between each pair of plates, sheet no. 1 is removed, sheets nos. 2, 3, 4 and 5 are moved up and a dry sheet is inserted in each section.

It is usually found that about 1000 litres of oil may be filtered for each kilogram of filter paper used.

In operating the filter the oil should preferably be warm, as oil which is warm and mobile may be filtered far more readily. The best working temperature varies for different oils, but satisfactory results are obtained if the temperature is not less than 25°C. If artificially heated before filtration, the temperature should not be raised above 70°C. It should be noted that as the temperature of the oil is increased, the water solubility of the oil also rises and under this condition the efficiency of the filter paper in drying the oil decreases so that a compromise between speed and quality must always be made.

CENTRIFUGAL OIL PURIFIERS

With the steadily increasing size and loading of transformers the importance of centrifugal oil purifiers as a means of maintaining insulating oil in good condition has also increased, and designs have been evolved to deal with most applications.

The centrifugal separator applies the principle of settling to a system in which gravitational acceleration is replaced by centrifugal accelerations of many thousand g. Centrifugal treatment of oil having a high percentage sludge content

Fig. 16.8   Motor driven stationary centrifugal oil purifier (Broadbent Hopkinson Ltd.)

will, for instance, remove matter which would very quickly clog a filter press, while on the other hand, large quantities of moisture which could not be absorbed in a filter press or would build up pressure in a filter, can be satisfactorily separated.

Figure 16.8 illustrates a Broadbent-Hopkinson oil purifier, the bowl being driven from the motor through helical gears enclosed in an oil chamber. The pump is gear driven from the motor shaft. A positive drive to the purifier ensures the constant speed of rotation, which is essential to efficient purification, as any variation in speed in centrifugal purifiers has the same adverse effect on the efficiency of purification as disturbance has to settlement by gravity.

The Broadbent-Hopkinson disc type bowl can be used for two distinct functions requiring different methods of assembly.

Purification: the separation of two immiscible liquids and suspended solids e.g. water and dirt from oil.

Clarification: the separation of solids from oil or any other liquid.

For purification of transformer oil, it is possible to set the bowl in two ways; one known as Standard Separation, in which the oil is given full access to the conical liners without contacting the sealing water, and the second Positive Washing Action. The first method is achieved by using a complete pack of

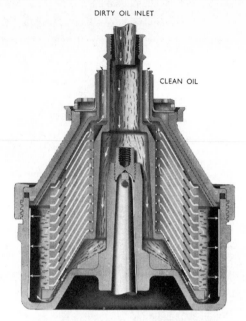

Fig. 16.9   Section of a purifier bowl

perforated conical liners placed on a perforated centre tube, while for positive washing action, an unperforated conical liner is fitted at the bottom of the pack.

Figure 16.9 shows a section of the bowl, which rotates at from 6500 to 12 000 rev/min depending on the size of the machine.

The disc type bowl has the advantage of preventing slip and remixing of the oil and water in the bowl, and also enables the minimum amount of water to be used, thus leaving the maximum surface area of the conical liners for purification purposes. It also aids the removal of dirt from the bowl when cleaning is necessary.

During operation the dirty oil is fed through a nozzle on top of the collecting trays, into the neck of the bowl centre tube where rotational acceleration of the liquid first takes place. With standard separation, immediate access to the conical liners is given to the lighter liquid—oil, while any water and impurities, being the heavier, are thrown to the outside, displacing some of the water already existing as a layer round the wall of the bowl. Entering the holes in the bottom of the centre tube and liners, the oil is divided into a number of thin films. Here the smaller and lighter particles of dirt have a radial distance of only 1 mm to travel before they are trapped on the underside of the disc, eventually sliding down and out into the water seal. The clean oil flows inwards and then upwards through the tubular portion of the top sealing cone, finally being spun off at the top edge into the clean oil tray.

When using the bowl with positive washing action, the oil cannot enter the pack of cones until it has actually passed through the seal or wall of water on the outside of the bowl, owing to the bottom unperforated conical disc. If, when using this method, a small trickle of clean water is allowed to enter the bowl along with the dirty oil, it is possible to effect some reduction in total acidity, as water soluble acids are worked out by the sealing water and so expelled.

The average transformer oil requires to be heated to about 60°C before being put through the machine, and this is most conveniently effected by means of an electric heater.

For the purification of oil in transformers on site, say up to 1000 kVA, special portable machines have been evolved, consisting of a purifier, heater, de-aerating tank, clean and dirty oil pumps, strainer and other accessories mounted on a metal-faced trolley so that the entire equipment can be run alongside each transformer in turn. The oil is circulated through the purifier by means of the pumps until the required dielectric strength is obtained, and there is no necessity to take the transformer out of commission during the operation. This type is illustrated in Fig. 16.10.

The contaminated-oil hose is connected to the base of the transformer, and the oil to be purified is drawn by the dirty oil pump through the strainer and thence flows through the heater to the purifier. Then clean oil is discharged from the purifier to a de-aerating tank, from which it is delivered by the clean-oil pump to the transformer.

Figure 16.11 shows a mobile oil treatment plant with a generator unit, on site.

The Alfa-Laval Co. Ltd. have two basic types of centrifugal oil purifier, the atmospheric type and the high vacuum type. Figure 16.13 illustrates a mobile high vacuum type.

Fig. 16.10 Portable motor driven centrifugal oil purifier (Broadbent-Hopkinson Ltd.)

Fig. 16.11   Mobile oil-treatment plant with its generator unit, in operation on site (Burmah-Castrol Industrial Ltd.)

The atmospheric purifiers are suitable for operating on transformers out of service or for treating oil in bulk storage tanks.

Under normal conditions the oil is heated to about 40°C and satisfactory results are obtained. Under very adverse conditions, *i.e.*, when handling oils which are badly sludged and/or very wet, or emulsified, it may be necessary to raise the temperature to approximately 60°C. This condition, however, would only arise when treating badly sludged oil from an old transformer.

With this equipment the oil is first heated to the required temperature in a continuous flow type electrical oil heater, with heating elements having very low surface loading to eliminate any possibility of overheating the oil. The warm oil is fed to the centrifugal oil purifier which removes all free moisture and solids and the clean dry oil is then returned either to the transformer, or to a clean oil tank.

The centrifuge bowl can again be set up either as a purifier or as a clarifier. When very wet oils are to be handled the purifier arrangement is used, from which any separated water is continuously discharged, the separated solids being retained in the centrifuge bowl. It will be appreciated, therefore, that extremely wet oils can be handled entirely satisfactorily and in addition, even if emulsions have formed these can be dealt with effectively. Also, when running the equipment as a purifier, water washing of the oil can be used to remove water soluble acids.

When treating oils which are not very wet and when it is not necessary to

356

Fig. 16.12   Vacuum centrifuge equipment installed at a h.v. substation (Alpha-Laval Co. Ltd.)

discharge the separated water continuously, the bowl is arranged as a clarifier when the separated water and the solids will be retained in the centrifuge bowl.

If the equipment is required to handle carbonised tapping switch oils as well as non-carbonised transformer oils, an auxiliary filter press is supplied to remove the last and very small traces of finely divided colloidal carbon which may not be taken out by the centrifuge itself as some of these minute particles will be lighter than the oil and, therefore, not centrifugable. These auxiliary filter presses, however, only have to handle very small traces of finely divided colloidal carbon and therefore they can operate for very long periods before it is necessary to change the filter papers. The pipework on the complete equipment is arranged so that the filter press can either be brought into circuit or by-passed, by means of a single three-way changeover valve.

Another Alfa-Laval purifier is the high vacuum type which, in addition to removing free water, solids, sludge etc., also removes dissolved moisture, air and gases, to a very low state of contamination. It is claimed that as a result of this the treated oil has a breakdown value considerably higher than that of oil treated in other types of equipment, and that this high breakdown strength is maintained for very much longer periods of time because the oil can subsequently be contaminated by greater quantities of moisture before any adverse effect takes place on the breakdown strength of the treated oil.

The dissolved moisture content of insulating oil can normally be reduced to 10 parts per million which represents an extremely effective removal of dissolved moisture and this can be used to great advantage in connection with the reconditioning of oil in transformers on load and also for the drying out of transformers on load. The very effective removal of dissolved air and gases from the oil also has considerable advantages, particularly when filling new transformers.

The main reason for the efficient results obtained with high vacuum equipment is that after all free water has been removed from the oil by centrifugal action in the centrifuge bowl, the oil is discharged from the bowl (which is rotating at 6000 rev/min) in the form of an extremely fine mist, into the collecting covers which are under high vacuum. The oil is raised to a temperature around 60°C which is above the boiling point of water at the vacuum under which the centrifuge is operating, i.e., at least 90% of absolute vacuum. Dissolved moisture in the oil, therefore, is vaporised rapidly and discharged through a 'dry' running high vacuum pump. Stress is laid on the fact that heating the oil to only 60°C has no deleterious effect on it whatsoever, particularly in view of the fact that the oil is under vacuum and not in the presence of air. The centrifuged and vacuum dried oil is then passed into a further vacuum and oil collecting tank, from which it is drawn out by an extraction pump and returned to the transformer or clean oil collecting tank as the case may be.

The oil spray from the centrifuge bowl is such that the surface area of the oil which is subjected to the vacuum is three times as great as the area which would result from passage through static nozzles under high pressure. Atomisation of the oil from the high-speed rotating centrifuge bowl gives oil droplets of very much smaller size than the droplets from any form of high-pressure spray nozzles.

This type of high-vacuum equipment is flexible in its applications and may be employed for:
- (a) drying, degassing and de-aerating transformer oil when treating transformers on site, either as new transformers or after repair
- (b) drying out new or old transformers which inadvertently may have become wet before they have initially been put on load, or after service
- (c) the improvement of the insulation resistance of transformers in service
- (d) the bulk treatment of either new or used oil in storage or make-up tanks.

The Alpha-Laval 'Delavac'

This oil-purifying equipment has been developed for the removal of dissolved moisture by vacuum treatment in addition to the separation of solids and free water by normal centrifugal action. This unit is designed to operate on a continuous bypass system, drawing off oil from the base of the tank and returning the treated oil to a connection at the top of the tank.

The centrifuge operates at atmospheric pressure on the water side enabling the unit continuously to separate and discharge away to a drain the water removed from the oil. The water separation is very effective and the oil discharging into the vacuum chamber contains only a trace of water. The vacuum drying section of the unit therefore operates at a high efficiency and provides clean oil from which all free water and dissolved moisture has been removed to less than 20 parts per million.

Fig. 16.13   Portable centrifugal oil purifying equipment of the high vacuum type (Alpha-Laval Co. Ltd.)

The unit incorporates a unique liquid vacuum seal which has no wearing parts. The top half of the seal is stationary and is secured to the outer collecting cover and to the centrifuge frame. The lower half of the seal rotates with the bowl and is filled with oil. This oil seal is under a high centrifugal force. A knife edge on the upper half protrudes into the oil forming a liquid seal between the upper vacuum section and the lower section which is at atmospheric pressure. A heating unit is incorporated around the outside of the vacuum chamber and either electrical or steam heating can be employed.

The clean oil, after the centrifugal separation of free water and solids, discharges from the rotating bowl, in the form of a fine spray, into the vacuum chamber. The oil is then drawn away from the vacuum chamber and discharged back to the tank by a motor extraction pump. Figure 16.12 shows a stationary vacuum centrifuge installation manufactured by the Alpha-Laval Co. Ltd.

The original Stream-Line* plant, Model 'T', was introduced some 30 years ago to serve the needs of the electrical industry. As a self-contained plant it combined the well-known edge filtration principle invented by Dr. H. S. Hele-Shaw, with dehydration and de-aeration of oil by the application of a vacuum of approximately 700 mm of mercury. This combined treatment filters insulating oil in a single passage to a dielectric strength above that specified for new oil.

During the last decade, advanced designs of oil treatment equipment using high-vacuum processing have been evolved for use with the higher voltage electrical equipment now being manufactured by the electrical industry. With

*Now Vokes Ltd, Stream-Line Filter Division.

Fig. 16.14   Stream-Line trailer-mounted high-vacuum filter unit (Vokes Ltd., Stream-Line Filter Division)

the wide experience gained over the years, from manufacturing these units, the Stream-Line Filters Division of Vokes Ltd. has designed and produced a completely new range of oil treatment plant incorporating both medium- and high-vacuum treatment of insulating oils. These are designated MT and HT respectively and have the advantages over earlier models of accessibility, compactness and simplicity. By developing a unit with a single vessel, the accessibility of pipework and pumps is improved. Simplification of the design enables many pipe connections and valves to be eliminated which results in a more compact unit with the risk of leakage reduced, and, since the size of the plant and number of components are reduced, the new unit is much lighter and can be towed and manoeuvred more easily. Furthermore, with this new design concept, additional accessories and items of equipment can be added to the plant as required. A trailer mounted Stream-Line high-vacuum unit is shown in Fig. 16.14.

This range of oil treatment plant consists of self-contained units especially developed for the treatment of insulating oil by the combined action of filtration and evaporation of moisture under vacuum conditions. The filter unit comprises a number of edge-type filter elements mounted in a chamber and totally enclosed in a vessel. Each element consists of a stack of thin paper discs compressed by springs on metal rods to form filter columns, as illustrated in Fig. 16.15. When cleaning becomes necessary it is achieved simply and quickly, without dismantling the plant, by passing compressed air through the filter in the reverse flow direction. The dislodged particles are then removed via a sludge drain valve.

To obtain maximum operating efficiency the temperature of the oil to be treated has to be raised. This is achieved in a heat exchanger incorporating electrical heaters thermostatically controlled and interlocked with the re-circulating pumps to prevent accidental overheating of the oil. The heater elements can be readily replaced without the need to drain the oil from the heat exchanger.

Basically the dehydration and de-aeration chamber is a vacuum receiver fabricated throughout from mild steel and fitted with a detachable cover to form a chamber capable of withstanding medium- or high-vacuum conditions. Air, water and any gaseous contaminants are evacuated from the top of the chamber by the vacuum pump and exhausted to atmosphere. A high-level float switch protects the vacuum pump against oil or foam rising to the level of the pump inlet and a vacuum gauge is fitted to the chamber. A typical flow diagram is given in Fig. 16.16.

The type of vacuum pump fitted depends upon the capacity of the plant and the degree of vacuum required. For medium plants (MT) a reciprocating vacuum pump is employed whilst for high-vacuum plant a single stage rotary vacuum pump is used. The pumps have gas ballast but are capable of producing a closed suction vacuum of 0·01 mm without gas ballast and 0·5 mm with gas ballast.

The unit is completely self-contained and oil entering the plant passes through a metering pump designed to give the specified rated flow of oil. Before entering the filter chamber the oil is heated to approximately 70°C in the heat exchanger which is located below the filter chamber in order to reduce heat losses. The oil being filtered passes from the outside to the inside edges through the minute interstices formed between the paper filter discs leaving even the smallest solids on the outside edges. As the oil flows through these very small spaces, it is dispersed into very thin films thereby providing the maximum area for moisture to evaporate and it is this film effect which produces the required filtration. The oil passing from the filter chamber directly into the vacuum chamber flows over

Fig. 16.15   Principles of the Stream-Line filter pack

FLOW DIAGRAM

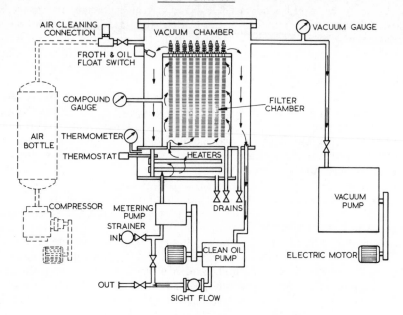

Fig. 16.16   Flow diagram of Stream-Line filter

the outside of the filter chamber allowing air and moisture to be extracted to atmosphere. The treated oil accumulates at the base of the vacuum chamber and is drawn off by a multi-stage centrifugal pump which maintains an oil seal between vacuum chamber and outlet pipeline.

For applications where the available power source is insufficient to supply the heat exchanger, filtration and removal of free moisture can be carried out at reduced temperatures, the moisture being absorbed and retained in the filter packs, which dry out automatically *in situ* when the plant is next operated and when heating can be applied. Cold filtration is recommended where heat soluble contaminants such as varnish are known to be present in the oil, since under this condition they remain out of solution and can more easily be removed during the passage through the filters.

Both versions of the plant are available in three standard forms, fixed; semi-mobile on castors; and fully mobile. If required a completely weatherproof cowling can be attached to any of the models.

Although designed primarily for the treatment of insulating oils, Stream-Line plants are equally suitable for use with other liquids where cleanliness and dryness are of major importance, including the synthetic insulant Aroclor. Plants are available with capacities ranging from 220 to 9000 litres per hour.

The standard unit is suitable for most applications such as the treatment of oil in transformers and switchgear in factories, substations and switching stations. Where special treatment is necessary on high-voltage transformers and the water content is required to be less than 10 parts per million, and gas content less than 0·5%, the high-vacuum unit should be employed.

The units can be fitted with various accessories to assist in the servicing and commissioning of transformers. Additional vacuum pumps or rotary exhausters can be fitted to the chassis to enable the transformer tank to be evacuated during oil treatment, and flowmeters, chart recorders for temperature, vacuum and resistivity measurements, to monitor the filling or drying-out operations are also available.

The resistivity test cell automatically meters off a small quantity of oil from the main filter outlet every three minutes and the sample is contained in the test cell for two minutes, during which time the resistivity is measured and recorded on a chart. If it is required, an estimate of the loss angle can be calculated from the resistivity reading.

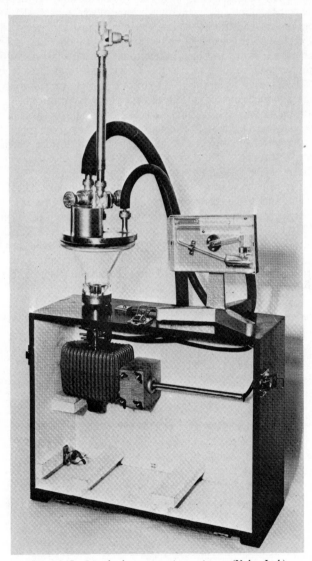

Fig. 16.17 Dissolved gas measuring equipment (Vokes Ltd.)

Another important consideration when treating transformer oil is the value of the dissolved air content. This value can be obtained by the Stream-Line Imaco gas measuring unit which provides a precise measurement. The unit is based on a development of the National Engineering Laboratory and operates by testing a sealed sample of known volume in an evacuated chamber, where it is then agitated by a vibration strainer. The resultant rise in pressure caused by the release of the dissolved gas is measured and the gas content ascertained. The de-aeration is complete in under two minutes and the test, including the filling of the sampler and the cleaning of the unit can be carried out in approximately fifteen minutes. Figure 16.17 illustrates this gas measuring equipment.

J. O. Buchanan (Engineering) Ltd., have also developed and manufacture fully mobile transformer oil treatment equipment which meets the requirements of oil filling and oil treatment for high-voltage transformers on site. A typical equipment is shown in Fig. 16.18. This plant can handle 7000 to 9000 litres of transformer oil per hour and has independent vacuum pumps with a capacity of 8500 litres per min, capable of evacuating a transformer tank down to 1·0 mm absolute. Oil delivered from the plant contains not more than 0·5 % dissolved gas, and has a water content of less than 5 parts per million.

This equipment has a self-contained Diesel generator and a patented oil burning boiler system to provide the heating to the heat exchanger, thus making the plant independent of site electrical supplies. It should be noted that at low ambient temperatures the power demand on site could be of the order of 250 kW. A line diagram of the equipment is given in Fig. 16.19.

Transformer oil when untreated will normally contain 20 to 30 parts per million of water and also approximately 10 to 11 % of air by volume at saturation. In order to bring the oil to the standard required for filling and servicing on site, it must be filtered to remove solids and vacuum treated to de-aerate and dehydrate the oil and the degree to which complete separation can be achieved depends upon the relative vapour pressures of oil-water-air system and also the temperature at which the process is being undertaken. It can be shown that optimum separation will occur when using a very high vacuum at a low temperature. However, in practice, a minimum temperature of 50°C is necessary in order to prevent the oil from foaming. At this temperature it will be found that

Fig. 16.18 Self-contained mobile oil-treatment plant (J. O. Buchanan (Engineering) Ltd.)

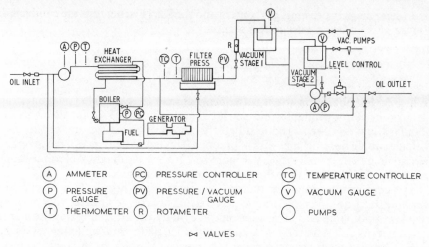

(A)	AMMETER	(PC)	PRESSURE CONTROLLER	(TC)	TEMPERATURE CONTROLLER
(P)	PRESSURE GAUGE	(PV)	PRESSURE / VACUUM GAUGE	(V)	VACUUM GAUGE
(T)	THERMOMETER	(R)	ROTAMETER	◯	PUMPS

▷◁ VALVES

Fig. 16.19   Line diagram of the mobile oil-treatment plant shown in Fig. 16.18

by applying Henry's and Rault's laws, providing that oil and vacuum are in equilibrium, a vacuum of 7·6 mm will be sufficient to ensure an air content of less than 0·10% and a water content of 1·5 parts per million (water solubility at 50°C = 150 parts per million).

It is therefore of great importance to achieve equilibrium between the oil and vacuum and in this equipment a compact two-stage vacuum unit has been incorporated with an overall equilibrium efficiency approaching 90%, and a second stage vacuum of approximately 1 mm. The use of a very high vacuum is avoided and therefore the attendant maintenance difficulties associated with operating a mobile plant, as distinct from a factory installation. The tendency to distil the lighter fractions from the transformer oil also does not arise under these conditions.

The design of this type of equipment is such that large surfaces of oil are provided during the passage through the plant and sufficient holding time allowed for the oil-vacuum system to come to equilibrium. This object has been achieved by allowing the oil to flow over a suitable packing and by designing the degassing unit so that the complete surface is wetted by the oil passing. The proportioning of the surface area is in fact a compromise. If too large a surface is provided there will be a marked reduction in the efficiency due to the oil channelling and, conversely, if insufficient surface is presented to the oil the film will become too thick and equilibrium will not be obtained. The shape of the oil film is also important, and should be as flat as possible. In this respect it should be noted that to spray the oil into the vacuum chamber is unsatisfactory. There are two reasons for this conclusion, firstly the rate of gas diffusion from the oil is extremely slow and with the oil in small droplets there is little internal mixing, and secondly, the effect of surface tension is to build up a positive pressure in the droplets which completely upsets the basic parameters of the oil-vacuum system.

In addition to the equipment described, the mobile oil treatment plant also has a combined control and laboratory section at the rear of the trailer. It is from here that the plant is operated and the processing continuously monitored and where dielectric testing, gas content and the Karl Fischer tests are conducted. Figure 16.20 shows a view of this section.

SITE TESTING

When drying out oil on site it is frequently found that suitable voltage testing equipment is not available, and a self-contained portable set for use with the standard gap has been produced, which can be designed for any of the low-voltage supplies usually met with in practice. This equipment gives a range on the high-voltage side from 0 to 50 kV. Figure 16.22 shows the diagram of connections while Fig. 16.21 shows the external view of the equipment.

The equipment comprises the testing transformer proper, a continuously variable auto transformer, double-pole circuit breaker, voltmeter with protective fuse, pilot lamp and oil testing cup. The oil immersed testing transformer is contained in an electrically welded plain boiler plate tank while the auxiliary gear is housed in the control panel mounted on the side of the tank. The oil testing cup is mounted upon an insulating board between the capacitor type high-voltage terminal and a metal column erected on the tank cover, and for visibility is made of glass. A large opening is provided at the top of the cup for the introduction and renewal of oil, and also to enable the electrodes to be cleaned after each discharge.

In order to provide smooth control of the testing voltage from zero to the

Fig. 16.20   Control and laboratory section of the mobile oil-treatment plant shown in Fig. 16.18
(J. O. Buchanan (Engineering) Ltd.)

Fig. 16.21 *3 kVA, single-phase, 50 Hz, 250/50 000 V testing equipment*

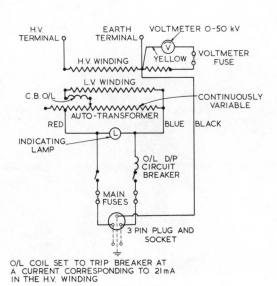

Fig. 16.22 *Diagram of connections of oil voltage testing equipment*

maximum value an auto transformer is connected in the primary circuit. The operating spindle of the auto transformer is brought out through the switch panel and fitted with a handwheel and indicator. The voltmeter which is scaled 0–50 kV is connected to a tertiary winding, one end of which is connected to the start of the secondary winding and to earth. A pilot lamp is included to indicate to persons in the vicinity that the testing set is alive.

Oil testing is carried out in accordance with the provisions of B.S.148, the method of operation being as follows:

First, the auto transformer handwheel is located in the minimum position hard against the stop. In order to determine the true condition of the oil it is essential that the oil cup and electrodes should be chemically clean and dry and free from adhering fibres, and special care is essential in obtaining this. Oil is then poured into the testing receptacle through the opening in the top to a minimum depth of 40 mm above the top of the spheres, care being taken to see that all air bubbles disperse. The oil should be allowed to stand for 10 minutes to ensure this. The circuit breaker may then be closed, and the handwheel rotated in a clockwise direction, thus increasing the voltage across the sphere gap. The voltage should be raised from approximately 15 kV to the B.S. site

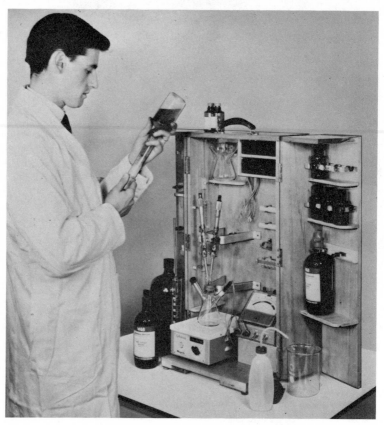

Fig. 16.23   Karl Fischer test equipment for the detection of the moisture content of transformer oil
(J. O. Buchanan (Engineering) Ltd.)

testing value at a uniform rate in 10–15 s, and then kept at that value for one minute. Three tests are carried out each on a separate filling. Complete breakdown is indicated by the establishment of sustained arcing, and when this occurs the handwheel should be rotated in the opposite direction until it comes up against its stop, and the switch opened before the oil testing cup is removed.

A higher minimum proof voltage may be required for certain high-voltage apparatus, or where recommended by the manufacturer.

After each breakdown of an oil sample all traces of carbonised matter should be removed from the oil testing cup, and the sphere electrodes themselves should be cleaned thoroughly. If in the course of time the electrodes become pitted, they should be renewed.

Care must be exercised when taking the oil samples for testing purposes either from steel containers or from a transformer tank. In every case the apparatus used for withdrawing the oil samples, and for subsequently holding them, must be chemically clean and dry as, if they are not, inaccurate test results on the oil will be obtained. As water and the majority of foreign matter settles to the bottom of the oil, samples should be withdrawn from the bottom of the container, but it is sometimes also advantageous to take samples from, say, the centre of the body of oil, as the results will usually indicate whether or not small loose fibres are present.

It may happen, however, that the necessary testing apparatus is not available, in which case rough and ready approximate indications must be resorted to. The crackle test for free water is performed by heating rapidly over a silent flame a sample of the oil in a test tube. If moisture is present in the sample a characteristic crackling will be detectable. Alternatively a metal rod, heated to a dull red heat may be lowered into the oil sample and used to stir it thoroughly. The characteristic crackling will again indicate the presence of moisture. Water can be detected if 50–60 parts per million are present. The Karl Fischer test may of course be used for the accurate determination of the water content of transformer oil. This test will detect values as low as 2 to 3 parts per million. Figure 16.23 illustrates the equipment which is comparatively robust and will enable an analysis to be made on site by staff having only limited academic training.

Chapter 17

Transformer oil

As in past editions of this book, this chapter aims mainly at the provision of a guide to the appreciation of the function of transformer oil in the performance of the electrical equipment which it fills.

The 1972 edition of B.S.148 brought the specification technically into accord with IEC Publication 296: 1969, 'Specification for new insulating oils for transformers and switchgear'; but the lower viscosity oil (Class II) of the two oils specified by the IEC was not covered, though details were given in an appendix.

Many changes were made in the new Standard. Saponification and mineral acidity tests were omitted, but a density test was included to conform with IEC requirements, and a new test was introduced for loss tangent (tan δ) to give an additional indication of the electrical quality of an insulating oil. Changes were made to the oxidation test, electric strength test and to the procedure for detection of corrosive sulphur, and the crackle test was replaced by a method giving precise determination of water content. The following characteristics are laid down in B.S.148:1972.

Characteristic	Limit
Sludge value (max.)	0·10%
Acidity after oxidation (max.)	0·40 mg KOH/g
Flash point (closed) (min.)	140°C
Viscosity at -15°C (max.)	800 cSt*
Viscosity at 20°C (max.)	40 cSt
Pour point (max.)	-30°C
Electric strength (breakdown) (min.)	
Oil delivered in bulk	30 kV
Oil received in Great Britain in drums	27 kV
Oil shipped overseas in drums	No set limit
Acidity (neutralisation value) (max.)	0·03 mg KOH/g
Corrosive sulphur	Shall be non-corrosive
Water content (max.)	
Oil delivered in bulk	35 ppm
Oil received in Great Britain in drums	50 ppm
Oil shipped overseas in drums	No set limit
Density at 20°C (max.)	0·895 g/cm^3
Loss tangent at 90°C (max.)	0·005
Resistivity	No set limit

*1 cSt = 10^{-6} m^2/s

PHYSICAL PROPERTIES OF TRANSFORMER OIL

In the foregoing specification, the tests governing the physical nature of transformer oil are viscosity, closed flash point, density and pour point.

The heat generated in the transformer itself is transferred by conduction through the solid insulation to the oil, which by convection effects, either alone or aided by forced circulation, flows over the cooling surfaces. These surfaces are either integral with the tank or in separate units, as discussed elsewhere. The efficacy of this process is essentially dependent upon the oil having a low viscosity.

Viscosity is measured in glass tube viscometers, which can be closely standardised and also allow use of the centistoke, which is based upon the absolute definition of viscosity. In the specification, the temperatures of measurement indicated for viscosity are $-15°C$ and $20°C$. A chart in Appendix E of the specification, here reproduced as Fig. 17.1, shows generally the extent to which an oil complying with the requirements of the Standard varies in viscosity with temperature.

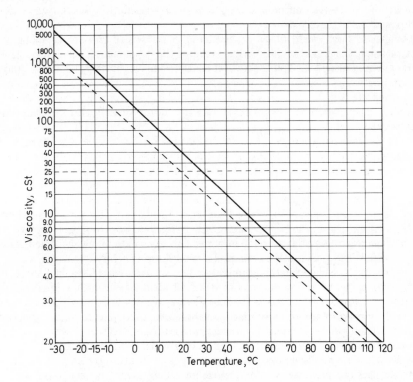

Fig. 17.1 Viscosity/temperature chart. The full black line relates to the maximum permitted viscosity of 40 cSt at 20°C as used in the UK, whilst the broken line relates to the second grade of oil specified in IEC Publication 296. The need for two grades of oil arose from differences in national requirements, largely dictated by climatic conditions. The characteristics of IEC Class II oils are identical with Class I except for the following: flash point (closed) (min.) 130°C; viscosity, at −30°C, 1800 cSt (max.) and at 20°C, 25 cSt (max.); pour point (max.) −45°C

With the increase of temperature, the viscosity of oil falls, at a rate dependent upon the chemical nature of the particular oil. An unduly high viscosity at low temperatures is guarded against by the introduction of a maximum viscosity limit at $-15°C$. B.S.148 does not lay down a lower limit for viscosity because the specification of a minimum for closed flash point prevents the use of the lowest viscosity fractions of oils. Similarly, because of the specification of a maximum value for viscosity, the closed flash point of transformer oil cannot be very much above the minimum requirement stated. The oil refiner gives due regard to these points in selecting a base oil for the manufacture of transformer oils so that a suitable compromise is obtained.

The practical reason for fixing a limit for closed flash point is to prevent the excessive loss of oil by vaporisation in the course of the normal cycle breathing action of a transformer, as a result of which the bulk of the oil would tend to decrease rapidly, and frequent topping-up would be necessary. This effect applies essentially to distribution transformers not fitted with conservators. Should proper attention not be given to the maintenance of the correct level of oil in the transformer tank, the cooling action of the oil by convective circulation may be reduced, or even part of the transformer core and windings may not be covered with oil. As a result, both the oil and the solid insulation may attain excessive temperatures, the winding insulation becoming brittle and possibly ultimately disintegrating.

The closed flash point of oil is measured by means of the Pensky-Martens apparatus. It is a guide to the temperature of the oil at which the combustible vapour in a confined air space above it accumulates sufficiently to 'flash' upon the application of a flame or other equivalent source of ignition, such as an arc or a spark. It will be recognised that fire and explosion are to some degree potential risks whenever petroleum oils are component parts of electrical equipment.

Since the operating temperature of the oil in service is very much lower than the allowable flash point of transformer oil, minor differences in flash point have no very great importance. On the other hand a change in the closed flash point may indicate the contamination of the oil by more volatile products which, even though present in only small quantities, may constitute an explosion hazard when the oil is heated in ordinary service, due to a high concentration of inflammable vapour above the oil surface. Such contamination has been known to occur upon removing oil from a transformer for inspection purposes, when using drums which had previously contained a volatile solvent.

Certain types of electrical faults cause the breakdown of oil in service to comparatively volatile low molecular weight hydrocarbons or to inflammable gases which, having dissolved in the transformer oil, result in a reduction of the closed flash point of the oil in service. It is therefore important to have on record the initial flash point of the oil at the commissioning of the transformer, so that comparatively small changes occurring during service can be detected.

A density limit has been introduced into the specification because at very low temperatures the increase in the density of an oil might be such that ice, if present, would float on top of the oil. The limiting density of 0.895 g/cm^3 (max.) at $20°C$ ensures that the temperature must fall to about $-20°C$ before the density of an oil, of the maximum permitted density at $20°C$, would exceed that of ice.

The remaining characteristic of a purely physical nature is the pour point, which is the lowest temperature at which an oil will flow under certain specified conditions of pressure difference. It gives a guide to the lowest ambient temperature at which transformers filled with oil can safely operate, and a maximum of $-30°C$ has been standardised as suitable for the majority of conditions.

CHEMICAL PROPERTIES OF TRANSFORMER OIL

The chemical tests listed in the Schedule of Characteristics fall into two classes: those having a bearing upon the basic type of oil, namely the tests for sludge formation and acidity development after oxidation, and those which ensure the freedom of the oil from small amounts of undesirable compounds, these being the tests specified for acidity (neutralisation value) and corrosive sulphur or copper discolouration. Water content, whilst not strictly a chemical property, may reasonably be considered within the second class.

The oxidation test is carried out by maintaining a sample of 25 g of the oil in the presence of metallic copper at 100°C whilst oxygen is bubbled through the sample for 164 hours. The oil is then cooled in the dark for one hour, diluted with normal heptane and allowed to stand for 24 hours, during which time the more highly oxidised products are precipitated as sludge, which is separated and weighed. The remaining solution of oxidised oil in n-heptane is used for the measurement of the acidity development. The combined two values satisfactorily define the oxidation stability of the oil.

The effectiveness of transformer cooling will be maintained only if the oil has a good oxidation stability, so that the viscosity is substantially unchanged during the many years of service life expected, and no solid materials, known by the general name 'sludge', are formed. Sludge formation reduces the efficacy of cooling in several ways. Deposition of sludge upon the insulation seriously impedes the conduction of heat from the conductors and other parts from which the removal of heat is essential. Sludge deposition may also occur in the oil ducts of windings, so that the rate of oil circulation over the heating surfaces is reduced. Further, sludge may accumulate in those parts of the unit employed for cooling the oil; this is not only equivalent to lagging the cooling surfaces, but also results in a loss of efficiency due to a reduction of the oil circulation rate within the transformer unit as a whole. The cumulative effect of sludging, therefore, is an increase of operating temperature within the transformer, the serious implications of which are only too well known.

It is to be concluded that oxidation stability is the most important single property of transformer oil and in this respect it is still true to say that sludging is the most undesirable type of oil deterioration, although the emphasis of more recent studies has been upon the relative tendencies of oils to develop acidity. In earlier days, when sludging was a real problem, large users of transformers, and also the oil suppliers, were not content merely to await the onset of operational troubles due to sludge formation. Samples of oil were drawn periodically for examination and the measurement of acidity was found to be the simplest test by which to follow the course of the early stages of oil deterioration. Later, the introduction of the so-called non-sludging oils, Class A of the B.S.148 current at that time, placed even more emphasis on acidity because, whilst rarely developing sludge in service, these oils were prone to relatively rapid acidity development.

It is to be understood that, concurrently or in sequence with the formation of acid bodies, other products such as peroxides, aldehydes, ketones, lactones, together with water and carbon dioxide, are formed, some of which by inter-action or further oxidation would ultimately become the solid material con-stituting sludge. However, further research found that acidity was more easily measurable than were the other initial products of oxidation, and this property became clearly important from another point of view. The occurrence of corrosion in some transformer tanks and on tank covers, for which volatile acids formed in the oil were naturally thought to be the cause, constituted a second reason for the decision to study acidity in later research work.

When corrosion occurred it was found not below the level of the oil surface, even with oils of high acidity, but on the walls of the tank above the level of the oil, and particularly on the undersides of tank covers. This naturally confined the corrosion troubles to the smaller distribution transformers as many of this type were not fitted with oil conservators.

The effects of corrosion could be serious because flakes of corroded tank covers would sometimes fall upon the core and windings, causing faults of various kinds. Unfortunately, the regular inspection of equipment and the periodic examination of oil had not become general practice, and reports were occasionally received of cases where the perforation of the cover by rusting had been the first intimation that all was not well with a transformer. In some of these cases the acidity of the oil was found to be unduly high, varying from 2 to more than 10 mg KOH/g, but this was by no means general and, on the other hand, many examples were quoted of high acidity not being associated with corrosion. It was evident that the trouble was due to various combinations of circumstances but no very clear pattern emerged. With improved oil maintenance, combined with the attention given to treatment of the steel work above oil level with anti-corrosive paints or other protectives, the operational difficulties which could be directly attributed to acidity became less serious. Nevertheless, oil technologists and others interested in the problem still concentrated their attention upon oil acidity because it can be the forerunner of sludge formation.

The result of research was the production by oil suppliers of a balanced transformer oil with low acid forming tendencies as well as low sludging. B.S.148:1959 covered a single oil of this type and this is also true of the current specification. In B.S.148:1972 though, the actual limiting values for sludge and acidity formation are much lower, because a more precise and convenient oxidation test at a lower temperature was introduced.

The acidity or neutralisation value of the oil as supplied, as distinct from that after the oxidation test, is not a property that can be regarded as a criterion of quality since, in the course of normal refining, it is possible to reduce the neutralisa-tion value to negligible proportions. It does, however, constitute a useful safeguard against adventitious contamination. The specification recognises the difficulty of attaining complete freedom, but a limit is set at the very low value of 0·03 mg KOH/g. The acidic materials in question are not capable of exact description as chemical entities; they may range in type from the so-called naphthenic acids, which are present naturally in unrefined petroleum, to organic acids resulting from the oxidative breakdown of the oil in processing. The method used for the quantitative estimation of acidity in oil is to find the volume of a standard solution of an alkali, usually caustic potash, required for neutralisation and, accordingly, a convenient mode of expression of acidity is the weight in milligrams of caustic potash (KOH) required per gram of oil. In practice, the 'titration' to the neutral point is carried out in the presence of a suitable solvent for organic acids, as described in Appendix B of the Specification, the point of neutralisation being shown by the colour change of an added indicator, this being an organic material of a type which experiences a constitutional change upon becoming alkaline, accompanied by a change in colour.

The test for corrosive sulphur, described in earlier editions as deleterious sulphur and copper discoloration, is more severe in B.S. 148:1972. It involves immersing a strip of polished copper in oil at a temperature of 140°C and in an atmosphere of nitrogen for a period of 19 hours, after which the copper is examined. An oil is failed if the copper strip, or part of it, is dark grey, dark brown or black. A pass does not mean that the oil is necessarily completely free from sulphur compounds but that any present are not of an active nature. In fact, trouble arising during service due to attack on copper is very rare indeed with modern transformer oil.

The need for cleanliness of oil is emphasised in a general manner by the statement in the specification that the oil shall be 'clean and free from matter likely to impair its properties'. However, the specification also includes tests that aim at securing freedom from free water and other contaminants, and as far as water is concerned a quantitative determination has been introduced for the first time in B.S.148, although it is not included in the IEC specification. The Karl Fischer method detailed in B.S.2511 is used. Nevertheless the specification notes that an approximate indication of free water (*i.e.* undissolved water) can be obtained by the crackle test. The crackle test is based upon the fact that the presence of quite minute amounts of free water can be detected by the audible crackle caused by its instantaneous expansion to vapour as a result of very rapid heating to well above its boiling point.

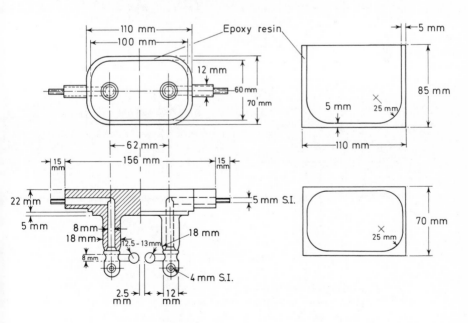

Fig. 17.2 Oil test cell

ELECTRICAL PROPERTIES OF TRANSFORMER OIL

The electrical tests included in the Schedule of Characteristics of B.S.148:1972 are those concerned with the use of oil as an insulant.

The electric strength test in the specification, although included here because it is directly associated with the use of oil as an insulant, is nevertheless not a test for the measurement of electrical quality in the normal sense, but mainly provides an indication of the physical condition of the oil. This emphasis is brought out in the specification itself, where it is stated that the electric strength is very greatly affected by the most minute traces of certain impurities. Those impurities most likely to influence the electric strength test in practice are moisture and fibre, and, in particular, a combination of the two.

In this test, oil is subjected to a steadily increasing alternating voltage until breakdown occurs. The breakdown voltage is the voltage reached at the time the first spark appears between the electrodes, whether it be transient or established. The test is carried out six times on the same cell filling, and the electric strength of the oil is the arithmetic mean of the six results obtained. The electrodes are mounted on a horizontal axis with a test spacing of 2·5 mm. Figures 17.2 and 17.3 illustrate standard oil test cells, of glass or transparent plastic

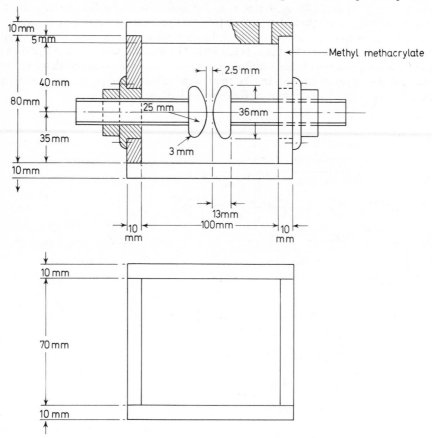

Fig. 17.3 Test cell with spherical surfaced electrodes

376

material, with effective volumes of 300 ml to 500 ml, preferably fitted with suitable lids. The electrodes, of copper, brass, bronze or stainless steel, are either spherical and of 12.5 to 13 mm diameter, as shown in Fig. 17.2, or spherical surfaced and of dimensions shown in Fig. 17.3.

The first application of the voltage is made as quickly as possible after the cell has been filled, provided that there are no longer any air bubbles in the oil, and no later than 10 minutes after filling. After each breakdown, the oil is gently stirred between electrodes with a clean, dry, glass rod, care being taken to avoid as far as possible the production of air bubbles. For the five remaining tests, the voltage is reapplied one minute after the disappearance of any air bubbles that may have been formed. If observation of the disappearance of air

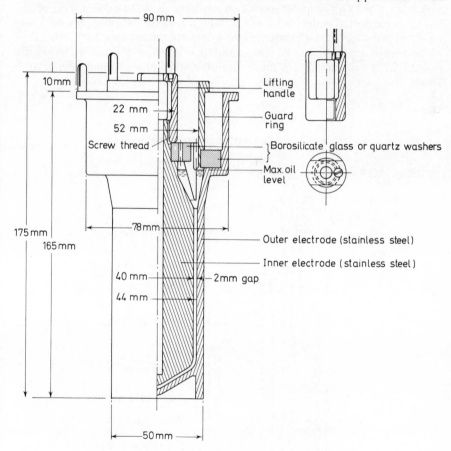

Fig. 17.4   Loss tangent test cell recommended by CIGRE
(Quantity of liquid required to fill cell, 45 ml approx.)

bubbles is not possible, it is necessary to wait five minutes before a new break-down test is commenced. It is generally recognised that the precision of the electric strength is not high, and it is not surprising that, in a test so susceptible to the presence of minute traces of impurities, variable results may be obtained, especially as any impurities may not be distributed uniformly over the bulk of the oil, or even the oil sample itself. Accordingly, care is essential in sampling, to avoid the introduction of contaminants not previously present in the oil. The maxima for electric strength in B.S.148:1972 are lower than previous editions. This does not reflect a lowering of standards; it arises only because a new method of testing has been introduced—especially because the gap between the electrodes has been reduced from 4 to 2·5 mm.

Electrical properties of a more fundamental character are loss tangent and resistivity, but only a test on the former is mandatory in B.S.148:1972. There is reference, however, to the value of resistivity as an indication of electrical quality, especially of used oils, and methods of testing for both properties are described in the specification. For the loss tangent test a specially designed test cell or capacitor is filled with oil under test which displaces air as the capacitor dielectric. The cell is connected in the circuit of a suitable a.c. bridge where its electric losses are directly compared with those of a low-loss reference capacitor.

The cell employed should be robust and have low loss; it must be easy to dismount, clean, reassemble and fill, without significantly changing the relative

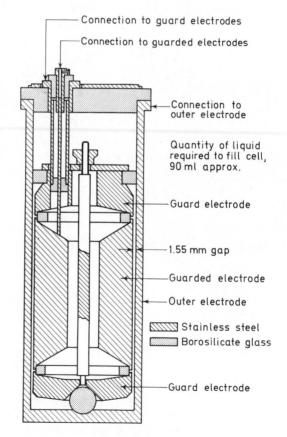

Fig. 17.5   A three-terminal loss tangent test cell

378

positions of the electrodes. Figure 17.4 is a diagram of a cell recommended by CIGRE*, and Fig. 17.5 illustrates a three-terminal test cell which is now widely used. For measurement of the loss tangent an alternating current (40 Hz to 62 Hz) bridge is used, which should be capable of measuring loss tangent down to 1×10^{-4} for normal applications, but preferably down to 1×10^{-5}, with a resolution of 1×10^{-5} in a capacitance of 100 pF. The voltage applied during the measurement must be sinusoidal. Measurement is made at a stress of 0·5 to 1·0 kV/mm at 90°C, and is started when the inner electrode attains a temperature within plus or minus 0·5°C of the desired test temperature.

For d.c. resistivity measurement, the current flowing between the electrodes is measured when a specified voltage, normally 500 V, is applied to the cell. The current is noted after the direct voltage has been applied for one minute, and again after reverse polarity has been applied for one minute, with the electrodes short circuited for five minutes between the two measurements. Average resistivity values are calculated from the readings taken after application of direct and reverse polarity. For measurement, an instrument capable of detecting 10^{-11} amperes is required.

INHIBITED TRANSFORMER OILS

In the 1972 edition of B.S.148 reference is made to the use of 'additives' as possible constituents of transformer oils: 'the oil shall be pure hydrocarbon mineral oil . . . . without additives. By arrangement between seller and buyer the oil may contain an oxidation inhibitor or other additive, in which case the oil, before inclusion of the additive, shall comply with this Standard. Oils complying with the requirements of this Standard are considered to be compatible with one another and can be mixed in any proportion; this does not necessarily apply to inhibited oils'.

The insertion of this clause has two objects, (i) to allay fears expressed by some users that the introduction of inhibitors would allow the use of under-refined oils, which are themselves basically unsuitable for use in transformers, and (ii), to ensure as far as possible that oils, after possible loss of inhibitor in service, would not be prone to unduly rapid deterioration, as might conceivably be the case if the base oils were not of the best modern type.

The use of additives in lubricating oils and other petroleum products is now familiar to most engineers. They are usually well defined chemical products and are of widely different natures according to the purposes of their introduction. Such purposes vary from the improvement of the properties of oils as lubricants to the modification of the ignition characteristics of petroleum fuels, but the special, if not sole, interest in the case of transformer oils is the improvement of oxidation stability.

To explain the behaviour and particular usefulness of anti-oxidant additives, it is necessary to expand further the subject of oxidation stability which has been discussed briefly earlier in this chapter. Reference was then made to certain intermediate products other than acid and sludge which occur as a result of oxidation of mineral oils but, in connection with the mechanism of action of anti-oxidant additives, the most important type of intermediate compound is

*Conférence Internationale des Grandes Reseaux Electriques

the initial oxidation product known as hydroperoxide. Such compounds are of an extremely short-lived nature, but are highly reactive and are rapidly transformed into one of the more stable intermediate oxidation products already mentioned. However, certain chemical groups have been found to be capable of direct interaction with hydroperoxides, thus breaking the chain of oxidation products, and compounds embodying such chemical groups may be specially added to the oils as oxidation inhibitors.

The oxidation processes discussed earlier inevitably occur with pure hydrocarbons in the presence of oxygen or air, the rate greatly increasing with elevation of temperature, but being affected to an even greater degree by the presence of pro-oxidant materials. Notably active in this regard are certain metals, and, of those normally present in transformers, copper and iron are specially important; metals acting in this manner are known as catalysts. In the case of copper the catalytic action may be shown by the metallic material, though the effect of oil-soluble copper compounds is even more pronounced; iron appears to be active mainly in the case of its compounds, especially those which are oil-soluble. Such catalytic action may be inhibited by certain types of anti-oxidant compounds entirely different from those used for the chain-breaking mechanism. In the case of transformer oils both types of anti-oxidant action may have useful application, according to the circumstances.

Actual operational experience in Great Britain with inhibited transformer oil extends back to the pre-war period, but there is available very little published information concerning it. The application of such oils in the United States has reached large proportions, although more recently some doubts have been expressed with regard to their universal use. The information available is somewhat conflicting, although this is possibly because the standardisation of transformer oil upon a national basis had not been attained in the United States to the same degree that it has in Great Britain. Probably the main reason why inhibited oils have not found wide usage in this country is that the uninhibited oils, which have been available for some years, coupled with the increased care with which they are maintained, has resulted in such long life in most transformers that users have been reluctant to meet the higher initial cost inhibited oil necessarily involves.

However, such oil tends to be more popular on the Continent as well as in the U.S.A., and studies in progress are likely to lead to the inclusion of a special oxidation test for inhibited oil in a future edition of the IEC specification and subsequently in an amendment to, or revision of, B.S.148.

The technical possibilities of inhibited oils are naturally most important in applications where the operating temperature of the oil may be higher than average in value, but due regard must necessarily still be given to the effects of such temperatures upon cellulose insulation.

OIL MAINTENANCE

It has been mentioned earlier in this chapter that the development of improved practices of oil maintenance led to a considerable reduction in operational difficulties directly due to acidity. Maintenance schemes were mainly introduced by the larger users, who were also interested in recording service histories of the various types of oils available, and there came into being a knowledge, on a broad basis, of the level of deterioration of oil that could be allowed without incurring serious dangers due to sludging. As a result, a substantial proportion of users have for all practical purposes been free of difficulties with sludge formation for many years. The importance of maintaining this happy state of affairs does not rest only upon the economic aspects of oil replacement costs, because a considerable saving in cost of electrical equipment over a period also results.

A very thin layer of sludge on the insulation inevitably increases the copper to oil temperature gradient, so that, with given conditions of loading, the winding temperature becomes somewhat higher than would otherwise be the case. If, under fully loaded conditions, the average winding temperature rise in the coil stack is, say, 60°C, it may not seem a matter of very great significance if, for example, it should rise to 64°C. Nevertheless, considered in terms of retention of strength of cellulose insulation on the copper, an increase of temperature of only 4°C has been reliably estimated to result in a decrease of insulation life of approximately 30%.

Further, it is extremely difficult to remove sludge from a transformer winding as some of the sludge will be deposited in places which are inaccessible. Accordingly, such a unit will thereafter be working continually at a temperature slightly higher than would otherwise arise for the load to be carried, with the inevitable loss of life of insulation and an increased rate of oil deterioration.

For the convenience of operating staff, oil maintenance has naturally been made a part of general transformer maintenance, the time cycle for the latter synchronising with that chosen for oil inspection or routine treatment, where this is practised. In addition to the examination of an oil sample for its degree of deterioration by oxidation, i.e., for its acidity or for the presence of sludge, the condition of the oil as an insulant is assessed, in particular with regard to its dryness. Further, it is sometimes possible to diagnose the existence of an electrical fault in the transformer, by the analysis of dissolved gases extracted from the oil. The technique is an important and valuable extension to the analysis of gases collected in Buchholz relays. It will readily be understood that the results obtained can sometimes guide the future pattern of transformer maintenance.

Detailed recommendations for the practice of oil maintenance have been issued by The British Standards Institution in a Code of Practice C.P.1009, having the title 'Maintenance of Insulating Oil', so that only the broad essentials will now be discussed.*

Oil sampling in the field

Inspection of oil should be made at least annually throughout the life of the transformer, and there are circumstances in which more frequent examination is advisable.

Cleanliness of the sampling equipment is essential, this beginning with the sampling cock itself and extending to all the vessels used, but especially to the sampling bottles. The size of the latter depends on the tests to be carried out; 500 ml should be adequate for all normal tests, but for acidity tests only, 100 ml is adequate.

*C.P.1009 is at present under revision

Protection from atmospheric contamination while sampling is in progress is important and the sample bottle must have a tight stopper, so that the oil is protected until the bottle is reopened at the place of test. Although this is often a chemical laboratory, it is not always so, as the routine tests are quite simple in nature and electrical maintenance staff can normally be trained to carry them out if provided with the necessary apparatus.

The point of sampling will normally be the drain cock at the bottom of the transformer tank. After cleaning the cock as far as possible with clean paper specially selected for its freedom from loose fibres, it should be flushed with a little oil from the tank before the test sample itself is taken. It is often necessary to draw the latter into an intermediate vessel, such as a large glass jar, with a larger opening than the sample bottle, the intermediate vessel, of course, also being scrupulously clean.

The oil sample will be more representative of the bulk if it is drawn whilst the transformer is still hot from recent loading and hence recent circulation.

Inspection of oil samples

The appearance of the sample will sometimes supply useful information. Thus, a lack of clarity will most likely be due to moisture, particularly if the cloudiness only appears after the oil has cooled. With oil which is very cloudy when taken warm from the transformer, it is advisable to heat a small quantity in a test tube as for the crackle test. Should it not then become clear the sample should be referred to an oil chemist, in case an adventitious contaminant is the cause.

Any sludge present will be noted. To distinguish sludge from iron oxide and other solid matter not derived from oil a chemist will use hot solvents such as carbon tetrachloride, in which sludge will be soluble, but such examination is not usually carried out by maintenance staff.

A very dark coloured oil may have suffered oxidation to a point where sludge formation is imminent; however, the coloration of the oil may be due to its contamination, for example, with bituminous compound.

A 'gassy' smell suggests that oil cracking may have occurred, and the closed flash point of the oil should then be determined.

A strong acid smell may indicate the presence of unduly high proportions of low molecular weight acids, which may result in tank corrosion above oil level, particularly if the oil contains moisture.

While no final conclusions can be reached from observations of the type described, it will be evident that they may provide useful information upon which to proceed.

Measurement of acidity of oil in service

A suitable method is that standardised by the Institute of Petroleum as I.P. 1/64, which is reproduced completely in the B.S. recommendations for maintenance. Two alternative procedures are given and, although the Method B, which uses the indicator phenolphthalein to show the neutral point, has been extensively used in the electrical industry for this purpose and is the more widely known, Method A, using alkali blue indicator is now often employed.

In Method B, ethyl alcohol, or industrial methylated spirit, is added to the measured oil sample and the mixture boiled together before titration of the acidity by standard alkali solution; as the spirit layer separates below the oil, and is to some degree coloured by it, the red colour developed by phenolphthalein at the neutral point cannot always easily be distinguished. On the other hand, Method A requires no heating and, with no separation of oil and solvent into layers, the colour change of the alkali blue is considered easier to see by many operators. Nevertheless, some workers find this colour change a little difficult and it is necessary to carry out the titration promptly as contamination with carbon dioxide from the air appreciably affects the result.

It is normally sufficient to test for acidity yearly, or even every two years, in the earlier years of the life of the transformer oil charge.

In Great Britain 1·0 mg KOH/g was considered a suitable acidity limit in normal transformer service, especially having regard to the importance of ensuring freedom from sludge formation; oil having reached this degree of acidity should be discarded. With the normal transformer loading usually practised in large distribution systems, with adequate reserves of transformer capacity available for units to be isolated for maintenance or other reasons, satisfactory service of 20 years was often obtained without troubles due to sludge formation.

However, it is increasingly recognised that with voltages and loadings rising, the earlier acidity limit no longer provides an adequate safeguard, and most users, in Great Britain and abroad, have been working to a limiting figure of 0·5 mg KOH/g for some years.

Appearance of sludge in oil

The very undesirable effects of sludge deposition in transformers have been referred to several times earlier in this chapter, so that it only remains to emphasise that, should oil show a tendency to form sludge before the normal limit for acidity is reached, it should be given immediate treatment. Testing methods are available to establish the presence of highly oxidised bodies in oil which have not actually precipitated as solids or, alternatively, are present as solids in such small amount and in such finely divided form as not to be readily visible; these methods are described in the B.S. recommendations for oil care and maintenance.

Electric strength and water tests of oil in service

Considered as tests for the presence of water there has been much discussion of the precise significance of electric strength and crackle tests, because it is certainly the case that free water in varying amounts is often found in the bottom of transformer tanks even though a previously drawn oil sample had shown a negative crackle test and a satisfactory breakdown value. However, a low breakdown value, especially when coupled with a positive crackle test, definitely points to the advisability of early treatment of the oil. It has been mentioned elsewhere that B.S.148:1972 introduces a quantitative determination of water content in place of the crackle test, and that water in conjunction with fibres is probably most damaging to the electric strength of oil. The actual determination of water content in used oil is likely to be recommended in the near future, with proposed action if quite low limits are exceeded, especially in very high-voltage transformers.

Resistivity

The use of resistivity measurement for the assessment of the overall condition of oil in service is growing. Relatively simple equipment for the determination of an approximate value is available, making the test suitable as a field test, and a guide to recommended action based on the resistivity level is likely to be embodied in the new maintenance guide which is in preparation by the British Standards Institution.

Treatment of oil

The alternative methods available for oil treatment and drying are discussed in some detail in the B.S. document already referred to, including the precautions to be observed in carrying out such work. It is useful to emphasise the importance of using equipment of adequate throughput. Treatment on site by continuous circulation through the oil cleaning unit without completely emptying the transformer tank is, in general, satisfactory for routine maintenance. However, where the presence of sludge is known, it may be necessary to remove the oil completely for filtration, so that the transformer itself may be inspected and cleaned. As sludge tends to harden when exposed to the atmosphere, it is recommended that the cleaning be carried out as soon as possible after the removal of the oil and of the core and windings from the tank.

Filtration of the oil for removal of sludge must be carried out in the cold, since, although sludge tends to dissolve in hot oil, it comes out of solution when the oil is cooled. Water, too, is much more soluble in hot oil than in cold, but this contaminant can be rapidly and virtually completely removed by vacuum treatment of the hot oil, and the most sophisticated modern oil treatment plants employ high-vacuum equipment in combination with fine filtration.

GENERAL RECOMMENDATIONS

It will have been noted from the earlier discussion that the factors most prominent in causing oil deterioration are the presence of air and the effect of temperature.

While the complete exclusion of air may be achieved by the use of sealed transformer tanks having suitable oil expansion arrangements, or by providing nitrogen or other inert gas cushions in the tanks, the extra capital cost involved makes such arrangements rather doubtful economic propositions except in special circumstances. However, a major exclusion of air is obtained by the use of oil conservators, particularly in the case of the larger transformers.

The rating of the transformer at which it becomes economical to add a conservator is naturally dependent upon the conditions of operation, the degree of loading, the duration of the load cycles, the ambient temperature and the degree of ventilation provided, in the case of a substation transformer.

Mention has been made earlier that an increase in temperature of 4°C has been estimated to result in a decrease in the effective life of cellulose insulation of approximately 30%. It is generally accepted that the rate of a chemical reaction approximately doubles with increases in temperature of 8°C to 10°C and, while it is difficult to demonstrate with any degree of precision that this applies to the oxidation of oil, because of the complication of the reactions and the number of separate chemical products involved, it is certainly the case that the individual reactions making up the process of oil deterioration will tend to follow the same general pattern with regard to increased temperature.

It is clearly evident that the life of oil can be prolonged by a reduction of the average temperature of even a few degrees. Changes in ambient temperature are fairly closely reproduced in transformer temperatures although the time lag for temperature change does not always make this evident when loads are fluctuating. Accordingly, reductions of a few degrees in the ambient temperatures, which could often easily be obtained by improvement of the ventilation of the transformer cell, would lead to worthwhile increases in coil life.

One further cause of acceleration of oxidation is the presence of catalysts. Those catalysts provided by the materials employed in the construction of transformers are beyond the control of the users, but traces of sludge and most forms of dirt are in varying degrees catalytic in action. Some of the larger supply authorities have adopted the practice of oil filtration at regular intervals as a routine, for example, every two years, and undoubtedly this has been one factor in reducing oil troubles.

Some of the products of oxidation of oil are also catalytic as regards further oxidation. It is therefore very important to avoid the mixing of oils having varying degrees of deterioration. The more oxidised oil may be improved somewhat, but any such advantage is outbalanced by the accompanying loss of effective life of the less oxidised oil. It is especially important to clean thoroughly any core and windings which have been immersed in severely deteriorated oil which is to be replaced. The old oil will generally contain soluble copper and iron compounds which, contaminating the new oil, will have a very strong catalytic effect upon its oxidation.

Oxidation cannot be prevented entirely so long as air has access to the oil but the greatest operating economy is in the long run secured by not allowing any oil to deteriorate unduly, whilst increased transformer life will be obtained by the adoption of routine procedures which minimise sludge formation in transformers.

As mentioned in the previous chapter, transformer oil will dissolve water, though the proportion is small, normally about 30 to 45 parts per million for new oil at atmospheric temperature. The atmosphere always contains some moisture, and it follows that the oil in contact with air will, under equilibrium conditions, hold a proportion of water in solution. This water is not objectionable in normal circumstances and does not prevent the preparation of oil with good dielectric strength. Similarly, oil dissolves air and other gases, the saturation proportion for air being about 11% by volume. Oil in contact with the atmosphere will contain that amount of air without loss to its quality

With the introduction of higher-voltage transformers, the need has arisen for the initial filling to be made with oil of better quality than hitherto. Excess moisture in the windings and air or the gas used to fill the transformer tank for transport, and which may also be trapped in the windings, can be removed by absorption into the oil but this is possible only if the oil is below the saturation condition in air and water content. Accordingly, de-aerated and dehydrated oil is delivered to the site in specially designed road tankers.

However, it is sometimes considered desirable to speed up the absorptive process by continuous oil circulation vacuum treatment on site. After an extensive study of the requirements and of various types of equipment, the Burmah-Castrol 'Ilovac' unit was introduced to give the same standard of oil filling and servicing on site as was previously available only at the manufacturer's works. The 'Ilovac' unit, shown in Fig. 17.6, incorporates a high-vacuum oil treatment plant and is capable of removing almost completely both dissolved air and water from transformer oil. It will also remove particles of free solids such as fibres, as well as any free water.

The use of this equipment for filling new transformers is only one of its applications. It will be realised that the quantity of water, gas or air that can be absorbed by oil, however efficiently dehydrated and de-aerated, is limited by the oil's capacity for water and air. However, if the air and water are continuously removed from the oil, its absorptive capacity is regenerated and only

Fig. 17.6 'Ilovac' mobile oil treatment plant in operation at a high-voltage substation (Burmah-Castrol Industrial Ltd.)

the servicing time available limits the quantities of water or air which can be removed from transformer windings, or other parts of the transformer in contact with the oil.

During a typical servicing period this type of equipment will reduce in a single pass, the dissolved air content of oil from 11% to under 1% by volume and dissolved moisture from 30 parts per million to less than 5. Simultaneously, the electric strength can be increased from under 30 kV to over 70 kV.

Chapter 18

Parallel operation

In the study of the parallel operation of transformers, polarity and phase sequence play important parts, and therefore it becomes very desirable to consider these characteristics in some detail first before passing on to the more general treatment of parallel operation. The points to consider are the relative directions of the windings and of the voltages in the windings; the relative positions of leads from coils to terminals also have a bearing upon the question. It is usual in the study of transformer polarity diagrams to base an explanation upon the instantaneous voltages *induced* in both windings, as this procedure avoids any reference to primary and secondary windings as such; this is desirable, as transformer polarity and phase sequence are independent of such a distinction.

As the voltages induced in the primary and secondary windings are due to a common flux, the induced voltages in each turn of both windings must be in the same direction for, considered alone, any individual turn cannot be said to possess one direction around the core any more than the opposite one. Direction is given to a complete winding, however, when a number of such turns are connected in series, one end of the winding being labelled 'start' and the other 'finish', or one being called, say, A_1 and the other end A_2. The directions of the total voltages induced in the primary and secondary windings will therefore depend upon the relative directions of the coil windings between terminals. It should clearly be understood that in considering directions of windings it is necessary to do so from similarly labelled or assumed similar terminals; that is, both primary and secondary windings should be considered in the direction from start to finish terminals (or even the reverse if desired), but they should not be considered one from start to finish and the other from finish to start. If the start and finish of the windings are not known, adjacent primary and secondary terminals may be assumed to correspond to similar ends of the respective windings.

TRANSFORMER TERMINAL MARKING, POSITION OF TERMINALS AND PHASOR DIAGRAMS

Terminal markings

A system of transformer terminal marking has been standardised in B.S.171 and agreed for both single and polyphase transformers. The employment of the various agreed standards facilitates the identification of the various external terminals on a transformer and assists installation, phasing and parallel operation of units. Phase windings are given descriptive letters and the same letter is used for all windings on one phase. The h.v. winding has been given a capital letter and the l.v. winding on the same phase a corresponding small letter. The following designations are used. For single-phase transformers:

> A : for the h.v. winding.
> 3A: for the third winding (if any).
> a   : for the l.v. winding.

For two-phase windings on a common core or separate cores in a common tank:

> A B: for the h.v. windings.
> a b : for the l.v. windings.

For three-phase transformers:

> A B C   : for the h.v. windings.
> 3A 3B 3C: for the third windings (if any).
> a b c   : for the l.v. windings.

Figure 18.1 shows an example of standard marking of a single-phase transformer.

Position of terminals

The physical position of all terminals in relation to each other has also been standardised and for three-phase transformers the position is such that when viewed from the h.v. side of the transformer the order is NABC, and when facing the l.v. side cban. Examples of both single- and three-phase terminal marking are shown in Fig. 18.2.

In addition to the letter marking of terminals, suffix numbers must be given to all tapping points and to the ends of the winding. These suffix numbers begin at unity and then with ascending numbers are ascribed to all tapping points, such that the sequence represents the direction of the induced e.m.f. at some instant of time. In the case of an h.v. winding without tappings for which the phase marking is A, the ends of the winding would be marked A_1, A_2. Similarly the l.v. winding would be marked a_1, a_2. As described later in this chapter, it is

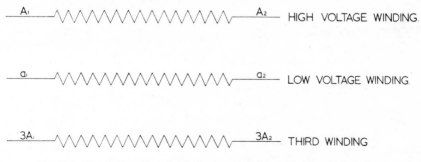

Fig. 18.1   Terminal marking of a single-phase transformer having a third winding

389

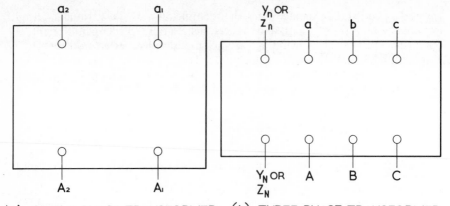

(a) SINGLE PHASE TRANSFORMER   (b) THREE PHASE TRANSFORMER

Fig. 18.2 Relative position of terminal of two-winding transformers

(a) SINGLE PHASE WINDING WITH TAPPINGS AT THE ENDS.

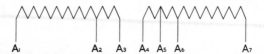

(b) SINGLE PHASE WINDING WITH TAPPINGS AT THE MIDDLE.

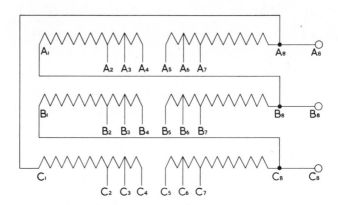

(c) THREE PHASE WINDING WITH TAPPINGS AT THE MIDDLE.

Fig. 18.3 Marking of tappings on phase windings

390

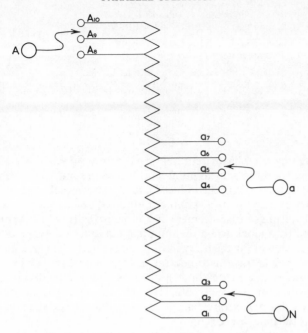

Fig. 18.4   Marking of tappings for an auto transformer

an easy matter to check the terminal marking (see Fig. 18.10). Typical examples of the marking of tappings are shown in Fig. 18.3.

The neutral connection, when brought out in the form of an external terminal is marked Y_N in the case of an h.v. winding and y_n in the case of an l.v. winding. No suffix number is required.

Auto transformer terminal marking includes the appropriate phase and suffix number and it should be noted that for tappings the higher suffix numbers correspond to the higher voltages. Figure 18.4 shows a typical terminal marking for an auto transformer.

Phasor diagrams

Phasors in transformer phasor diagrams represent the induced e.m.f.s and the counter-clockwise direction of rotation of the phasor is employed. The phasor representing any phase voltage of the l.v. winding is shown parallel to that representing the corresponding phase voltage of the h.v. winding.

Certain conventional notational standards are now agreed and have been in general use for a considerable period and these conventions are now explained in some detail for those who may not be familiar with the details involved. Various types of interphase connections for three-phase transformers having the same phase displacement between the h.v. and l.v. windings are grouped together and the four groups are shown in Table 18.1.

In Table 18.1 it will be seen that the phase displacement has a corresponding clock hour number. The phase displacement is the angle of phase advancement

Table 18.1
GROUP NUMBERS

Group No.	Phase displacement	Clock hour number
I	0°	0
II	180°	6
III	minus 30°	1
IV	plus 30°	11

turned through by the phasor representing the induced e.m.f. between a high-voltage terminal and the neutral point which may, in some cases, be imaginary, and the phasor representing the induced e.m.f. between the l.v. terminal having the same letter and the neutral point. An internationally adopted convention used to indicate phase displacement is to use a figure which represents the hour indicated by a clock when the minute hand takes the place of the line to neutral voltage phasor for the h.v. winding and is set at 12 o'clock and when the hour hand represents the line to neutral voltage phasor for the l.v. winding. It therefore follows that the clock hour number is obtained by dividing the phase displacement angle in degrees by 30. The phase angles of the various windings of three-phase transformers are determined with reference to the highest voltage being taken as the phasor of origin.

The phasor diagram, the phase displacement and the terminal marking are all identifiable by the use of symbols which for transformers having two windings if taken in order have the following significance:

First symbol   : h.v. winding connection.
Second symbol: l.v. winding connection.
Third symbol   : phase displacement expressed as the clock hour number (see Table 18.1, Column 3).

The interphase connections of the h.v. and l.v. windings are indicated by the use of the initial letters as given in Table 18.2 and the terms high and low voltage used in this table are employed in a relative sense only.

Table 18.2
WINDING CONNECTION DESIGNATIONS

Winding connection		Designation
high voltage	delta	D
	star	Y
	interconnected star	Z
low voltage	delta	d
	star	y
	interconnected star	z

A transformer having a delta connected high-voltage winding, a star connected lower-voltage winding and a phase displacement of plus 30° (corresponding to a clock hour number of 11), therefore has the symbol Dy11.

The following standard phasor diagrams which are frequently encountered in practice are included for single-, two- and three-phase transformers.

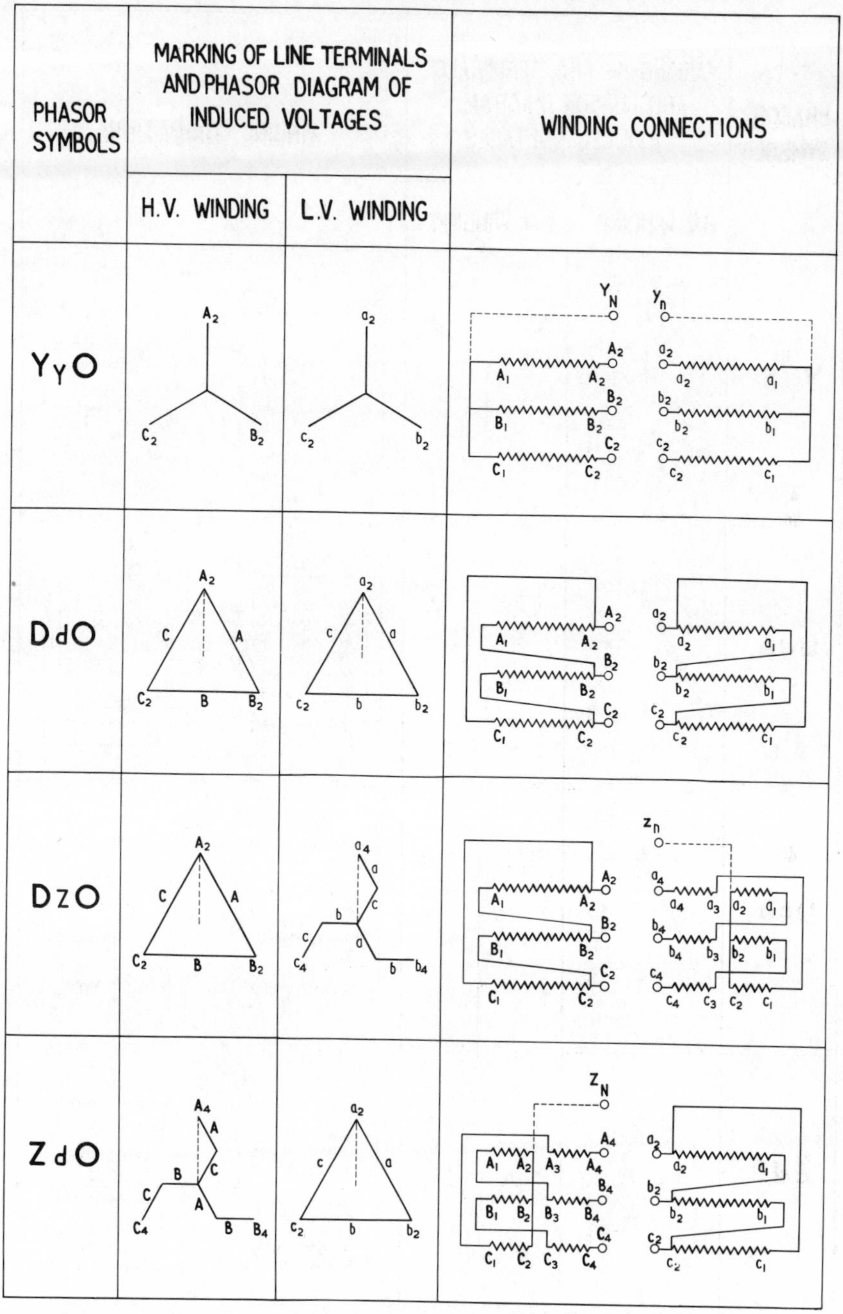

PHASOR SYMBOLS	MARKING OF LINE TERMINALS AND PHASOR DIAGRAM OF INDUCED VOLTAGES		WINDING CONNECTIONS
	H.V. WINDING	L.V. WINDING	
YᵧO			
DdO			
DzO			
ZdO			

Fig. 18.5   Phasor diagrams for three-phase transformers. Group No. I: phase displacement $= 0°$

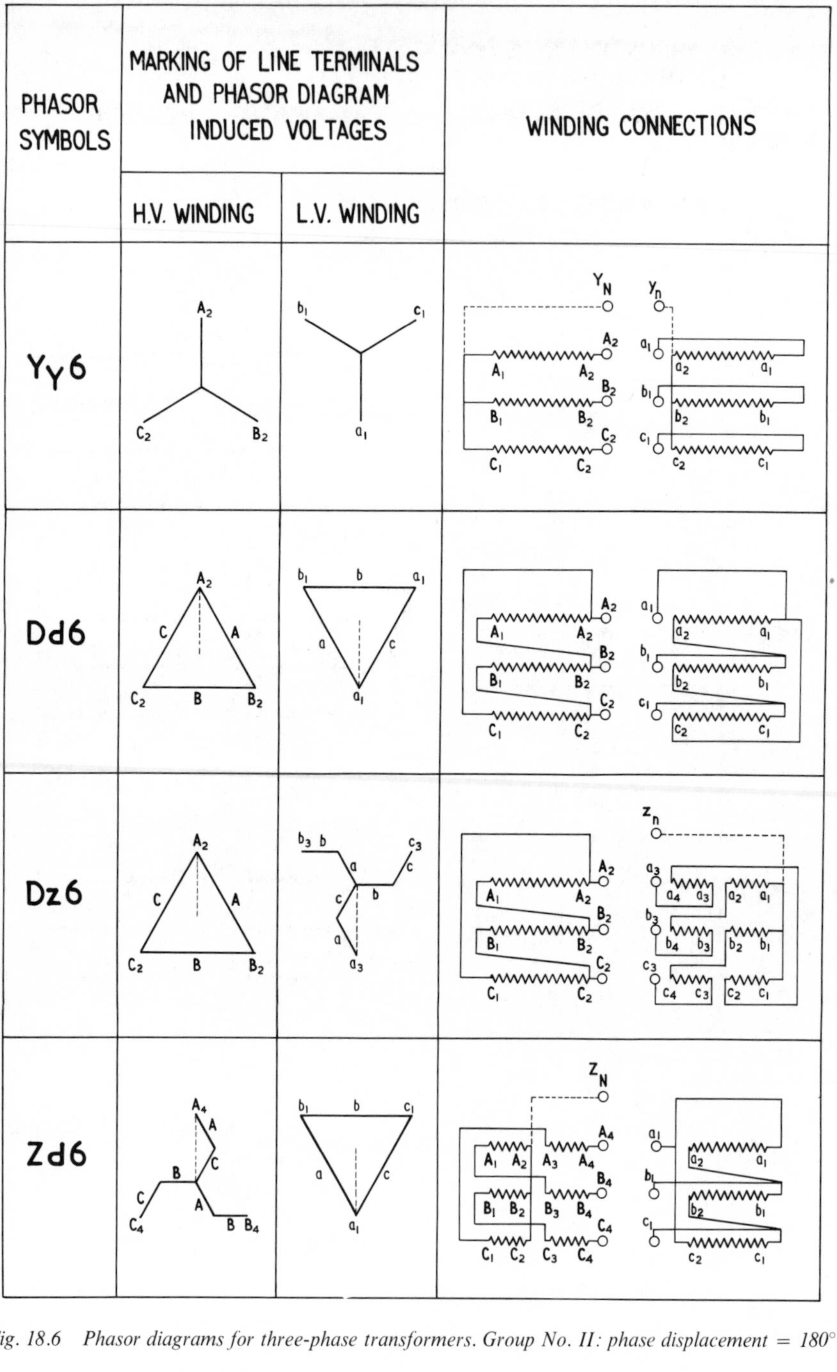

PHASOR SYMBOLS	MARKING OF LINE TERMINALS AND PHASOR DIAGRAM INDUCED VOLTAGES		WINDING CONNECTIONS
	H.V. WINDING	L.V. WINDING	
Yy6			
Dd6			
Dz6			
Zd6			

Fig. 18.6   Phasor diagrams for three-phase transformers. Group No. II: phase displacement = 180°

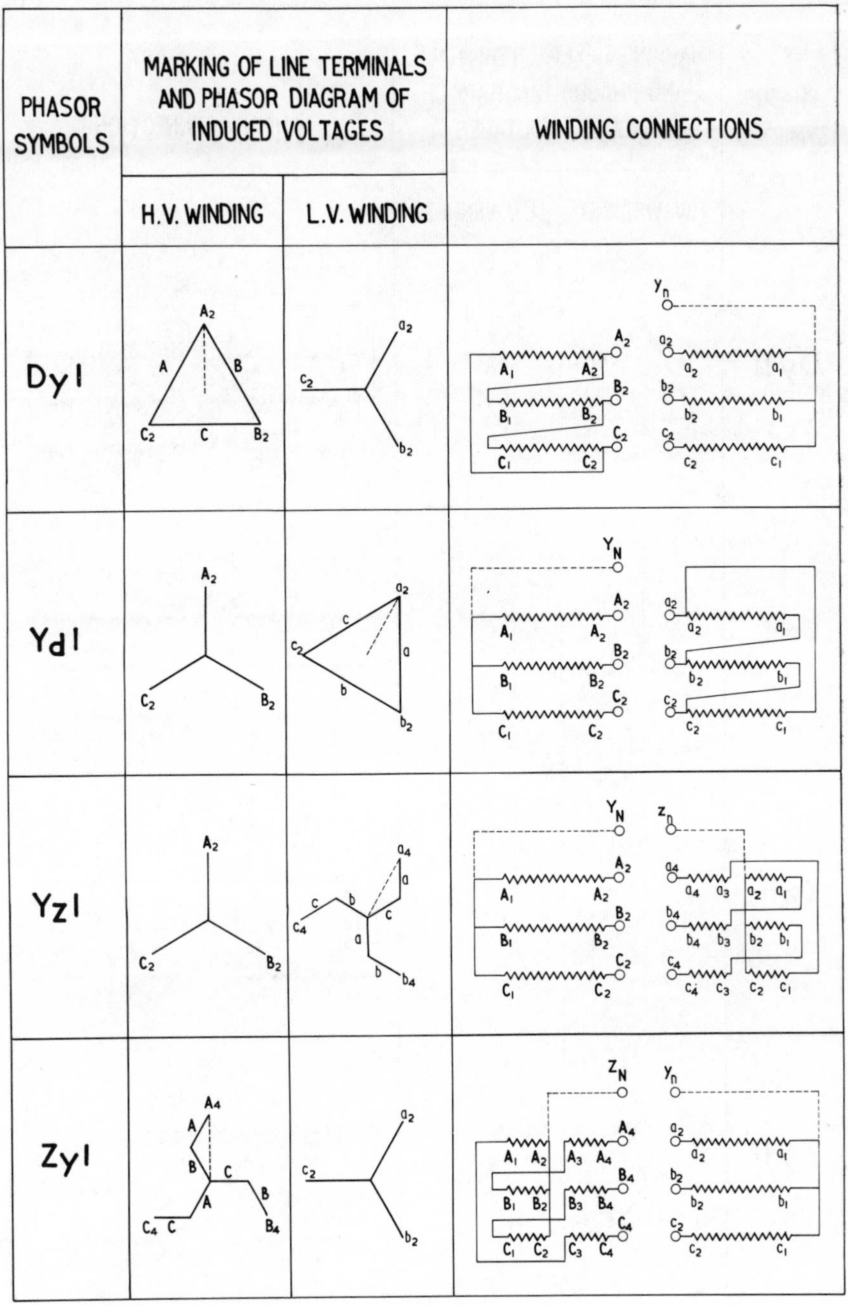

PHASOR SYMBOLS	MARKING OF LINE TERMINALS AND PHASOR DIAGRAM OF INDUCED VOLTAGES		WINDING CONNECTIONS
	H.V. WINDING	L.V. WINDING	
Dy1			
Yd1			
Yz1			
Zy1			

Fig. 18.7 Phasor diagrams for three-phase transformers. Group No. III: phase displacement = minus 30°

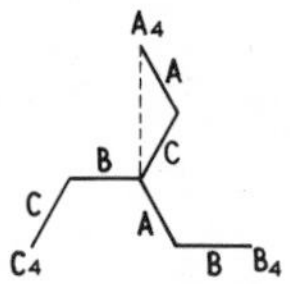

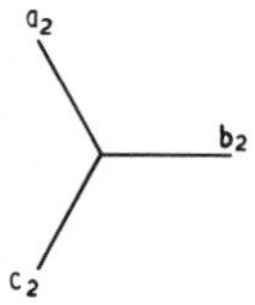

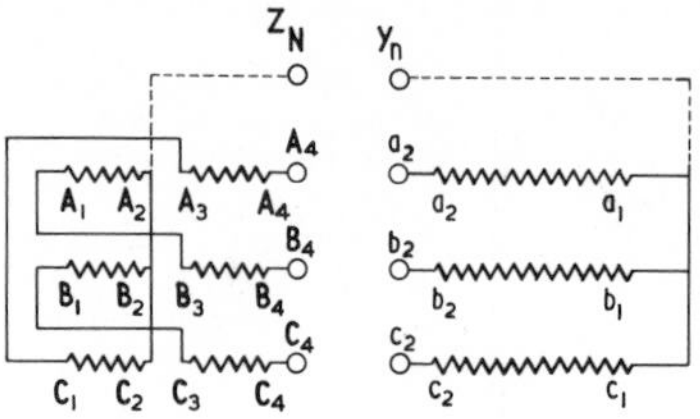

Fig. 18.8 *Phasor diagrams for three-phase transformers. Group No. IV : phase displacement = plus 30°*

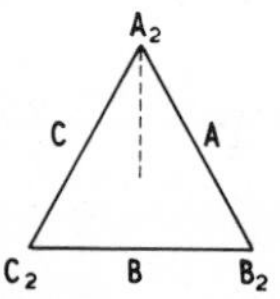

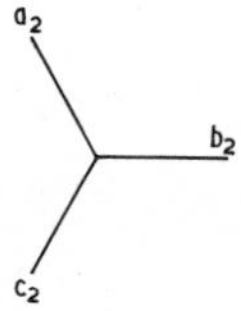

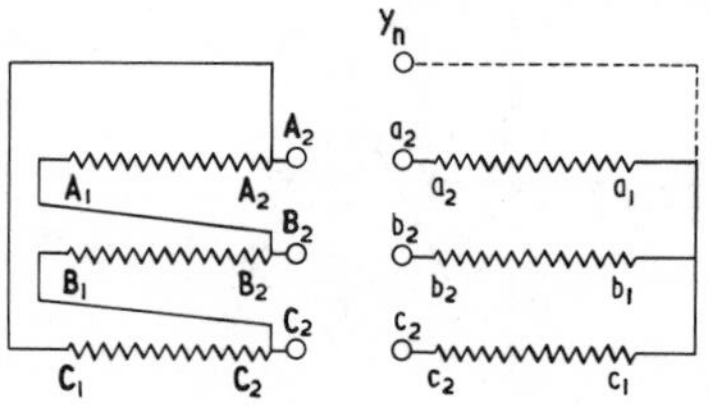

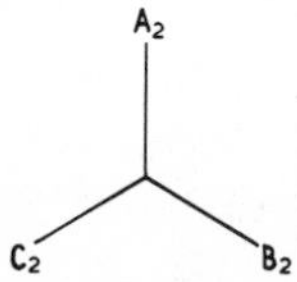

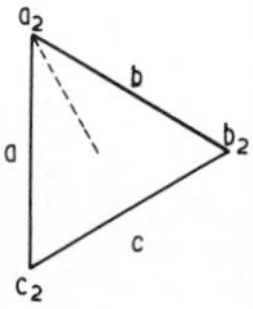

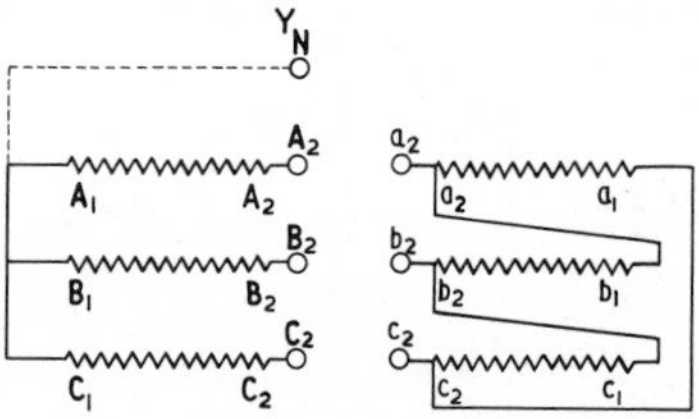

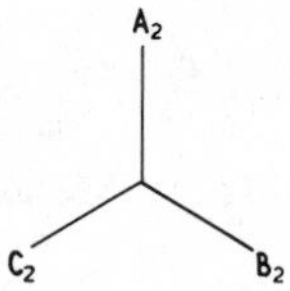

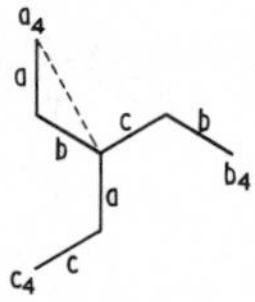

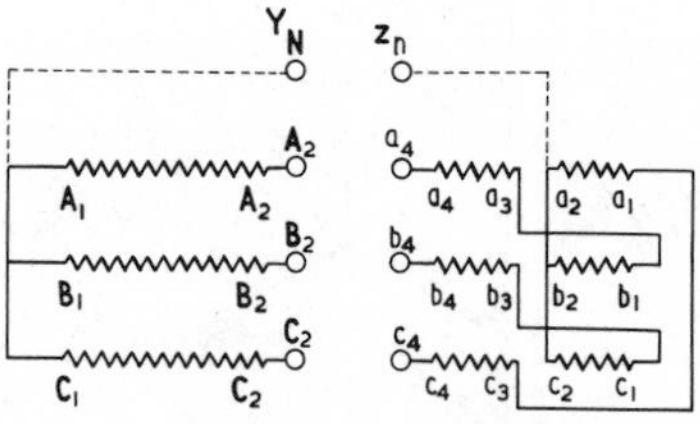

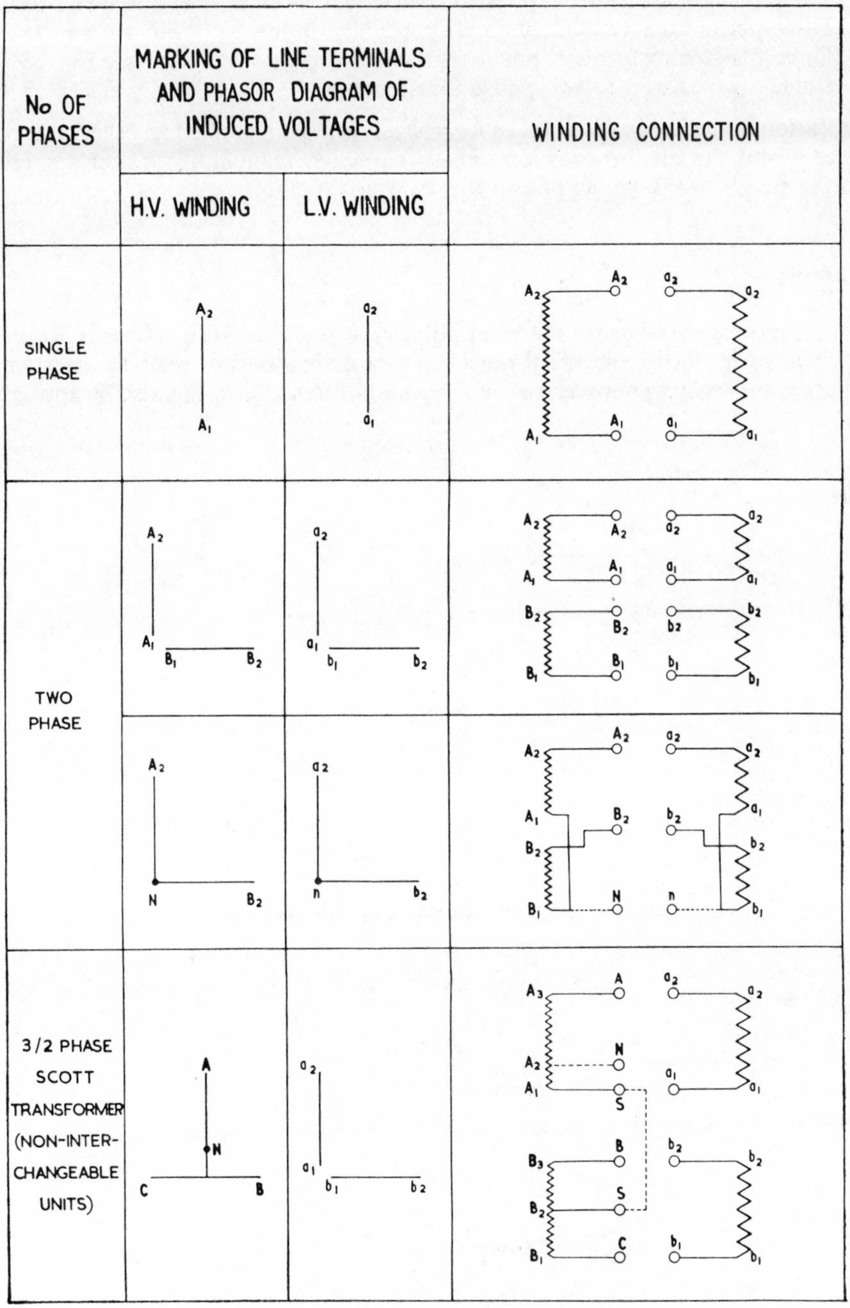

No OF PHASES	MARKING OF LINE TERMINALS AND PHASOR DIAGRAM OF INDUCED VOLTAGES		WINDING CONNECTION
	H.V. WINDING	L.V. WINDING	
SINGLE PHASE			
TWO PHASE			
3/2 PHASE SCOTT TRANSFORMER (NON-INTER-CHANGEABLE UNITS)			

Fig. 18.9   Phasor diagrams for single-, two-, and three- to two-phase transformers

Three-phase transformers, phase displacement   0° see Fig. 18.5.
Three-phase transformers, phase displacement   180° see Fig. 18.6.
Three-phase transformers, phase displacement minus   30° see Fig. 18.7.
Three-phase transformers, phase displacement plus   30° see Fig. 18.8.
Single-, two-, three- to two-phase transformers   see Fig. 18.9.

Various other combinations of interphase connections having other phasor relationships occur but they are only infrequently manufactured and it is left to the reader to evolve the phasor diagram and symbol.

Polarity

In the more general sense the term polarity, when used with reference to the parallel operation of electrical machinery, is understood to refer to a certain relationship existing between *two or more* units, but the term can also be applied

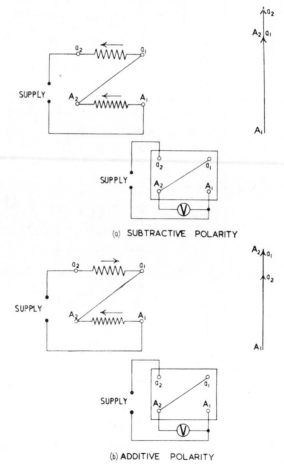

(a) SUBTRACTIVE POLARITY

(b) ADDITIVE POLARITY

Fig. 18.10   *Test connections for determining single-phase transformer winding polarity*

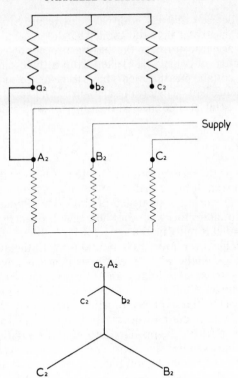

Fig. 18.11   Test connections for determining three-phase transformer winding polarity

to two separate windings of any individual piece of apparatus. That is, while two separate transformers may, under certain conditions of internal and external connection, have the same or opposite polarity, the primary and secondary windings of any individual transformer may, under certain conditions of coil winding, internal connections, and connections to terminals, have the same or opposite polarity. In the case of the primary and secondary windings of an individual transformer when the respective induced terminal voltages are in the same direction, that is, when the polarity of the two windings is the same, this polarity is generally spoken of as being subtractive; when on the other hand, the induced terminal voltages are in the opposite direction, the windings are of opposite polarity, which is usually referred to as being additive.

This style of designating the polarity of individual transformers has really arisen from the particular test conditions usually adopted for determining the relative voltage directions.

Figure 18.10 shows the test connections at (*a*) and (*b*) respectively for single-phase transformers having subtractive and additive polarity. Dealing first with (*a*), if in the h.v. winding the induced e.m.f. at a given instant is in the direction A_1 to A_2, the direction of the induced e.m.f. generated in the l.v. winding at the same instant will also be in the sequence a_1 to a_2. If the two windings are joined in series by connecting A_2 to a_1 and a single-phase e.m.f. applied to terminals A_1 and a_2, the e.m.f. measured between the h.v. terminals A_1 A_2 will be less than the applied voltage. Using the same connections and considering case (*b*) it will be seen that for additive polarity the e.m.f. measured between terminals $A_1 A_2$ is greater than the applied e.m.f.

For three-phase transformers the testing procedure is similar, except that the windings must, of course, be excited from a three-phase supply, and considerably more voltage measurements have to be made before the exact polarity and phase sequence can be determined. Figure 18.11 shows the test connections and results for a star/star connected transformer with subtractive polarity.

Phase sequence

Phase sequence is the term given to indicate the angular direction in which the voltage and current phasors of a polyphase system reach their respective maximum values during a sequence of time. This angular direction may be clockwise or counter-clockwise, but for two transformers to operate satisfactorily in parallel it must be the same for both. Phase sequence of polyphase transformers is, however, intimately bound up with the question of polarity.

It should be remembered that phase sequence is really a question of the sequence of *line terminal* voltages, and not necessarily of the voltages across individual windings. While the actual phase sequence of the supply is fixed by the generating plant, the sequence in which the secondary voltages of a transformer attain their maximum values can be in one direction or the other, according to the order in which the primary terminals are supplied.

Figure 18.12 shows four instances of a delta/star connected transformer under different conditions of polarity and phase sequence, and a comparison of these diagrams shows that interchanging any one pair of the supply leads to the primary terminals reverses the phase sequence. If, however, the internal connections on the secondary side of the transformer are reversed, the interchanging of any two primary supply leads will produce reverse phase sequence and non-standard polarity. If with reverse internal connections on one side the primary leads are not interchanged, the resulting phase sequence will be the same and the polarity will be non-standard. The above remarks apply strictly to transformers in which the primary and secondary windings have different connections, such as delta/star, but where these are the same, such as star/star, the polarity can only be changed by reversing the internal connections on one side of the transformer. The phase sequence alone may, however, be reversed by interchanging two of the primary supply leads.

If tests indicate that two transformers have the same polarity and reverse phase sequence, they may be connected in parallel on the secondary side simply by interchanging a certain pair of leads to the busbars of one of the transformers. Referring to Fig. 18.12, for instance, transformers to diagrams (*a*) and (*d*) can be paralleled so long as the secondary leads from a_1 and c_1 to the busbars are interchanged.

The satisfactory parallel operation of transformers is dependent upon five principal characteristics; that is, any two or more transformers which it is desired to operate in parallel should possess:

(1) the same inherent phase angle difference between primary and secondary terminals
(2) the same voltage ratio
(3) the same percentage impedance

The supply phase sequence is RYB in each case and the conventional counter-clockwise phasor rotation is observed

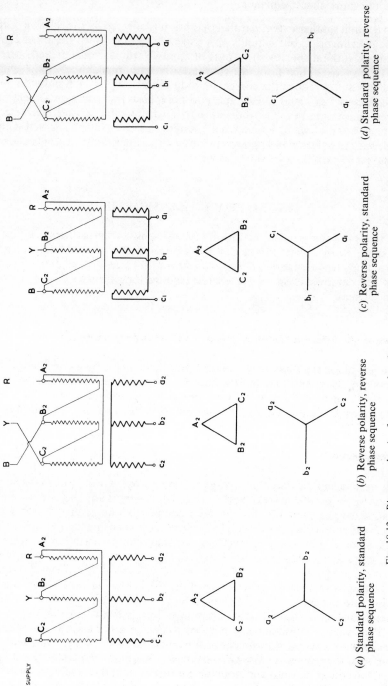

(a) Standard polarity, standard phase sequence

(b) Reverse polarity, reverse phase sequence

(c) Reverse polarity, standard phase sequence

(d) Standard polarity, reverse phase sequence

Fig. 18.12   Diagrams showing four examples of a three-phase delta/star connected transformer having differing polarity and phase sequence conditions

(4) the same polarity

(5) the same phase sequence.

To a much smaller extent parallel operation is affected by the relative outputs of the transformers, but actually this aspect is reflected into the third characteristic as, if the disparity in outputs of any two transformers exceeds three to one it may be difficult to incorporate sufficient impedance in the smaller transformer to produce the correct loading conditions for each individual unit.

Characteristics 1 and 5 only apply to polyphase transformers. A very small degree of latitude may be allowed with regard to the second characteristic above mentioned, while a somewhat greater tolerance may be allowed with the third, but the polarity and phase sequence, where applicable, of all transformers operating in parallel *must* be the same.

SINGLE-PHASE TRANSFORMERS

The theory of the parallel operation of single-phase transformers is essentially the same as for three phase, but the actual practice for obtaining suitable connections between any two single-phase transformers is considerably simpler than the determination of the correct connections for any two three-phase transformers.

Phase angle difference between primary and secondary terminals

In single-phase transformers this point does not arise, as by the proper selection of external leads any two single-phase transformers can be connected so that the phase angle difference between primary and secondary terminals is the same for each. Consequently the question really becomes one of polarity.

Voltage ratio

It is very desirable that the voltage ratios of any two or more transformers operating in parallel should be the same, for if there is any difference whatever a circulating current will flow in the secondary windings of the transformers which are connected in parallel, but before they are connected to any external load. Such a circulating current may or may not be permissible, according, first, to its actual magnitude, and second, to whether the load to be supplied is less than or equal to the sum of the rated outputs of the transformers operating in parallel. As a rule, however, every effort should be made to obtain identical ratios, and particular attention should be given to obtaining these at all ratios when transformers are fitted with tappings. In passing, it may be well to point out that when a manufacturer is asked to design a transformer to operate in parallel with existing transformers, the actual ratio of primary and secondary *turns* should be given, as this ratio can easily be obtained exactly. Such figures, would of course, be obtained from the manufacturer of the existing transformers.

Equations 18.1 to 18.19 inclusive show how the values of these circulating currents can be calculated when certain of the transformer characteristics

402

differ. Equations 18.1 to 18.5 show how to derive the circulating currents when two single- or three-phase transformers, having different ratios, operate in parallel, while eqns. 18.6 to 18.10 apply to the case of three single- or three-phase transformers.

It is to be noted that this flow of circulating current takes place before the transformers are connected up to any external load. A circulating current in the transformer windings of the order of, say, 5% of the full-load current may generally be allowed in the case of modern transformers without any fear of serious overheating occurring. It is sometimes very difficult to design new transformers to give an identical turns ratio on, say, four tappings as an existing one may possess, and while it is desirable that these ratios should be the same, it is not necessary to *insist* on their being identical.

Equation 18.1: The circulating current in amperes at no-load in two single- or three-phase transformers A and B connected in parallel, having different voltage ratios, the same or different outputs, the same or different impedances, and the impedances having the same ratios of resistance to reactance, is equal to:

$$\frac{V_A - V_B}{Z_A + Z_B} \qquad (18.1)$$

where

V_A = secondary line terminal voltage of transformer A having the lower ratio, *i.e.*, the higher secondary voltage.

V_B = secondary line terminal voltage of transformer B having the higher ratio, *i.e.*, the lower secondary voltage.

*Z_A, Z_B = ohmic impedances of transformers A and B respectively, and are obtained from the equations:

$$Z_A = \frac{V_{ZA} V_A}{100 I_A} \qquad Z_B = \frac{V_{ZB} V_B}{100 I_B} \qquad (18.2)$$

where

V_{ZA}, V_{ZB} = percentage impedance voltage drops at full-load ratings of transformers A and B respectively.

I_A, I_B = full-load line currents in amperes of transformers A and B respectively.

Equation 18.3. The circulating current in amperes at no-load in two single- or three-phase transformers A and B connected in parallel, having different voltage ratios, the same or different outputs, the same or different impedances, but the impedances having different ratios of resistance to reactance, is equal to:

$$\frac{V_A - V_B}{Z} \qquad (18.3)$$

where

V_A = secondary terminal voltage of transformer A, having the lower ratio, *i.e.*, the higher secondary voltage.

*These quantities are the transformer resistances and reactances between two secondary line terminals.

403

V_B = secondary line terminal voltage of transformer B, having the higher ratio, *i.e.*, the lower secondary voltage.

*Z = vector sum of the ohmic impedances of transformers A and B, and is obtained from the equation:

$$Z = \sqrt{\{(R_A + R_B)^2 + (X_A + X_B)^2\}} \qquad (18.4)$$

where

*R_A = ohmic resistance of transformer A, and equals

$$\frac{V_{RA} V_A}{100 I_A}$$

*R_B = ohmic resistance of transformer B, and equals

$$\frac{V_{RB} V_B}{100 I_B}$$

*X_A = ohmic reactance of transformer A, and equals $\qquad (18.5)$

$$\frac{V_{XA} V_A}{100 I_A}$$

*X_B = ohmic reactance of transformer B, and equals

$$\frac{V_{XB} V_B}{100 I_B}$$

V_{RA}, V_{RB} = percentage resistance voltage drops at normal full-load ratings of transformers A and B respectively.

V_{XA}, V_{XB} = percentage reactance voltage drops at normal full-load ratings of transformers A and B respectively.

I_A, I_B = normal full-load line currents in amperes of transformers A and B respectively.

Equations 18.6–18.8: The circulating currents in amperes at no load in three single- or three-phase transformers A, B and C connected in parallel, each having different voltage ratios, the same or different impedances, the same or different outputs, and the impedances having the same ratio of resistance to reactance, are given by:

In transformer A $\qquad \dfrac{V_A - M}{Z_A} \qquad\qquad (18.6)$

In transformer B $\qquad \dfrac{V_B - M}{Z_B} \qquad\qquad (18.7)$

and in transformer C $\qquad \dfrac{V_C - M}{Z_C} \qquad\qquad (18.8)$

*These quantities are the transformer resistances and reactances between two secondary line terminals.

where

V_A = secondary line terminal voltage of transformer A, having the lowest ratio, *i.e.*, the highest secondary voltage.

V_B = secondary line terminal voltage of transformer B, having the next higher ratio, *i.e.*, the next lower secondary voltage.

V_C = secondary line terminal voltage of transformer C, having the highest ratio, *i.e.*, the lowest secondary voltage.

And where for transformers A, B and C respectively:

*Z_A, Z_B, Z_C = ohmic impedances and are obtained from the equations:

$$
\left.
\begin{aligned}
Z_A &= \frac{V_{ZA} V_A}{100 I_A} \\[6pt]
Z_B &= \frac{V_{ZB} V_B}{100 I_B} \\[6pt]
Z_C &= \frac{V_{ZC} V_C}{100 I_C}
\end{aligned}
\right\}
\tag{18.9}
$$

where

V_{ZA}, V_{ZB}, V_{ZC} = percentage impedance voltage drops at full-load ratings.

I_A, I_B, I_C = full-load line currents in amperes.

$$
M = \frac{V_A Z_B Z_C + V_B Z_A Z_C + V_C Z_A Z_B}{Z_A Z_B + Z_B Z_C + Z_C Z_A}
\tag{18.10}
$$

Equations 18.11–18.13: The circulating currents in amperes at no load in three single- or three-phase transformers A, B and C connected in parallel, having different voltage ratios, the same or different outputs, the same or different impedances, but the impedances having different ratios of resistance to reactance, are given by:

In transformer A
$$
\frac{100 I_A}{V_A V_{ZA}} \sqrt{\{(V_A - S)^2 + T^2\}}
\tag{18.11}
$$

In transformer B
$$
\frac{100 I_B}{V_B V_{ZB}} \sqrt{\{(V_B - S)^2 + T^2\}}
\tag{18.12}
$$

and in transformer C
$$
\frac{100 I_C}{V_C V_{ZC}} \sqrt{\{(V_C - S)^2 + T^2\}}
\tag{18.13}
$$

where

$$
T = \frac{\left(\sum \dfrac{I V_R}{V_Z^2 V} \sum \dfrac{I V_X}{V_Z^2}\right) - \left(\sum \dfrac{I V_X}{V_Z^2 V} \sum \dfrac{I V_R}{V_Z^2}\right)}{\left(\sum \dfrac{I V_X}{V_Z^2 V}\right)^2 + \left(\sum \dfrac{I V_R}{V_Z^2 V}\right)^2}
\tag{18.14}
$$

*These quantities are the transformer ohmic impedances between two secondary line terminals.

$$S = \frac{\left(\sum \dfrac{IV_X}{V_Z^2 V}\sum \dfrac{IV_X}{V_Z^2}\right) + \left(\sum \dfrac{IV_R}{V_Z^2 V}\sum \dfrac{IV_R}{V_Z^2}\right)}{\left(\sum \dfrac{IV_X}{V_Z^2 V}\right)^2 + \left(\sum \dfrac{IV_R}{V_Z^2 V}\right)^2} \qquad (18.15)$$

The symbol 'Σ' has the usual mathematical significance, *i.e.*,

$$\sum \frac{IV_R}{V_Z^2 V} = \frac{I_A V_{RA}}{V_{ZA}^2 V_A} + \frac{I_B V_{RB}}{V_{ZB}^2 V_B} + \frac{I_C V_{RC}}{V_{ZC}^2 V_C} \qquad (18.16)$$

$$\sum \frac{IV_X}{V_Z^2} = \frac{I_A V_{XA}}{V_{ZA}^2} + \frac{I_B V_{XB}}{V_{ZB}^2} + \frac{I_C V_{XC}}{V_{ZC}^2}$$

The angle of lag* between the circulating current and the normal secondary line terminal voltages of transformers A, B and C respectively is equal to:

$$\text{for transformer A: } \tan^{-1} \frac{V_{XA} V_A - V_{XA} S - V_{RA} T}{V_{RA} V_A - V_{RA} S + V_{XA} T} \qquad (18.17)$$

$$\text{for transformer B: } \tan^{-1} \frac{V_{XB} V_B - V_{XB} S - V_{RB} T}{V_{RB} V_B - V_{RB} S + V_{XB} T} \qquad (18.18)$$

$$\text{and for transformer C: } \tan^{-1} \frac{V_{XC} V_C - V_{XC} S - V_{RC} T}{V_{RC} V_C - V_{RC} S + V_{XC} T} \qquad (18.19)$$

where T and S have the same values as before. The remaining symbols used have the following meanings for transformers A, B and C respectively:

V_A, V_B, V_C = secondary line terminal voltages.

I_A, I_B, I_C = normal full-load line currents.

V_{ZA}, V_{ZB}, V_{ZC} = percentage impedance voltage drops at full-load rating.

V_{RA}, V_{RB}, V_{RC} = percentage resistance voltage drops at full-load rating.

V_{XA}, V_{XB}, V_{XC} = percentage reactance voltage drops at full-load rating.

Percentage impedance voltage drop

The percentage impedance voltage drop is a factor inherent in the design of any transformer, and is a characteristic to which particular attention must be paid when designing for parallel operation. The percentage impedance drop is determined by the formula

$$V_Z = \sqrt{(V_R^2 + V_X^2)} \qquad (18.20)$$

where V_Z is the percentage impedance drop, V_R the percentage resistance drop and V_X the percentage reactance drop, corresponding to the full-load rating of the transformer. Assuming that all other characteristics are the same, the percentage impedance drop determines the load carried by each transformer, and in the simplest case, viz., of two transformers of the same output operating in parallel, the percentage impedances must also be identical if the transformers

*The angle of lag is taken as being positive. If the sign of any of these expressions is negative the angle is leading.

are to share the total load equally. If, for instance, of two transformers connected in parallel having the same output, voltage ratio, etc., one has an impedance of 4% and the other an impedance of 2%, the transformer having the larger impedance will supply a third of the total bank output and the other transformer will supply two-thirds, so that the transformer having the higher impedance will only be carrying 66% of its normal load while the other transformer will be carrying 33% overload.

Equations 18.21 to 18.41 inclusive show how the division of load currents can be calculated when certain of the transformer characteristics differ. Equations 18.21 to 18.29 show how to derive the transformer load currents when two single- or three-phase transformers having different impedances operate in parallel, while eqns. 18.30 to 18.41 apply to the case of three single- or three-phase transformers.

When there is a phase displacement between transformer and total load currents, the phase angles can also be calculated from the equations.

Equations 18.21 and 18.22. The division of total load current I_L amperes between two single- or three-phase transformers A and B connected in parallel, having the same or different outputs, the same voltage ratios, the same or different impedances, and the same ratios of resistance to reactance, is given by

$$I_A = \frac{I_L N_A}{N_A + N_B} \tag{18.21}$$

$$I_B = \frac{I_L N_B}{N_A + N_B} \tag{18.22}$$

where, for transformers A and B respectively

$$I_A, I_B = \text{line currents in amperes.}$$

$$\left. \begin{array}{l} N_A = \dfrac{K_A}{V_{ZA}} \\[2ex] N_B = \dfrac{K_B}{V_{ZB}} \end{array} \right\} \tag{18.23}$$

and

K_A, K_B = normal rated outputs in kVA.

V_{ZA}, V_{ZB} = percentage impedance voltage drops at full-load ratings.

Note. The load currents in transformers A and B are in phase with each other and with the total load current.

Equations 18.24 and 18.25. The division of total load current I_L amperes between two single- or three-phase transformers A and B connected in parallel, having the same or different outputs, the same voltage ratios, the same or different impedances, but different ratios of resistance to reactance, is given by:

$$I_A = \frac{I_L N_A}{\sqrt{(N_A{}^2 + N_B{}^2 + 2N_A N_B \cos \theta)}} \tag{18.24}$$

$$I_B = \frac{I_L N_B}{\sqrt{(N_A{}^2 + N_B{}^2 + 2N_A N_B \cos \theta)}} \tag{18.25}$$

where

$$N_A = \frac{K_A}{V_{ZA}}$$

$$N_B = \frac{K_B}{V_{ZB}}$$ (18.26)

$$*\theta = \left(\tan^{-1}\frac{V_{XB}}{V_{RB}}\right) - \left(\tan^{-1}\frac{V_{XA}}{V_{RA}}\right)$$ (18.27)

$$\beta = \sin^{-1}\frac{I_A\sin\theta}{I_L}$$ (18.28)

$$*\alpha = \theta - \beta$$ (18.29)

and where for transformers A and B respectively:

I_A, I_B = line currents in amperes.

K_A, K_B = normal rated outputs in kVA.

V_{ZA}, V_{ZB} = percentage impedance voltage drops at full-load ratings.

V_{XA}, V_{XB} = percentage reactance voltage drops at full-load ratings.

V_{RA}, V_{RB} = percentage resistance voltage drops at full-load ratings.

$*\theta$ = phase angle difference between the load currents I_A and I_B.

$*\beta$ = phase angle difference between I_L and I_B.

$*\alpha$ = phase angle difference between I_L and I_A.

For the diagram in Fig. 18.13

θ is positive.

I_A is leading I_L.

I_B is lagging I_L.

Transformer A has the smaller value of V_X/V_R.

Transformer B has the greater value of V_X/V_R.

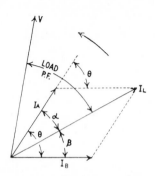

Fig. 18.13 *Phasor diagram showing current distribution with two transformers in parallel having different ratios of resistance to reactance*

*See Fig. 18.13.

408

When θ is negative:

I_A is lagging I_L.

I_B is leading I_L.

Transformer A has the greater value of V_X/V_R.

Transformer B has the smaller value of V_X/V_R.

Equations 18.30–18.32. The division of total load current I_L between three single- or three-phase transformers A, B and C connected in parallel, having the same or different outputs, the same voltage ratio, the same or different impedances, and the same ratios of resistance to reactance, is given by:

$$I_A = \frac{N_A I_L}{N_A + N_B + N_C} \tag{18.30}$$

$$I_B = \frac{N_B I_L}{N_A + N_B + N_C} \tag{18.31}$$

$$I_C = \frac{N_C I_L}{N_A + N_B + N_C} \tag{18.32}$$

where for transformers A, B and C respectively

$$I_A, I_B, I_C = \text{line currents in amperes}$$

$$\left. \begin{array}{l} N_A = \dfrac{K_A}{V_{ZA}} \\[2ex] N_B = \dfrac{K_B}{V_{ZB}} \\[2ex] N_C = \dfrac{K_C}{V_{ZC}} \end{array} \right\} \tag{18.33}$$

and where

K_A, K_B, K_C = normal rated outputs in kVA.

V_{ZA}, V_{ZB}, V_{ZC} = percentage impedance voltage drops at full-load ratings.

Equations 18.34–18.36. The division of total load current I_L between three single- or three-phase transformers A, B and C connected in parallel, having the same or different outputs, the same voltage ratios, the same or different impedances, but different ratios of resistance to reactance, is given by:

$$I_A = \frac{I_L}{\sqrt{\{1 + k_1^2 + 2k_1 \cos(\theta_2 + \beta)\}}} \tag{18.34}$$

$$I_B = \frac{I_A N_B}{N_A} \tag{18.35}$$

$$I_C = \frac{I_A N_C}{N_A} \tag{18.36}$$

where $\quad I_A, I_B, I_C$ = line currents in amperes

k_1 is a constant and equals:

$$\frac{1}{N_A}\sqrt{(N_B^2+N_C^2+2N_BN_C\cos\theta_1)} \qquad (18.37)$$

$$\left.\begin{aligned} N_A &= \frac{K_A}{V_{ZA}} \\ N_B &= \frac{K_B}{V_{ZB}} \\ N_C &= \frac{K_C}{V_{ZC}} \end{aligned}\right\} \qquad (18.38)$$

*β is an angular constant and equals:

$$\sin^{-1}\left(\frac{N_C\sin\theta_1}{N_Ak_1}\right) \qquad (18.39)$$

$$*\theta_1 = \tan^{-1}\left(\frac{V_{XB}}{V_{RB}}\right)-\tan^{-1}\left(\frac{V_{XC}}{V_{RC}}\right) \qquad (18.40)$$

$$*\theta_2 = \tan^{-1}\left(\frac{V_{XA}}{V_{RA}}\right)-\tan^{-1}\left(\frac{V_{XB}}{V_{RB}}\right) \qquad (18.41)$$

and where for transformers A, B and C respectively

K_A, K_B, K_C = normal rated outputs in kVA.

V_{ZA}, V_{ZB}, V_{ZC} = percentage impedance voltage drops at full-load ratings.

V_{XA}, V_{XB}, V_{XC} = percentage reactance voltage drops at full-load ratings.

V_{RA}, V_{RB}, V_{RC} = percentage resistance voltage drops at full-load ratings.

*θ_1 = phase angle difference between the load currents I_A and I_B.

*θ_2 = phase angle difference between the load currents I_B and I_C.

From the geometry of the figure,

$$\alpha = \sin^{-1}\left\{\frac{I_C}{I_L}\sin(\beta+\theta_2)\right\}$$

and the phase angle difference between the load current I_A in transformer A and the total load current I_L is $(\theta_1-\beta+\alpha)$. Having fixed the phase relationship of the total load current to the load current in one transformer, it is a simple matter to determine the angles between the total load current and the load currents in the remaining two transformers. If β is greater than α, the load current I_B in transformer B is lagging with respect to I_L: if β is smaller than α, I_B is leading with respect to I_L.

For the diagram in Fig. 18.14:

θ_1 and θ_2 are positive.

I_A is leading I_L.

I_C is lagging I_L.

Transformer A has the smallest ratio of V_X/V_R.

Transformer C has the greatest ratio of V_X/V_R.

*See Fig. 18.14.

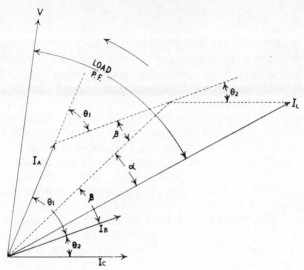

Fig. 18.14   Phasor diagram showing current distribution with three transformers in parallel having different ratios of resistance to reactance

I_B may lead or lag I_L, according to the inter-relationship of its value of V_X/V_R with the values of V_X/V_R of the other two transformers.

When θ_1 and θ_2 are negative:

I_A is lagging I_L.

I_C is leading I_L.

Transformer A has the greatest ratio of V_X/V_R.

Transformer C has the smallest ratio of V_X/V_R.

As before, I_B may lead or lag I_L, depending upon the various values of V_X/V_R.

When dealing with transformers having different outputs and different impedances which are to operate in parallel, it should be remembered that the impedance drop of a single transformer is based on its own rated full-load current, and this point should not be overlooked when determining the current distribution of two such transformers operating in parallel. If the *ohmic* values of the impedances of the individual transformers are deduced from the impedance drop and normal full-load current of each and the results inserted in the usual formula for resistances in parallel, the same final results for current distribution are obtained by already well known and simple methods. In using this ohmic method care should be taken to notice whether the ratio of resistance to reactance is the same with all transformers, for if it is not, the value of the impedance voltage drop as such cannot directly be used for determining the current distribution, but it must be split up into its power and reactive components.

When operating transformers in parallel the output of the smallest transformer should not be less than one-third of the output of the largest, as otherwise it is extremely difficult, as mentioned above, to incorporate, the necessary impedance in the smallest transformer.

411

Polarity

The term polarity when used with reference to the parallel operation of electrical machinery is generally understood to refer to a certain relationship existing between *two or more* units, though, as stated previously, it can be applied so as to indicate the directional relationship of primary and secondary terminal voltages of a single unit. Any two single-phase transformers have the same polarity when their instantaneous *terminal* voltages are in phase. With this condition a voltmeter connected across similar terminals will indicate zero.

Single-phase transformers are essentially simple to phase in, as for any given pair of transformers there are only two possible sets of external connections, one of which must be correct. If two single-phase transformers, say X and Y, have to be phased in for parallel operation, the first procedure is to connect both primary and secondary terminals of, say, transformer X, to their corresponding busbars, and then to connect the primary terminals of transformer Y to their busbars. If the two transformers have the same polarity, corresponding secondary terminals will be at the same potential, but in order to ascertain if this is so it is necessary to connect one secondary terminal of transformer Y to what is thought to be its corresponding busbar. It is necessary to make the connection from one secondary terminal of transformer Y, so that when taking voltage readings there is a return path for the current flowing through the voltmeter. The voltage across the disconnected secondary terminal of transformer Y and the other busbar is then measured, and if a zero reading is obtained the transformers have the same polarity, and permanent connections can accordingly be made. If, however, the voltage measured is twice the normal secondary voltage, then the two transformers have opposite polarity. To rectify this it is only necessary to cross-connect the secondary terminals of transformer Y to the busbars. If, however, it is more convenient to cross-connect the primary terminals, such a procedure will give exactly the same results.

Phase sequence

In single-phase transformers this point does not arise, as phase sequence is a characteristic of polyphase transformers.

POLYPHASE TRANSFORMERS

Phase angle difference between primary and secondary terminals

The determination of suitable external connections which will enable two or more polyphase transformers to operate satisfactorily in parallel is more complicated than is a similar determination for single-phase transformers, largely on account of the phase angle difference between primary and secondary terminals of the various connections. It becomes necessary, therefore, to study carefully the internal connections of polyphase transformers which are to be operated in parallel before attempting to phase them in.

Transformers made strictly to comply with later editions of B.S.171 have certain fixed terminal positions and terminal markings and two transformers of

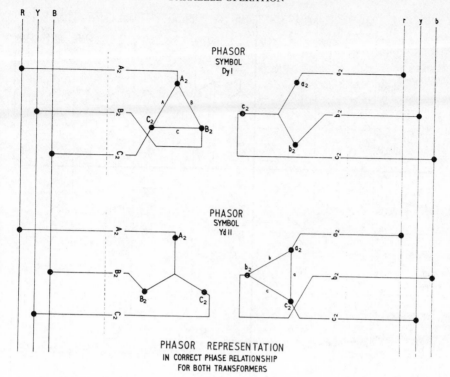

Fig. 18.15   Example of parallel operation of transformers in B.S. 171, Groups 3 and 4.

Note: The phasor diagram of the transformer DyI is identical with Fig. 18.7, but that for the transformers YdII, for which the phase sequence has been reversed from A-B-C to A-C-B, differs from Fig. 18.8

similar characteristics and having the same group number can be operated in parallel by connecting together terminals which correspond both physically and alphabetically. In addition it is possible to arrange the external connections of a transformer from group number 3 to enable it to operate in parallel with another transformer connected to group number 4 without changing any internal connections. Figure 18.15 indicates how this can be achieved, and it will be seen that two of the h.v. connections and the corresponding l.v. connections are interchanged.

Transformers connected in accordance with phasor groups 1 and 2 respectively *cannot* be operated in parallel with one another without altering the internal connections of one of them and thus bringing the transformer so altered within the other group of connections.

Figure 18.16 shows the range of three to three-phase connections met with in practice, and it will be noticed that the diagram is divided up into four main sections. The pairs of connections in the groups of the upper left-hand section may be connected in parallel with each other, and those in the lower right-hand section may also be connected in parallel with one another, but the remaining pairs in the other two groups cannot so be connected, as there is a 30° phase displacement between corresponding secondary terminals. This displacement is indicated by the dotted lines joining the pairs of secondaries. Figure 18.17 is

413

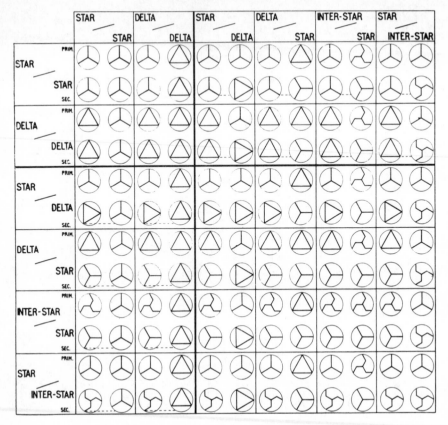

Fig. 18.16 Diagram showing the pairs of three- to three-phase transformer connections which will and which will not operate together in parallel

also given to show the range of three- to six-phase connections which may be used, and the same remarks apply.

It should be noted that this question of phase displacement is one of displacement between the *line terminals*, and not necessarily of any internal displacement which may occur between the phasors representing the voltages across the individual phase windings.

Voltage ratio

With polyphase transformers, exactly the same remarks apply as outlined for single-phase transformers. Equations 18.1 to 18.19 inclusive, also apply in the same way, but the currents, voltages and impedances should all be based on the line values.

Percentage impedance

The treatment given in eqns. 18.21 to 18.41 inclusive, applies exactly for polyphase transformers, the currents, voltages and impedances being based on line values.

Polarity and phase sequence

When phasing in any two or more transformers it is essential that both their polarity and phase sequence should be the same. The phase sequence may be clockwise or counter-clockwise, but so long as it is the same with both transformers, the direction is immaterial. It is generally advisable, when installing two or more transformers for parallel operation, to test that corresponding secondary terminals have the same instantaneous voltage, both in magnitude and phase.

With regard to the actual procedure to be followed for determining the correct external connections, there are two ways in which this may be done. The first one is to place the two transformers in parallel on the primary side and take voltage measurements across the secondaries, while the other is to refer to the manufacturer's diagram, such a diagram being illustrated in Fig. 18.18. From a diagram of this kind, together, if necessary, with the key diagrams which are given in Fig. 18.19, it is an easy matter to obtain precisely the correct external connections which will enable the transformers to operate in parallel.

Dealing first with the method in which a series of voltage readings are taken for the purpose of determining how the transformers shall be connected, assume

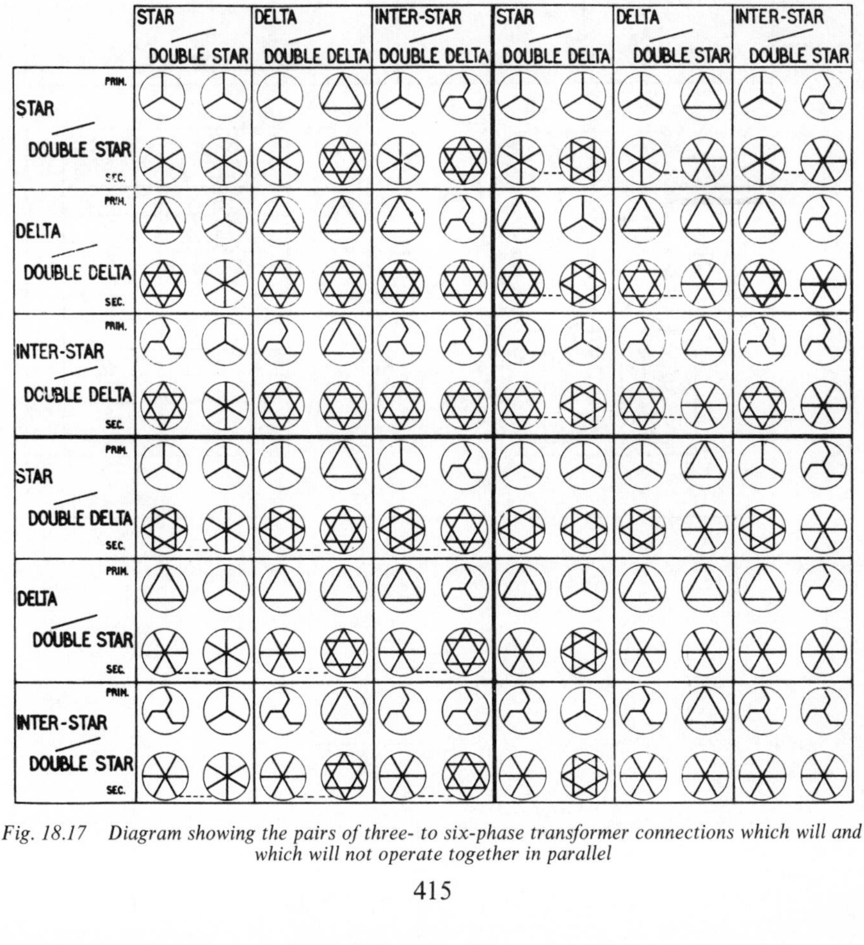

Fig. 18.17   Diagram showing the pairs of three- to six-phase transformer connections which will and which will not operate together in parallel

415

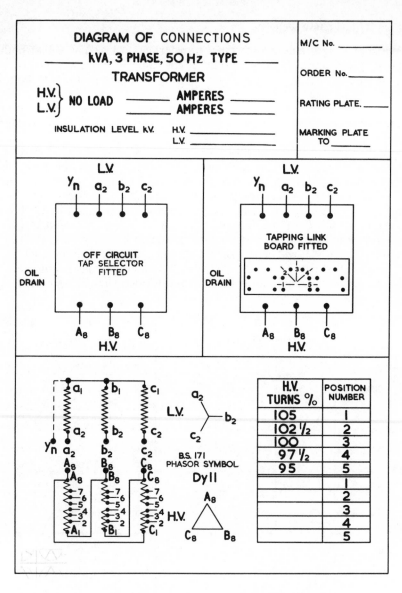

Selector position	Tappings connected
1	4, 5
2	6, 4
3	3, 6
4	7, 3
5	2, 7

Fig. 18.18   Manufacturer's typical diagram of connections

416

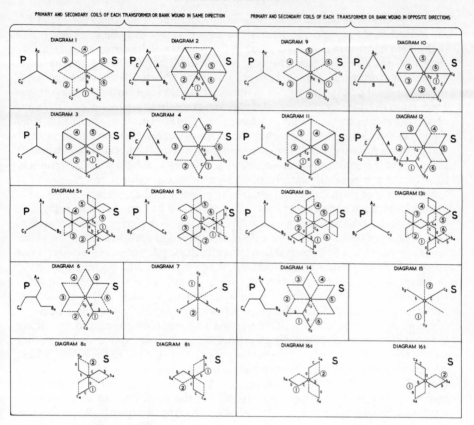

Fig. 18.19   Key diagrams for the phasing-in of three- to three-phase transformers

two transformers X and Y having the same voltage ratios and impedances and with their internal connections corresponding to any one pair of the permissible combinations given in Fig. 18.16. The first procedure is to connect all the primary terminals of both transformers to their corresponding busbars, and to connect all the secondary terminals of *one* transformer, say X, to its busbars. Assuming that both secondary windings are unearthed, it is next necessary to establish a link between the secondary windings of the two transformers, and for this purpose any one terminal of transformer Y should be connected, via the busbars, to what is thought to be the corresponding terminal of the other transformer. These connections are shown in Fig. 18.20. Voltage measurements should now

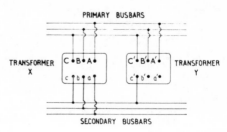

Fig. 18.20   Phasing-in a three-phase transformer

417

be taken across the terminals aa′ and bb′, and if in both instances zero readings are indicated, the transformers are of the same polarity and phase sequence, and permanent connections may be made to the busbars. If, however, such measurements do not give zero indications, it is sometimes helpful to take, in addition, further measurements, that is, between terminals ab′ and ba′, as such measurements will facilitate the laying out of the exact phasor relationship of the voltages across the two transformer secondary windings. Figure 18.19 gives key diagrams of the different positions that the secondary voltage phasors of a transformer could take with respect to another transformer depending upon their relative connections, polarity, phase sequence and the similarity or not of those terminals which form the common junction, and this will serve as a guide for determining to what the test conditions correspond on any two transformers.

In the case of transformers of which the primary and secondary connections are different, such as delta/star, it is only necessary when one of the transformers is of opposite polarity, to change over any two of the primary *or* secondary connections of *either* transformer. As such a procedure also reverses the phase sequence, care must be exercised finally to join those pairs of secondary terminals across which zero readings are obtained. When, however, the connections on the primary and secondary sides are the same, such as, for instance, delta/delta, transformers of opposite polarity cannot be phased in unless their *internal* connections are reversed. When the phase sequence is opposite, it is only a question of changing over the lettering of the terminals of one transformer, and, providing the polarity is correct, connecting together similarly lettered terminals; in other words, two of the secondary connections of one transformer to the busbars must be interchanged. With two transformers both having star connected secondaries, the preliminary common link between the two can be made by connecting the star points together if these are available for the purpose, and this leaves all terminals free for the purpose of making voltage measurements. As a rule, this procedure makes the result much more apparent at first glance owing to the increased number of voltage measurements obtained.

Table 18.3 shows all the voltage readings which may be obtained between the secondary terminals of two transformers which are being phased in for parallel operation, under different conditions of connected terminals, polarity and phase sequence. The table refers to voltage measurements taken across

Table 18.3

STAR CONNECTED SECONDARIES, NEUTRALS JOINED, VOLTAGE
MEASUREMENTS ACROSS A TO A′, B TO B′, C TO C′

Same polarity, same phase sequence.
(a) Three zero readings . . . . . .
(b) Three line voltage readings. . . . .

Same polarity, reverse phase sequence.
(a) { One zero reading . . . . .
{ Two line voltage readings . . . .

Reverse polarity, same phase sequence.
(a) Three double phase voltage readings . . .
(b) Three phase voltage readings . . . .

Reverse polarity, reverse phase sequence.
(a) { One double phase voltage reading . . .
{ Two phase voltage readings . . . .

418

Table 18.3 (*continued*)

STAR, INTERCONNECTED STAR, OR DELTA CONNECTED SECONDARIES, LINE TERMINALS JOINED, VOLTAGE MEASUREMENTS ACROSS A TO A', B TO B', C TO C'.

Same polarity, same phase sequence.
  Like phases joined:
    (a) Three zero readings . . . .
    (b) Three line voltage readings . . . .
  Unlike phases joined:
    (a) Three line voltage readings . . . .
    (b) { One zero reading . . . .
        { Two line voltage × 1·73 readings .
    (c) { Two line voltage readings . . .
        { One double-line voltage reading . .

Same polarity, reverse phase sequence.
  Like phases joined:
    (a) { One zero reading . . . .
        { Two line voltage readings . . .
  Unlike phases joined:
    (a) { Two line voltage readings . . .
        { One line voltage × 1·73 reading . .
          One zero reading . . . .
    (b) { One line voltage reading . . .
        { One double-line voltage reading . .

Reverse polarity, same phase sequence.
  Like phases joined:
    (a) { One zero reading . . . .
        { Two double-line voltage readings .
    (b) { Two line voltage readings . . .
        { One line voltage × 1·73 reading . .
  Unlike phases joined:
    (a) { Two line voltage readings . . .
        { One line voltage × 1·73 reading . .
    (b) { Two line voltage readings . . .
        { One zero reading . . . .

Reverse polarity, reverse phase sequence.
  Like phases joined:
    (a) { One zero reading . . . .
        { Two line voltage × 1·73 readings .
    (b) { Two line voltage readings . . .
        { One double-line voltage reading . .
  Unlike phases joined:
    (a) Three line voltage readings . . . .
    (b) { Two zero readings . . . .
        { One line voltage × 1·73 reading . .

similarly lettered terminals, that is, across A to A', B to B', and C to C', but the different readings obtainable, as set out in the table, take care of the fact that any one phase may be lettered A, B or C, the other phases being lettered accordingly. The small diagrams simply show, in each case, an example of star and delta connected secondaries, with one relative set of letterings only, but in every instance there are really three of these for each diagram shown in the table corresponding to the different conditions of terminals connected, polarity and

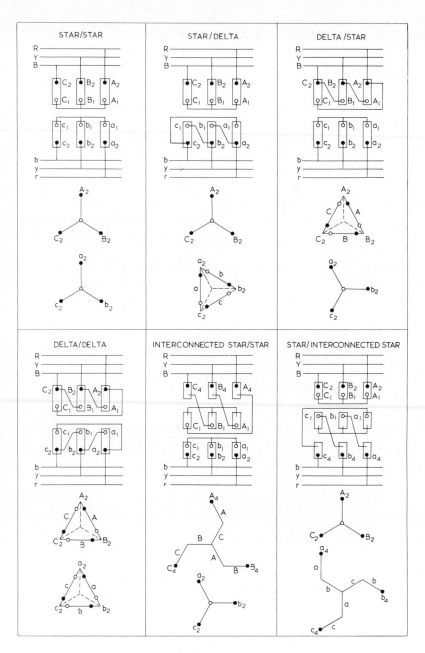

Fig. 18.21   Standard connections and polarities for three- to three-phase transformers
Note: Primary and secondary coils wound in the same direction
  ● indicates start of windings
  ○ indicates finish of windings

420

phase sequence. Having obtained voltage measurements across similarly lettered secondary terminals of two transformers which are being phased in, it will be an easy matter to determine exactly what their relative polarities and phase sequences are by referring to this table.

Dealing next with the method in which the transformer manufacturer's diagram is used for obtaining the correct external connections, Fig. 18.21 shows the six most common combinations of connections for three- to three-phase transformers. This diagram illustrates the standard internal connections between phases of the transformers, and also gives the corresponding polarity phasor diagrams. It is to be noted that the phasors indicate instantaneous induced voltages, as by arranging them in this way the phasor diagrams apply equally well irrespective of which winding is the primary and which the secondary.

Both primary and secondary coils of the transformers are wound in the same direction, and the diagrams apply equally well irrespective of what the actual direction is. With the standard polarities shown in Fig. 18.21, it is only necessary to join together similarly placed terminals of those transformers which have connections allowing of parallel operation, to ensure a choice of the correct external connections. That is, there are two main groups only, the first comprising the star/star and the delta/delta connection, while the other consists of the star/delta, delta/star, interconnected star/star and star/interconnected star. As a matter of interest, Fig. 18.22 is included to show the connections and polarity of three- to six-phase transformers, and the remarks previously made regarding three- to three-phase transformers also generally apply. The procedure subsequently outlined for three- to three-phase transformers can so easily be applied to three- to six-phase transformers that the latter are not again specifically dealt with in this chapter.

When phasing in any two transformers having connections different from the star or the delta, such as, for instance, two Scott-connected transformer groups to give a three- to two-phase transformation, particular care must be taken to connect the three-phase windings symmetrically to corresponding busbars. If this is not done the two-phase windings will be 30° out of phase, and Fig. 18.23 shows the correct and incorrect connections together with the corresponding phasor diagrams.

If, on the other hand, two tee-connected groups for three- to three-phase transformation are to be connected in parallel, then a perfectly symmetrical connection of windings to corresponding busbars is not essential, such a case being illustrated in Fig. 18.24.

The polarity, of course, is as important with this connection as with any other, and phasing in is carried out in the same way as for ordinary three-phase transformer groups.

A further point to bear in mind when phasing in Scott-connected transformer banks for two- to three-phase transformation, is that similar ends of the teaser windings on the primary and secondary sides must be connected together. This applies with particular force when the three-phase neutrals are to be connected together for earthing. If the connection between the teaser transformer and the main transformer of one bank is taken from the wrong end of the teaser winding, the neutral point on the three-phase side of that bank will be at a potential above earth equal to half the phase voltage to neutral when the voltage distribution of the three-phase line terminals is symmetrical with respect to earth.

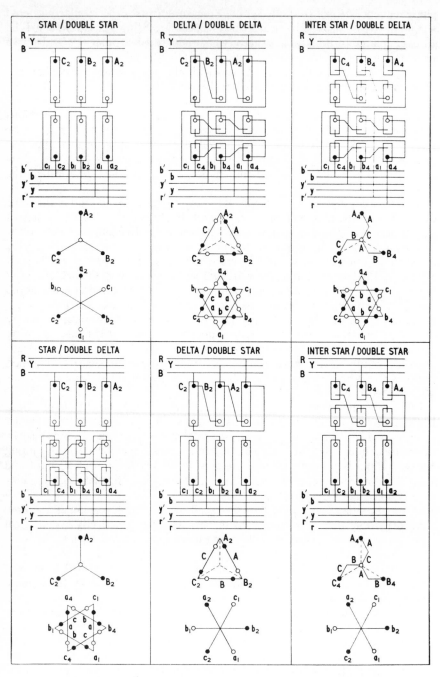

Fig.18.22 Standard connections and polarities for three- to six-phase transformers.
Note: Primary and secondary coils wound in the same direction.
    ● indicates start of windings
    ○ indicates finish of windings

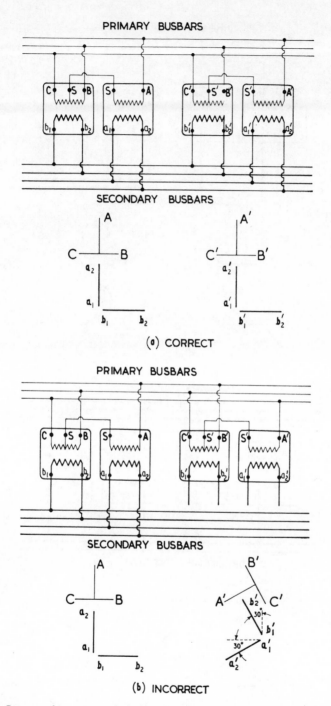

Fig. 18.23 Correct and incorrect method of paralleling two Scott-connected transformer groups for three- to two-phase transformation

423

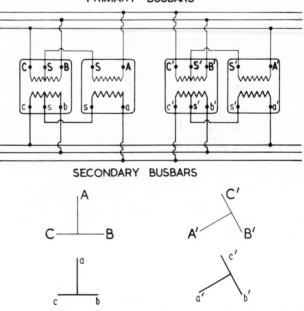

PRIMARY BUSBARS

SECONDARY BUSBARS

PRIMARY BUSBARS

SECONDARY BUSBARS

Fig. 18.24   Correct alternative methods of paralleling two tee-connected transformer groups for three-to three-phase transformation

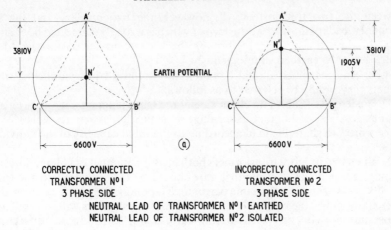

CORRECTLY CONNECTED
TRANSFORMER N° I
3 PHASE SIDE

INCORRECTLY CONNECTED
TRANSFORMER N° 2
3 PHASE SIDE

NEUTRAL LEAD OF TRANSFORMER N° I EARTHED
NEUTRAL LEAD OF TRANSFORMER N° 2 ISOLATED

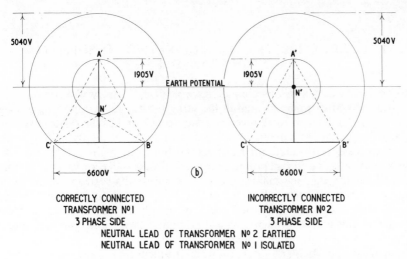

CORRECTLY CONNECTED
TRANSFORMER N° I
3 PHASE SIDE

INCORRECTLY CONNECTED
TRANSFORMER N° 2
3 PHASE SIDE

NEUTRAL LEAD OF TRANSFORMER N° 2 EARTHED
NEUTRAL LEAD OF TRANSFORMER N° I ISOLATED

Fig. 18.25   Phasing-in Scott-connected transformer groups

Figure 18.25 shows the phasor conditions on the three-phase side when the teaser transformer of one bank is correctly connected and the teaser transformer of the other bank is reverse connected: (*a*) when the neutral point of the correctly connected bank is earthed, the other neutral being isolated, and (*b*) when the neutral of the incorrectly connected bank is earthed, the other neutral, in turn, being isolated. The diagrams at (*a*) show that if the true neutral on the correctly connected bank is earthed, the so-called 'neutral' lead on the other bank attains a voltage above and below earth equal to half the phase voltage to neutral, although in both banks the voltage distribution of the windings with respect to earth is normal and symmetrical. On the other hand, if the 'neutral' lead of the incorrectly connected bank is earthed, the neutral point of the other bank attains a voltage below and above earth also equal to half the phase voltage to neutral, this condition being shown at (*b*). Such conditions are dangerous both to human life and to the transformer insulation, while further, the two neutral points cannot be earthed together, as heavy circulating currents would flow

425

between the two transformers. If, however, the two primary and the two secondary leads of the teaser transformer which is wrongly connected be changed over, the three-phase neutral is brought back to its proper position and both neutrals may be earthed simultaneously.

Other features which should be taken into account when paralleling transformers may briefly be referred to as follows:

(1) The length of cables on either side of the main junction should be chosen, as far as possible, so that their percentage resistance and reactance will assist the transformers to share the load according to the rated capacity of the individual units.

(2) When two or more transformers both having a number of voltage adjusting tappings are connected in parallel, care should be taken to see that the transformers are working on the same percentage tappings. If they are connected on different tappings, the result will be that the two transformers will have different ratios, and consequently a circulating current will be produced between the transformers on no-load.

THE PARALLEL OPERATION OF NETWORKS SUPPLIED THROUGH TRANSFORMERS

The previous sections in this chapter have dealt exclusively with the parallel operation of transformers located in the same substation or supplying a common circuit. As the loads on a given system increase and as the system extends, due to new load requirements in more distant areas of supply, it frequently becomes necessary to interconnect either, or both, the high-voltage and low-voltage networks at different points, in order to produce an economical distribution of load through the mains, and to minimise voltage drops at the more remote points of the networks. This problem of network interconnection due to increasing loads and extended areas of supply becomes, perhaps, most pressing in the case of systems which originally have been planned, either partially or wholly, as radial systems.

, In such cases, particularly, perhaps, when the problem is one of interconnecting higher-voltage supplies to extensive low-voltage networks, it may be found that the different circuits between the common source of supply and the proposed point, or points of interconnection, contain one or more transformers, which may, or may not, have the same combinations of primary and secondary connections, the same impedances, etc. The different circuits, moreover, may not contain the same number of transforming points.

It has been stated previously that two delta/star or star/delta transformers, for instance, may be paralleled satisfactorily simply by a suitable choice of external connections to the busbars, provided their no-load voltage ratios are the same, and such transformers will share the total load in direct proportion to their rated outputs provided their percentage impedances are equal. When, however, two or more compound circuits each comprising, say transformers and overhead lines, or underground cables, are required to be connected in parallel at some point remote from the source of supply, the question of permissible parallel operation is affected by the combined effect of the numbers of transformers in the different circuits and the transformer connections.

A typical and practical instance of what might be encountered is shown in

426

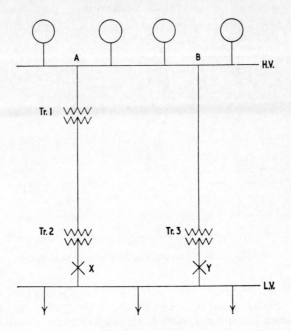

Fig. 18.26   Network layout

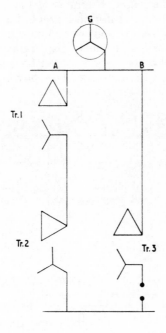

Fig. 18.27   Connections not permitting parallel operation

427

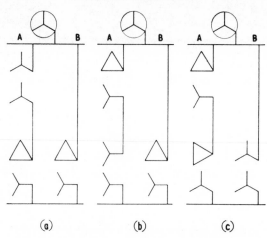

Fig. 18.28   Connections permitting parallel operation

Fig. 18.26 where a common l.v. network is fed from a power station through two parallel h.v. circuits A and B, one of which, A, contains a step-up transformer and a step-down transformer, both having delta connected primaries and star connected secondaries, while the other, B, contains one transformer only, having its primary windings delta connected and its secondary windings in star. From such a scheme it might be thought at first that the switches at the points X and Y might be closed safely, and that successful parallel operation would ensue. Actually, this would not be the case.

Figure 18.27 shows the phasor diagrams of the voltages at the generating station and at the different transforming points for the two parallel circuits lying between the power station and the common l.v. network, and it will be seen from these that there is a 30° phase displacement between the secondary line of neutral voltage phasors of the two transformers, (2) and (3), which are connected directly to the l.v. network. This phase displacement cannot be

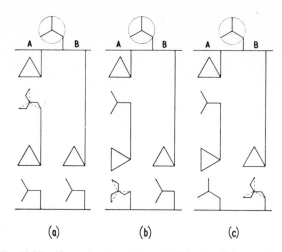

Fig. 18.29   Alternative connections permitting parallel operation

428

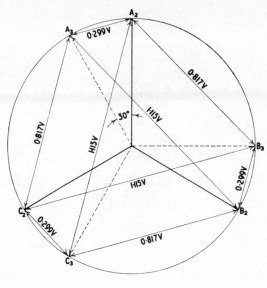

V· LINE VOLTAGE.

Fig. 18.30   Phasor diagram of l.v. voltages corresponding to Fig. 18.27

eliminated by any alternative choice of external connections to the busbars on either primary or secondary sides of any of the delta/star transformers, nor by changing any of the internal connections between the phase windings. The difficulty is created by the double transformation in circuit A employing delta/star connections in both cases, and actually the sum total result is the same as if the two transformers concerned were connected star/star. As mentioned earlier in this chapter, it is impossible to connect a star/star transformer and a delta/star in parallel.

The two circuits could be paralleled if the windings of any of the three transformers were connected star/star as shown at (*a*), (*b*) and (*c*) in Fig. 18.28, or delta/interconnected star as shown in (*a*), (*b*) and (*c*) of Fig. 18.29.

Apart from the fact that the delta/interconnected star transformer would be slightly more expensive than the star/star, the advantage lies with the former, as it retains all the operating advantages associated with a primary delta winding.

The phasor diagram in Fig. 18.30 shows the relative voltage differences which would be measured between the secondary terminals of the two transformers, (2) and (3), assuming their neutral points were temporarily connected together for the purpose of taking voltmeter readings, and that all three transformers were delta/star connected as in Fig. 18.27.

With properly chosen connections, as shown in Figs. 18.28 and 18.29 the loads carried by the two parallel circuits A and B will, of course, be in inverse proportion to their respective sum total ohmic impedances.

Thus, when laying out a network supplied through the intermediary of transformers, the primary and secondary connections of the latter, at the different transforming centres, should be chosen with a view to subsequent possible network interconnections, as well as from the other more usual considerations governing this question.

429

Chapter 19

The minimum total loss loading of substation transformers operating in parallel

Transformer substations, especially those serving large consumers or areas, are usually equipped with several transformers operating in parallel, or they are designed for accommodating additional transformers to cope with the demands of growing loads. In some cases it is the suppliers' practice to switch in additional transforming plant only when the load increases to such an extent that the transformers already in circuit would be overloaded seriously without further assistance. From the viewpoint of minimum total transformation losses, this procedure is not always an economical one, nor is it financially economical in those cases where iron and copper losses are lumped together and costed or charged for at a flat rate. If the separate losses are priced upon different bases, such as, for instance, that which takes account of the transformer load factor, the financial aspect introduces a modifying influence upon the values of the substation loads at which transformers should be switched in or out of commission, although the minimum total loss feature is not affected. It is not the purpose of this study to discuss the effects of different price rates for transformer losses based upon different load factors, but only to show how maximum overall operating economy may be effected by reducing the total transformation losses to a minimum under all the loading conditions which the equipment may be called upon to handle.

It is realised that the economical switching in and out of transformers can only be carried out in attended substations or in unattended substations if provided with automatic switching equipment specially designed for the purpose.

In this chapter it is assumed that transformers operating in parallel are designed for the same voltage ratios, the same percentage impedances, and the same ratios of resistance to reactance voltages. Of these three requirements for perfect parallel operation, only the first two are of very great importance, as considerable differences between the various ratios of resistance to reactance voltages may exist without materially affecting the distribution of current between transformers operating in parallel.

The general problem is to determine at what load or loads one or more transformers shall be switched in or out of circuit in order to obtain minimum total losses, that is, iron plus copper losses, at all loads.

430

IDENTICAL TRANSFORMERS

The simplest case is that involving only two transformers of the same output. The load at which any given transformer possesses its highest efficiency, *i.e.*, its minimum percentage losses, is that at which the iron and copper losses are equal and is given by the expression

$$K_1 = WK \text{ where } W = \sqrt{\frac{P_f}{P_c}} = \sqrt{\frac{'P_f}{'P_c}} \qquad (19.1)$$

$$K = \text{rated kVA}$$

where K_1 = load at which maximum efficiency occurs expressed as a fraction of the normal full-load rating.

P_f = iron loss expressed in watts or as a percentage of the normal full-load rating.

P_c = normal full-load copper loss expressed in watts or as a percentage of the normal full-load rating.

At loads higher than K_1, the copper loss, which increases as the square of the load, operates to reduce the efficiency, while at loads lower than K_1, the iron loss which remains substantially constant and independent of the load, reduces the efficiency from that given by eqn. 19.1.

The foregoing general principle may now be applied to two identical transformers operating in parallel.

Suppose in a certain substation there are three identical transformers, each of normal full-load rating K kVA, one of which operates alone while the other two are connected in parallel. At some particular, definite, and equal value of the load given by each supply, the total transformation losses of the two equipments are equal. That is, at a certain fixed output K_x kVA given by the one transformer, on the one hand, and at the same output K_x kVA given by the two transformers in parallel, on the other hand, the total losses in the transformer operating singly are the same as the sum total losses of the two transformers operating in parallel. The critical output K_x kVA at which this equalisation of total losses occurs obviously represents a higher loading on the transformer operating alone compared with the load on each of the other two transformers, and as all three are identical the load on each of the two transformers in parallel is $K_x/2$ kVA.

Suppose, now, in some other substation two identical transformers, each of normal full-load rating K kVA, are installed for parallel operation, each capable of being switched into circuit independently of the other. Assume one transformer only is necessary at first to supply the load demand, which latter, however, subsequently increases until it clearly becomes desirable to switch in the second transformer. The question arises, at what load on the substation should this be done? From the viewpoint of substation minimum total transformation losses at any load the answer is, at the total substation load equal to the critical output K_x kVA, when, as already stated in the case of the three-unit substation, the total losses of one transformer equal the sum total losses of two transformers in parallel. Below this critical output the sum total losses of two transformers in parallel would be greater than the total losses of one transformer alone, while above the critical output the total losses of one transformer would

be greater than the sum total losses of two transformers in parallel, even suppos-
ing that one transformer alone could carry any appreciable additional load above
the critical value K_x kVA without serious overheating. Thus we arrive at the
ruling that *at the critical group output* the total transformation losses are a
minimum, as the sum total losses of the two transformers in parallel are the same
as the total losses of one transformer alone. At the critical output the load carried
by each of the two transformers in parallel is, of course, one-half the load on the
one transformer working singly. Thus,

$$P_f + \left(\frac{K_x}{K}\right)^2 P_c = 2\left[P_f + \left(\frac{K_x}{2K}\right)^2 P_c\right] \qquad (19.2)$$

gives the equality of the total losses of one transformer with the sum total
losses of two transformers working together, which occurs at the critical group
output and which gives minimum total losses at that output.

In eqn. 19.2,

   P_f = iron loss of one transformer in watts

   P_c = normal full-load copper loss of one transformer in watts

   K_x = critical output in kVA at which the second transformer should be
   switched in or out of circuit with rising or falling load in order
   to keep the total group losses to a minimum at any load up to the
   critical output corresponding to K_x for two transformers in parallel.

For the case cited above, of the changeover from one to two identical transformers, the kVA value of the group critical loading is,

$$K_x = K\sqrt{\left(\frac{2P_f}{P_c}\right)} \qquad (19.3)$$

It is interesting to note that when changing over from one to two transformers
of the same characteristics, the value of the critical output is such that an individual transformer would have the same efficiency when carrying a load equal
to the critical load as when carrying half that load. Figure 19.1 shows typical
curves of total kW losses and percentage efficiencies plotted against total group
outputs for one, two and three 500 kVA identical transformers, each transformer
having an iron loss of 2·7 kW and a full-load copper loss at 60°C of 5·7 kW. It
will be seen from curves A and B, diagram I of Fig. 19.1, that the critical output,
which is given by the point at which the curves intersect, is 486 kVA for one
and two transformers. Reference to diagram II of Fig. 19.1 shows that the
corresponding efficiency curves A′ and B′ intersect at the same group output,
namely, 486 kVA, and that below this output the efficiency of two transformers
in parallel would be lower than that obtaining at the critical output, while above
this output the efficiency of one transformer operating alone would also be lower.

Extending the study to a consideration of changing over from two to three
identical transformers, each of normal full-load rating K kVA, it will be found
that the critical group output which preserves the total transformation losses
at a minimum at all loads is obtained from the equation,

$$2P_f + \left[\left(\frac{K_x}{2K}\right)^2 2P_c\right] = 3P_f + \left[\left(\frac{K_x}{3K}\right)^2 3P_c\right] \qquad (19.4)$$

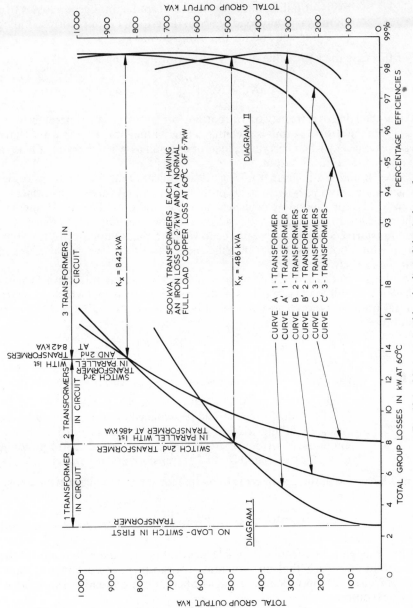

Fig. 19.1 Minimum total loss loading of identical transformers

433

where P_f = iron loss of one transformer expressed in watts

P_c = normal full-load copper loss of one transformer expressed in watts

K_x = critical output in kVA for two transformers in parallel at which the third transformer should be switched in or out of circuit with rising or falling load in order to keep the total group losses to a minimum at any load between the critical outputs corresponding to K_x for one and for three transformers.

For this case the kVA value of the group critical loading is,

$$K_x = K \sqrt{\left(\frac{6P_f}{P_c}\right)} \qquad (19.5)$$

When changing over from two to three transformers of the same characteristics, the value of the critical output is such that an individual transformer would have the same efficiency when carrying a load of one half as when carrying a load of one third of the critical output.

Curves B and C (diagram I of Fig. 19.1) indicate that for the transformer characteristics involved, the critical output for two and three transformers is 842 kVA. The corresponding efficiency curves B′ and C′ of diagram II intersect at the same group output, namely, 842 kVA. Below this output the efficiency of three transformers in parallel would be lower than that obtaining at the critical output, while above this output the efficiency of two transformers operating in parallel would also be lower.

For the general case of changing over from n to $n+1$ identical transformers each of normal full-load rating K kVA, the critical group output which preserves the total transformation losses at a minimum at all loads is obtained from the equation,

$$nP_f + \left[\left(\frac{K_x}{nK}\right)^2 nP_c\right] = \left\{(n+1)P_f\right\} + \left[\left\{\frac{K_x}{(n+1)K}\right\}^2 \left\{(n+1)P_c\right\}\right] \qquad (19.6)$$

where P_f and P_c have the same significance as in eqn. 19.4.

K_x = critical output in kVA for n transformers in parallel at which the $(n+1)$th transformer should be switched in or out of circuit with rising or falling load in order to keep the group losses to a minimum at any load between the critical outputs corresponding to K_x for $(n-1)$ and for $(n+1)$ transformers in parallel.

The kVA value of the group critical loading for the general case given above is,

$$K_x = K \sqrt{\left\{\frac{n(n+1)P_f}{P_c}\right\}} \qquad (19.7)$$

When changing over from n to $(n+1)$ identical transformers, the value of the critical output is such that an individual transformer would have the same efficiency when carrying a load of $1/n$th as when carrying a load of $1/(n+1)$th of the critical output.

In the foregoing treatment the critical loads, at which additional transformers should be switched in or out of circuit, have been determined solely from considerations of maximum efficiency. There is, however, a further aspect of the problem which must be considered. If the ratio of normal full-load copper

434

loss to iron loss in the individual transformers is very low, the equations which have been evolved for determining the values of the critical loads will result in values of the latter involving considerable overloads on the transformers already in commission before extra transformer capacity is provided. If it be postulated that no transformer shall be overloaded, then the limiting loss ratios for which eqns. 19.3, 19.5 and 19.7 apply may be found by equating K_x to the number of transformers already in circuit. If this calculation be performed it will be found that the equations for critical loads will hold good, provided the ratio of iron loss to normal full-load copper loss be not more than the ratio between the number of transformers already in circuit to that number plus one. When the ratio of losses is such that eqns. 19.3, 19.5 and 19.7 give values of the critical loads in excess of the combined normal full-load ratings of the transformers already in circuit, this combined normal full-load rating should be substituted for the calculated critical load in determining the transition point. Arising from these considerations the following conclusions may be drawn.

For changing from a single transformer to two in parallel, eqn. 19.3 holds good if the ratio of normal full-load copper loss to iron loss is not less than 2 to 1. If the loss ratio is less than this the change should be made when the total sub-station load reaches the normal full-load rating of one transformer.

Similarly, eqn. 19.5 holds good for ratios of normal full-load copper loss to iron loss not less than 1·5 to 1.

If the iron loss is equal to or greater than the normal full-load copper loss, extra transformers should only be switched into circuit when those already in circuit are fully loaded.

It should be noted that this limitation will rarely apply except in the case of changing from one to two transformers, as modern transformers rarely have ratios of normal full-load copper to iron loss less than 1·5 to 1 at normal working temperatures.

DISSIMILAR TRANSFORMERS

The ruling has now been established that the critical group output of a transformer substation is the total output at which the total losses are a minimum at any load between the critical output corresponding to K_x for $(n-1)$ and for $(n+1)$ identical transformers in parallel and at which the total losses of the $(n-1)$ transformers are equal to the total losses of the $(n+1)$ transformers. This ruling applies with exactly the same force to transformers of different individual outputs operating in parallel, although, of course, voltage ratios and percentage impedances must be the same to ensure satisfactory parallel operation under any condition.

For the purpose of further analysis of the problem we will consider three transformers of equal voltage ratios and percentage impedances and having the following full-load ratings and losses.

Transformer A

$P_{fA} = 1\cdot4$ kW = iron loss in watts.
$P_{cA} = 2\cdot7$ kW = normal full-load copper loss at 60°C in watts.
$K_A = 200$ kVA = normal full-load rating of transformer A.

435

Transformer B

P_{fB} = 1·8 kW = iron loss in watts.
P_{cB} = 4·1 kW = normal full-load copper loss at 60°C in watts.
K_B = 300 kVA = normal full-load rating of transformer B.

Transformer C

P_{fC} = 2·7 kW = iron loss in watts.
P_{cC} = 5·7 kW = normal full-load copper loss at 60°C in watts.
K_C = 500 kVA = normal full-load rating of transformer C.

Consider first the determination of the critical substation output above which it is not economical, from the point of view of total minimum transformation losses, to continue working the smallest transformer, namely, the 200 kVA alone.

The equation for determining the kVA value of the critical output K_x at which the 200 kVA transformer A should be superseded by the 300 kVA transformer B is,

$$P_{fA} + \left[\left(\frac{K_x}{K_A} \right)^2 P_{cA} \right] = P_{fB} + \left[\left(\frac{K_x}{K_B} \right)^2 P_{cB} \right] \tag{19.8}$$

where K_x = critical output in kVA at which transformer B should be substituted for transformer A with rising load or, vice versa, with falling load in order to keep the substation total losses to a minimum at any load up to the critical output corresponding to the value of K_x, at which transformer B would be superseded.

Substituting the respective numerical values in eqn. 19.8 we get,

$$1·4 + \left[\left(\frac{K_x}{200} \right)^2 2·7 \right] = 1·8 + \left[\left(\frac{K_x}{300} \right)^2 4·1 \right]$$

from which, K_x = 135 kVA.

The critical output K_x at which the 200 kVA transformer A should be superseded by the 500 kVA transformer C is determined from the equation,

$$P_{fA} + \left[\left(\frac{K_x}{K_A} \right)^2 P_{cA} \right] = P_{fC} + \left[\left(\frac{K_x}{K_C} \right)^2 P_{cC} \right] \tag{19.9}$$

where K_x = critical output in kVA at which transformer C should be substituted for transformer A with rising load or, vice versa, with falling load in order to keep the substitution total losses to a minimum at any load up to the critical output corresponding to the value of K_x, at which it would be necessary to switch two transformers in parallel.

Substituting the respective values in eqn. 19.9 we get,

$$1·4 + \left[\left(\frac{K_x}{200} \right)^2 2·7 \right] = 2·7 + \left[\left(\frac{K_x}{500} \right)^2 5·7 \right]$$

from which, K_x = 170 kVA.

Considering next the critical point at which the 500 kVA transformer C should be substituted for the 300 kVA transformer B, the equation for determining the kVA value of the critical output K_x is,

$$P_{fB} + \left[\left(\frac{K_x}{K_B} \right)^2 P_{cB} \right] = P_{fC} + \left[\left(\frac{K_x}{K_C} \right)^2 P_{cC} \right] \tag{19.10}$$

where K_x = critical output in kVA at which transformer C should be substituted for transformer B with rising load or, vice versa, with falling load in order to keep the substation total losses to a minimum at any load up to the critical output corresponding to the value of K_x, at which transformer C would be superseded if a fourth larger transformer were available.

Substituting the respective numerical values in eqn. 19.10 we get,

$$1 \cdot 8 + \left[\left(\frac{K_x}{300} \right)^2 4 \cdot 1 \right] = 2 \cdot 7 + \left[\left(\frac{K_x}{500} \right)^2 5 \cdot 7 \right]$$

from which, $K_x = 199$ kVA.

Having studied the respective critical outputs at which changeovers should be made from one transformer to another in a three-transformer substation, for example, we are confronted with the problem of ascertaining the respective critical outputs at which different transformers should be grouped for parallel running.

The general form of equations from which the critical outputs K_x may be determined is similar to that of eqns. 19.2 and 19.4, except that the terms embodying the constants of 2 and 3 must be replaced by other more comprehensive terms which correspond to the summations of the different values of the characteristics of the transformers concerned.

The equation for determining the respective values of the critical outputs K_x at which certain specified transformers should be switched in or out of parallel with, or other transformers substituted for, specified transformers already on load are as follows. The loss and output symbols have the significance already given, and in each subsequent case cited the value of K_x is that of the critical load in kVA at which the additional specified transformer or transformers should be switched in or out of parallel with, or other transformers substituted for, those already in circuit with rising or falling load in order to keep the substation total losses to a minimum at any load between the immediate lower and higher critical outputs at which a further change would be necessary economically.

(1) Critical output K_x at which the 300 kVA transformer B should be switched in or out of parallel with the 200 kVA transformer A.

$$P_{fA} + \left[\left(\frac{K_x}{K_A} \right)^2 P_{cA} \right] = (P_{fA} + P_{fB}) + \left[\left(\frac{K_x}{K_A + K_B} \right)^2 (P_{cA} + P_{cB}) \right] \qquad (19.11)$$

Substituting the respective numerical values in eqn. 19.11,

$$1 \cdot 4 + \left[\left(\frac{K_x}{200} \right)^2 2 \cdot 7 \right] = (1 \cdot 4 + 1 \cdot 8) + \left[\left(\frac{K_x}{200 + 300} \right)^2 (2 \cdot 7 + 4 \cdot 1) \right]$$

from which, $K_x = 212$ kVA.

(2) Critical output K_x at which the 500 kVA transformer C should be switched in or out of parallel with the 200 kVA transformer A.

$$P_{fA} + \left[\left(\frac{K_x}{K_A} \right)^2 P_{cA} \right] = (P_{fA} + P_{fC}) + \left[\left(\frac{K_x}{K_A + K_C} \right)^2 (P_{cA} + P_{cC}) \right] \qquad (19.12)$$

Substituting the respective numerical values in eqn. 19.12,

$$1\cdot4+\left[\left(\frac{K_x}{200}\right)^2 2\cdot7\right] = (1\cdot4+2\cdot7)+\left[\left(\frac{K_x}{200+500}\right)^2 (2\cdot7+5\cdot7)\right]$$

from which, $K_x = 232$ kVA.

(3) Critical output K_x at which the 300 and 500 kVA transformers B and C in parallel should be substituted for the 200 kVA transformer A.

$$P_{fA}+\left[\left(\frac{K_x}{K_A}\right)^2 P_{cA}\right] = (P_{fB}+P_{fC})+\left[\left(\frac{K_x}{K_B+K_C}\right)^2 (P_{cB}+P_{cC})\right] \qquad (19.13)$$

Substituting the respective numerical values in eqn. 19.13,

$$1\cdot4+\left[\left(\frac{K_x}{200}\right)^2 2\cdot7\right] = (1\cdot8+2\cdot7)+\left[\left(\frac{K_x}{300+500}\right)^2 (4\cdot1+5\cdot7)\right]$$

from which, $K_x = 243$ kVA.

(4) Critical output K_x at which the 200 kVA transformer A should be switched in or out of parallel with the 300 kVA transformer B.

$$P_{fB}+\left[\left(\frac{K_x}{K_B}\right)^2 P_{cB}\right] = (P_{fA}+P_{fB})+\left[\left(\frac{K_x}{K_A+K_B}\right)^2 (P_{cA}+P_{cB})\right] \qquad (19.14)$$

Substituting the respective numerical values in eqn. 19.14,

$$1\cdot8+\left[\left(\frac{K_x}{300}\right)^2 4\cdot1\right] = (1\cdot4+1\cdot8)+\left[\left(\frac{K_x}{200+300}\right)^2 (2\cdot7+4\cdot1)\right]$$

from which, $K_x = 276$ kVA.

(5) Critical output K_x at which the 200 and 500 kVA transformers A and C in parallel should be substituted for the 300 kVA transformer B.

$$P_{fB}+\left[\left(\frac{K_x}{K_B}\right)^2 P_{cB}\right] = (P_{fA}+P_{fC})+\left[\left(\frac{K_x}{K_A+K_C}\right)^2 (P_{cA}+P_{cC})\right] \qquad (19.15)$$

Substituting the respective numerical values in eqn. 19.15,

$$1\cdot8+\left[\left(\frac{K_x}{300}\right)^2 4\cdot1\right] = (1\cdot4+2\cdot7)+\left[\left(\frac{K_x}{200+500}\right)^2 (2\cdot7+5\cdot7)\right]$$

from which, $K_x = 285$ kVA.

(6) Critical output K_x at which the 300 and 500 kVA transformers B and C should be switched in or out of parallel with the 200 kVA transformer A.

$$P_{fA}+\left[\left(\frac{K_x}{K_A}\right)^2 P_{cA}\right] = (P_{fA}+P_{fB}+P_{fC})+\left[\left(\frac{K_x}{K_A+K_B+K_C}\right)^2\right.$$
$$\left. (P_{cA}+P_{cB}+P_{cC})\right] \qquad (19.16)$$

Substituting the respective numerical values in eqn. 19.16,

$$1\cdot4+\left[\left(\frac{K_x}{200}\right)^2 2\cdot7\right] = (1\cdot4+1\cdot8+2\cdot7)+\left[\left(\frac{K_x}{200+300+500}\right)^2\right.$$
$$\left. (2\cdot7+4\cdot1+5\cdot7)\right]$$

from which, $K_x = 286 \, \text{kVA}$.

(7) Critical output K_x at which 500 kVA transformer C should be switched in or out of parallel with the 300 kVA transformer B.

$$P_{fB} + \left[\left(\frac{K_x}{K_B} \right)^2 P_{cB} \right] = (P_{fB} + P_{fC}) + \left[\left(\frac{K_x}{K_B + K_C} \right)^2 (P_{cB} + P_{cC}) \right] \qquad (19.17)$$

Substituting the respective numerical values in eqn. 19.17,

$$1 \cdot 8 + \left[\left(\frac{K_x}{300} \right)^2 4 \cdot 1 \right] = (1 \cdot 8 + 2 \cdot 7) + \left[\left(\frac{K_x}{300 + 500} \right)^2 (4 \cdot 1 + 5 \cdot 7) \right]$$

from which, $K_x = 298 \, \text{kVA}$.

(8) Critical output K_x at which the 200 and 500 kVA transformers A and C in parallel should be substituted for the 200 and 300 kVA transformers A and B in parallel.

$$(P_{fA} + P_{fB}) + \left[\left(\frac{K_x}{K_A + K_B} \right)^2 (P_{cA} + P_{cB}) \right] = (P_{fA} + P_{fC}) + \left[\left(\frac{K_x}{K_A + K_C} \right)^2 \right.$$
$$\left. (P_{cA} + P_{cC}) \right] \qquad (19.18)$$

Substituting the respective numerical values in eqn. 19.18,

$$(1 \cdot 4 + 1 \cdot 8) + \left[\left(\frac{K_x}{200 + 300} \right)^2 (2 \cdot 7 + 4 \cdot 1) \right] = (1 \cdot 4 + 2 \cdot 7) + \left[\left(\frac{K_x}{200 + 500} \right)^2 \right.$$
$$\left. (2 \cdot 7 + 5 \cdot 7) \right]$$

from which, $K_x = 299 \, \text{kVA}$.

(9) Critical output K_x at which the 300 and 500 kVA transformers B and C in parallel should be substituted for the 200 and 300 kVA transformers A and B in parallel.

$$(P_{fA} + P_{fB}) + \left[\left(\frac{K_x}{K_A + K_B} \right)^2 (P_{cA} + P_{cB}) \right] = (P_{fB} + P_{fC}) + \left[\left(\frac{K_x}{K_B + K_C} \right)^2 \right.$$
$$\left. (P_{cB} + P_{cC}) \right] \qquad (19.19)$$

Substituting the respective numerical values in eqn. 19.19,

$$(1 \cdot 4 + 1 \cdot 8) + \left[\left(\frac{K_x}{200 + 300} \right)^2 (2 \cdot 7 + 4 \cdot 1) \right] = 1 \cdot 8 + 2 \cdot 7) + \left[\left(\frac{K_x}{300 + 500} \right)^2 \right.$$
$$\left. (4 \cdot 1 + 5 \cdot 7) \right]$$

from which, $K_x = 331 \, \text{kVA}$.

(10) Critical output K_x at which the 200 and 500 kVA transformers A and C should be switched in or out of parallel with the 300 kVA transformer B.

$$P_{fB} + \left[\left(\frac{K_x}{K_B} \right)^2 P_{cB} \right] = (P_{fA} + P_{fB} + P_{fC}) + \left[\left(\frac{K_x}{K_A + K_B + K_C} \right)^2 \right.$$
$$\left. (P_{cA} + P_{cB} + P_{cC}) \right] \qquad (19.20)$$

Substituting the respective numerical values in eqn. 19.20,

$$1{\cdot}8 + \left[\left(\frac{K_x}{300}\right)^2 4{\cdot}1\right] = 1{\cdot}4 + 1{\cdot}8 + 2{\cdot}7) + \left[\left(\frac{K_x}{200 + 300 + 500}\right)^2 \right.$$

$$\left. (2{\cdot}7 + 4{\cdot}1 + 5{\cdot}7)\right]$$

from which, $K_x = 352$ kVA.

(11) Critical output K_x at which the 500 kVA transformer C should be switched in or out of parallel with the combined 200 and 300 kVA transformers A and B.

$$(P_{fA} + P_{fB}) + \left[\left(\frac{K_x}{K_A + K_B}\right)^2 (P_{cA} + P_{cB})\right] = (P_{fA} + P_{fB} + P_{fC})$$

$$+ \left[\left(\frac{K_x}{K_A + K_B + K_C}\right)^2 (P_{cA} + P_{cB} + P_{cC})\right] \quad (19.21)$$

Substituting the respective numerical values in eqn. 19.21,

$$(1{\cdot}4 + 1{\cdot}8) + \left[\left(\frac{K_x}{200 + 300}\right)^2 (2{\cdot}7 + 4{\cdot}1)\right] = (1{\cdot}4 + 1{\cdot}8 + 2{\cdot}7) +$$

$$\left[\left(\frac{K_x}{200 + 300 + 500}\right)^2 (2{\cdot}7 + 4{\cdot}1 + 5{\cdot}7)\right]$$

from which, $K_x = 428$ kVA.

(12) Critical output K_x at which the 300 and 500 kVA transformer B and C in parallel should be substituted for the 200 and 500 kVA transformers A and C in parallel.

$$(P_{fA} + P_{fC}) + \left[\left(\frac{K_x}{K_A + K_C}\right)^2 (P_{cA} + P_{cC})\right] = (P_{fB} + P_{fC}) + \left[\left(\frac{K_x}{K_B + K_C}\right)^2 \right.$$

$$\left. (P_{cB} + P_{cC})\right] \quad (19.22)$$

Substituting the respective numerical values in eqn. 19.22,

$$(1{\cdot}4 + 2{\cdot}7) + \left[\left(\frac{K_x}{200 + 500}\right)^2 (2{\cdot}7 + 5{\cdot}7)\right] = (1{\cdot}8 + 2{\cdot}7) + \left[\left(\frac{K_x}{300 + 500}\right)^2 \right.$$

$$\left. (4{\cdot}1 + 5{\cdot}7)\right]$$

from which, $K_x = 468$ kVA.

(13) Critical output K_x at which the 300 kVA transformer B should be switched in or out of parallel with the 500 kVA transformer C.

$$P_{fC} + \left[\left(\frac{K_x}{K_C}\right)^2 P_{cC}\right] = (P_{fB} + P_{fC}) + \left[\left(\frac{K_x}{K_B + K_C}\right)^2 (P_{cB} + P_{cC})\right] \quad (19.23)$$

Substituting the respective numerical values in eqn. 19.23,

$$2{\cdot}7 + \left[\left(\frac{K_x}{500}\right)^2 5{\cdot}7\right] = (1{\cdot}8 + 2{\cdot}7) + \left[\left(\frac{K_x}{300 + 500}\right)^2 (4{\cdot}1 + 5{\cdot}7)\right]$$

440

from which, $K_x = 491$ kVA.

(14) Critical output K_x at which the 200 kVA transformer A should be switched in or out of parallel with the 500 kVA transformer C.

$$P_{fC} + \left[\left(\frac{K_x}{K_C} \right)^2 P_{cC} \right] = (P_{fA} + P_{fC}) + \left[\left(\frac{K_x}{K_A + K_C} \right)^2 (P_{cA} + P_{cC}) \right] \qquad (19.24)$$

Substituting the respective numerical values in eqn. 19.24,

$$2 \cdot 7 + \left[\left(\frac{K_x}{500} \right)^2 5 \cdot 7 \right] = (1 \cdot 4 + 2 \cdot 7) + \left[\left(\frac{K_x}{200 + 500} \right)^2 (2 \cdot 7 + 5 \cdot 7) \right]$$

from which, $K_x = 498$ kVA.

(15) Critical output K_x at which the 200 and 300 kVA transformers A and B in parallel should be substituted for the 500 kVA transformer C.

$$(P_{fA} + P_{fB}) + \left[\left(\frac{K_x}{K_A + K_B} \right)^2 (P_{cA} + P_{cB}) \right] = P_{fC} + \left[\left(\frac{K_x}{K_C} \right)^2 P_{cC} \right] \qquad (19.25)$$

Substituting the respective numerical values in eqn. 19.25,

$$(1 \cdot 4 + 1 \cdot 8) + \left[\left(\frac{K_x}{200 + 300} \right)^2 (2 \cdot 7 + 4 \cdot 1) \right] = 2 \cdot 7 + \left[\left(\frac{K_x}{500} \right)^2 5 \cdot 7 \right]$$

The solution to this equation is imaginary.

The reason for this is that as the ratings of the two groups are the same there would be no change in the total normal full-load group rating by the proposed substitution, and the total iron losses and total copper losses of transformers A and B are both greater than the respective iron and copper losses of transformer C at all total loads on the substations. Figure 19.2 shows that the output loss curves of the two combinations are approximately parallel over the output range given, although they are diverging with increasing output. The efficiency of transformer C thus would be higher at all loads and there can be no critical output.

(16) Critical output K_x at which the 200 and 300 kVA transformer A and B should be switched in parallel with the 500 kVA transformers C.

$$P_{fC} + \left[\left(\frac{K_x}{K_C} \right)^2 P_{cC} \right] = (P_{fA} + P_{fB} + P_{fC}) + \left[\left(\frac{K_x}{K_A + K_B + K_C} \right)^2 \right.$$
$$\left. (P_{cA} + P_{cB} + P_{cC}) \right] \qquad (19.26)$$

Substituting the respective numerical values in eqn. 19.26,

$$2 \cdot 7 + \left[\left(\frac{K_x}{500} \right)^2 5 \cdot 7 \right] = (1 \cdot 4 + 1 \cdot 8 + 2 \cdot 7) + \left[\left(\frac{K_x}{200 + 300 + 500} \right)^2 \right.$$
$$\left. (2 \cdot 7 + 4 \cdot 1 + 5 \cdot 7) \right]$$

from which, $K_x = 557$ kVA.

(17) Critical output K_x at which the 300 kVA transformer B should be switched in and out of parallel with the combined 200 and 500 kVA transformers A and C.

441

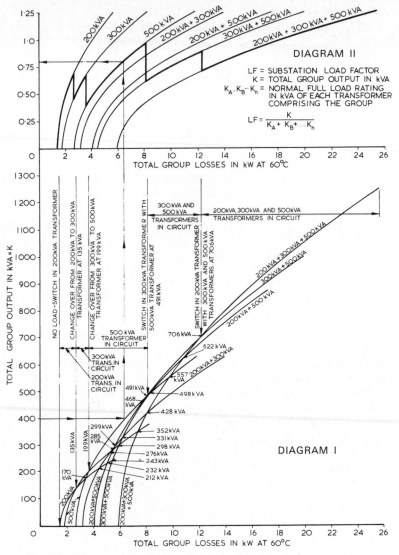

Fig. 19.2 Substation losses and load factors for different transformer groupings

$$(P_{fA}+P_{fC})+\left[\left(\frac{K_x}{K_A+K_C}\right)^2\ \ (P_{cA}+P_{cC})\right]=(P_{fA}+P_{fB}+P_{fC})$$

$$+\left[\left(\frac{K_x}{K_A+K_B+K_C}\right)^2(P_{cA}+P_{cB}+P_{cC})\right] \quad (19.27)$$

Substituting the respective numerical values in eqn. 19.27,

$$(1\cdot4+2\cdot7)+\left[\left(\frac{K_x}{200+500}\right)^2(2\cdot7+5\cdot7)\right]=(1\cdot4+1\cdot8+2\cdot7)$$

$$+\left[\left(\frac{K_x}{200+300+500}\right)^2(2\cdot7+4\cdot1+5\cdot7)\right]$$

from which, $K_x = 622$ kVA.

(18) Critical output K_x at which the 200 kVA transformer A should be switched in or out of parallel with the combined 300 and 500 kVA transformers B and C.

$$(P_{fB}+P_{fC})+\left[\left(\frac{K_x}{K_B+K_C}\right)^2(P_{cB}+P_{cC})\right]=(P_{fA}+P_{fB}+P_{fC})$$

$$+\left[\left(\frac{K_x}{K_A+K_B+K_C}\right)^2(P_{cA}+P_{cB}+P_{cC})\right] \qquad (19.28)$$

Substituting the respective numerical values in eqn. 19.28,

$$(1\cdot8+2\cdot7)+\left[\left(\frac{K_x}{300+500}\right)^2(4\cdot1+5\cdot7)\right]=(1\cdot4+1\cdot8+2\cdot7)$$

$$+\left[\left(\frac{K_x}{200+300+500}\right)^2(2\cdot7+4\cdot1+5\cdot7)\right]$$

from which, $K_x = 706$ kVA.

Summarising the foregoing solutions for the different group combinations of the three transformers, we have the following data in the order of ascending critical output.

Substitution of 300 kVA transformer B for 200 kVA transformer A.
$$K_x = 135 \text{ kVA}.$$

Substitution of 500 kVA transformer C for 200 kVA transformer A.
$$K_x = 170 \text{ kVA}.$$

Substitution of 500 kVA transformer C for 300 kVA transformer B.
$$K_x = 199 \text{ kVA}.$$

Switching in of 300 kVA transformer B with 200 kVA transformer A.
$$K_x = 212 \text{ kVA}.$$

Switching in of 500 kVA transformer C with 200 kVA transformer A.
$$K_x = 232 \text{ kVA}.$$

Substitution of 300 + 500 kVA transformers B and C for 200 kVA transformer A.
$$K_x = 243 \text{ kVA}.$$

Switching in of 200 kVA transformer A with 300 kVA transformer B.
$$K_x = 276 \text{ kVA}.$$

Substitution of 200 + 500 kVA transformers A and C for 300 kVA transformer B.
$$K_x = 285 \text{ kVA}.$$

Switching in of 300 + 500 kVA transformers B and C with 200 kVA transformer A.
$$K_x = 286 \text{ kVA}.$$

Switching in of 500 kVA transformer C with 300 kVA transformer B.
$$K_x = 298 \text{ kVA}.$$

Substitution of $200+500$ kVA transformers A and C for $200+300$ kVA transformers A and B.

$$K_x = 299 \text{ kVA}.$$

Substitution of $300+500$ kVA transformers B and C for $200+300$ kVA transformers A and B.

$$K_x = 331 \text{ kVA}.$$

Switching in of $200+500$ kVA transformers A and C with 300 kVA transformer B.

$$K_x = 352 \text{ kVA}.$$

Switching in of 500 kVA transformer C with $200+300$ kVA transformers A and B.

$$K_x = 428 \text{ kVA}.$$

Substitution of $300+500$ kVA transformers B and C for $200+500$ kVA transformers A and C.

$$K_x = 468 \text{ kVA}.$$

Switching in of 300 kVA transformer B with 500 kVA transformer C.

$$K_x = 491 \text{ kVA}.$$

Switching in of 200 kVA transformer A with 500 kVA transformer C.

$$K_x = 498 \text{ kVA}.$$

Substitution of $200+300$ kVA transformers A and B for 500 kVA transformer C.

$$K_x \text{ is imaginary}.$$

Switching in of $200+300$ kVA transformers A and B with 500 kVA transformer C.

$$K_x = 557 \text{ kVA}.$$

Switching in of 300 kVA transformer B with $200+500$ kVA transformers A and C.

$$K_x = 622 \text{ kVA}.$$

Switching in of 200 kVA transformer A with $300+500$ kVA transformers B and C.

$$K_x = 706 \text{ kVA}.$$

Diagram 1 of Fig. 19.2 shows the total group losses at 60°C. plotted against total group outputs in kVA and the points of intersection of the various losses —output curves correspond to the different values of the critical group outputs K_x given in the above summary. The curves are drawn for maximum loads of 125% of the normal full-load ratings of the transformers, so that the respective values of K_x correspond to substation loadings permitting not more than 25% overload on any individual transformer.

Of the foregoing summary of the critical outputs at which changes in the groupings of the transformers take place, some represent points of substation overall economic changes, while others are merely possible groupings. The heavy curve in diagram I of Fig. 19.2 shows the relationship between the total group outputs and the most economical total group losses in kW at all loads from zero to 1250 kVA. This curve, it will be seen, is a composite one combining portions of certain of the different output-losses curves which represent the maximum possible efficiency obtainable at any given substation load. The curve comprises those portions of the individual curves lying between the critical

Table 19.1

Transformer Groupings, kVA	200	300	500	200 + 300	200 + 500	300 + 500	200 + 300 + 500
200		135	170	*212*	232	243	286
300	135		199	276	285	298	*352*
500	170	199		*· ·*	498	491	*557*
200 + 300	*212*	276	· ·		299	331	428
200 + 500	232	285	498	299		468	622
300 + 500	243	298	**491**	331	468		**706**
200 + 300 + 500	286	*352*	*557*	428	622	**706**	

The figures in italics involve overloads on the smaller of the two group rated outputs concerned. The bold figures represent the points of economical changes in groupings.

outputs corresponding to the upper and lower limits for each uppermost section of the composite curve and each of which gives the lowest total group losses for any given load lying within the particular range of outputs embraced.

A tabular statement of all the various critical outputs involved is given in Table 19.1. The output at which the change should be made from any one grouping to any other will be found at the junction of the appropriate row and column. The sequence of switching operations may be followed readily from this table. For very small outputs it is evident that the 200 kVA transformer would be used alone as it has the lowest iron loss.

Referring to Table 19.1, it is seen that the 300 kVA transformer should be substituted at 135 kVA. Continuing in the row or column for 300 kVA the next change is at 199 kVA, at which the 500 kVA transformer becomes more economical. Again referring to the appropriate row the next change is at 491 kVA, at which the 300 kVA transformer is switched in parallel. Finally, from the row for the 300 kVA and 500 kVA transformers in parallel, it is seen that the 200 kVA transformer should also be switched in parallel at 706 kVA.

The changes in groupings corresponding to the heavy composite curve in diagram 1 of Fig. 19.2 thus are as follows:

Switching in of 200 kVA transformer A at no-load.

Substitution of 300 kVA transformer B for 200 kVA transformer A,

$$K_x = 135 \text{ kVA.}$$

Substitution of 500 kVA transformer C for 300 kVA transformer B,

$$K_x = 199 \text{ kVA.}$$

Switching in of 300 kVA transformer B with 500 kVA transformer C,

$$K_x = 491 \text{ kVA.}$$

Switching in of 200 kVA transformer A with 300 + 500 kVA transformers B and C,

$$K_x = 706 \text{ kVA.}$$

The general formula for the critical output K_x at which it is economical to make substitutions for, or additions to, the transformers already in circuit is,

$$P_f + \left[\left(\frac{K_x}{K} \right)^2 P_c \right] = P_f{}^1 + \left[\left(\frac{K_x}{K^1} \right)^2 P_c{}^1 \right] \qquad (19.29)$$

445

where P_f = sum total iron losses of all transformers already in circuit, expressed in watts.

P_c = sum total normal full-load copper losses of all transformers already in circuit, expressed in watts.

K^1 = sum total normal full-load rating in kVA of the transformers already in circuit.

$P_c{}^1$ = sum total normal full-load copper losses of all transformers in circuit after the proposed change, expressed in watts.

$P_f{}^1$ = sum total iron losses of all transformers in circuit after the proposed change, expressed in watts.

K^1 = sum total normal full-load rating in kVA of the transformers in circuit after the proposed change.

K_x = critical output in kVA at which the proposed change should be made.

Solving eqn. 19.29 for K_x gives,

$$K_x \sqrt{\frac{(P_f{}^1 - P_f)}{(P_c/K^2) - (P_c{}^1/K^{1^2})}} \qquad (19.30)$$

The expansion of the general expression given by eqn. 19.29, from which the value of the critical output K_x may be determined, at which it is economical to substitute any given transformers W, X, Y, Z for any other transformers A, B, C, D, already on load is:

$$(P_{fA} + P_{fB} + P_{fC} + P_{fD}) + \left[\left(\frac{K_x}{K_A + K_B + K_C + K_D} \right)^2 \right.$$

$$\left. (P_{cA} + P_{cB} + P_{cC} + P_{cD}) \right] = (P_{fW} + P_{fX} + P_{fY} + P_{fZ})$$

$$+ \left[\left(\frac{K_x}{K_W + K_X + K_Y + K_Z} \right)^2 (P_{cW} + P_{cX} + P_{cY} + P_{cZ}) \right] \qquad (19.31)$$

The expansion of the general expression given by eqn. 19.29, from which the value of the critical output K_x may be determined, at which it is economical to switch in parallel any given transformers W, X, Y, Z, with any other transformers A, B, C, D, already on load is:

$$(P_{fA} + P_{fB} + P_{fC} + P_{fD}) + \left[\left(\frac{K_x}{K_A + K_B + K_C + K_D} \right)^2 \right.$$

$$\left. (P_{cA} + P_{cB} + P_{cC} + P_{cD}) \right] = (P_{fA} + P_{fB} + P_{fC} + P_{fD} + P_{fW} + P_{fX} + P_{fY} + P_{fZ}) -$$

$$+ \left[\left(\frac{K_x}{K_A + K_B + K_C + K_D + K_W + K_X + K_Y + K_Z} \right)^2 \right.$$

$$\left. (P_{cA} + P_{cB} + P_{cC} + P_{cD} + P_{cW} + P_{cX} + P_{cY} + P_{cZ}) \right] \qquad (19.32)$$

If certain units only of a number of transformers operating in parallel are superseded by other transformers, the expansion of eqn. 19.29 for determining

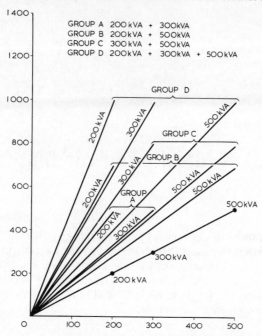

GROUP A   200 kVA + 300 kVA
GROUP B   200 kVA + 500 kVA
GROUP C   300 kVA + 500 kVA
GROUP D   200 kVA + 300 kVA + 500 kVA

Fig. 19.3   Division of loads among individual transformers operating in various paralleled groups

the value of the critical output K_x takes the following form, in which, for example, three transformers X, Y, Z replace two transformers A and B of a group of four, of which transformers C and D remain on load.

$$(P_{fA}+P_{fB}+P_{fC}+P_{fD})+\left[\left(\frac{K_x}{K_A+K_B+K_C+K_D}\right)^2\right.$$

$$\left.(P_{cA}+P_{cB}+P_{cC}+P_{cD})\right] = (P_{fC}+P_{fD}+P_{fX}+P_{fY}+P_{fZ})$$

$$+\left[\left(\frac{K_x}{K_C+K_D+K_X+K_Y+K_Z}\right)^2(P_{cC}+P_{cD}+P_{cX}+P_{cY}+P_{cZ})\right] \qquad (19.33)$$

Diagram II of Fig. 19.2 shows substation load factors for the different transformer groupings plotted against total group losses at all loads up to the maximum substation output. This diagram presents a clear picture of the relation between substation outputs, losses, and load factors at all loadings and groupings. The heavy composite curve in diagram II shows the substation load factors corresponding to those transformer groupings which maintain the total substation losses at a minimum at any given load from zero to the full output the plant is able to carry. Figure 19.3 is a simple diagram which shows very effectively how the loads taken by individual transformers vary according to those grouped in parallel

447

Chapter 20

Neutral point earthing

The subject of transformer neutral point earthing is divisible into two distinct sections, that is, earthing of the h.v. neutral and the l.v. neutral.

The h.v. neutral is earthed chiefly for the sake of the protection afforded to the electrical system, while the l.v. neutral is earthed mainly in order to reduce to a minimum the possible danger to human life, should contact accidentally be made with any live low-voltage conductor.

H.V. NEUTRAL POINT EARTHING

On account of the limitations imposed in Great Britain upon multiple earthing of any one electrical system, transformer h.v. neutral point earthing is really usually a question of system earthing, so that the issues involved are broader than those of a consideration of the transformer alone.

It is therefore not proposed to discuss the problem very fully here, but the chief aspects may be elaborated slightly as follows:

(1) Persistent arcing grounds are eliminated, and an earth fault definitely becomes converted into a short circuit from line to neutral. The high-voltage oscillations which are so prevalent in systems having isolated neutrals, and which have caused so much damage on such systems in the past, are reduced to a minimum, and consequently the factor of safety of the system against earth faults is largely increased. This reasoning applies to systems having overhead lines or underground cables, though to a greater extent with the former.

(2) Earthing the neutral point secures the maximum effectiveness of automatic protective gear immediately an earth fault occurs on the system. In h.v. networks most of the line faults take place to earth, particularly where such networks consist of underground cable, and on a system employing an isolated neutral these faults would generally take the form of an arcing ground, which in the case of multicore cables, would result ultimately in a severe short circuit between phases. The earthed neutral in conjunction with automatic protective gear, results in the faulty section becoming isolated in the incipient stages of the fault.

(3) If the neutral point is solidly earthed, the voltage of any live conductor

cannot exceed the voltage from line to neutral. As under such conditions the neutral point will be at zero potential, it is possible to effect appreciable reductions in the insulation to earth of cables and overhead lines, which produce a corresponding saving in cost. It is also possible to make similar insulation reductions in the transformers, the major insulation being graded down from the line terminals to the neutral point. In Great Britain, the neutral end insulation is graded for system voltages above 66 kV.

A stable earth fault on one line of a system having an isolated neutral raises the voltage of the two sound lines to full line voltage above earth, which is maintained so long as the fault remains. The insulation of all apparatus connected to the lines is subjected to this high voltage, and although it may be able to withstand the first few shocks it will eventually fail from cumulative weakening action. In very high voltage systems, on account of their electrostatic capacitance, the voltage of the two sound lines may, at the first instant of the fault, reach a value approaching twice the normal line voltage by the same phenomenon as the double overcharging which takes place when switching a pure capacitance into circuit, and the insulation of the system will correspondingly be affected.

(4) When earthing the neutral point it is not essential that this should be done direct, and in fact in h.v. systems having large generating capacities it is usual to insert some form of current limiting device between the neutral point and earth By this means the current flowing upon the occurrence of an earth fault can be limited to such a value as will, with a certain margin, operate the protective relays, while at the same time lightening the duty on the oil switches and reducing the electromagnetic stresses on those parts of the system which would be subjected to the fault current.

(5) On an unearthed system the voltage to earth of any line conductor may have any value up to the breakdown voltage of the insulation to earth, even though the normal voltage between lines and from lines to neutral is maintained. Such a condition may easily arise through electrostatic induction on systems employing overhead lines, as these are particularly subject to induced static charges from adjacent charged clouds, dust, sleet, fog and rain, and to changes in altitude of the lines. If provision is not made for freeing the line from these induced charges, gradual accumulation takes place, and the line and apparatus connected to it may reach a high 'floating' potential above earth until the abnormal condition is relieved by a breakdown to earth of the line or machine insulation or by the flashover of surge diverter gaps.

If, however, the neutral point is earthed either directly or through a current limiting device, the induced static charges are conducted to earth as they appear, and all danger to the insulation of the line and apparatus is removed. Therefore, no part of a solidly neutral system can reach a voltage above earth greater than the normal voltage from line to neutral.

Neutral earthing apparatus

The choice of suitable apparatus for inserting in the neutral connection for the purpose of limiting earth fault currents is dependent partly upon the voltage of the system to be earthed, that is whether it is h.v. or l.v., and partly upon the kVA output of the system. The neutral point may, in the case of h.v. systems, be earthed directly or through a resistor or reactor, and while the insertion of some kind of current limiting device is beneficial from the point of view of reducing fault currents, it should be remembered that the sound lines may, for a short period of time, be subjected to voltages higher than the normal phase voltage. However, if the automatic protective gear operates with sufficient rapidity, the effects of such increases in voltage are not likely to be serious.

Dealing first with the h.v. neutral, this may be earthed directly if the short circuit fault current of one phase is only of the order of the current necessary to operate the protective relays, or if the flow of the fault current only produces such electromagnetic stresses in the system apparatus as can easily be provided against.

The most common device used in connection with h.v. neutral point earthing is some form of resistor. For smaller systems, and up to 11 kV, this may take the form of cast or pressed grids, while for larger systems, where a larger fault current flows, liquid resistors are often used. The resistors are generally designed to carry for 30 s a fault current equal to the transformer rated full load, or a proportional value if two or more transformers are connected in parallel. The actual ohmic value of the resistor is a function of the system voltage to earth and of the permissible fault current.

Alternatively, the connection between the neutral point and earth may be made through an arc suppression coil. This form of earth connection reduces the number of interruptions of the supply under line to earth fault conditions where the fault is self clearing. If the fault is intermittent but not self clearing, the coil allows the system to be operated with one line earthed for comparatively short periods until the fault can be located and cleared.

In the event of a line to earth fault on an overhead line isolated neutral system, the electrostatic capacitance of the two sound lines to earth results in a capacitance current which is usually insufficient to operate protective relays, but which helps to sustain the arc across the faulty insulators. As a result the sound lines are subjected to an abnormal voltage above earth. If a variable reactor (arc suppression coil) is connected between the system neutral and earth, it may be tuned so that the current flowing in it, at approximately zero lagging power factor substantially neutralises the total fault current to earth at approximately zero leading power factor. The small residual earth current is sufficient to sustain the arc, and since it is substantially in phase with the voltage of the faulty conductor, both pass through zero at the same instant, and the arc is unlikely to restrike. A feature of the method is the slow reappearance of the system voltage after the arc has been extinguished.

The insulation level of all the plant and apparatus on a system on which arc suppression coils are installed must be adequate to allow operation for a period with one line earthed, and it is generally found uneconomic to install them on systems operating above 33 kV. Up to this voltage, the standard insulation level, without grading, should be employed for all transformers. B.S.171 recommends that a higher insulation level should be considered if operation of the system with one line earthed is likely for more than 8 hours in any 24, or more than 125 hours in any year.

Provision should be made for automatic short circuiting of the coil with subsequent operation of the protective relays if the fault is not cleared within the designed time rating.

Earth connection

When dealing with the question of neutral point earthing it is important to devote the utmost care to the earth connection itself, that is, to the apparatus buried in the earth (or otherwise located) for the purpose of obtaining a point at or near the potential of the earth. If the earthing system itself is not carefully considered, installed and maintained, it may be, under fault conditions, a distinct danger to human life.

The apparatus used for obtaining a direct earth contact may consist of copper or cast iron plates, iron pipes of small or large diameters, driven copper rods, or copper or galvanised iron strips.

It is not always appreciated that it is very difficult to obtain resistance values of less than about 2 Ω from a single earth plate or other piece of apparatus, and often it is still more difficult to *maintain* the ohmic value after the earthing system has been installed for some time. On account of this it is usual to install several earth plates, pipes, etc., in parallel, so that the combined resistance of the installation is reduced to a reasonably low value of, say, 1 Ω or less. Where a parallel arrangement is employed, each plate, rod, etc., should be installed outside the resistance area of any other. Strictly, this requires a separation of the order of 10 metres which, however, can often be substantially reduced without increasing the total resistance by more than a few per cent.

The chief points to be borne in mind when installing an earthing equipment are, that it must possess sufficient total cross-sectional area to carry the maximum fault current, and it must have a very low resistance in order to keep down to a safe value the potential gradient in the earth surrounding the plates, etc., under fault conditions. As most of the resistance of the earthing system exists in the immediate vicinity of the plates, etc., the potential gradient in the earth under fault conditions is naturally similarly located, and in order that this shall be kept to such a value as will not endanger life, the current density in the earth installation should be kept to a low figure either by using a number of the plates, pipes, etc., in parallel, or else by burying the devices to a considerable depth, making the connection to them by means of insulated cable. The former arrangement is one which can best be adopted where there are facilities for obtaining good earths, but in cases where, on account of the nature of the ground, it has been difficult to obtain a good earth, driven rods have been sunk to a depth of 10 metres or more. The maximum current density around an electrode is, in general, minimised by making its dimensions in one direction large with respect to those in the other two, as is the case with a pipe, rod or strip.

Earth plates are usually made of galvanised cast iron not less than 12 mm thick, or of copper not less than 2·5 mm in thickness, the sizes in common use being between 0·6 and 1·2 m^2. If an earth of greater conductivity is required, it is preferable to use two or more such plates in parallel.

Earth pipes may be of cast iron up to 100 mm diameter, 12 mm thick and 2·5 to 3 m long, and they must be buried in a similar manner to earth plates. Alternatively, in small installations, driven mild steel pipes of 30 to 50 mm diameter are sometimes employed.

Where the driving technique is adopted, copper rods are more generally used. These consist of 12–20 mm diameter copper in sections of 1–1·5 m, with screwed couplers and a driving tip. Deeply driven rods are effective where the soil resistivity decreases with depth but, in general, a group of shorter rods arranged in parallel is to be preferred.

In cases where high-resistivity soil (or impenetrable strata) underlies a shallow surface layer of low resistivity soil an earthing installation may be made up of untinned copper strip of section not less than 25 by 1·5 mm or of bare stranded copper conductor.

If a site can be utilised which is naturally moist and poorly drained, it is likely to exhibit a low soil resistivity. A site kept moist by running water should, however, be avoided. The conductivity of a site may be improved by chemical treatment of the soil, but it should be verified that there will be no deleterious effect on the electrode material. To ensure maximum conductivity, earth electrodes must be in firm direct contact with the earth.

It is most important that the connections from the neutral or auxiliary apparatus to the earth installation itself should be of ample cross-sectional area, so that there is adequate margin over the maximum fault current, and so that no abnormal voltage drop occurs over their length. The connections to the earthing structure should be of ample surface contact, and they should be permanently brazed.

Multiple earthing

The Acts and Regulations relating to electrical installations are framed in such a way that, consistent with the maximum practicable safeguard of the installations and of life, and of minimum interference with telecommunication systems, any system of earthing is permissible.

In Great Britain the Department of Trade and Industry and the Secretary of State for Scotland impose limitations on the earthing of systems at more than one point: that is, certain specified conditions must be complied with before two points, which are more or less widely apart, on the same system may be earthed. Extracts from Regulation 8 of the Electricity Supply Regulations, 1937, read:

'8. The following provisions shall apply to the connection with earth of systems for use at high voltage:

- (i) Unless otherwise allowed by the * Electricity Commissioners and subject as hereinafter provided, a point of every such system shall be connected with earth.
- (ii) The connection with earth shall, subject as hereinafter provided, be made at one point only in each system, and the insulation of the system shall be efficiently maintained at all other parts.
- (iii) In the case of a system as aforesaid comprising electric lines having concentric conductors, the external conductor shall be the one to be connected with earth.
- (iv) Where the Undertakers propose to connect with earth at one point only an existing system for use at high voltage which has not hitherto been so connected with earth, the Undertakers shall give notice and particulars to the Postmaster-General of the proposed connection with earth, and such notice shall be deemed to be a notice of works served upon the Postmaster-General within the meaning and for the purposes of Section 14 of the Schedule to the Electric Lighting (Clauses) Act, 1899, or

*Now the Department of Trade and Industry and the Secretary of State for Scotland.

corresponding provision in any Act or Order relating to the undertaking of the Undertakers.

(v) It shall not be permissible for the Undertakers to interconnect electrically systems for use at high voltage which are each connected with earth at one point, or, except as hereinbefore provided, to connect any such system with earth at more than one point, unless electrical interconnection as aforesaid or connection with earth at more than one point is for the time being approved by the Electricity Commissioners with the concurrence of the Postmaster-General† and is made in accordance with the conditions, if any of that approval.

Nothing in this Regulation or in any approval given by the Electricity Commissioners thereunder shall affect any rights or remedies of the Postmaster-General in relation to injury to or injurious affection of his telegraphic lines, or confer any exemption from any liability or penalty in respect of any such injurious affection.'

The object of these Regulations, and in particular para. (ii), is to avoid the short circuit paths which result if more than one point on the system is earthed, and hence to prevent the risks of interference with telecommunication circuits, of voltage gradients in the earth, and of electrolysis.

The practice of 'protective multiple earthing' has, however, much to commend it and, under specified conditions, is commonly adopted on high voltage systems with the approval of the Department of Trade and Industry or the Secretary of State for Scotland, and the concurrence of the Post Office. This method of earthing is adopted for the British Grid.

Earthing of delta connected transformers

It is sometimes desired to earth the neutral point of a circuit supplied by a transformer having delta connected windings, in which case a neutral point must

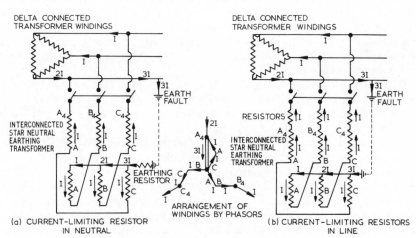

Fig. 20.1 Interconnected star neutral earthing transformer

†Now the Minister of Posts and Telecommunications.

be derived artificially by the inclusion of auxiliary apparatus specially designed for the purpose. The apparatus usually installed takes the form of what is known as an interconnected star neutral transformer, or alternatively a star/delta transformer. The two schemes are shown diagrammatically in Figs. 20.1 and 20.2. The interconnected star neutral earthing transformer is similar in construction to a three-phase core-type transformer, but it has a single winding only on each limb which is split up into two equal portions and interconnected, as shown in Fig. 20.1. The apparatus is therefore a 1 to 1 auto transformer with the windings so arranged that, while the voltages from each line to earth are maintained under normal operating conditions, a minimum impedance is offered to the flow of single phase fault current, such as is produced by an earth fault on one

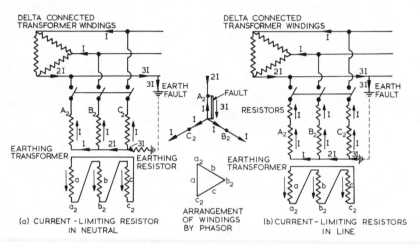

Fig. 20.2   Three-phase, star/delta neutral earthing transformer

line of a system having an earthed neutral. Under normal operating conditions the currents flowing through the windings are the magnetising currents of the earthing transformer only, but the windings are designed to carry the maximum possible fault current to which they may be subjected for a period of thirty seconds. The apparatus is built exactly as a three-phase core-type transformer, and is immersed in a tank of oil.

While the interconnected star earthing transformer is the type of apparatus most often used for providing an artificial neutral point, an alternative may be adopted in the form of an ordinary three-phase core-type transformer having star connected primary windings, the neutral of which is earthed and the line ends connected to the three phase lines, while the secondary windings are connected in closed delta, but otherwise isolated. Normally, the current taken by the transformer is the magnetising current only, but under fault conditions the closed delta windings act to distribute the fault currents in all three phases on the primary side of the transformer, and as primary and secondary fault ampereturns balance each other, there is no transient choking effect. The transformer is rated on the same basis as outlined for the interconnected star earthing transformer and it is constructed exactly the same as an ordinary power transformer.

For the purpose of fault-current limitation, resistors may be used in conjunction with either of the above types of apparatus, and they may be inserted between the neutral point and earth, or between the terminals of the apparatus and the lines. In the former case one resistor is required, but it must be designed to pass the total fault current, while it should be insulated for a voltage equal to the phase voltage of the system. On the other hand, the neutral point of the earthing transformer windings will rise to a voltage above earth under fault conditions equal to the voltage drop across the earthing resistor, and the earthing transformer windings will have to be insulated for the full line voltage above earth. While in any case this latter procedure may be adopted, it is not desirable to subject the earthing transformer windings to sudden voltage surges any higher than can be avoided, as the insulated windings are the most vulnerable part of the equipment. If, however, suitably proportioned resistors are placed between the terminals of the earthing apparatus and the lines instead of between neutral and earth, exactly the same purpose is served so far as fault-current limitation is concerned, while, in addition, the neutral point of the earthing transformer always remains at earth potential, and the windings are therefore not subject to any high voltages. On the other hand, the resistors must now be insulated for full line voltage, but this is a relatively easy and cheap procedure. For the same fault current and voltage drop across the resistors the ohmic value of each of those placed between the earthing transformer terminals and lines is three times the ohmic value of the single resistor connected between the neutral and earth, but the current rating of each resistor in the line is one third of the current rating of a resistor in the neutral, as under fault conditions the three resistors in the lines operate in parallel to give the desired protection.

L.V. NEUTRAL POINT EARTHING

L.V. neutral point earthing may be considered from two different points of view, depending upon whether the so called l.v. winding is designed for a voltage which is really a high voltage, such, for instance, as 6000 V or above, or for a low voltage of the order of 440 V. In the former case the transformer will be part of a general h.v. distribution system, and the considerations involved would almost be identical with those outlined in connection with h.v. neutral earthing. In addition, there is the question of induced electrostatic voltages to be taken into account, and this is dealt with later. Transformers wound for low voltages of the order of 440 V would be supplying consumers direct, and in such cases l.v. neutral point earthing is adopted chiefly as a precaution to the consumer. In this connection it is therefore pertinent to consider l.v. neutral point earthing with respect to the *Electricity Commissioners, Factories and Coal Mines Electricity Regulations.

Extracts from Regulation No. 4 of the Electricity Supply Regulations, 1937, as amended, read as follows:

'The following provisions shall apply to the connection with earth of alternating current systems at low voltage in cases where the voltage normally exceeds 125 volts, and of systems at medium voltage employed for giving a general supply:

    (i) Unless otherwise allowed by the *Electricity Commissioners, a point of every such system shall be connected with earth.

    (ii) The connection with earth shall, subject as hereinafter provided, be

*Now the Department of Trade and Industry and the Secretary of State for Scotland.

made at one point only in each system and the insulation of the system shall be efficiently maintained at all other parts.

(iii) In the case of a system comprising electric lines having concentric conductors, the external conductor shall be the one to be connected with earth.

(v) In the case of an alternating current system, there shall not be inserted in the connection with earth any impedance (other than that required solely for the operation of switchgear or instruments), fusible cut-out, or automatic circuit-breaker, and the result of any test made to ascertain whether the current (if any) passing through the connection with earth is normal shall be duly recorded by the Undertakers.

Provided that for the purpose of operating relays for the remote control of switches, the Undertakers may insert in the connection with earth the secondary winding of a high frequency transformer, the ohmic resistance of the said secondary winding not to exceed 2000 microohms at a temperature of $60°F$ ($15°C$) and its inductance not to exceed 10 microhenries.

(vi) Alternating current systems which are connected with earth at one point as aforesaid may be electrically interconnected subject to the following conditions and qualifications:

(a) Each connection with earth shall be bonded to the metal sheathing and metallic armouring (if any) of the electric lines concerned; and the Undertakers shall serve a notice on the Postmaster-General at least seven days prior to the making of any such interconnection, specifying the location of the point or points at which such interconnection is to be made and at which the interconnected systems are connected with earth, and the date on which such interconnection is to be made.

(b) Overhead lines forming a system or part of a system shall not be electrically interconnected with other electric lines (including other overhead lines) unless the neutral conductor or conductors of the overhead lines is or are of the same material and cross-sectional area as the corresponding phase conductors of the overhead lines.

(c) Where a system includes a generator or a transformer not having a mesh winding of low impedance, it shall not be electrically interconnected with another system if the neutral point of such generator or transformer is connected with earth.

(viii) Except as hereinbefore provided, it shall not be permissible for the Undertakers to connect any system or interconnected systems with earth at any further point unless such additional connection with earth is for the time being approved by the *Electricity Commissioners with the concurrence of the Postmaster-General and is made in accordance with the conditions, if any, of that approval.

Nothing in this regulation, or in any approval given by the *Electricity Commissioners thereunder shall:

(a) relieve the undertakers from the obligation to comply with the requirements of Regulation 21(a); or

*Now the Department of Trade and Industry and the Secretary of State for Scotland.

(b) affect any rights or remedies of the Postmaster-General in relation to injury to or injurious affection of his telegraphic lines, or confer any exemption from any liability or penalty in respect of any such injurious affection.'

Regulation No. 20 of the Electricity (Factories Act) Special Regulations, 1908 and 1944, reads: 'Where a high pressure or extra high pressure supply is transformed for use at a lower pressure, or energy is transformed up to above low pressure, suitable provision shall be made to guard against danger by reasons of the lower pressure system becoming accidentally charged above its normal pressure by leakage or contact from the higher pressure system.'

The requirements of the Coal Mines Act 1911 (M & Q Form No. 11) concerning l.v. neutral point earthing are specified in Regulation 126(b), which reads: 'Where energy is transformed, suitable provision shall be made to guard against danger by reason of the lower pressure apparatus becoming accidentally charged above its normal pressure by leakage from or contact with the higher pressure apparatus.'

The above Regulations have been framed with two principal ideas in mind: first, that the earthing of the l.v. neutral will prevent the presence of any voltage above the normal appearing in the l.v. circuit, and therefore the possible danger to human life will be reduced to a minimum; and second, the earthing of the neutral point eliminates the possibility of an arcing fault to earth, and therefore of fire risk, while it also ensures the rapid disconnection of faulty apparatus from the system without undue delay. The Regulations and the Explanatory Memoranda issued by the Electrical Inspector of Factories show that the l.v. neutral point is earthed chiefly for reasons somewhat different to those which influence the earthing of the h.v. neutral. As pointed out earlier, in the latter case the neutral is earthed mainly for the sake of the protection afforded to the electrical system, while in the former case the neutral is earthed chiefly in order to reduce to a minimum the possible danger to human life should contact accidentally be made with any live low voltage conductor.

The l.v. neutral points of three-phase transformers are earthed in a variety of ways, in some cases direct without the inclusion of any auxiliary apparatus, and in others through a resistor or through a fuse which is shunted by an ammeter and link. The ammeter and link provide means for periodically determining the amount of leakage current to earth, while the fuse will automatically open the neutral connection should a heavy sustained earth fault develop, thereby giving an indication of the presence of such a fault. The advantage of direct earthing is that no voltage higher than the normal l.v. phase voltage can ever appear in the l.v. circuit, and consequently the risk of shock is reduced to a minimum. This applies irrespective of whether the h.v. winding breaks down to the l.v. or an electrostatically induced voltage is liable to appear on the secondary winding. It is sometimes urged that with the neutral point directly earthed the danger from shock is increased, as contact with any one of the line wires produces a more or less high resistance short circuit across one phase, and that as a result of this the fault current is relatively large. In reality, however, no definite ruling can be laid down upon this point, as the whole question is largely dependent upon the health of the individual concerned and the particular local conditions existing at the time. There are many cases known to operating engineers of individuals having survived shocks of alternating voltages as high as 20 kV,

457

others succumbing to voltages as low as 110 V.

High voltages may be present in the l.v. windings if the l.v. neutral is unearthed when a breakdown between high- and low-voltage windings occurs, or under certain fault conditions of the h.v. circuit when the various electrostatic capacitances of the transformer have certain ratios.

Dealing first with the question of a breakdown between h.v. and l.v. windings, Fig. 20.3 shows how the voltages in the two windings are distributed with reference to the potential of the earth when the l.v. neutral is isolated and when it is solidly earthed. A breakdown between h.v. and l.v. windings has been assumed at the point X on phase A, as shown in diagram (a). Diagram (b) shows the normal voltage distribution, and diagram (c) shows the voltage distribution when a breakdown occurs at the point X on phase A with the l.v. neutral isolated. As the l.v. neutral is isolated the voltage distribution on the h.v. side would be the same as under normal operating conditions, as the h.v. neutral point would be held stable by the generating or transforming plant at the other end of the line which would have an earthed neutral. The loci of the points A, B and C are one and the same circle with N as centre, while the locus of the point X_H is also a circle with N as centre. X_H and X_L are directly connected by the fault, and therefore the locus of X_L is also a circle with Y (which is the same as N) as centre and radius YX_L. The loci of the points a, b, c and n are similarly circles having the same centre marked Y in diagram (c), and radii Ya, Yb, Yc and Yn respectively. Diagram (d) shows the relative positions of h.v. and l.v. phasors at three different instants, indicating the accuracy of the above statements, and also shows that the normal rotation of both h.v. and l.v. phasors about their respective neutral points is maintained unaltered even with the direct connection between h.v. and l.v. windings at X on phase A. The diagrams show that very high voltages would be present in the l.v. windings and the insulation would be liable to failure, while the danger to human life would be very considerable. Diagram (e) shows how the voltages would be distributed with the l.v. neutral directly earthed, and in this case it will be seen that the l.v. voltage distribution is the same as under normal operating conditions, while the voltage balance on the h.v. side with respect to earth is entirely upset. Two of the h.v. lines would be subjected to considerably higher voltages above earth than the normal phase voltage, while the remaining phase would have a correspondingly lower voltage above earth. The loci of the points, a, b and c are one and the same circle with n as centre, while the locus of the point X_L is also a circle with n as centre. X_L and X_H are directly connected by the fault, and therefore the locus of X_H is also a circle with Y (which is the same as n) as centre and radius YX_H. The loci of the points A, B, C and N are similarly circles having the same centre marked Y in diagram (e), and radii YA, YB, YC and YN respectively. Diagram (f) gives the relative positions of h.v. and l.v. phasors at three different instants, showing that the normal rotation of both h.v. and l.v. phasors about their respective neutral points is maintained unaltered even with the direct connection between h.v. and l.v. windings at X on phase A. It should be remembered, however, that if the neutral point of the generating or transforming plant at the other end of the h.v. line is earthed, the conditions shown in diagram (e) are transitory only, as the transformer would very soon be tripped out of circuit by circulating fault current which would flow from the plant at the far end of the line through the fault X, returning via the l.v. neutral earth and the neutral earth of the generating or transforming plant.

3000/440V STAR/STAR CONNECTED, THREE PHASE TRANSFORMER
VOLTS PER PHASE = 1730/254   V/T = 10·6
TURNS PER PHASE = 163/24   RATIO = 6·8/1

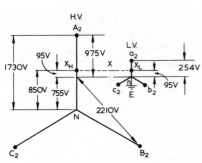

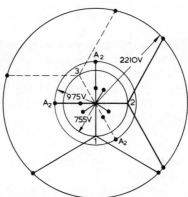

(a)

(b)

NORMAL VOLTAGE CONDITIONS
NEUTRALS UNEARTHED OR EARTHED

(c)

VOLTAGE CONDITIONS WITH BREAKDOWN
BETWEEN H.V. & L.V. WINDINGS ON PHASE A
AT POINT X. L.V. NEUTRAL UNEARTHED

(d)

PHASOR DIAGRAM OF THE CONDITIONS
AT THREE PARTICULAR INSTANTS
CORRESPONDING TO DIAGRAM (c)

(e)

VOLTAGE CONDITIONS WITH BREAKDOWN
BETWEEN H.V. & L.V. WINDINGS ON PHASE A
AT POINT X. L.V. NEUTRAL EARTHED

(f)

PHASOR DIAGRAM OF THE CONDITIONS
AT THREE PARTICULAR INSTANTS
CORRESPONDING TO DIAGRAM (e)

Fig. 20.3   Diagrams illustrating breakdown between h.v. and l.v. windings of one phase of a three-phase, star/star-connected transformer

459

Figure 20.4 shows how the voltage distribution on the h.v. side is temporarily affected with respect to earth when a breakdown occurs on the l.v. windings at different points along the limb, the l.v. neutral being earthed. Diagram (*a*) shows the coil arrangement, and diagram (*b*) gives curves showing the maximum instantaneous voltages of the h.v. phases and neutral point above and below earth with varying positions of the fault on phase A. Diagram (*c*) gives two

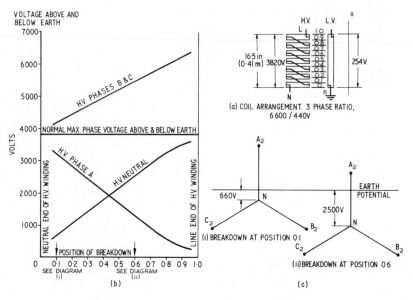

Fig. 20.4   *Curves showing the maximum voltage of the h.v. phases and of the neutral point above and below earth with varying positions of the fault on phase A.*

Phasor diagrams illustrating the instantaneous voltage distribution in the h.v. windings with respect to earth when a breakdown occurs on phase A. Phase A at its maximum.

Diagrams illustrating the influence of the position of the fault upon the voltage distribution in the h.v. windings with respect to earth when a breakdown occurs between the h.v. and l.v. windings on one phase of a three-phase star/star connected transformer with l.v. neutral directly earthed

phasor arrangements corresponding to breakdown at the positions marked 0·1 and 0·6 on the two previous diagrams, and it will be seen that the further the breakdown between h.v. and l.v. windings is located from the earthed end of the l.v. winding, the more is the high-voltage balance disturbed.

Figure 20.5 similarly shows how the voltage distribution on the l.v. side is temporarily affected with respect to earth when a breakdown between h.v. and l.v. windings occurs at different points along the limb, the l.v. neutral being isolated. As would be expected, the nearer the fault occurs to the h.v. line terminal, the more severe is the voltage to which the l.v. windings become subjected.

Under certain operating conditions abnormally high voltages may be present in low-voltage windings due to electrostatic induction, and these high voltages may result in insulation failures in the l.v. circuit.

Consider first, a single-phase transformer having windings entirely symmetrical as regards impedance, insulation resistance and capacitance; in such a piece of apparatus considerable metallic surfaces exist relatively adjacent to

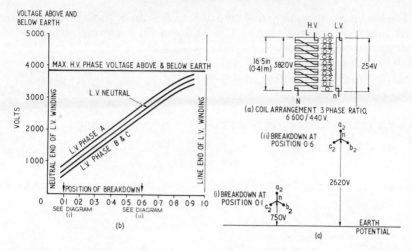

Fig. 20.5 Curves showing the maximum voltages of the l.v. phases and of the neutral point above and below earth with varying positions of the fault on phase A.

Phasor diagrams illustrating the instantaneous voltage distribution in the l.v. windings with respect to earth when a breakdown occurs on phase A. Phase A at its maximum.

Diagrams illustrating the influence of the position of the fault upon the voltage distribution in the l.v. windings with respect to earth when a breakdown occurs between the h.v. and l.v. windings on one phase of a three-phase, star/star connected transformer with l.v. neutral isolated

one another and interspersed with these, insulating materials forming the dielectric. The metallic surfaces are formed by the h.v. and l.v. windings and by the cores and yokes, while the dielectrics consist chiefly of Bakelite, oil and fullerboard. As a result of the disposition of the various essential parts of the transformer, electrostatic capacitors are formed which influence the voltage distribution between windings and to earth particularly under certain fault conditions.

Figure 20.6 shows how this capacitance effect is distributed throughout a typical single-phase transformer, and the capacitances involved are:

C_1, the capacitance between h.v. and l.v. windings.

C_2, the capacitance between l.v. winding and earthed core.

C_3, the capacitance between h.v. winding and earthed core.

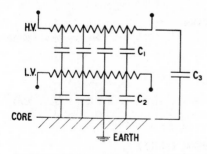

Fig. 20.6 Major capacitance distribution in a single-phase transformer

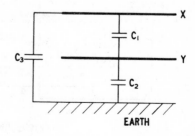

Fig. 20.7 Simple plate capacitor system

461

The transformer windings and core of Fig. 20.6 can be replaced by the simple plate capacitor system shown in Fig. 20.7, and the following general rule for voltage distribution between the plates holds good. If plate X is maintained at a certain voltage V_1 above earth and the ratio

$$\frac{C_1}{C_2} = \frac{1}{n}$$

(20.1)

then the voltage above earth of the plate Y is given by the equation

$$V_2 = \frac{V_1}{n+1}$$

(20.2)

where n is the number derived from the known capacitance ratio of eqn. 20.1 above.

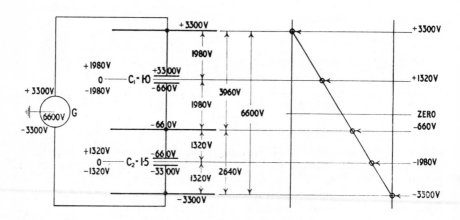

(a) SUPPLY NEUTRAL EARTHED.

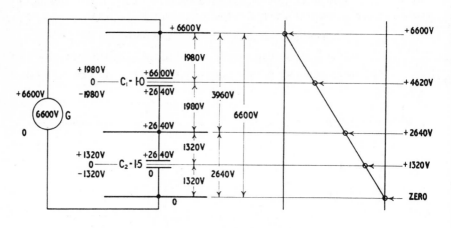

(b) ONE SUPPLY TERMINAL EARTHED.

Fig. 20.8   Diagrams showing voltage distribution across series capacitors

462

For instance, if V_1 is 10000 V and C_1 equals one quarter of C_2, then n is 4 and V_2 is 2000 V. Or if V_1 is 30000 V and C_1 equals one-half of C_2, then n is 2 and V_2 is 10000 V. Transformer l.v. windings designed for voltages of the order of 440 V may therefore fail unless steps are taken to eliminate these high induced static voltages.

Figure 20.8 illustrates in detail at (a) and (b) how the total voltage applied to a system of series-connected capacitors divides up across the individual units when the supply neutral and when one terminal is earthed respectively. These diagrams show that the resultant voltage differences across the individual capacitors are in inverse proportion to their capacitance, and that this *relative* distribution of voltage difference remains unaffected by the conditions of earthing. The *actual* voltages of the different parts of the system above and below earth are, however, fixed by the earthing conditions, as will be seen from the two diagrams of Fig. 20.8. It is also shown in the diagram how, while the electrical mid points *between* the capacitor plates have certain voltages above and below earth, the plates themselves have equal and opposite charges imposed upon them with respect to their electrical midpoint axes. This is shown by the columns of figures given between the capacitor plates and the generator, and equating the voltages of the midpoint axes to zero it is clear that the electrodes of C_1 have equal and opposite charges corresponding to 1980 V above and below the voltage of its electrical midpoint axis, and the electrodes of C_2 have equal and opposite charges corresponding to 1320 V above and below the voltage of its electrical midpoint axis.

For symmetrically designed single-phase transformers operating as such, the following equation holds good:

$$V_2 = V_1 \frac{C_1}{C_1 + C_2} \tag{20.3}$$

where V_1 is the voltage above or below earth of the midpoint of the h.v. winding, V_2 is the voltage above or below earth of the midpoint of the l.v. winding.

C_1 is the capacitance between h.v. and l.v. windings.

C_2 is the capacitance between l.v. winding and earthed core.

Figure 20.9 shows at (a) and (b) how the normal steady voltages are distributed in the windings of a symmetrical single-phase transformer under normal conditions when the transformer is excited from the h.v. side, the l.v. windings being open circuited. Both h.v. and l.v. windings are unearthed. As the voltages of each h.v. and l.v. terminal are at their respective equal values above and below earth, the voltage of the midpoints is zero. If, however, the windings are not symmetrical, the points of zero voltage will move towards that end corresponding to the halves of the windings having the lower impedance, as shown by the dotted curves A' and B' of Fig. 20.9. For the conditions corresponding to curves A and B, the capacitance distribution has no effect as the h.v. midpoint is at zero or earth potential; for the conditions corresponding to curves A' and B', the effects of the capacitance distribution will be to raise the voltage of the l.v. midpoint to the value V_2 above earth, and therefore to modify the voltages of the l.v. terminals with respect to earth as shown in Fig. 20.9. Diagram (a) of Fig. 20.9 shows the conditions in a transformer of subtractive polarity, and

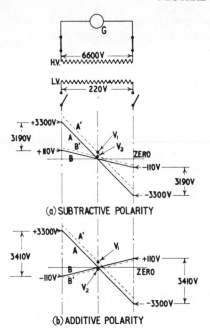

Fig. 20.9 *Normal steady voltage distribution across the windings and to earth of a single-phase transformer on open circuit*

diagram (*b*) shows the conditions corresponding to additive polarity. While Fig. 20.9 shows that under normal steady conditions the l.v. winding of a symmetrically designed single-phase transformer operating as such is not subjected to a high induced static voltage, certain conditions external to the transformer may have serious results.

If, for instance, when switching the h.v. winding on to the supply one pole of the switch makes contact before the other, then during that interval existing between partial and complete connection of the transformer the whole of the h.v. winding will assume a voltage above and below earth corresponding to that of the switch pole which first establishes the connection. If the midpoint of the supply is earthed, or if the supply is entirely isolated, the transformer h.v. winding assumes a voltage above earth equal to half the full line voltage, while if one side of the supply is earthed and the other side first makes contact with the transformer, the h.v. winding assumes the full line voltage above earth. The voltage assumed by the h.v. winding is the same throughout its entire length, that is, there is no *voltage difference* between any parts of the winding, as the windings are assumed to be symmetrical, and consequently the mean voltage, or the voltage of the midpoint is the same as that of the switch pole first making contact. On account of the capacitance distribution as shown in Fig. 20.6 the whole of the l.v. winding, which is open-circuited and unearthed, assumes a definite high static voltage above earth as given by eqn. 20.3. For illustration, take a symmetrically designed single-phase transformer with a voltage ratio of 6600/220 V and with a capacitance distribution such that

$$\frac{C_1}{C_1 + C_2} = 0.5 \qquad (20.4)$$

464

In the first case it is assumed that the midpoint of the supply is earthed so that the voltage V_1 of the h.v. winding above earth is consequently 3300 V and therefore V_2 equals $3300 \times 0.5 = 1650$ V above earth, these conditions being shown in Fig. 20.10 in which C and D represent the voltage above earth of the h.v. and l.v. windings respectively. The voltage of the whole of the l.v. winding above earth is, of course, the same as that of its midpoint. If one end of the supply is solidly earthed and the live pole of the switch first makes contact with the transformer, the h.v. winding is immediately raised to a voltage of 6600 V above earth, and therefore the voltage V_2 of the whole of the l.v. winding equals $6600 \times 0.5 = 3300$ V above earth. These conditions are shown in Fig. 20.11 at C' and D', and it will be seen in both Figs. 20.10 and 20.11 that the l.v. winding has been raised

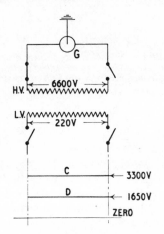

Fig. 20.10   Initial transient voltage distribution on the windings of a single-phase transformer on open circuit due to unipolar switching. Supply neutral earthed

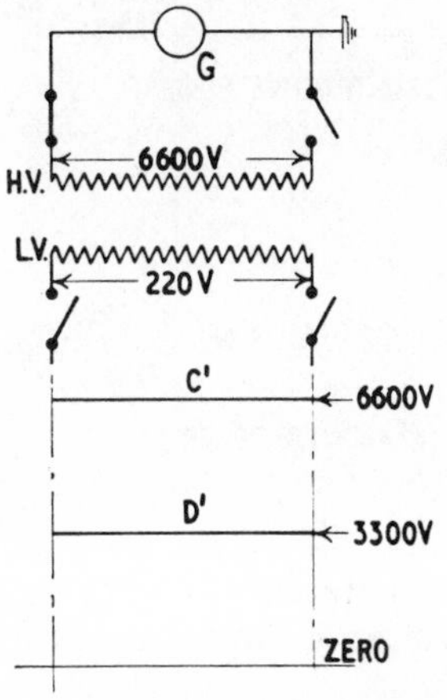

Fig. 20.11   Initial transient voltage distribution on the windings of a single-phase transformer on open circuit due to unipolar switching when one h.v. terminal is earthed

to a dangerously high voltage above earth on account of the distribution of electrostatic capacitance throughout the transformer and uneven switching.

The conditions shown in Figs. 20.10 and 20.11 are, of course, transient only, and the abnormal stresses are relieved as soon as the other pole of the switch makes firm contact. At the same time the insulation of the l.v. winding is undoubtedly strained, and repeated occurrences of this high static voltage in the l.v. winding may ultimately lead to insulation failure.

High induced static voltages may also appear in the l.v. winding when this is open circuited and unearthed if one side of the h.v. winding is earthed. As soon as one h.v. terminal is earthed the other is raised to the full line voltage above earth, the voltage distribution across the h.v. winding being shown by curve E of Fig. 20.12. The midpoint of the h.v. winding, therefore, has a voltage above earth of half of the full line voltage, and the voltage above earth of the l.v. midpoint is half that of the h.v. midpoint. If it is assumed that the ratio of transformation is 6600/220 V, the voltage above earth of the h.v. midpoint is 3300 V and that of the l.v. above earth is 1650 V. The l.v. terminals will therefore have voltages above earth of 1650 V plus and minus half the normal l.v. terminal voltage, that is, one end of the l.v. winding will be at a voltage above earth of

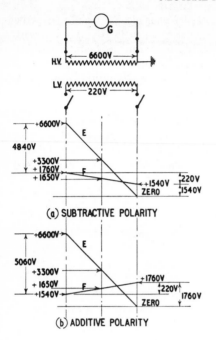

(a) SUBTRACTIVE POLARITY

(b) ADDITIVE POLARITY

Fig. 20.12 Normal steady voltage distribution across the windings and to earth of a single-phase transformer on open circuit with one h.v. terminal earthed

1760 V and the other end will be at a voltage of 1540 V above earth. These conditions are shown by the curves E and F in Fig. 20.12, (a) corresponding to transformers of subtractive polarity and (b) to those of additive polarity.

Three-phase transformers

For studying the electrostatic voltages induced in the l.v. windings of three-phase transformers, consider a core-type star/star connected transformer having a line voltage ratio of 11 000/440 V which gives a voltage ratio to neutrals of 6360/254 V. The capacitance ratio corresponding to eqn. 20.4 is assumed to be 0·5. The principles involved are identical with those outlined for single-phase transformers, but before passing on to a consideration of true capacitance effects the study will materially be assisted by indicating the voltage distribution in the windings under normal operating conditions and when one h.v. terminal becomes earthed.

Figure 20.13 shows the winding and phasor diagrams for a transformer operating on open circuit and excited on the h.v. side from a three-phase supply having an isolated neutral, while both the neutral points of the transformer are also unearthed. If both h.v. and l.v. transformer windings are symmetrically designed and the connections between the transformer and the supply are not subject to any extraneous influences, the h.v. and l.v. transformer neutrals will be at earth potential, as also will be the core. The resultant voltage of the transformer windings will therefore also be zero and the electrostatic capacitance between different parts of the transformer have no effect, so that the total voltage across the l.v. windings is only that due to the winding ratio. If the transformer and the supply are balanced as regards insulation resistance, electrostatic

466

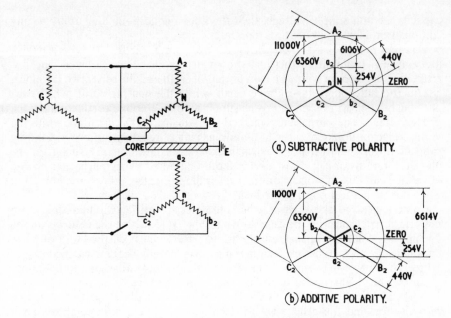

Fig. 20.13 Normal steady voltage distribution in a symmetrical three-phase star/star connected transformer on no-load and excited on the h.v. side. All neutrals may be earthed or unearthed. Core earthed

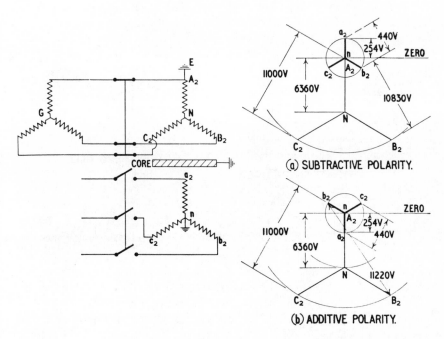

Fig. 20.14 Voltage distribution in a symmetrical three-phase star/star connected transformer on no-load with one h.v. terminal earthed and l.v. neutral and core earthed. Supply and transformer primary neutrals unearthed

capacitance and supply voltage, then the above conditions also apply to any combination of earthed neutrals and core.

Consider next the case where the l.v. neutral point is earthed, the core earthed, and one h.v. terminal earthed, as shown in Fig. 20.14. As soon as one h.v. terminal is earthed its voltage becomes that of earth and the other two terminals rise to the full line voltage above earth, while the neutral point, which was previously at earth potential, now assumes a voltage above earth equal to the phase voltage. The core, being earthed, remains at earth potential, while the voltage distribution across the l.v. windings is that due only to normal winding ratio as the l.v. neutral point is earthed. These conditions are shown in the phasor diagrams of Fig. 20.14, and it should be noted particularly that no rise of voltage can possibly occur on the l.v. windings due to electrostatic induction when the l.v. neutral point is earthed.

Coming now to those cases where the electrostatic capacitance plays an important part, consider the case shown in Fig. 20.15, where the neutrals on the h.v. and l.v. sides are unearthed, the transformer core is earthed and one h.v. terminal is earthed. One h.v. terminal may accidentally become earthed due to a breakdown to earth of one core of an underground cable, or by the failure of an overhead line insulator, or by the actual fracture of an overhead wire. The h.v. terminal connected to the earthed line immediately falls to earth or zero voltage, and the other two h.v. terminals assume a voltage above earth equal to the full line voltage, while the h.v. neutral point has a voltage above earth equal to the phase voltage, these values being 11 000 and 6360 V respec-

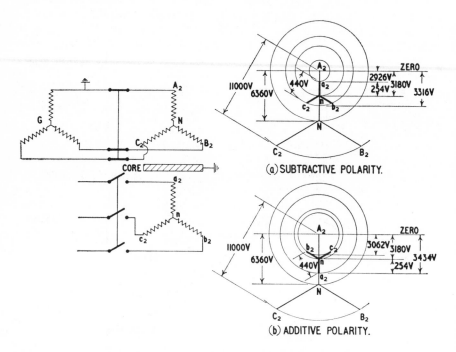

Fig. 20.15   *Voltage distribution in a symmetrical three-phase star/star connected transformer on no-load with one h.v. line earthed and l.v. neutral unearthed. Supply and transformer h.v. neutrals unearthed. Transformer core earthed*

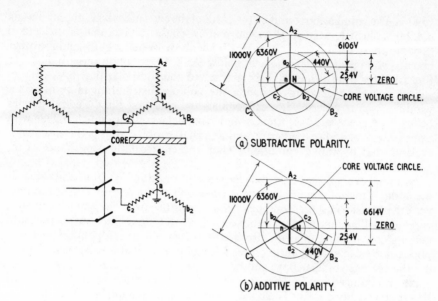

(a) SUBTRACTIVE POLARITY.

(b) ADDITIVE POLARITY.

Fig. 20.16   Voltage distribution in a symmetrical three-phase star/star connected transformer on no-load with the transformer core earthed. Transformer l.v. neutral earthed. Supply and transformer h.v. neutral unearthed

tively. As the average voltage of the l.v. windings is one half the average voltage of the h.v. windings, the voltage of the l.v. neutral point is one half that of the h.v. neutral, and in this case is 3180 V above and below earth, as shown in the phasor diagrams of Fig. 20.15. As the l.v. neutral point is subject to this voltage above and below earth, the whole of the l.v. windings are severely strained and in all probability the insulation to earth on the l.v. side would fail. The *voltage difference* between l.v. line terminals, of course, remains unaltered.

It is the usual practice to earth the transformer core, and in all modern transformers means are provided for doing this by the provision of an earthing strip, usually of copper, from the core to the tank while a terminal socket is fitted on the outside of the tank for taking a lead to the earthing structure. If, for any reason, the earth connection from the core has been left open ended, the core, and possibly the tank, may assume high voltages above earth on account of the electrostatic capacitance between the windings and core. Figure 20.16 illustrates the conditions that may arise if the earth connection to the core is omitted, and the core has been known to reach voltages above earth as high as half the h.v. line voltage. If the l.v. neutral is earthed the l.v. voltages to earth are not abnormal.

Consider next the case of a three-phase transformer in which the low-voltage phases are disconnected from each other, as shown in Fig. 20.17, such as would be the case for a transformer supplying a six-phase rotary converter. In this case the average voltage of each l.v. phase is the same as if each phase was operating a single-phase transformer, and the induced electrostatic voltage on the secondary windings is determined by the voltage of the midpoint of each h.v. phase, and not by the voltage of the h.v. neutral. The average voltage of each h.v. phase above earth is 3180 V, and as the capacitance ratio corresponding to eqn. 20.4 is 0·5, the voltage of each l.v. midpoint above and below earth is

1590 V. The voltage above and below earth of the ends of each l.v. phase winding is 1590 V plus and minus half the normal l.v. phase voltage; that is, one end of each l.v. phase will be at a voltage of 1717 V above or below earth and the other end at a voltage of 1463 V above or below earth. These conditions are shown in Fig. 20.17 for transformers of subtractive and additive polarities.

The preceding diagrams showing the voltage distribution in transformer windings due to electrostatic capacitance indicate that the dielectric stresses between adjacent ends of primary and secondary coils are much higher with transformers of additive polarity, and as a matter of fact, this is the case under even normal operating conditions when the electrostatic capacitance is inoperative.

As a final example, take the case previously shown in Fig. 20.17 and consider the initial transient conditions arising when one switch pole establishes contact with the transformer h.v. windings before the other two. The whole of the h.v. windings will assume a voltage above earth equal to that of the switch pole first making contact, and this will also be the voltage of the midpoint of each h.v. phase. If the capacitance ratio corresponding to eqn. 20.4 is still assumed to be 0·5, the average voltage above and below earth of the l.v. phases will be 3180 V, and as no induced e.m.f.s are active this will be the voltage of the whole of the l.v. windings above earth. The conditions are illustrated in Fig. 20.18.

From the foregoing review of the voltages induced in transformer windings due to electrostatic capacitance, it will be gathered that these mainly arise when certain fault conditions occur in the h.v. circuit. It has been shown that high voltages may be impressed upon the l.v. circuit when a direct breakdown occurs between h.v. and l.v. windings, and it has also been shown that under this kind of failure the l.v. circuit may be safeguarded by directly earthing the l.v. neutral

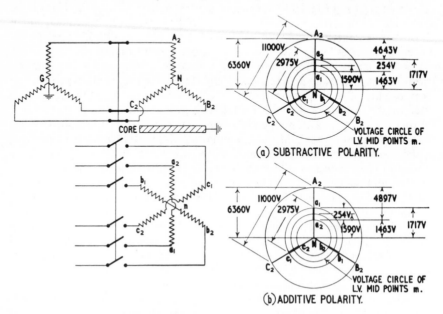

Fig. 20.17 Voltage distribution in a symmetrical three-phase transformer on no-load with star connected primary and open-circuited secondary windings. Supply and transformer primary neutrals may be earthed or unearthed. Transformer secondary windings unearthed and core earthed

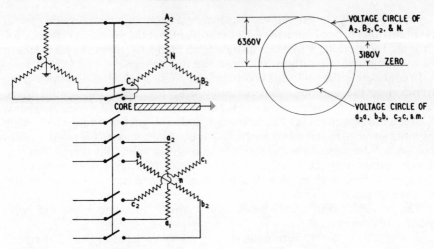

Fig. 20.18   Voltage distribution in a symmetrical three-phase transformer on no-load with star connected primary and open-circuited secondary windings. Transformer core earthed. Supply and transformer primary neutrals may be earthed or unearthed. Transformer secondary windings unearthed

point. In the case of electrostatically induced voltages the l.v. circuit may similarly be freed from these high voltages if the l.v. neutral point is directly earthed, and this is clearly indicated, for instance, in Figs. 20.14 and 20.16. In this connection it should be borne in mind that high induced electrostatic voltages can only occur in transformer l.v. windings when these are open circuited, as a load connected to the transformer very considerably increases the capacitance of the l.v. circuit to earth, and this correspondingly reduces the maximum voltage that can be induced electrostatically on the l.v. windings. As these high voltages are generally induced on the l.v. windings when an earth fault occurs on one h.v. line it will be appreciated that earthing the h.v. neutral will also prevent the presence of high voltages in the l.v. windings, as an earth fault would produce a short circuit across one phase and the transformer would automatically be tripped out of circuit.

Solid v. impedance earthing of transformer neutral points

The neutral points of transformers having star connected windings are earthed (*a*) to facilitate system operation and protection, and/or (*b*) to decrease the hazard to human life, according to service requirements of the plant. Thus in the case of a main three-phase step-up transformer having delta/star connected windings and supplying a high-voltage transmission line, the secondary h.v. neutral would be earthed for reason (*a*) above; a star/delta step-down transformer at the receiving end of the line might also have its primary h.v. neutral earthed for the same reason. A three-phase step-down transformer with delta/star connected windings, for supplying a consumer direct at low or medium voltage, would have its secondary l.v. neutral earthed to comply with the requirements of both (*a*) and (*b*).

In this treatment we are concerned only with category (*a*).

Now, not only do the service requirements governing transformer neutral point earthing depend upon the duty of the plant, but the provisions adopted for carrying the earthing into effect also depend, indirectly, upon the same factor, because this involves both rated output and line voltage.

In considering the problems involved in earthing the neutral point of a given transformer the questions that naturally arise are: (1) is the neutral to be earthed solidly or through a current limiting device? and (2) if the latter, what is the required ohmic value? The following study is confined strictly to showing how the answers to these two fundamental questions may be obtained. It does not set out definitely to recommend certain ohmic values for current limiting devices inserted in the neutrals of transformers of given rated outputs and voltages as these depend upon the individual characteristics of the system and its protective gear.

The principles enumerated herein apply equally to single-phase and poly-phase transformers.

It is first necessary to determine upper and lower limits of the short circuit current permissible, or required for relay operation, which will flow upon the occurrence of a line fault to earth at various points of the system. These will be determined by the extent and characteristics of the system supplied by the transformer and by the type and characteristics of the relay or other automatic gear used. In this study the effects of neutral current values ranging from $1 \cdot 0$ to $5 \cdot 0$ times full-load current are investigated.

When designing a neutral earthing equipment it is customary to assume that the highest allowable neutral current is produced at the point of earthing when one transformer terminal is solidly earthed. Upon this basis the following formula for the ohmic value of the neutral current limiting device for three-phase transformers can be derived as shown. The current limiting device may consist of a resistor, an inductor, a capacitor or any combination of these.

$$I = \frac{kVA1000}{V1 \cdot 73} \text{ amperes} \tag{20.5}$$

$$Z_N = \frac{(V/1 \cdot 73)}{nI} \text{ ohms}$$

$$= \frac{VV1 \cdot 73}{1 \cdot 73nkVA1000} \text{ ohms} \tag{20.6}$$

$$= \frac{V^2}{nkVA1000} \text{ ohms}$$

where 　I = transformer rated full load line current, amperes
　　　　V = transformer rated line, volts
　　kVA = transformer rated output, kVA
　　　　n = neutral short circuit current in terms of normal full-load line current
　　　Z_N = impedance in neutral connection, ohms.

Thus it will be seen the neutral impedance varies (1) as the square of the line voltage for a given transformer output and ratio of neutral fault current to normal line current, (2) inversely as the transformer output for a given line voltage and ratio of neutral current to line current, and (3) inversely as the ratio of neutral current to line current for a given line voltage and transformer output.

Equation 20.6 may be transformed to give the following expression for n, the ratio of neutral fault current to normal full-load line current.

$$n = \frac{V^2}{Z_N k \text{VA} 1000} \tag{20.7}$$

Equation 20.7 shows that the ratio of neutral fault current to normal line current varies, (1) with the square of the line voltage for a given neutral impedance and transformer rated output, (2) inversely as the neutral impedance for a given line voltage and transformer output, and (3) inversely as the transformer rated output for a given line voltage and neutral impedance.

For single-phase and other polyphase transformers, eqns. 20.6 and 20.7 become as follows:

Single-phase

$$Z_N = \frac{V^2}{n k \text{VA} 2000} \text{ ohms} \tag{20.8}$$

$$n = \frac{V^2}{Z_N k \text{VA} 2000} \tag{20.9}$$

Two-phase, three-wire

$$Z_N = \frac{V^2}{n k \text{VA} 500} \text{ ohms} \tag{20.10}$$

$$n = \frac{V^2}{Z_N k \text{VA} 500} \tag{20.11}$$

Two-phase, four-wire

$$Z_N = \frac{V^2}{n k \text{VA} 1000} \text{ ohms} \tag{20.12}$$

$$n = \frac{V^2}{Z_N k \text{VA} 1000} \tag{20.13}$$

Six-phase, diametric

$$Z_N = \frac{V^2}{n k \text{VA} 667} \text{ ohms} \tag{20.14}$$

$$n = \frac{V^2}{Z_N k \text{VA} 667} \tag{20.15}$$

V being the diametric line voltage.

Figure 20.19 shows, for three values of n, how Z_N varies with the line voltage for a given three-phase transformer output of 2000 kVA. The chain dotted curve shows the ohmic impedance per phase winding of the transformer (constant impedance voltage of 5 %) which, it will be noticed, is extremely small compared to the total impedance Z_N required to confine the neutral fault current within the stated limits, and hence often may be neglected in fixing the value of Z_N in practice.

Figure 20.20 shows the curves of Fig. 20.19, up to Z_N equals 11, to an enlarged scale.

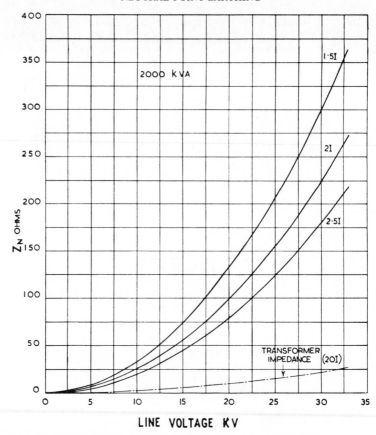

Fig. 20.19   Variation of Z_N with line voltage for a 2000 kVA three-phase transformer

When considering whether or not to insert a current-limiting impedance in the neutral connection to earth, it must be remembered that the total impedance limiting the flow of short circuit earth fault current includes the resistance of the buried earth electrodes and the surrounding body of earth. In particularly favourable circumstances the resistance of the earth connection proper may be of the order of 2 Ω or even less, and under slightly less favourable conditions around 5 Ω, although, unfortunately, considerably higher values than these are often encountered.

The two values mentioned may, however, be regarded as representing the approximate limits of good practice, and these levels are marked on Fig. 20.20 to intercept the voltage/impedance curves. The points at which the earth resistance level lines cut the curves thus give the approximate lines of demarcation between the line voltage at which solid earthing is permissible and at which impedance earthing should be adopted for the particular size of three-phase transformer involved. Thus for a 2 Ω earth resistance and neutral fault current equal to twice normal full-load line current solid earthing is permissible up to 2800 V, and above this line voltage, impedance earthing should be resorted to. For a 5 Ω earth resistance and the same neutral current, the change from solid to impedance earthing is at 4400 V approximately.

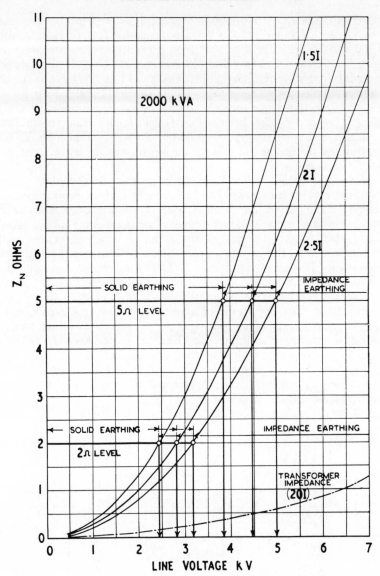

Fig. 20.20   Variation of Z_N with line voltage for a 2000 kVA three-phase transformer

The following conclusions can be drawn from Figs. 20.19 and 20.20. For a transformer of given rated output and for a given ratio of neutral current to line current solid earthing may be adopted up to higher line voltages with increasing values of actual earth resistance. For a transformer of given rated output and for a given earth resistance solid earthing may be adopted up to higher line voltages with increasing values of the ratio of neutral fault current to line current.

The remaining variable factor affecting the earthing problem is that of transformer rated output. Figure 20.21 shows how the total neutral impedance Z_N

is affected by the rated output of three-phase transformers for three typical line voltages, viz. 3·3 kV, 11 kV and 33 kV, when the neutral fault currents are limited to 1·5, 2·0 and 2·5 times the normal full-load currents of the transformers.

Figure 20.22 shows the 3·3 kV and 11 kV curves of Fig. 20.21 up to Z_N equals 20, to an enlarged scale. The chain dotted curve in Fig. 20.21 shows the ohmic impedances per phase winding of the 33 kV transformers, and those in Fig. 20.22 the winding impedances of the 3·3 kV and 11 kV transformers. A constant impedance voltage of 5 % has been assumed over the entire range of transformer rated outputs, and while this figure would be somewhat on the high side for the smaller outputs and on the low side for the larger ones, it may be regarded as a reasonable average.

These curves show clearly how the earthing problem is affected by consideration of transformer rated output. The 2 Ω and 5 Ω levels of earth resistance are marked on Fig. 20.22 to intercept the output/impedance curves, and the points of intersection give the approximate lines of demarcation between the rated outputs at which solid earthing is permissible and at which impedance earthing should be adopted for the particular line voltages and ratios of neutral currents to line currents concerned. Thus for a 2 Ω earth resistance and neutral fault current equal to twice normal full-load current solid earthing of 3·3 kV three-phase windings (for instance) is permissible above 2750 kVA, and below this output impedance earthing should be adopted. For a 5 Ω earth resistance, the same ratio of neutral current to line current and the same line voltage, the change from solid to impedance earthing is at 1100 kVA three-phase, approximately.

Figure 20.22 shows that with a neutral earth current equal to twice normal full-load current 11 kV windings (for instance) should not be earthed solidly above a three-phase transformer rated output of about 5000 kVA when the earth resistance is around 12 Ω.

The following conclusions can be drawn from Figs. 20.21 and 20.22.

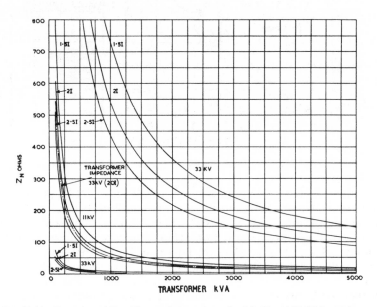

Fig. 20.21   Variation of Z_N with transformer output for 3·3, 11 and 33 kV line voltages

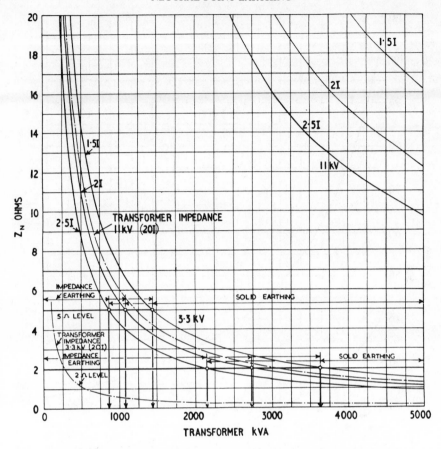

Fig. 20.22 Variation of Z_N with transformer output for 3·3 and 11 kV line voltages

For a transformer of given rated line voltage and for a given ratio of neutral current to line current solid earthing may be adopted down to lower rated outputs, with increasing values of actual earth resistances. For a transformer of given rated line voltage and for a given earth resistance solid earthing may be adopted down to lower rated outputs, with increasing values of the ratio of neutral fault current to line current.

Figures 20.23 and 20.24 show how the total neutral impedance Z_N is affected by the ratio of neutral fault current to full-load line current for 3·3 kV and 11 kV three-phase transformers respectively of outputs ranging from 100 kVA to 5000 kVA. In Fig. 20.23 the points of intersection of the 2 Ω and 5 Ω earth resistance levels with certain of the curves mark the lines of demarcation between the ratios of neutral current to line current at which solid earthing is permissible and at which impedance earthing should be adopted for the particular transformer outputs concerned. The curves of Fig. 20.24 all lie above the 5 Ω level.

From Figs. 20.23 and 20.24 it is evident that for a transformer of given rated line voltage and rated output solid earthing may be adopted down to lower ratios of neutral current to line current, with increasing values of actual earth resistance. For a given ratio of neutral current to line current and for a given

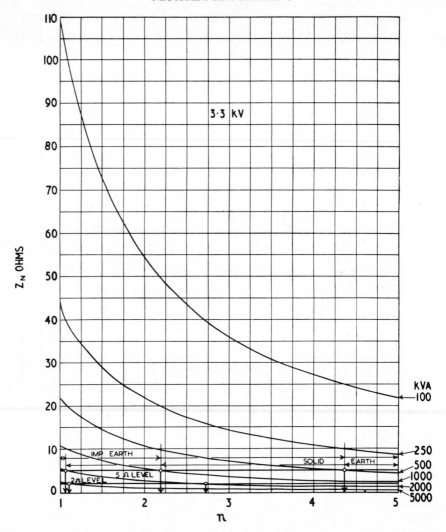

Fig. 20.23   Variation of Z_N with neutral fault current for 3·3 kV transformers of 100 to 5000 kVA

earth resistance solid earthing may be adopted down to lower rated outputs, with decreasing values of line voltage.

The data given in the foregoing individual illustrations are summarised in Fig. 20.25 for three-phase transformers up to 5000 kVA rated output and wound for standard sending end line voltages from 0·44 kV to 33 kV. From this fault chart the total neutral impedance can be determined for limiting the neutral fault current to any value between $n = 1$ and $n = 5$. The basis of the output/impedance curves on the chart is $n = 2$, so that the value of total neutral impedance Z_N required to limit the neutral fault current to twice normal full-load current in any instance is read direct from the particular line voltage curve for the output concerned. For other values of n the procedure is exemplified by the two cases (a) and (b) shown on the chart.

478

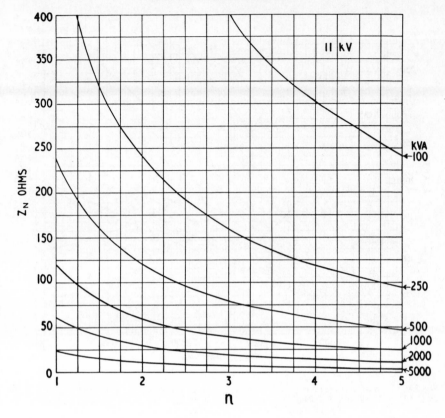

Fig. 20.24   Variation of Z_N with neutral fault current for 11 kV transformers of 100 to 5000 kVA

Case (a) 1000 *kVA*, 11 *kV transformer,* n *being limited to* 1·0

Trace upwards from the abscissa of 1000 kVA to meet the 11 kV curve, and from the point of intersection proceed horizontally to the left to cut the oblique line marked $n = 2$. From this point of intersection trace upwards to cut the oblique line marked $n = 1$, and thence, horizontally to the left again, reading the required value of $Z_N = 121\ \Omega$ on the ordinate.

Case (b) 2000 *kVA*, 33 *kV transformer,* n *being limited to* 3·0

Trace upwards from the abscissa of 2000 kVA to meet the 33 kV curve, and from the point of intersection proceed horizontally to the right to cut the oblique line marked $n = 2$. From this point of intersection trace downwards to cut the oblique line marked $n = 3$, and thence, horizontally to the left, reading the required value of $Z_N = 182\ \Omega$ on the ordinate.

The results of this investigation may now be summarised as in Table 20.1 and Figs. 20.26 and 20.27.

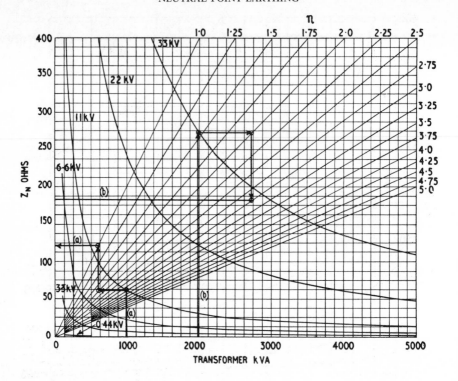

Fig. 20.25   Chart for determining values of the total earthing impedance Z_N for three-phase transformers up to 5000 kVA rated output and 33 kV rated line voltage when the ratio of neutral fault current to full-load line current ranges between 1 and 5

The validity of the statements in Table 20.1 will be realised when comparing them with eqn. 20.6.

The neutral points of star connected windings of all transformers wound for rated line voltages up to medium voltages (i.e., 650 V) generally may be earthed solidly as the resistance of the buried earth electrodes and surrounding earth usually is more than sufficient to limit the neutral fault current to the desired value.

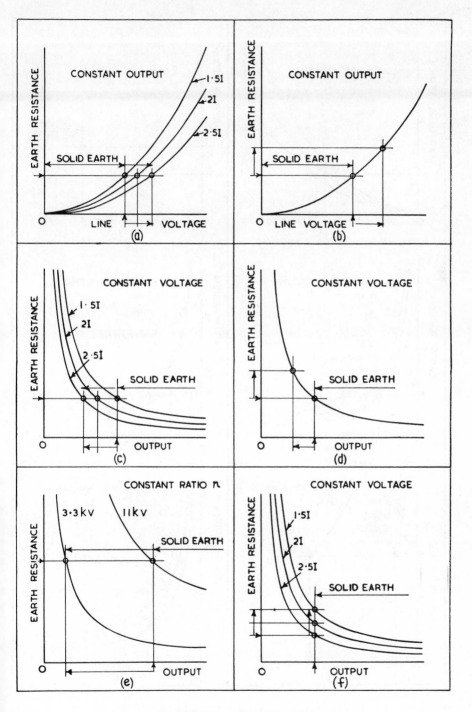

Fig. 20.26 Summary: solid earthing

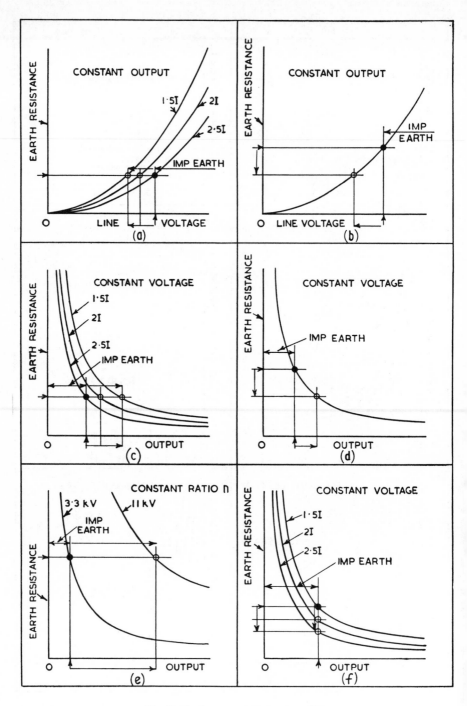

Fig. 20.27   Summary: impedance earthing

482

Table 20.1

For given:	Solid earthing is permissible:	See Fig.	Impedance earthing is necessary:	See Fig.
Output and earth resistance	Up to higher voltages with increasing ratio of neutral current to line current	20–26(a)	At lower voltages with decreasing ratios of neutral current to line current	20–27(a)
Output and ratio of neutral current to line current	Up to higher voltages with increasing earth resistance	20–26(b)	At lower voltages with decreasing earth resistance	20–27(b)
Voltage and earth resistance	Down to lower outputs with increasing ratio of neutral current to line current	20–26(c)	At higher outputs with decreasing ratios of neutral current to line current	20–27(c)
Voltage and ratio of neutral current to line current	Down to lower outputs with increasing earth resistance	20–26(d)	At higher outputs with decreasing earth resistance	20–27(d)
Ratio of neutral current to line current and earth resistance	Down to lower outputs with decreasing voltage	20–26(e)	At higher outputs with increasing voltage	20–27(e)
Output and voltage	Down to lower ratios of neutral current to line current with increasing earth resistance	20–26(f)	At higher ratios of neutral current to line current with decreasing earth resistance	20–27(f)

483

Chapter 21

The three-phase, interconnected star neutral earthing transformer and static balancer

EARTHING TRANSFORMERS

From time to time the necessity may arise for earthing the neutral of one part or another of an interconnected transmission or distribution network at a place where no natural neutral point is available. In such cases auxiliary apparatus is required for the purpose, and recourse must be had either to an interconnected star neutral earthing transformer or to a double-wound star/delta connected transformer. In many instances the star connected windings of a star/delta transformer have been used for deriving a neutral point for the system, but when this has been done it has been because an ordinary power transformer of suitable size has been available from the Supply Authority's stock. In such cases the ratio of transformation has no importance, although the transformer earthing windings must be suitable for the line voltage concerned. If, however, such transformers can be used to give power supplies on other parts of the system, the procedure of utilising them as earthing transformers is distinctly uneconomical, and in most cases it is better to install interconnected star neutral earthing transformers designed specially for the purpose.

In construction, the transformer is generally the same as an ordinary three-phase, core-type power transformer, but having a single winding on each limb, which is split up into two parts, the halves of the windings on the three limbs being interconnected, as shown in Fig. 21.1. The three-phase, core-type construction is, however, usual.

The neutral point of the earthing transformer is connected to earth either direct or through a current-limiting impedance, while the terminals of the apparatus are connected to the three-phase lines. The rating of the earthing transformer is, of course, totally different from that of a power transformer, as the latter is designed to carry its total load continuously, while the former has only to be supplied with the iron loss, whilst the copper loss due to the passage of the short-circuit fault current occurs only for a fraction of a minute.

Figure 21.2 shows at (a) and (b) the connection of the earthing transformer to the lines with earthing resistances in between the transformer neutral and earth, and in between the transformer terminals and the line wires respectively. The relative merits of the two arrangements have been stated in Chapter 20.

Neutral earthing transformers are normally designed to carry the maximum estimated fault current for thirty seconds or other periods, depending upon the

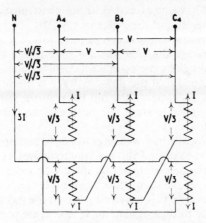

Fig. 21.1   Three-phase interconnected star neutral earthing transformer

time settings of the automatic protective gear on the system and the location of the earthing transformer with respect to the generating plant. It is more usual to specify the single-phase earth fault current that the earthing transformer must carry rather than the equivalent short circuit kVA of the apparatus, as this avoids possible misinterpretations of the requirements. If $3I$ is the total earth fault current and V the line voltage, the earthing transformer short time rating is equal to $\sqrt{3}VI$.

The earthing transformer must be designed for two conditions, namely, when the system is normal and when an earth fault occurs on one line. In the former case the only current flowing in the earthing transformer is that required to provide the necessary magnetisation and to supply the iron loss. The tank cooling surface must be sufficient to dissipate the steady iron loss without a

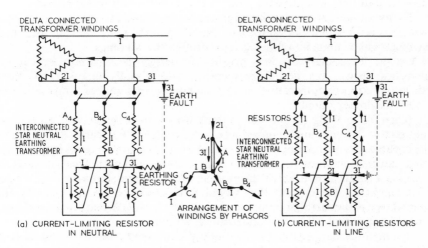

Fig. 21.2   Interconnected star neutral earthing transformer. Arrows indicate instantaneous directions of fault currents

485

temperature rise of 60°C being exceeded, although in most cases the temperature rise will be lower than this, of the order of say 20°C.

In the latter case the earth fault current only flows for a very short time, and earthing transformers are designed on the assumption that none of the heat generated in the windings by the earth fault current enters the cooling medium; in addition, it is assumed that the phase voltage is maintained at its full value across the fault, thus maintaining the full value of the fault current. Neither of these assumptions is strictly true, but both can be considered as giving an additional safety factor.

Neutral earthing transformer windings are designed upon these assumptions and current practice is to permit the winding temperature to exceed 250°C, the two factors stated in the preceding paragraph giving some measure of additional protection against excessive temperature rise under fault conditions. An initial winding temperature of 75°C is assumed prior to the passage of any earth fault current, this being due to the ambient air temperature plus the rise in temperature when the tank surface is dissipating the iron loss. The permissible temperature rise of the windings is therefore 250°C less 75°C.

The two most common time intervals specified for the duration of the earth fault current are 30 and 60 seconds, current practice tending to favour the shorter one. It will be appreciated that the longer the time specified, the larger and more costly will be the neutral earthing transformer due to the fact that more active materials must be employed in its construction. Shorter time intervals can of course be allowed for in the design of earthing transformers but this is for the user to decide, as he has all the relevant information relating to the system to which the earthing transformer is to be connected, including the method of system protection and the protective relay characteristics. In certain cases users have stated that the windings are to be designed to carry specified currents continuously, the current being either the rated current of the earthing transformer or a definite percentage of this current. This practice permits earthing transformer windings to carry the fault currents safely in the event of any failure of the protective devices but it necessitates the purchase of a larger unit than would be the case if a short time rating was specified.

The maximum current density employed for the design of earthing transformer windings is 23 A/mm^2 for 30 seconds, based upon the r.m.s. value of the initial symmetrical component of the current through the windings.

On the foregoing basis it is possible to design the windings for current densities to give any desired temperature rise. Usually this is 175°C from an initial copper temperature of 75°C, giving a final copper temperature of 250°C as already stated.

When an earthing transformer is designed for the neutral point to be connected to a current limiting impedance in the connection to earth, it should also be capable of withstanding, for a period of 5 s, the maximum earth fault current that can flow without the additional impedance in circuit. This safeguard is necessary should, for instance, the bushing of an earth resistor flash over. Under this condition, the current that would flow would be limited only by the impedance of the earthing transformer itself. The current density in the windings should not exceed 62 A/mm^2 under this condition.

It sometimes happens that an l.v. supply is required at an h.v. substation. A 415/240 V supply could be obtained by installing a conventional step-down transformer, but if it is intended to employ an earthing transformer it is possible to incorporate a star connected auxiliary winding of, say, 100 to 150 kVA rating and hence, a supply is available for the local l.v. load.

As in the case of standard double-wound transformers, the maximum temperature rise of the oil is limited to 60°C when the system is normal and no earth fault current is flowing. Under this condition only the iron loss will contribute to the temperature rise of the oil. The smaller ratings of earthing transformer have ample plain surface for dissipating this loss and in many cases have a temperature rise appreciably less than 60°C, and it is only on the larger units that cooling tubes are required to be fitted to the tank.

The price of earthing transformers will theoretically vary approximately inversely as the cube root of the temperature rise and approximately as the cube root of the time rating. In practice this is often a very rough approximation only, due to the exigencies of standardisation of component parts.

In operation the interconnected star earthing transformer is really the acme of simplicity. The total fault current to earth divides up when reaching the earthing transformer neutral point into approximately equal parts in each phase, so that the current in the windings with a single line earth fault is approximately one-third of the total fault current to earth. The current distribution under fault conditions, assuming equal currents in all windings, is shown in Figs. 21.1 and 21.2, and it will be seen that the currents in the halves of the windings on the same limb flow in opposite directions so that they introduce no choking effect, thus permitting a free flow of current from the earthing transformer neutral to each line wire. This, of course, is the reason for interconnecting the windings, as a star connection would produce an additional single-phase magnetic flux in each limb, which in turn would give exceedingly high single-phase voltages operating to reduce the fault current to values which would render the automatic protective gear inoperative.

With regard to the voltage distribution in the earthing transformer windings, it is interesting to see how this is disturbed from the normal, upon the occurrence

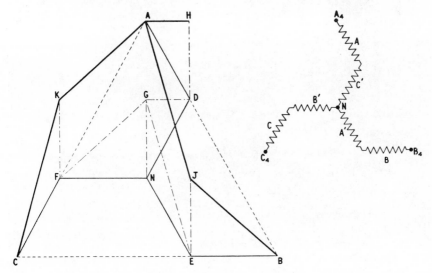

Fig. 21.3   Phasor diagram showing voltage distribution across the windings of a three-phase interconnected star neutral earthing transformer under normal and under fault conditions

of a single-phase earth fault. Figure 21.3 shows the normal and the fault conditions in phasor form, assuming for simplicity that a short circuit occurs on phase A and that no earthing resistors are present.

The voltage distribution under normal conditions is represented by the phasors ADN, BEN and CFN, the terminals of the earthing transformer corresponding to the points A, B and C while N is the neutral point. The phasors AB, BC and CA represent the system line voltages. If an earth fault occurs on line A the windings A and C′ represented by the phasors AD and DN, are short circuited and N then coincides with A. That is, in the winding A there is a voltage drop DH equal to half the phase voltage, and in the winding C′ there is a voltage drop NG of the same magnitude, both these voltage drops having the same phase relationship as the normal phase voltage AN. The resultant voltages under the fault conditions across the windings A and C′ are, therefore, the resultants of DA-DH and ND-NG, namely, AH and GD respectively, these two being, of course, equal. The voltages AH and GD are in quadrature with the phasor AN, which is the phase of the fault current.

In the windings A′ and B′ voltage drops occur as shown by NG, these voltage drops being the same as already discussed. NE combined vectorially with NG gives the resultant voltage EG across the winding A′ while NF combined with NG gives the resultant voltage FG across the winding B′.

The same voltage drops occur in the windings B and C these being shown by

Fig. 21.4   Three-phase 33 000 V, 50 Hz earthing transformer designed for a fault current of 1050 A for 30 seconds and having a star connected 415/240 V auxiliary winding continuously rated at 220 kVA. This winding has all ends brought to a link board, to enable the l.v. phasor group to be changed (Bonar Long & Co. Ltd.)

EJ and FK respectively. Combining phasors EB and EJ gives the resultant voltage BJ across the winding B, while FC combined with FK gives the resultant voltage CK across the winding C. We thus now have six resultant voltages, AH, GD, BJ, EG, CK, and FG across the windings A, C′, B, A′, C and B′ respectively, and these phasors must fit in with the fixed points A, B and C. By inspection, the phasors AH, BJ and CK already terminate, at one end, at the points A, B and C respectively, while GD really coincides with AH on account of the short circuit from A to N. If the phasors EG and FG be now transferred to parallel positions such that they are extensions of BJ and CK respectively, it will be found that in both instances their other ends terminate at A. That is, JA is equal and parallel to EG, and KA is equal and parallel to FG. We now have the complete final phasor diagram of the voltages across different parts of the earthing transformer windings under fault conditions, assuming the line voltages are maintained at their full normal values. That is, across A the voltage AH and across C′ the same voltage, but acting in an opposite direction so that the resultant voltage across AN is zero; across B the voltage BJ and across A′ the voltage JA; across C the voltage CK and across B′ the voltage KA. The voltages from terminals B and C to neutral are the same as the line voltages.

It will be understood that Fig. 21.3 is drawn upon the assumption that the line voltages are maintained at their full values by the generating plant. A fall of line voltage will distort the resultant voltages across the earthing transformer windings, according to the distribution of line voltages between the earthing transformer terminals. In any case, the abnormal voltage conditions in the earthing transformer are transient only, as the circuit becomes interrupted by the automatic switchgear.

Figure 21.4 illustrates a three-phase neutral earthing transformer and Fig. 21.5 shows a typical earthing transformer for providing an artificial neutral on a large three-phase h.v. network.

STATIC BALANCERS

A proportion of existing three-phase distribution cable networks is three-wire, designed for supplying balanced three-phase loads or single-phase loads across lines. The necessity frequently arises for giving a distant consumer a four-wire supply from such a distributor in order to permit the connection of three-phase motor loads at, say, 415 V, and single-phase lighting or domestic loads at 240 V. The motor load is, of course, balanced, and the lighting or domestic load is generally unbalanced. The problem therefore arises as to how a neutral point can be derived to which a fourth wire may be connected. This can be solved by installation, on the consumer's premises, of a three-phase interconnected star static balancer. The general scheme is illustrated in Fig. 21.6, which indicates that the existing three-wire distributor is tapped by a similar short three-wire line running to the consumer's premises and there supplying a balancer which provides the neutral point for a fourth-wire connection. It is thus unnecessary to open up the ground to lay a fourth wire alongside the existing three-wire distributor from the substation, and consequently the cost saved may be considerable.

A three-phase interconnected star static balancer is simply a one to one auto transformer, and in construction is identical with a normal power transformer.

Fig. 21.5   Three-phase, outdoor, core-type neutral earthing transformer with interconnected star windings for neutral point earthing of a 33 kV, 50 Hz system. Disc coils; two concentric coil stacks per limb. Maximum fault current in the neutral, 788 A for 30 seconds. This earthing transformer is also wound with a three-phase 130 kVA, 415 V auxiliary winding

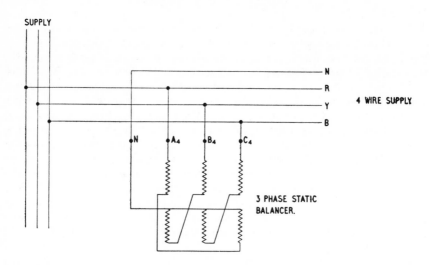

Fig. 21.6   Application of balancer to three-wire distributor

490

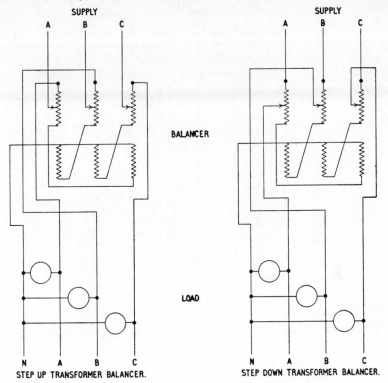

Fig. 21.7   Step-up and step-down transformer balancers

The most common form consists of an ordinary three-phase core-type magnetic circuit wound with two coils on each limb which are connected in the interconnected star manner as described elsewhere for double-wound transformers. There is thus a single three-phase winding only, from the ends of which are taken suitable terminals for connection to the supply and to the load. A terminal is also taken from the neutral point for connection to the fourth wire. The balancer may be oil immersed or simply air insulated and fitted with a protective ventilated casing.

The three-phase core-type form of construction is preferable, as other forms introduce third-harmonic voltages in the windings, and while these would not be injected into the supply system or the load on account of their opposition in those halves of the windings which are connected in series to form each phase, their presence may produce undesirable dielectric stresses. The neutral point of the balancer is connected to a fourth wire to give the requisite four-wire supply.

When the consumer's three-phase line voltage is the same as the voltage of the existing three-wire distributor and the single-phase voltage is 58 % of this, the balancer only carries the net out of balance current of the supply, so that if the whole of the load is balanced no current flows through the balancer. The balancer can easily be designed to supply a consumer at voltages higher or lower than that immediately available, in which case the windings carry currents greater than the net out of balance current, the values of the currents depending upon the ratio of transformation. The principle of this is shown in Fig. 21.7.

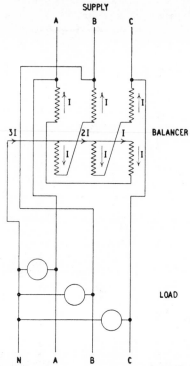

Fig. 21.8   Distribution of out of balance current in balancer windings

The total out of balance current $3I$ flowing in the fourth wire divides up, when reaching the balancer neutral point, into equal parts in each phase, so that the out of balance current in the windings is one third of the total current in the fourth wire.

Figure 21.8 shows the distribution of out of balance current, and it will be seen that the currents in the halves of the windings on the same limb flow in opposite directions so that they introduce no choking effect, thus permitting a free flow of current from the balancer neutral to each line wire.

The balancers can be built for indoor and outdoor service and for any system voltage.

On account of the unbalanced four-wire loads to be supplied from existing three-wire distributors, the currents on the line side of the balancer, that is, in the three-wire distributor, are also unbalanced. Due to the nature of the loads as regards phase difference, the magnitude of the line currents cannot be determined by simple arithmetical calculations, but instead resort must be made to phasor analysis. This is often a laborious process, and in order to shorten it charts have been devised to enable a rapid determination to be made of the unbalanced line currents which flow in the existing three-wire distributor*.

The current distribution can be determined, however, for individual cases, as shown graphically by the following examples.

Journal I.E.E., Supplement to Vol. 57, 1919, pp. 201–208.

Conditions of loading

The loads supplied (which are throughout assumed to be at unity power factor) may be any of the following, as shown in Fig. 21.9:

(a) One single-phase load only from one line to neutral.

(b) Two single-phase loads from lines to neutral; the loads may be equal or unequal.

(c) Three single-phase loads from each line to neutral; the loads may be all equal or all unequal.

Case (a). The top diagram in Fig. 21.9 illustrates case (a) where an assumed load of 100A is supplied. As the load is single-phase, arithmetical quantities only have to be dealt with. The problem can most easily be handled by assuming, to commence with, that the outgoing full-load current of 100A traverses line A passing on to the load to be supplied. The neutral wire carries the full value of the load current, which divides up, on reaching the neutral point of the balancer, into 33·3A in each phase winding and returns to the generator in equal amount through the three line wires A, B and C. Line A, however, already carries 100A in an opposite direction, so that the net result in this line is $100 - 33·3 = 66·7A$. The other two lines each carry 33·3A. With one single-phase load this ratio between the actual currents in the three lines always remains the same, *i.e.*, in the line from which the load is connected to the neutral equals 66·7% and in each of the other two lines 33·3% of the normal full-load current.

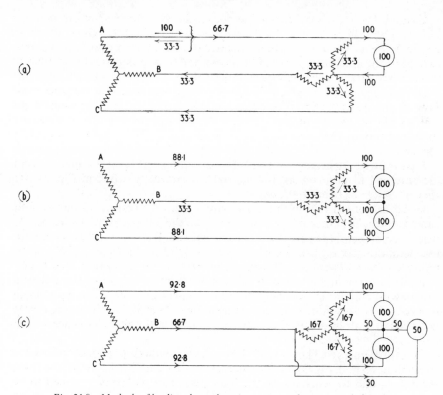

Fig. 21.9 *Methods of loading three-phase interconnected star, static balancers*

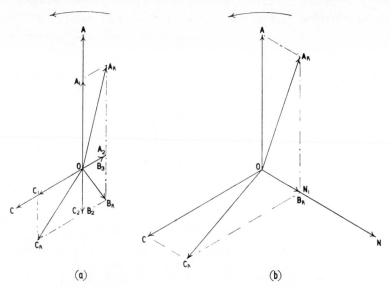

Fig. 21.10   Phasor diagrams for determining the line and neutral current when loading a three-phase interconnected star balancer as shown in Fig. 21.9(b)

Case (b). There are two different methods to adopt, both of which are, however, closely allied. The one is to treat each load separately throughout the entire circuit, knowing from the method indicated in case (a) the proportion of current carried by the unloaded phases in each case, and to combine the current phasors (in each line wire) which result from the two loads. The neutral current must then be found separately, by phasor addition of the separate load currents which flow in the neutral wire. The other method is first to find the neutral current as above, and combine one third of this with each of the normal full-load currents in the three line wires A, B and C. The results obtained are identical.

Figure 21.10 illustrates the two methods; in Fig. 21.10(a) the load currents are assumed to be unequal, 100 and 60A respectively, while in Fig. 21.10(b) they are taken as being 100A each.

It will be noticed that in the case of equal loads the neutral current is of the same value as the currents on the load side of the balancer and displaced 120° from each of them; the currents in the line wires A, B and C bear the fixed ratios to one another of 88·1 % in each of the two loaded phases and 33·3 % in the unloaded phase. These percentages are in terms of the current in each load.

Relative maximum loading of the neutral wire (and also of the balancer) occurs with either one load only from line to neutral or with two equal loads.

As case (c) offers more scope for explanations of the phasor diagrams, case (b) is not dealt with in detail. Precisely the same remarks apply in general as for case (c).

Case (c). Figures 21.11 and 21.12 show the two methods previously referred to. In the first method, Fig. 21.11, the complete path of each load current is followed separately, and the currents on the load side of the balancer assumed to be: phase A = 100A, phase B = 50A and phase C = 100A. These currents are

494

represented by OA, OB and OC respectively in the phasor diagram, being in phase with the voltage to the neutral in each case and therefore spaced 120° apart.

Then due to the load across phase A, the current in line A is equal to two thirds of OA and is represented by OA_1, while the currents in lines B and C equal one third of OA and, flowing in opposite direction to OA_1, are represented by OB_2 and OC_2. Due to the load across phase B the current in line B equals two thirds of OB and is represented by OB_1, and in lines A and C the current equals one third of OB and is represented by OA_2 and OC_3. Similarly with phase C the current in line C due to its load is represented by OC_1, and in lines A and B by OA_3 and OB_3 respectively.

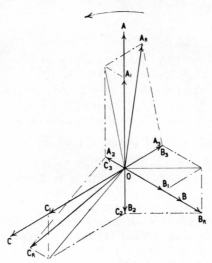

Fig. 21.11   Phasor diagram for determining the line currents when loading a three-phase intercon-nected star balancer as shown in Fig. 21.9(c)

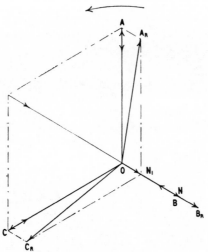

Fig. 21.12   Phasor diagram for determining the line and neutral currents when loading a three-phase interconnected star balancer as shown in Fig. 21.9(c)

495

Tabulated, the currents so far involved are:

In line wire A

$$OA_1 = 66{\cdot}7A$$
$$OA_2 = 16{\cdot}7A \quad \text{leading } OA_1 \text{ by } 60°$$
$$OA_3 = 33{\cdot}3A \quad \text{lagging } OA_1 \text{ by } 60°$$

In line wire B

$$OB_1 = 33{\cdot}3A$$
$$OB_2 = 33{\cdot}3A \quad \text{lagging } OB_1 \text{ by } 60°$$
$$OB_3 = 33{\cdot}3A \quad \text{leading } OB_1 \text{ by } 60°$$

In line wire C

$$OC_1 = 66{\cdot}7A$$
$$OC_2 = 33{\cdot}3A \quad \text{leading } OC_1 \text{ by } 60°$$
$$OC_3 = 16{\cdot}7A \quad \text{lagging } OC_1 \text{ by } 60°$$

Combining OA_1, OA_2 and OA_3 as shown, gives $OA_R = 92{\cdot}8A$
Combining OB_1, OB_2 and OB_3 as shown gives $OB_R = 66{\cdot}7A$
Combining OC_1, OC_2 and OC_3 as shown, gives $OC_R = 92{\cdot}8A$

$$\cos \phi_A = 0{\cdot}987; \cos \phi_B = 1; \cos \phi_C = 0{\cdot}987$$

The neutral current is obtained in a similar manner by adding the three load current phasors OA, OB and OC assumed to be flowing towards the neutral point. The value so obtained is represented by ON as shown in Fig. 21.12, and equals 50A. In the second graphical method, Fig. 21.12, the neutral current is first determined as above. This current, on reaching the neutral point of the balancer, divides up into one third of its value in each phase, being represented by ON_1, and flows back along lines A, B and C. In this case the actual current in line A is the resultant of OA and ON_1 which, as in the previous graphical case, is represented by OA_R. Similarly OB_R is the resultant of OB and ON_1 while OC_R is the resultant of OC and ON_1.

Chapter 22

Transient phenomena occurring in transformers

There is, perhaps, no phase of electrical engineering which possesses a more absorbing interest than does the study of transient phenomena. This interest is doubtlessly stimulated by the very elusiveness of the subject and also by the difficulty often met with in producing, in the laboratory or the test room, just the identical conditions which occur in practice. Quantitative calculations are sometimes rendered very difficult since, under extremely abnormal conditions (for instance, when dealing with voltages at lightning frequencies and with supersaturation of magnetic circuits), the qualities of resistance, inductance and capacitance undergo very material temporary apparent changes compared with their values under normal conditions. A considerable amount of connected investigation has been carried out on transient phenomena of different kinds, by many brilliant investigators, and it is largely to these that we owe our present knowledge of transients. A number of individual papers have been presented before the technical engineering institutions of Great Britain, the United States and the Continent, and these have formed valuable additions to the literature of the subject. We cannot hope, in a single chapter, to cover anything approaching the whole field of the subject, but we have here endeavoured to present a brief survey of the chief disturbances to which transformers are particularly liable.

The transients to which transformers are mainly subjected are:
(1) impact of high-voltage and high-frequency waves arising from various causes, including switching in
(2) switching out voltage rises
(3) switching in current rushes
(4) short-circuit current rushes.

It is not intended to discuss specifically the results of faulty operations, such as paralleling out of phase or the opening on load of a system isolator link, as the resulting transients would be of the nature of one or more of those mentioned above.

IMPACT OF HIGH-VOLTAGE AND HIGH-FREQUENCY WAVES

Transformer primary windings may be subjected to the sudden impact of high-frequency waves (which may consist, in effect, of a single half cycle uni-directional impulse) arising from switching in operations, atmospheric lightning

497

discharges, arcing grounds and short circuits, and, in fact, from almost any change in the electrostatic and electromagnetic conditions of the circuits involved. An appreciable number of transformer failures has occurred in the past, particularly in the earlier days of transformer design, owing to failure of the insulation between adjacent turns, principally of those end coils directly connected to the line terminals, though similar insulation failures have also occurred at other places within the windings, notably at points at which there is a change in the winding characteristics. The failures which have occurred on the line end coils have been due chiefly to the concentration of voltage arising on those coils as a result of the relative values and distribution of the inductance and of capacitance between the turns of the coils.

In the early stages when these breakdowns were being manifested, considerable discussion arose as to the relative merits of external choke coils and reinforced insulation of the end coils, but actual experience with external choke coils showed that in many cases the provision of these did not eliminate the necessity for reinforcement of the end coils, while, on the other hand, added reinforcement of the end coils was itself still occasionally subject to failure, and more frequent breakdown of the insulation between turns occurred beyond the reinforcement. It was for many years the custom of manufacturers to reinforce various sections of the higher-voltage windings in the manner recommended in B.S.422. This practice has today been largely discontinued and the Specification is no longer regarded as representative of the best current practice. Instead, it is now considered far more satisfactory to design transformer windings to meet a specified impulse voltage test level, and this test level is related to the system voltage. The interturn and other parts of the transformer insulation are thus proportioned on this basis to give the most satisfactory overall performance instead of being tied to rather arbitrary standards as hitherto.

While transformer windings are generally considered merely as large inductances, they also contain capacitance distributed throughout the windings in different ways, depending upon the type of coils and the arrangement of the windings. At normal operating frequencies the effect of electrostatic capacitance between turns and layers of individual coils is negligible, and as a result of this the windings act as simple concentrated inductances giving uniformly distributed voltages. When the windings are subjected to the sudden impact of high-voltage and high-frequency or steep-fronted waves, the effect of electrostatic capacitance in determining the initial voltage distribution becomes important, due to the fact that capacitances which are unimportant at low frequencies may have very low impedances or even become virtually short circuits when subjected to high-frequency waves or to steep-fronted impulses.

Moreover, at high frequencies conditions of resonance may be reached for the various combinations of inductances and capacitances.

This will be clear if we consider, first, the effect of an alternating voltage impressed upon an inductance and a capacitance in parallel. With a constant applied voltage the current taken by the capacitance is directly proportional to the frequency, while the current taken by the inductance is inversely proportional to the frequency, and that particular frequency at which these two currents are equal is termed the resonant frequency for the combination. As, at the resonant frequency, the currents are equal and opposite, the combination draws no resultant current from the external circuit no matter how high the voltage may be, and it therefore acts, at the resonant frequency, like an open circuit.

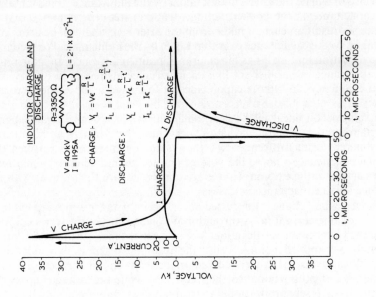

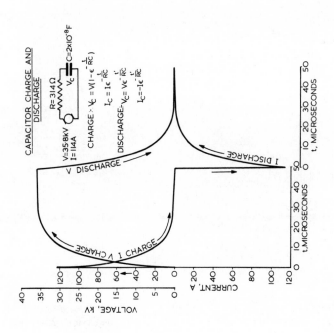

Fig. 22.1 Typical capacitor and inductor charge and discharge curves under direct current excitation

499

At frequencies below that of resonance the resultant current taken from the external circuit will be the excess of that taken by the inductance over that taken by the capacitance at the applied voltage, while at frequencies above that of resonance the resultant current taken from the external circuit will be the excess of that taken by the capacitance over that taken by the inductance. We therefore have the condition that, so far as the external circuit is concerned, an inductance and a capacitance in parallel act like an inductance at frequencies below the resonant frequency, and like a capacitance at frequencies above the resonant frequency.

From this brief outline it will be clear that if we have two parallel arrangements of inductance and capacitance in series with each other, the two arrangements having different resonant frequencies, the results, for frequencies between the resonant frequencies, so far as the voltage across the individual combinations and the current in the external circuit are concerned, are the same as with an inductance and a capacitance in series.

If an alternating voltage is impressed across an inductance and a capacitance in series, the same current flows through both on account of the series connection, but the voltages across the inductance and capacitance are in opposition to each other, the applied voltage being the resultant of these two individual voltages. With a constant value of current in the series circuit the voltage across the inductance is proportional to the frequency, while the voltage across the capacitance is inversely proportional to it, and the resonant frequency is that at

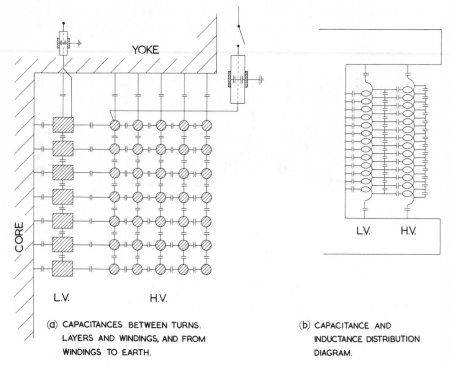

YOKE

CORE

L.V. H.V.

(a) CAPACITANCES BETWEEN TURNS.
LAYERS AND WINDINGS, AND FROM
WINDINGS TO EARTH.

L.V. H.V.

(b) CAPACITANCE AND
INDUCTANCE DISTRIBUTION
DIAGRAM.

Fig. 22.2   Distribution of electrostatic capacitance and inductance in a single-phase core-type transformer

which these two voltages are equal. Except for the effects of the losses in the series circuit the voltages across the inductance and capacitance would be infinite, although a finite voltage was applied across the combination, and therefore at its resonant frequency the combination acts as a short circuit. At frequencies lower than that of resonance the voltage across the capacitance will be greater than that across the inductance, while at frequencies higher than that of resonance the voltage across the inductance will be the greater. The combination will therefore act like a capacitance at frequencies below the resonant frequency, and like an inductance at frequencies above that of resonance.

Consider now the effects of a sudden application of voltage to typical combinations of inductance and capacitance. In the simple case of a pure capacitance only, the current at the first instant is limited only by the characteristics of the external supply circuit, and on account of the inductance of this circuit the current cannot instantly build up in it, and therefore the voltage across the capacitance will commence at zero at the first instant, and as the capacitor becomes charged it will build up to the full value and the current will cease to flow. At the first instant, therefore, the capacitor acts as a short circuit, but finally like an open circuit.

The response of a pure inductance to a sudden application of voltage is the reverse of that with a capacitor. At the first instant the current will be zero with the full value of the applied voltage, but ultimately the current is limited by the characteristics of the supply circuit only, and the voltage across the inductance is zero. At the first instant the inductance acts like an open circuit, but finally like a short circuit. When considering the effects of the *sudden* application of alternating voltages to inductances and capacitances it should be remembered that the initial transient distribution of voltages and currents are the same as when applying steady undirectional voltages, as during the charge and discharge periods the amplitudes of the incident waves remain practically constant. Figure 22.1 shows typical charge and discharge curves for a capacitor and an inductor.

With an inductance and capacitance in parallel the combination acts as a short circuit at the first instant of impact of the exciting wave due to the presence of the capacitor, and finally it also acts as a short circuit due to the presence of the inductance. During the intermediate period a certain voltage will grow and then disappear, due to the combined action of the capacitance and the inductance, but this voltage will never reach the value which would appear with a true open circuit as current exists during this period in both the inductance and the capacitance.

With an inductance and capacitance in series the combination acts wholly as an open circuit at the first instant because current cannot flow instantly through the inductance, and finally, also, because current cannot flow continuously through the capacitance. At the first instant the total voltage acts across the inductance, while finally it all acts across the capacitance. During the interval existing between the first instant and the final condition, a voltage oscillation occurs with a maximum voltage across the inductance equal to the applied voltage, and a maximum voltage across the capacitance equal to double the applied voltage.

In a transformer, capacitance and inductance are distributed as shown in Fig. 22.2, from which it will be seen that capacitances between portions of the same winding act in parallel with the inductances of the same portions. The capacitances of the windings to earth act in parallel with the inductances between the points where the capacitances are located and the earthed point of the winding if such an earth exists. We have, therefore, various parallel combinations of inductance and capacitance in series with various other similar combinations which give opportunities for resonance and excessive internal voltages at various points inside the windings occurring respectively at different frequencies.

A consideration of the theories outlined will show that it is possible to obtain high transient voltages between the end turns of transformer windings as a result of the impact of disturbing voltage waves arising from such causes as switching, lightning discharges, and arcing grounds, and the severity of the dielectric stresses on the insulation between turns is determined by the frequency or steepness of the front and the amplitude of the disturbing wave and by the ratios between the various capacitances of the windings. For instance, if the capacitance of the end turns is low as compared with the capacitances between the remaining turns and to earth, the voltage concentration on the line end turns is most intense, and the risk of insulation failure is correspondingly great. If, on the other hand, the capacitances between the end turns are high, these capacitances act as a short circuit to the front of the disturbing wave, and the voltage of the wave is instantly distributed across a number of the turns so that the voltage across adjacent turns is considerably reduced. Although the voltage does not distribute itself evenly over the turns, the concentration on the extreme end turns is very much lower than when the capacitances between these turns are negligible. It will be seen, therefore, that from this point of view insulation

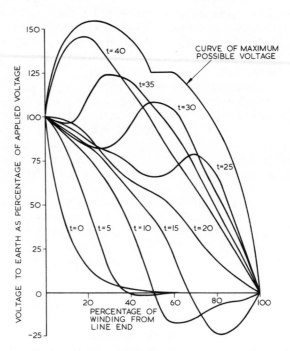

Fig. 22.3   *Voltage to earth distribution throughout a winding with earthed neutral point, after impact of an uninterrupted rectangular wave.*
Times (t) shown against the curves are time in microseconds after the instant of impact

502

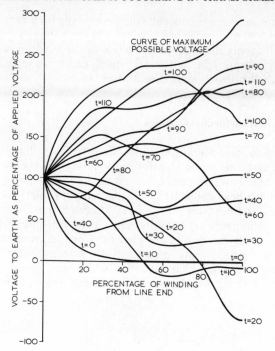

Fig. 22.4   Voltage distribution throughout a winding with isolated neutral point, after impact of an un-interrupted rectangular wave.
Times (*t*) shown against the curves are time in microseconds after the instant of impact

reinforcement of the end turns constitutes a disadvantage, as such reinforcement increases the distance between the copper conductors, so decreasing the capacitances between adjacent turns. Also it must be borne in mind that the reinforcement of the end turn insulation causes increased stresses to be imposed upon the turns comprising the main body of the winding.

It is possible, however, that a disturbing wave reaching a winding may give rise to oscillations of high values which are particularly dangerous to the insulation at points where there is a change in the circuit constants, such as at tappings, open-circuited sections, and intercoil spaces, or by internal resonance with individual coils which comprise the whole winding, and failures of this description have occurred in the past. The voltage concentration on the end turns becomes intensified if the capacitances of the turns to earth are high compared with the capacitances between turns, but in practice the degree to which this occurs depends upon the design of the winding.

The shunt capacitances of the terminals to earth, even of capacitor terminals, have been shown experimentally to be insufficient to produce any noticeable smoothing down of the front of the voltage wave transmitted into the windings, and thus they do not relieve the dielectric stress on the end turns under abnormal voltage conditions. For example, the capacitance of a 66 kV capacitor is of the order only of 150 μF and of a 132 kV bushing, 250 μF, and are far too small to effect any appreciable charge of surge wave form.

When considering the time element of these transient effects, it should be remembered that the voltage concentration exists only for a very short period of time, of the order of millionths of a second, and the voltage wave subsequently distributes itself evenly throughout the whole of the winding, so that in a very short space of time, approximately one ten-thousandth of a second, the electrical stresses reach their normal cyclic conditions.

Surge voltages in transformer windings

The following notes give a qualitative idea of the typical voltage to earth magnitudes which may be set up in transformer windings by the incidence of rectangular surge voltage waves originating on the line. The curves of Figs. 22.3 to 22.7 inclusive have been calculated from the equations derived by Blume and Boyajian (*Trans. A.I.E.E.*, vol. 38, p. 577), although the illustrations and the accompanying comments are actually reproduced from Report Ref. S/T2* of the E.R.A., in which the results are presented in a convenient manner. The curves which are given apply specifically to a transformer having certain capacitance and inductance characteristics (which need not be given here), but they are typical of what may take place in any transformer winding under severe

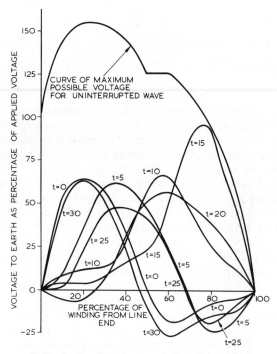

Fig. 22.5   *Voltage to earth distribution throughout a winding with earthed neutral point, due to the abrupt termination of a rectangular wave.*
As Fig. 22.3. Wave abruptly terminated after 10 μs.
Times (*t*) shown against the curves are time in microseconds after the termination of the wave

*See Appendix 8.

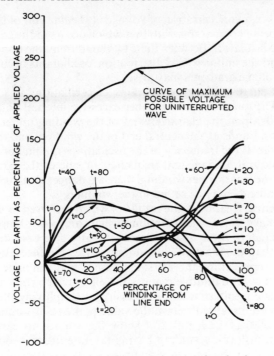

Fig. 22.6   Voltage to earth distribution throughout a winding with isolated neutral point, due to the abrupt termination of a rectangular wave.
As Fig. 22.4.   Wave abruptly terminated after 20 μs seconds.
Times (t) shown against the curves are time in microseconds after the termination of the wave.

surge conditions. The effects in transformers having an earthed and an isolated neutral are illustrated for the two cases of an uninterrupted surge wave and an interrupted (or chopped) wave. It will be noted that, with both types of surges, the internal winding surge voltages to earth are more severe when the neutral is isolated.

At normal frequencies such as 50 Hz, and neglecting resistance, a transformer winding behaves as though it were a pure inductance, the effects of the small capacitances present being negligible. At much higher frequencies, however, the reactance due to the inductance of the winding becomes very high, while the reactance due to the interturn capacitances becomes progressively lower, with the result that currents set up by high frequency waves tend to take the capacitance path. If, for instance, a rectangular travelling wave reaches a transformer, the initial distribution of voltage depends entirely upon the capacitances present, as the front of a rectangular wave represents a quarter cycle of infinite frequency. Subsequently, however, the distribution depends entirely upon the inductance, as the wave, apart from its front, represents uniform voltage, that is, zero frequency. Now the distribution of inductance in a transformer is practically uniform and, therefore, the final distribution of voltage also is uniform. The initial distribution of voltage, however, is not uniform owing to the fact that there is appreciable capacitance between the windings and the core and tank in addition to the interturn capacitances. The capacitance currents from the windings to earth (core and tank) must flow through the interturn capacitances,

505

and this results in a concentration of voltage at the line end of the winding, the phenomenon being comparable with that which occurs in the case of a string of suspension line insulators. Between these two conditions, the non-uniform initial distribution and the uniform final distribution, oscillations occur and excessive voltages and voltage gradients may arise.

The case of surges consisting of trains of oscillations is somewhat more complicated. The question as to whether excessive internal voltages are likely to arise depends upon the characteristics of the winding under consideration, the terminal conditions at the neutral end of the winding, and the frequency of the applied surge. If the frequency of the oscillations is equal or approximately equal to the natural fundamental frequency, or one of the natural harmonic frequencies, of the transformer winding, there is a danger that high local voltages and voltage gradients may be set up in the windings.

In the theoretical treatment, the results of which are given in curve form, it is necessary to make certain assumptions which are not wholly justifiable in practice. Fortunately these assumptions generally give conditions which are more onerous than those actually obtaining, so that the theoretical results may be regarded as representing the worst conditions that can arise. It is also possible to indicate in a qualitative manner the effect which each assumption has upon the ultimate result, so that the necessity for simplifying the mathematical treatment does not greatly diminish the value of the results obtained.

The curves of Figs. 22.3 and 22.4 show the voltage to earth distribution throughout a transformer winding having (a) an earthed neutral, and (b) an isolated neutral, after the impact of an uninterrupted rectangular wave in each example.

The time intervals between the successive curves of Fig. 22.3 are 5 μs, and the total period covered by the curves is a little more than one half cycle of the fundamental, the time for one complete cycle of the latter being 75 μs.

In the case of Fig. 22.4 the time intervals between successive curves are 10 μs and the total period covered by the curves is again a little more than one half cycle of the fundamental, the time for one complete cycle of the latter being, in this case, 215 μs.

The reason that none of the various curves of instantaneous voltage distribution, shown in Figs. 22.3 and 22.4, approaches very closely to the curves of maximum possible voltage is that these instantaneous curves are shown for a total period of only one half cycle of the fundamental, and consequently, at none of the instants considered are the fundamental and the principal harmonics even approximately in phase. If the series of curves had been continued at regular intervals over a much longer period, it is likely that some of them would approach much more closely to the maximum curves shown.

The curves of Figs. 22.5 and 22.6 show the voltage to earth distribution throughout a transformer winding having (a) an earthed neutral, and (b) an isolated neutral after the abrupt termination of a rectangular wave applied to the winding. In Fig. 22.5 the wave is assumed to have been interrupted abruptly 10 μs after its application, and the various curves shown are obtained by taking differences between curves of this time interval in Fig. 22.3; thus the curve for $t = 15$, of Fig. 22.5 for instance, is the difference between the curves of Fig. 22.3 for $t = 25$ and $t = 15$.

The curve showing the maximum possible voltage to earth due to the impact of a rectangular wave is repeated (from Fig. 22.3) for purposes of comparison,

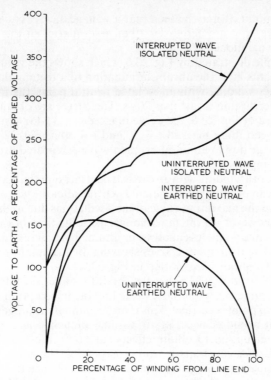

Fig. 22.7   Maximum possible voltages to earth obtainable in a winding with rectangular incident waves. The curves for interrupted waves apply only after the interruption

although it is not applicable to the new conditions. The new maximum, under the worst possible conditions, theoretically is equal to twice the maximum amplitude of the oscillations as the maximum positive voltage above the axis of oscillation, due to the impact of the wave, may at some instant coincide with the maximum negative voltage below the axis of oscillation, due to the interruption of the wave. These oscillations take place about the zero line as an axis, as this represents the uniform distribution line once the wave has been interrupted. The two curves representing the maximum possible voltage to earth before and after interruption are shown in Fig. 22.7.

The maximum voltage to earth after the interruption of a wave is unlikely to be attained in practice owing to the fact that there is a definite and constant time interval between the two sets of oscillations. In order that any particular harmonic shall reach its maximum possible amplitude under these conditions, the time interval between the impact and the interruption must be an odd integral multiple of one half of the periodic time of the harmonic. The chance of this condition being fulfilled in the case of the fundamental and even the first few harmonics is evidently somewhat remote.

Referring to Fig. 22.5, a noticeable feature is that the curve for $t = 15$ reaches a higher voltage at a point in the winding between 70 and 80% from the line end, than any of the curves of Fig. 22.3. This is typical of abruptly terminated waves, which generally produce very onerous conditions in the neighbourhood

507

of the neutral point. It may be seen that a voltage, nearly equal to that of the impressed wave, appears across less than one quarter of the winding in the particular case mentioned.

In the example illustrated by Fig. 22.6, which shows the voltage distribution at various instants after the abrupt termination of a rectangular wave applied to a transformer winding with an isolated neutral point, the interval between impact and interruption is taken as 20 μs. The curves are obtained in a similar manner to those of Fig. 22.5, the curve marked $t = 30$, for instance, being the difference between those marked $t = 50$ and $t = 30$ in Fig. 22.4. The curve of maximum voltage due to an uninterrupted wave is repeated (from Fig. 22.4) in Fig. 22.6 for comparison.

As in the case of a winding with an earthed neutral point the maximum possible voltage which can occur, with an isolated neutral, after the wave has terminated is equal to twice the maximum value of the oscillations due to the impact of the wave. Owing, however, to the much longer periodic time of the fundamental, the maximum value is still less likely to be attained in practice than in a winding with an earthed neutral point. Curves showing the maximum possible voltages under the two conditions are plotted in Fig. 22.7.

It will be seen from Fig. 22.6 that the conditions after the interruption of the wave are far less onerous than those preceding the interruption than is the case when the neutral point is earthed. This is on account of the longer periodic times involved, and it would be necessary to assume a much longer interval between impact and interruption to obtain effects similar to those of Fig. 22.5. With suitable transformer constants it would be possible, however, to obtain such effects. Low values of leakage inductance and capacitance to earth would give high values for the fundamental and lower harmonic frequencies with correspondingly short periodic times. Such values would obtain in small high-voltage transformers designed with relatively short windings and high values of volts per turn.

In the theoretical study presented by Figs. 22.3 to 22.7, certain assumptions were made which were necessary in order to simplify the mathematical processes. These assumptions are as follows:

(1) that the wave is rectangular in shape
(2) that the interturn capacitance is uniform throughout the winding
(3) that the capacitance to earth per unit length of winding is uniform throughout the winding
(4) that the time taken to charge the various capacitances is so short that no appreciable current flows through the inductance of the winding during this period.

Reference should be made to the E.R.A. report for a more detailed discussion of the effects of these assumptions upon the results shown in the figures.

In the theoretical study of the initial voltage to earth distribution, it has been assumed that the wave front is vertical, so that the maximum wave voltage is applied instantaneously. In practice the capacitances of the winding are charged to within 90 % of the ultimate value in a time of the order of a few hundredths of a microsecond, and the assumption that no appreciable current flows in the inductance of the winding during the time required to charge the capacitance system is thus justified. With a vertical wave front, it is this very short time which alone determines whether any appreciable current will flow through the inductance of the winding before the voltage distribution, as determined by the capacitance system, is established.

If, however, the slope of the wave front departs appreciably from the vertical there will be a definite time interval between the arrival of the foremost part of the wave at the transformer terminal and the attainment of the maximum voltage. This time interval will then determine the time for charging the capacitance system, and, if the time is sufficiently long, the assumption that no appreciable current will flow in the inductance during the charging period is no longer justified.

A long wave front may be regarded as a quarter cycle of relatively low frequency. Since at low frequency the inductance of the winding represents a low reactance and the capacitance a high reactance, the initial voltage distribution is determined mainly by the inductance. The assumption that the inductance is uniformly distributed is justified in the case of iron-cored apparatus, such as a transformer, and consequently a long fronted wave produces an approximately uniform initial distribution of voltage throughout the winding. As the final distribution is also uniform, there is thus no subsequent oscillation.

These considerations show that calculations based upon the assumption of a rectangular wave give results which represent the worst that can happen in practice, in so far as wave shape is concerned. The simplification of the calculations by the assumption of a rectangular wave is therefore justified.

The phenomena attendant on the termination of a travelling wave are similar to those occurring on the arrival of the wave front. Wave tails are not generally

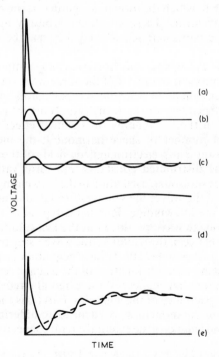

Fig. 22.8 *Composition of the surge voltage wave at the l.v. terminal of a transformer due to the impact of a surge wave on the h.v. terminal*

very steep unless the wave is interrupted by an insulator flashover. Under this latter condition, however, the wave tail approximates to the rectangular shape. This is equivalent to the sudden application of a wave having a rectangular front and an amplitude equal to that of the original wave, at the instant of its termination, but of opposite polarity.

The distribution of voltage, due to such an abrupt termination, is calculated by treating it as the result of the application of a second wave of opposite polarity. In this case the calculated voltages are superposed on those already existing in the windings due to the original wave, in order to determine the resultant effects, and Figs. 22.5 and 22.6 were obtained in this way.

The order of merit of coil types in respect of minimising internal voltage gradients and voltages to earth, whether oscillatory or not, is spiral, continuous disc, crossover, and it is significant that in general this is the reverse of the order of coil susceptibility to interturn insulation failure. Gaps between coils cause somewhat violent oscillations in their neighbourhood, the oscillations having a more serious effect upon the voltage gradients in the vicinity of the gaps than upon the corresponding voltages to earth. Here again the coil order in which the gaps represent proportionately more violent discontinuities is crossover, continuous disc and spiral types.

It is now an established fact that surge voltage waves occurring in the h.v. windings are transformed just as much as 50 Hz waves are transformed, so that they have their counterparts in the l.v. windings. An interesting way of considering the composition of the surge voltage wave produced at the l.v. terminal of a transformer by a rectangular wave appearing at the h.v. terminal has been presented,* in which the former is regarded as the result of superposing four component waveforms. These are shown typically by Fig. 22.8, which is reproduced from the paper referred to, and they may be explained briefly as follows:

Part (a) shows the exponential electrostatic voltage impulse produced at the l.v. terminal at the moment of impact of the surge wave on the h.v. terminal; it depends upon the relation of the distributed capacitances between windings and to earth, and is independent of the turns ratio. Part (b) is the free sinusoidal oscillation induced in the l.v. winding by a similar effect in the h.v. winding, which consists of a number of space harmonics; the induction is achieved through the electrostatic and electromagnetic fields of these harmonics, being dependent upon the distributed constants and turns ratio of the windings. Part (c) shows the free sinusoidal oscillation in the l.v. winding which immediately follows the impact of the wave; the voltages produced depend upon the distributed constants of the l.v. winding. Part (d) is typical of the exponential unidirectional surge voltage wave produced in the l.v. winding by purely electromagnetic induction between the windings, the wave rising from zero to a certain maximum and then falling to zero; the voltage magnitude is directly proportional to the turns ratio, being a simple function of the leakage reactance of the transformer and of the surge impedances of the external circuits; it is independent of the distributed capacitance of the windings. Part (e) is the resultant l.v. surge voltage wave. All four component waves depend, to differing degrees, upon the surge impedances and lengths of the external circuit connected to the transformer.

---

*'Effect of Transient Voltages on Power Transformer Design: IV., Transition of Lightning Waves from One Circuit to Another Through Transformers', by K. K. Palueff and J. H. Hagenguth. *Trans. A.I.E.E.*, Vol. 51, Sept. 1932, pp. 601–615.

The surge wave transformation mainly is effected electromagnetically, although the initial, very short, voltage transients in the l.v. windings, which are determined chiefly by the capacitance relations between windings and to earth, become of increasing importance as the rated voltage of the secondary circuit is reduced. Thus, in general, the maximum amplitude of the surge voltage wave in an l.v. winding is approximately equal to that of the surge wave in the corresponding h.v. winding divided by the turns ratio of the two windings; the capacitance effects between windings and to earth, producing initial high voltage transients, may predominate, however, in high-voltage transformers having large ratios, e.g., 33 000/400 V.

The extent to which the connections of three-phase transformers affect the relative maximum l.v. terminal surge voltage to earth (neglecting the initial, extremely short, electrostatic transient) has been investigated,* and on the basis of a 100 % surge amplitude in single-phase transformers and in three-phase star/star connected transformers with earthed neutrals, in both of which types the ratio of h.v. and l.v. crest surge voltages is approximately the same as the winding turns ratio, the following were found to apply. For equal line voltage ratios, the amplitude is 67 % when the connections are the same on both sides and when only one neutral is earthed or both neutrals are isolated (in the case of star/star connections); where h.v. and l.v. connections are different, the amplitude is 57 %. For equal line to neutral voltages (i.e., equal phase winding turns ratio) the amplitude is again 67 % when the connections are the same on both sides and when only one neutral is earthed or both are isolated; where h.v. and l.v. connections are different, the amplitude is 100 % when the h.v. windings are delta connected and 33 % when star connected with earthed or isolated neutral. The lengths of the wave fronts are 100 % for all star/star connected transformers for all combinations of earthed and isolated neutrals and for equal line voltages and equal line to neutral voltages; for delta/delta transformers the front length is 33 % for equal line and equal phase voltages; for delta/star transformers it is 300 % for equal line voltages and 100 % for equal phase voltages, the star neutral being earthed or isolated; for star/delta transformers the front length is 11 % for equal line voltages and 33 % for equal phase voltages, the star neutral being earthed or isolated. The wave front length is taken as the time required to reach 95 % of ultimate crest value. The foregoing figures take no account of possible wave reflection from both ends of the l.v. circuits. The exciting incident surge wave is, of course, assumed to originate in the line connected to the h.v. windings of the transformer.

The foregoing figures hold good when identical surge voltage waves are set up on one, or simultaneously, on two lines only. If identical surge waves are present simultaneously on all three h.v. lines and a delta winding is present on one or both sides, no surge voltage waves appear on the l.v. lines as the surge voltages in the three delta connected phase windings are equal and cophasal; but, of course, certain surge voltages are present in the l.v. windings immediately following wave impact on the h.v. windings, due to initial electrostatic transference and to free oscillations of space harmonics in both h.v. and l.v. windings.

The front of the wave in the l.v. winding generally is less steep than is that of the exciting wave in the h.v. winding, as the former is smoothed down by the

*'Effect of Transient Voltages on Power Transformer Design: IV., Transition of Lightning Waves from One Circuit to Another Through Transformers', by K. K. Palueff and J. H. Hagenguth. Trans. A.I.E.E., Vol. 51, Sept. 1932, pp. 601–615.

normal leakage reactance of the transformer. The dielectric stress on the l.v. interturn insulation is correspondingly reduced.

SWITCHING IN VOLTAGE RISES

An appreciable number of transformer failures has occurred in the past, due to the failure of the insulation between adjacent turns, principally of those end coils directly connected to the line terminals, and these failures have been due chiefly to the concentration of voltage which occurs on the line end coils when switching the transformer on to a live circuit.

The effects of the inductance and capacitances of the transformer windings in concentrating and distributing the initial voltage and current waves are exactly the same as outlined earlier in the chapter, but when the disturbing waves are due to switching a live circuit on to an unexcited transformer, the severity of the phenomenon referred to above is dependent upon the point of the voltage wave at which switching in occurs, the presence of arcing at the switch contacts, chattering of switch contacts, and by the phenomenon which occurs when one pole of the switch makes contact by an advancing spark.

It will doubtless be appreciated that if contact is first made at a point on the voltage wave equal, say, to half the maximum amplitude, the consequent voltage concentration due to reflection cannot (ignoring the effects of arcs and chattering switch contacts) reach a value higher than the normal maximum peak value of the wave; if, however, initial contact is made at the peak of the voltage wave, the maximum voltage concentration will theoretically equal double this amount.

Arcing at and chattering of the switch contacts produces trains of high frequency waves, which impinge upon the transformer windings and represent a distinct menace to the insulation.

The phenomenon of advancing switch spark connection may, so far as switching is concerned, produce the highest voltage rises possible. Consider, for instance, the case of a three-phase star connected transformer winding being switched into circuit; as the switch contacts approach each other the voltage of one phase reaches its maximum at such an instant as to cause a spark to pass between its pair of contacts. In a very short space of time the whole of the transformer windings become charged to a voltage above earth equal to the maximum line voltage. At a short period of time later, however, the remaining switch contacts have made connection, and on the other two phases there is therefore a sudden transition from the maximum peak voltage, say, above earth, to a voltage below earth of some value between zero and the maximum according to the length of time which has elapsed between the initial spark connection and the final metallic connection of all three contacts. While normal stable conditions are eventually reached, high-voltage, high-frequency waves are present in the transformer windings during the transition period, and they may overstress the insulation of the windings. This phenomenon may also be aggravated by any intermittent arcing that may take place at the switch contacts.

These switching surges produce in the first place an excessive dielectric strain on the insulation of the coils; in the second place punctures of the insulation between turns which are sometimes self-sealing (particularly in oil immersed transformers); and in the third place a short circuit between turns or a breakdown of the windings to earth. Cases have also been known in practice of the flashing over of transformer terminals and the failure of insulated leads at open-ended tappings. Usually the short circuit between turns is the only serious effect, but, of course, the others enumerated mainly lead up to this. Such a short circuit between turns may cause very considerable damage if the transformer protective gear allows the fault to be maintained for any length of time, and faults of this description have been known subsequently to result in short circuits between phases.

In practice it is usually very difficult, if not impossible, to eliminate entirely the voltage rises which occur when switching a transformer into circuit, and in fact the only obvious cause which can easily be eliminated is that of chattering switch contacts, and the remedy here is obvious. While, however, the other causes are inherent in the switching operation, their effects can very considerably be subdued by the provision of suitable protective apparatus such as surge diverters and by careful design of the transformer windings and insulation.

SWITCHING OUT VOLTAGE RISES

In practice, voltage rises occur when switching out of circuit an unloaded transformer. The phenomenon is typical of the condition arising when switching out a highly inductive winding irrespective of whether the winding is excited by alternating or direct currents. As soon as the switch breaks contact, the magnetising current, and therefore the magnetic flux, *tends* to collapse instantaneously, and consequently to produce high voltage rises. Actually, however, an instantaneous collapse of the magnetic flux does not occur, as the switch does not break contact instantaneously, owing to the inertia of the switch mechanism and to the small arcs which in general cannot be eliminated entirely from the switch contacts. However, the flux does sometimes drop at a much more rapid rate than that corresponding to the normal cyclic rate of change, and this increased rate of cutting the conductors of the transformer windings produces in them high voltage rises. While the usual tendency of modern oil switches is to break circuit at zero current, this does not always happen, and voltage rises occur accordingly. It should, of course, be borne in mind that the induced voltage rises are dependent upon the *rate of change* of magnetic flux and not upon the actual value, so that severe voltage rises may occur when the flux is broken at any point on the wave.

The rapid cooling of the interrupting arc, particularly during the last half cycle, has been found to augment these voltage rises considerably.

These switching out voltage rises generally become manifested by flashing over of the transformer terminals and by short circuits between turns of the windings resulting from the puncture of the insulation between them.

It can be said that, generally speaking, the effect of switching out voltage rises on the transformer windings is the same as switching in voltage rises, that is, an excessive dielectric strain is placed on the insulation of the coils.

While voltage rises do occur in transformers when switching them out of circuit, they are by no means very prevalent and, generally speaking, it is perhaps only desirable to contemplate the use of auxiliary apparatus in connection with high-voltage transformers where the electrostatic and electromagnetic energies are high. When a transformer working on no-load is switched out of circuit, the electromagnetic energy stored therein becomes rapidly dissipated, giving high voltage rises as a result. If, however, the secondary side of the transformer is closed either through a load or a short circuit, the magnetic flux cannot collapse so rapidly, as there is a tendency for it to be maintained by the action of the induced voltage in the closed secondary circuit. Voltage rises can therefore be reduced by switching the transformer off together with its load or with an artificial one.

An ideal method is, of course, to allow the speed of the rotating machinery supplying a transformer to fall gradually so that the transformer can be switched out of circuit unexcited. This, of course, is ideal and can seldom be realised in practice.

General

There are various other points which have not been discussed but which to some extent have important bearings upon the transient phenomena arising when switching in or out. The chief of these are:

(a) h.v. versus l.v. switching
(b) the effect of switching in and out through overhead lines and cables
(c) air break versus oil break switches
(d) no-load versus load switching.

(a) The question of h.v. versus l.v. switching is really one of the permissibility of switching in current rushes. In earlier days operating engineers were faced with having to make a decision whether to risk failures due to voltage rises when switching in on the h.v. side, or to incur the risk of dislocation of the windings resulting from heavy current rushes when switching in on the l.v. side. There was also the attendant probable annoyance of the tripping of circuit breaker relays or of the blowing of fuses. Today, however, none of these phenomena gives cause for anxiety, and therefore modern transformers may be switched in on either side as may be the more convenient.

(b) If a transformer is switched in through a length of overhead line or cable the results are somewhat similar to those occurring when using a reactor or capacitor protective device. That is, the overhead line or cable affords a certain degree of protection to the transformer against voltage rises, while either has the effect of reducing the current rush by virtue of the impedance of the line. The protective action, however, is largely dependent upon the switch making a clean contact, for otherwise the capacitance of the line or cable is likely to make any switch arcs which may occur unstable, and so set up serious voltage oscillations.

(c) In Great Britain much of the switching of power transformers is done by oil circuit-breakers and to some extent air break switches are also used. With the former of these, arcing at the contacts is reduced to a minimum, and so long as the contacts are suitably designed, no very considerable amount of arcing will occur when switching. In addition, the tendency of an oil circuit-breaker is to break circuit when the current is passing through its zero value, and while this does not always happen, it does so in a large percentage of cases. Switching out voltage rises are, of course, more likely to occur the nearer the current is in phase with the voltage, for then zero current interrupted would not correspond to zero magnetic flux and voltage rises would therefore be bound to occur. With air breakers this tendency to interrupt at zero current does not arise, and the arcs forming even when breaking small currents at moderately high voltages may involve short circuits between phases, while on account of the instability of arcs they may be productive of high-frequency voltage oscillations.

(*d*) Generally speaking, a transformer is usually switched on at no-load, the load subsequently being applied, and the reverse procedure is adopted when switching out of circuit. The practical advantage of doing this is that should there be a fault in the transformer when switching on, the disturbance arising from such a fault is not transferred to the load circuit, and consequently there is no danger of breakdown there. There does not, however, appear to be the same justification for switching off separately a transformer and its load, providing the transformer is supplying one load only, and in this case the switching off of the two would be provision against switching out voltage rises.

It will be seen from this general review that transformer switching gives rise to a variety of transient phenomena against which precautions must be taken in order to avoid subsequent damage. At the same time, so long as these abnormal conditions are recognised and understood, and suitable provision made to guard against them, the switching of modern transformers need not give rise to undue anxiety. It can be said, however, that the smaller the amount of switching carried out on a system the better, for transient conditions of some kind or other must arise to connect the unexcited with the normal running conditions.

SWITCHING IN CURRENT RUSHES

It is often noticed when switching in a transformer on no-load that the ammeter registers an initial current rush (which, however, rapidly dies down) greatly in excess of the normal no-load current and sometimes even greater than the normal full-load current of the transformer. In the latter case it may seem at first glance that there is a fault in the transformer. Upon considering the problem fully, however, and bearing in mind the characteristics of iron-cored apparatus, the true explanation of the transient current rush will become clear. The initial value of the current taken on no-load by the transformer at the instant of switching in is principally determined by the point of the voltage wave at which switching in occurs, but it is also partly dependent on the magnitude and polarity of the residual magnetism which may be left in the core after previously switching out. There are six limiting conditions to consider, namely:

(*a*) switching in at zero voltage—no residual magnetism

(*b*) switching in at zero voltage—with maximum residual magnetism having a polarity opposite to that to which the flux would normally attain under equivalent normal voltage conditions

(*c*) switching in at zero voltage—with maximum residual magnetism having the same polarity as that to which the flux would normally attain under equivalent normal voltage conditions

(*d*) switching in at maximum voltage—no residual magnetism

(*e*) switching in at maximum voltage—with maximum residual magnetism having a polarity opposite to that to which the flux would normally attain under equivalent normal voltage conditions

(*f*) switching in at maximum voltage—with maximum residual magnetism having the same polarity as that to which the flux would normally attain under equivalent normal voltage conditions.

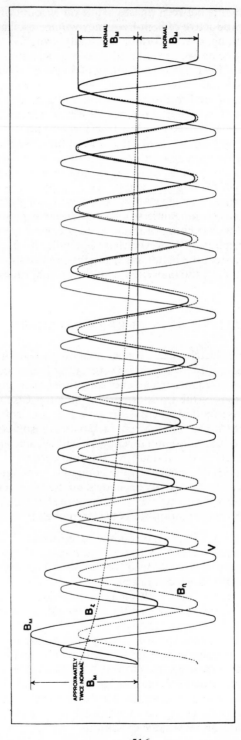

Fig. 22.9   Transient flux density conditions when switching in a transformer at the instant $V = 0$. No residual magnetism.

Dotted lines represent the normal steady flux density B_n, and the transient component B_t

(a) Switching in at zero voltage—no residual magnetism

Under normal conditions the magnetic flux in the core, being 90° out of phase with the voltage, reaches its peak value when the voltage passes through zero and vice versa. Due to this phase displacement it is necessary for the flux to vary from a maximum in one direction to a maximum in the opposite direction in order to produce one half cycle of the required back e.m.f. in the primary winding, so that a total flux is embraced during the half cycle corresponding to twice the maximum flux density.

At the instant of switching in, there being no residual magnetism in the core, the flux must start from zero, and to maintain the first half cycle of the voltage wave it must reach a value corresponding approximately to twice the normal maximum flux density.

This condition, together with the succeeding voltage and flux density waves, is shown in Fig. 22.9, and it will be seen that the rate of change of flux (upon which the magnitudes of the induced voltages depend) is nearly the same, throughout each cycle, as the normal flux density which is symmetrically placed with regard to the zero axis and which corresponds to the steady working conditions. The maximum values of the flux density, upon which the magnitude of the no-load current depends, vary gradually from a figure initially approaching twice the normal peak value in one direction only, down to the normal peak value disposed symmetrically on each side of the zero axis. As the magnitude of the no-load current is dependent upon the flux density, it follows that the current waves also will initially be unsymmetrical, and that they will gradually settle down to the steady conditions. While, however, in the case of the flux density the transient value cannot exceed twice the normal, the transient current reaches a value very many times the normal no-load current and can exceed the full-load current. The reason for this current rush is to be found in the characteristic shape of the B/H curve of transformer core steel, which is shown in Fig. 22.10, and from this

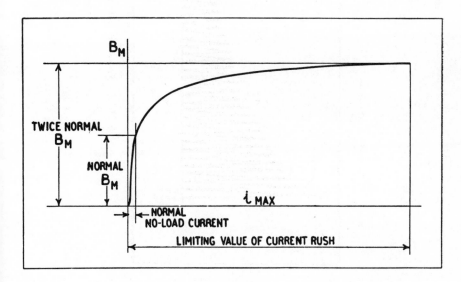

Fig. 22.10   Typical B/H curve showing relationship between maximum flux density and no-load current

517

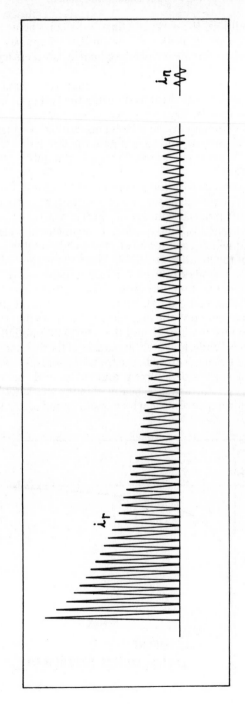

Fig. 22.11   Typical transient current rush when switching in a transformer at the instant V = 0.
i_n = normal no-load current
i_r = switching in current rush

it will be seen that the no-load current at twice the normal flux density is increased out of all proportion as compared with the current under steady conditions.

Figure 22.11 illustrates this current rush phenomenon, and the total current may be considered to consist of the normal no-load current and a drooping characteristic transient current superimposed upon it. Due to the initial high saturation in the core, the current waves may be extremely peaked and contain prominent third harmonics.

In practice, the transient flux does not actually reach a value corresponding to twice the normal flux density, as the voltage drop, due to the heavy rush of current flowing through the resistance of the entire primary circuit during the flux variation from zero to twice the maximum, is greater than the drop occurring with the normal flux distribution. Consequently a somewhat smaller back e.m.f. is to be generated by the varying flux, so that the latter does not reach a value corresponding to $2B_{max}$, but remains below this figure, the more so the higher the resistance of the primary circuit.

(b) Switching in at zero voltage—with maximum residual magnetism having a a polarity opposite to that to which the flux would normally attain under equivalent normal voltage conditions

If there is residual magnetism in the core at the instant of switching in and the residual magnetism possesses an opposite polarity to that which the varying flux would normally have, the phenomena described under (a) will be accentuated. That is, instead of the flux wave starting at zero it will start at a value corresponding to the polarity and magnitude of the residual magnetism in the core, and in the first cycle the flux will reach a maximum higher than outlined in (a) by the amount of residual magnetism. The theoretical limit is a flux which corresponds to a value approaching three times the normal maximum flux density, and at this value the initial rush of current will be still greater.

Figure 22.12 illustrates the resulting transient flux/time distribution, while the current rush will be similar to that shown in Fig. 22.11, except that the maximum values will be much higher and the rush will take a longer time to reach steady conditions. In this case also the drop in voltage, due to the resistance of the primary circuit, operates to reduce the maximum flux density and consequently the current rush, but to a greater extent than in the case of (a).

(c) Switching in at zero voltage—with maximum residual magnetism having the same polarity as that to which the flux would normally attain under equivalent normal voltage conditions

The converse of (b) where the residual magnetism possesses the same polarity as that which the changing flux would normally attain, results in a diminution of the initial maximum values of the flux, and consequently of the current rush.

If the value of the residual magnetism corresponds to maximum flux density the flux will follow its normal course and the no-load current rush will be avoided. Figure 22.13 illustrates the flux/time distribution. If, however, the residual magnetism corresponds to a flux density lower than the maximum, the initial flux waves are unsymmetrically disposed about the zero axis, the more so

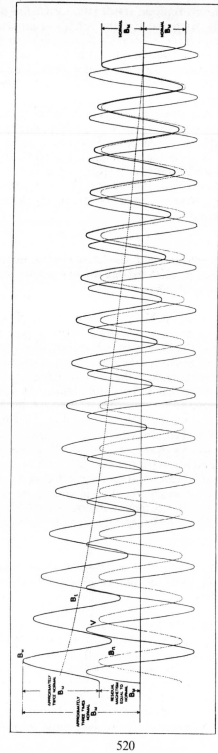

Fig. 22.12   Transient flux density conditions when switching in a transformer at the instant $V = 0$ and with residual magnetism in the core equal and opposite to the normal flux density. Dotted lines represent the normal steady flux B_n and the transient component B_t

the lower the value of the residual magnetism. Figure 22.14 illustrates this, and a current rush occurs according to the maximum value of the flux.

(*d*) Switching in at maximum voltage—no residual magnetism

In this case at the instant of switching in, the flux should be zero, due to its 90° phase displacement from the voltage, and as we have assumed there is no residual magnetism in the core, the desired conditions are obtained which produce the normal steady time distribution of the flux. That is, at the instant of switching in the flux starts from zero, rises to the normal maximum in one direction, falls to zero, rises to the normal maximum in the opposite direction and again reaches zero, the wave being symmetrically disposed about the zero axis. The no-load current, therefore, pursues its normal course and does not exceed the magnitude of the normal no-load current.

(*e*) Switching in at maximum voltage—with maximum residual magnetism having a polarity opposite to that to which the flux would normally attain under equivalent normal voltage conditions

In this case the residual magnetism introduces the transient components, so that the initial flux waves are unsymmetrically disposed about the zero axis, high initial maximum flux values are attained, and in the case where the residual magnetism has the same value as corresponds to the normal maximum flux density the current rush will have a value corresponding approximately to twice the normal maximum flux density. This is shown in Fig. 22.15.

(*f*) Switching in at maximum voltage—with maximum residual magnetism having the same polarity as that to which the flux would normally attain under equivalent normal voltage conditions

This is the converse of the foregoing case, and the initial flux waves will again be unsymmetrically disposed about the zero axis. For the same value of residual magnetism the total maximum flux would be the same as in case (*e*), but both flux and current waves would initially be disposed on the opposite side of the zero axis. Figure 22.16 illustrates this case.

The foregoing remarks are strictly applicable to single-phase transformers operating as such, but the principles can also be applied to polyphase transformers or banks so long as one considers the normal magnetic relationship between the different phases. That is, curves have been given which relate to one phase only, but the principles apply equally well to polyphase transformers, providing each phase is treated in conjuction with the remaining ones.

A single instance will suffice to show what is meant, and for this purpose consider a three-phase core-type star/star connected transformer switched in under the same conditions as (*b*) which is illustrated in Fig. 22.12. Figure 22.17 shows the normal main flux space and time distribution at intervals of 30°, and the number of lines in the cores indicate the relative flux density in each. Due to the usual three-phase relationship, the transformer can only be switched in

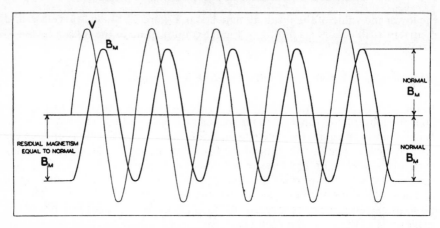

Fig. 22.13   *Flux density conditions when switching in a transformer at the instant V = 0 and with residual magnetism in the core equal to and of the same polarity as the normal flux density. No transient condition*

when any *one* phase, say A, is at zero voltage, so that the remaining two phases B and C will each give a voltage at the instant of switching in equal to 86·6 % of the maximum of each phase, one being positive and the other negative. Similarly, if the transformer has previously been switched out so that the residual magnetism in phase A of the core has a value corresponding to the maximum flux density and a polarity opposite to that which the flux would normally attain to under equivalent normal voltage conditions, the residual magnetism in each phase B and C will have a value corresponding to half the normal flux density in each phase, and a polarity opposite to the residual magnetism in phase A. The current rushes in the three phases will therefore not be equal, but they will be modified by the flux conditions, which are shown in Fig. 22.18. It is only necessary to refer to the B/H curve and hysteresis loop to obtain an approximation of the current value corresponding to each value of flux density.

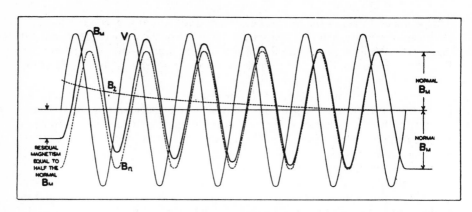

Fig. 22.14   *Flux density conditions when switching in a transformer at the instant V = 0 and with residual magnetism in the core equal to half the normal flux density and of the same polarity as the normal flux density.*
Dotted lines represent the normal steady flux B_n and the transient component B_t

Fig. 22.15 Transient flux density conditions when switching in a transformer at $V = V_{max}$ and with residual magnetism in the core equal to but in the opposite direction to the normally increasing flux density. Dotted lines represent the normal steady flux B_n and the transient component B_t

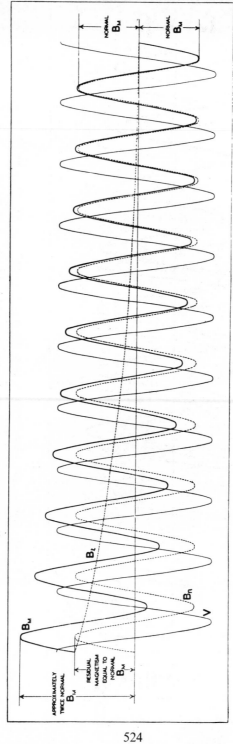

Fig. 22.16   Transient flux distribution conditions when switching in a transformer at $V = V_{max}$ and with residual magnetism in the core equal to and in the same direction as the normally increasing flux density. Dotted lines represent the normal steady flux B_n and the transient component B_t

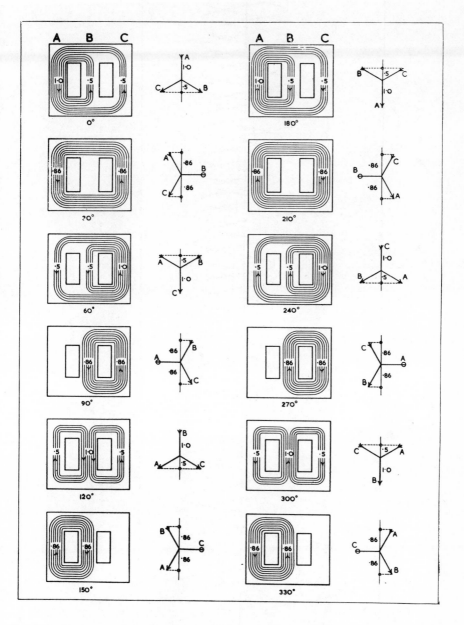

Fig. 22.17 Flux space and time distribution at 30° intervals in a three-phase core-type transformer with star/star connected windings

525

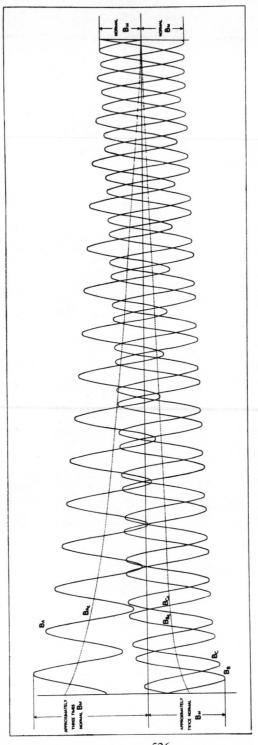

Fig. 22.18 Transient flux density conditions when switching in a three-phase core-type transformer with star/star connected windings at the instant $V_a = 0$, $V_b = 0.866\ V_{max}$ and $V_c = 0.866\ V_{max}$ and with residual magnetism in phase A equal to twice the normal flux density and in phases B and C equal to half the normal flux density.

Dotted lines represent the transient components B_{At}, B_{Bt}, and B_{Ct}

The flux waves have been drawn sinusoidal in order to present the illustrations in the clearest manner, but the actual shape of the flux and current waves will be determined by the connections of the transformer windings and the type of magnetic circuit. In a three-phase core-type transformer with star/star connected windings the normal flux wave may contain small third harmonics, and may therefore be flat topped, while the no-load current will be a sine wave. With a delta connected winding on either primary or secondary side the normal flux wave will be sine shaped, while the no-load current may contain third harmonics and be peaked.

Figures 22.19 and 22.20 show the method of obtaining the non-sinusoidal wave shapes of flux and no-load current from the hysteresis loop of the core material when the no-load current and flux respectively are sine waves. In the initial transient stages the saturation of the cores will accentuate the higher harmonics so that current rushes will have much higher peak values than can

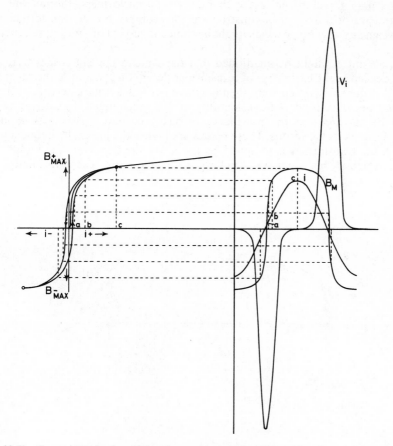

Fig. 22.19 Construction for determining the non-sinusoidal waveform of flux density with a sine wave of no-load current
N.B. Induced voltage wave determined by differentiation of the flux wave

527

be deduced from ammeter readings. The *B/H* curve and hysteresis loops at various maximum flux densities would have to be available if more accurate theoretical determinations of the current values were to be made, but even then further difficulties would arise from the unequal form factors of the current waves in the three phases. These would particularly be marked in star/star connected transformers and, due to the tendency to balance out the currents, the neutral point would be deflected and the phase voltages unbalanced.

From observations made over a considerable period of time it would seem that new transformers are more subject to transient current rushes than are transformers which have been in service for a number of years. This may *possibly* be accounted for by the fact that with the latter transformers the core plates have become somewhat further annealed due to the temperatures they have been subjected to while in service, and as increased annealing renders the iron less able to hold residual magnetism, the maximum flux densities possible at the instant of switching in are correspondingly reduced.

It is also worthy of note that transformers with butt type yokes do not retain residual magnetism so much as if the yokes and cores are interleaved. Current rushes may therefore be less with butt yoke transformers, though the disadvantages of this form of construction for ordinary power transformers far outweigh any advantage which might be gained in respect of a minimised current rush.

In passing it might be mentioned that these heavy current rushes were not experienced in the early days of transformer design on account of the relatively low flux densities which were then employed. The loss characteristic of transformer steel has improved considerably so that much higher flux densities can now be allowed and the prevalence of heavy current rushes with modern transformers is due to this. These rushes are higher the lower the frequency for which the transformer is designed, as the lower the frequency the higher can the flux density be, which will still keep the iron loss to a reasonable figure.

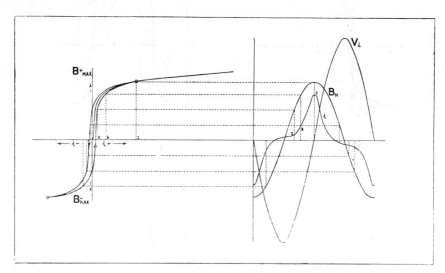

Fig. 22.20   Construction for determining the non-sinusoidal waveform of no-load current with sine waves of flux and induced voltage

For ordinary power and lighting transformers it has been suggested that residual magnetism may be greatly minimised if the load on the transformer is switched off before the primary circuit is opened. In this case when the transformer is finally switched out of circuit the only current flowing will be the normal no-load current which will be lagging behind the applied voltage by an angle usually between 70° and 90°. As it is generally found that a circuit-breaker opens a circuit at or near zero current, this will correspond to a point at or near the maximum point on the voltage wave so that the flux in the core will be nearly zero. If the transformer is switched out of circuit on load, zero current will, in the case of a non-inductive load, correspond nearly to zero voltage, so that the residual magnetism left in the core would be a maximum; but the more inductive the load, the less likelihood is there of switching out at zero voltage.

Theoretically the residual magnetism may almost be eliminated by gradually reducing the applied voltage before switching the transformer out of circuit, while a further possible method would be to provide some kind of contact mechanism which would ensure the switch opening the circuit at maximum voltage. It should be remembered, however, that in any case with polyphase transformers it is not possible to switch out all phases at the maximum voltage, and consequently this last method would, at the best, only result in zero magnetism in one phase and something between zero and the maximum in the remaining phases. Both of the last two methods are objectionable, however, in so far as they involve additional apparatus.

It has also been suggested at various times that switching in current rushes may be minimised by slowly closing the switch in the exciting circuit. The idea underlying this suggestion is that as the switch contacts approach one another, a point will be reached just prior to actual closing at which a spark will bridge the contacts of one phase at the maximum peak voltage. At this instant the normally varying flux will have zero value and consequently current rushes will be avoided. This effect could, of course, only occur with one phase, and consequently the method could theoretically only be applied with perfect success to single-phase transformers. In the case of polyphase transformers, the remaining phases will each have a voltage at some value between zero and the maximum, and therefore abnormal flux distribution occurs in these phases, which will produce current rushes similar to those previously described. When considering this method, however, the fact should not be lost sight of that arcing at the switch contacts is liable to produce high frequency voltage oscillations, the more so the more slowly the switch is closed. As this type of disturbance is generally very liable to produce a breakdown in the transformer windings, the method of slowly switching in for the purpose of avoiding current rushes is not one which can easily be recommended.

At one time circuit breakers controlling transformers were sometimes fitted with buffer resistors and auxiliary contacts so that the resistors were connected in series with the transformer when switching in, being subsequently short circuited upon completion of the switching operation. These buffer resistors are, however, no longer employed but as a matter of academic interest Fig. 22.21 illustrates the effect of a buffer resistor on the transient switching in current rush of a 20 kVA transformer, the resistor having such a value that it takes 5 % of the normal supply voltage at no-load.

It has been the practice with some transformer makers actually to provide air gaps in the magnetic circuit in order to reduce the residual magnetism.

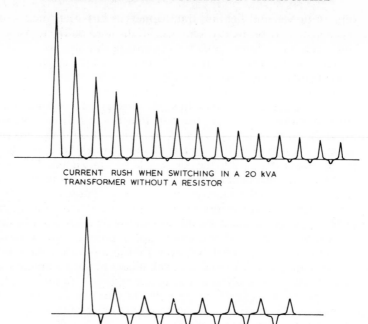

CURRENT RUSH WHEN SWITCHING IN A 20 kVA
TRANSFORMER WITHOUT A RESISTOR

CURRENT RUSH WHEN SWITCHING IN A 20 kVA
TRANSFORMER THROUGH A RESISTOR TAKING
5 % OF THE NORMAL LINE VOLTAGE

Fig. 22.21   Current rush when switching in a 20 kVA transformer

This is definitely not to be recommended, as the normal no-load current and iron loss of such a transformer would be much higher than in a transformer without gaps, while the eddy currents which may be produced in the vicinity of the gaps frequently cause local heating and burning of the cores.

An objection to these current rushes is the mechanical forces which are exerted between coils at the instant of switching in. While these certainly die down more or less rapidly, the conductors are strained to some extent and the insulation between individual conductors may become compressed in places, while in other places the normal mechanical pressure due to the winding process may be released, so that the mechanical rigidity of the coils as a whole becomes entirely altered. That is, in some parts adjacent conductors may be slack, while in others they may be compressed tightly, and with repeated switching in operations there may be a grave risk of failure of the insulation between turns of the windings.

Cases have been known in practice in which a transformer switched in under particularly adverse conditions has moved in its tank, and this introduces the possibility of damage to connections between coils and connections from coils to terminals, resulting in one or more open circuits in the windings concerned.

Among the minor disadvantages are the tripping of main switches, blowing of fuses, and the operation of relays, but while these are often annoying, they are not serious.

SHORT-CIRCUIT CURRENT RUSHES

It has already been shown that abnormal current rushes may occur in the primary windings under certain adverse conditions when switching in a transformer on no-load, but much heavier currents may flow in both primary and secondary windings when a transformer momentarily supplies its heavies load; that is, when a short circuit occurs across the secondary terminals. We thus have four distinct current conditions to which a transformer may be subjected, these being:

    (a) transient switching in no-load current rush
    (b) steady no-load current
    (c) steady normal load current
    (d) transient short-circuit current rush.

The currents which represent a danger to the windings are the transient rushes only, viz., (a) and (d), and of these two the latter is the one against which special precaution must be taken, as the resulting currents set up very severe mechanical stresses in the windings.

If a conductor carries current a magnetic field is set up round the conductor in the form of concentric circles, the density of the field at any point being directly proportional to the current in the conductor and inversely proportional to the distance between the conductor and the point considered. If two conductors both carrying current are in close proximity to each other they will each be subjected to the influence of the magnetic field surrounding the other, and in the case of adjacent conductors carrying currents in the same direction the magnetic fields will coalesce and a force of attraction between the two conductors will arise, while with currents flowing in opposite directions the magnetic fields mutually repel each other and a repulsion force is set up between the two conductors. For a given current and spacing between the two conductors the value of the forces is the same, irrespective of whether they are attractive or repulsive.

If, now, the above principles are applied to transformer coils it will be seen that any one coil, either primary or secondary, carries current so that the currents in opposite sides flow in opposite directions, and repulsion forces are thus set up between opposite sides so that the coil tends to expand radially outwards in just the same way as does a revolving ring or other structure on account of centrifugal force. The coil thus tends to assume a circular shape under the influence of short circuit stresses, and therefore it is obvious that a coil which is originally circular is fundamentally the best shape, and is one which is least liable to distortion under fault conditions. From this point of view the advantages of the circular core type of construction are obvious.

In a coil composed of a number of wires arranged in a number of layers, each having a number of turns per layer, such as is often the case with h.v. windings, the wires situated in the same sides of the coil carry current in the same direction, and therefore attract one another and tend to maintain the homogeneity of the coil.

In addition to the radial forces set up in the individual windings tending to force the coils into a circular shape, other repulsion forces exist between primary and secondary windings, as these windings carry currents flowing in opposite

531

directions. The directions of these repulsion forces are shown in Fig. 22.22 for circular and rectangular coils under the conditions of

(i) coincident electrical centres
(ii) non-coincident centres.

When the electrical centres coincide it will be seen that if the coils are of the same dimensions, repulsion forces normal to the coil surfaces only exist, but if the electrical centres do not coincide, a component at right angles to the force normal to the coil surfaces is introduced which tends to make the coils slide past one another. A similar component at right angles to the normal component is introduced, even if the electrical centres are coincident if the dimensions of the coils are different. In this case the system consisting of the primary and secondary coils is balanced as a whole, but adjacent sides of primary and secondary coils are liable to distortion on account of the sliding components introduced. In actual practice, both with core-type and shell-type transformers, the sliding component of the mechanical forces between primary and secondary coils is the one which has been responsible for many failures under external short-circuit conditions, particularly of some of the older transformers having low impedances. In passing it should be remembered that often it is not possible to preserve the coincidence of electrical centres at all ratios when transformers are fitted with voltage adjusting tappings.

The value of the current flowing under external short-circuit conditions is inversely proportional to the impedance of the entire circuit up to the actual fault, and, so far as the transformer itself is concerned, the most onerous condition which it has to withstand is a short circuit across the secondary terminals. The question of how the transformer impedance is affected by matters of design is dealt with elsewhere in this book, and it is only necessary to point out here that as the impedance voltage is that voltage required to circulate *full-load* current in the transformer windings, the short-circuit current bears the same relation to the normal full-load current as does the normal full-line voltage to the impedance voltage, the latter being expressed in terms of the full line voltage. Expressed in equation form, the connection between short-circuit current and impedance voltage is as follows:

$$I_{SC} = I_{FL}\frac{100}{V_z} \tag{22.1}$$

where I_{SC} = primary or secondary short-circuit current
I_{FL} = primary or secondary full-load current
V_z = percentage impedance voltage.

It must be remembered that the short-circuit current derived from the above equation presumes maintenance of the full line voltage under fault conditions, but as a matter of fact the line voltage is generally maintained for the first few cycles only after the first instant of short circuit, and then only on the larger systems having sufficient kVA of generating plant behind the fault. Therefore, at the first instant of short circuit the current reaches a value given by eqn. 22.1, but as the line voltage drops the value of the short-circuit current similarly falls until the transformer is automatically switched out of circuit.

The initial value of the short-circuit current rush may be further modified by the normal conditions existing at the instant of short circuit, and in the worst

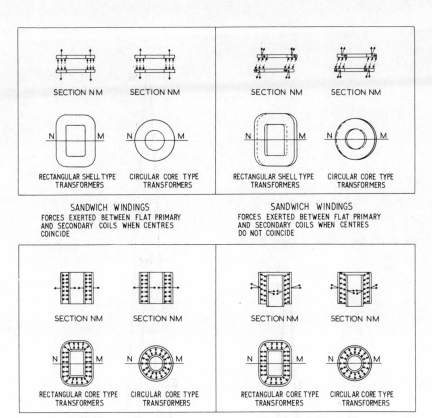

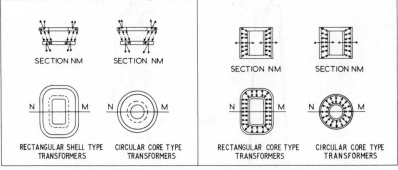

Fig. 22.22   Mechanical forces on transformer coils

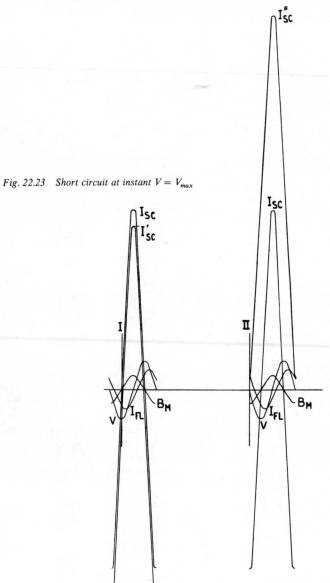

Fig. 22.23   Short circuit at instant $V = V_{max}$

Fig. 22.24   Short circuit at instant $V = zero$

Curves illustrating short circuit doubling effect

534

case the initial value of the short-circuit current reaches twice the amount given by eqn. 22.1 by what has been termed the 'doubling effect'. This doubling effect occurs when the actual short circuit is made at the instant when the voltage of the circuit is zero, and we will consider the two extreme cases when the short circuit occurs at the instant (a) when the voltage is passing through its maximum value, and (b) when the voltage is passing through zero. In Fig. 22.23, V represents the voltage wave, B_M the flux wave, and I_{FL} the wave of normal full-load current. If a short circuit takes place at the instant marked I in the diagram, the flux leading the voltage by 90° is zero, and as on short circuit the resulting current is in phase with the flux or nearly so, the short-circuit current should have a similar phase relationship. At the instant I the short-circuit current should therefore be zero, and if no current existed in the circuit at this instant the short-circuit current would pursue its normal course, reaching an initial maximum value corresponding to eqn. 22.1, and it would be disposed symmetrically on either side of the zero axis, gradually and symmetrically dying down until the transformer was tripped out of circuit. This condition is shown by the current wave I_{SC} in Fig. 22.23. On account of the presence of the normal full-load current which has a definite value at the instant I, the short-circuit current must initially start from that point and the resulting wave will be somewhat unsymmetrical, depending upon the ratio between the full-load and short-circuit currents and upon their relative power factors. This wave is shown at I_{SC}' in Fig. 22.23.

If, on the other hand, the short circuit occurs at the instant marked II in Fig. 22.24, the voltage is zero and the flux has a negative maximum value, so that the initial short circuit-current should also be at or near its negative maximum value. Actually this cannot occur, as the short-circuit current cannot *instantly* attain the value corresponding to the position and value of the flux wave, but instead it must start from a value corresponding in sign and magnitude to the current already in the circuit at the particular instant, viz., the normal full-load current at the instant II. During the first cycle immediately following the short circuit the full voltage is generally active in producing abnormal short-circuit currents, and in order to maintain this voltage during the first half cycle, the short-circuit current must vary from a maximum negative to a maximum positive value, that is, the total change is twice that occurring when a short circuit takes place at maximum voltage and zero flux. This abnormal current wave, therefore, commences from the value of the full-load current in the circuit at the instant II, and from this point rises to a value approaching twice that obtained with a symmetrical short circuit, as shown at I_{SC}'' in Fig. 22.24. This explains the so-called 'doubling effect', though, as a rule, the short-circuit current does not reach the full double value on account of resistance voltage drops. This highly abnormal wave is, of course, unsymmetrical, but dies down rapidly, giving ultimately the same symmetrical current distribution as when a short circuit takes place at the instant corresponding to maximum voltage.

The intensity of the mechanical stresses set up in transformer windings varies as the square of the current flowing, and it will be seen, therefore, that the doubling effect may have very serious consequences. For instance, in a transformer having an impedance of 5%, the initial stresses under short-circuit conditions would be 400 times as great as those in the transformer under normal full-load conditions when making the short circuit at maximum voltage, but when making the short circuit at zero voltage the resulting stresses in the windings would be approximately 1600 times as great as those under normal full-load conditions, on account of the doubling effect.

In actual practice these high mechanical stresses have been responsible for the severe rupture of h.v. end coils of core-type transformers with concentrically disposed windings, though in such transformers the radial bursting tendency on the outer windings is not usually high enough to reach the elastic limit of the copper conductors. Similarly, the compressive stress on the inner windings is amply combated by the mechanical rigidity inherent in such windings. With rectangular shell type transformers employing flat rectangular coils arranged in sandwiched fashion, severe distortion and subsequent rupture of the ends of the coils projecting beyond the core has occurred from the same cause. With low reactance transformers particularly, the h.v. end coils and also the coil clamping structure have been severely distorted, but as the forces which come into play have become more and more appreciated, transformer reactance has been increased and the coil clamping structure better designed and more adequately braced. Such coil clamps have been applied to compress the coils in an axial direction and to retrain them from moving under short-circuit conditions. Except in very special cases, radial coil supports are not necessary, though there is more justification for their use with rectangular type of coils such as are used with rectangular shell-type transformers. In the circular core-type transformer the coils also are circular, and as this is inherently the best possible shape, the conductors are best able to stand the short-circuit stresses.

While on short circuit the phenomenon most to be feared is the mechanical stresses to which the windings and structure are subjected, it must be remembered as somewhat of a paradox, that this is so on account of the rapidity with which modern circuit-breakers automatically disconnect such a fault from the supply. If for any reason the automatic means provided did not operate, the transformer would rapidly become overheated, and it would exhibit all the appearances associated with severe overloads. In a very short space of time short circuits between turns would take place and the windings would become destroyed.

The reader should refer to Chapter 27 for more detailed information regarding electromagnetic forces in transformer windings.

Chapter 23

Transformer protection

The subject of transformer protection falls naturally under two major headings, which are:

Protection of the transformer against the effects of faults occurring on any parts of the system; and

Protection of the system against the effects of faults arising in the transformer.

(1) Considering first the means to be adopted for protecting the transformer itself against the effects of system faults, three distinct types of disturbances (apart from overloads) have to be provided for, these being: (*a*) short circuits; (*b*) high-voltage, high-frequency disturbances; (*c*) pure earth faults.

(*a*) System short circuits may occur across any two lines, or, if the neutral point is earthed, between any one line and the earthed neutral. As pointed out elsewhere,* the effect of a system short circuit is to produce overcurrents, the magnitudes of which are dependent upon the system kVA feeding the fault, the voltage which has been short circuited, and upon the impedance of the circuit up to the fault. The short-circuit currents produce very high mechanical stresses in the apparatus through which they flow, these stresses being proportional to the square of the currents. The necessary protection may be afforded to the transformer by the insertion of additional reactance in the circuit and by adequate bracing of the transformer windings. The inclusion of additional reactance operates directly to reduce the magnitudes of the short-circuit currents, and therefore considerably of the short-circuit stresses. This reactance may be incorporated in the transformer itself by suitable design, or it may be provided in the form of separate reactors connected in series with the primary of the transformer. As a rule, when reactance is incorporated in the transformer for the purpose of protecting it against short-circuit stresses, the desired value of the reactance is obtained by a suitable design of the h.v. and l.v. windings. External reactors for short-circuit protection are usually, though not always, air-cored coils designed to withstand the stresses which may occur on short circuit.

In addition to providing the reactance sufficient to give the necessary degree of protection, transformers connected to large power systems have their winding very securely braced in order to minimise the effects of the mechanical

*Chapter 22.

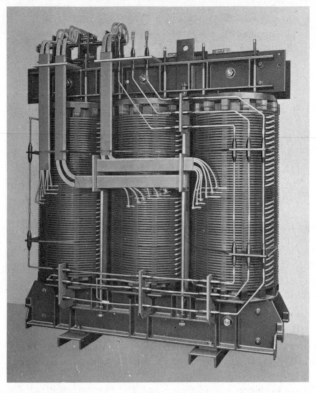

Fig. 23.1   10 MVA, 33 000/300 V, 50 Hz transformer. Continuous disc h.v. windings, spiral l.v. windings. View shows transformer out of its tank and looking at the h.v. side

forces to which they may be subjected. A typical transformer which has been designed specially to withstand severe short circuit stresses and which is provided with hand-adjustable axial coil supports is shown in Fig. 23.1, and this illustration clearly shows the great care which is taken in the mechanical design of modern power transformers which are liable to be subjected to the stresses arising on large power systems under fault conditions. In this connection it should also be noted that it is equally necessary to brace the leads from the coils in an effective manner, as otherwise these may become distorted under short-circuit conditions, and whilst the damage to the transformer may not be so extensive, the unit would almost certainly be put out of commission for some length of time.

For the purpose of protection against short-circuit stresses, it is recommended that transformers should have the minimum impedance values given in Table 23.1.

These figures are average values and they can be varied to suit the requirements of the system.

Modern transformers designed to comply with the current B.S. 171 are capable of withstanding without damage whilst in service, the electromagnetic forces arising under short-circuit conditions, as determined by the asymmetrical peak value of the current in the windings, which can be taken as 2·55 times the short-circuit current.

(b) High-voltage, high-frequency surges may arise in the system due to arcing grounds if the neutral point is isolated, to switching operations, and to the effects of atmospheric disturbances. These disturbances principally take the form of travelling waves having high amplitudes and steep wave fronts, and often successive surges may follow rapidly upon one another. On account of both their high amplitudes and frequencies these surges may, upon reaching the windings of a transformer, break down the insulation between turns, particularly of turns adjacent to the line terminals, causing short circuits between turns and producing extensive damage to the transformer windings. The effects of these surge voltages may, however, largely be minimised by designing the windings to withstand the application of a specified surge test voltage and then ensuring that this test value is not exceeded in service by the provision of a suitable surge diverter mounted adjacent to the transformer terminals. Regarding surge diverters the old combined choke coil and horn gap arrester has been superseded by more modern diverters which possess improved characteristics. All types of surge diverters aim at attaining the same results, viz., of shunting disturbing surges from the lines to earth to prevent their reaching the transformer. In essence, the different kinds of valve type surge diverters employ several spark gaps in series with a nonlinear resistor. The diverters are connected from each line to earth, and they are used without any auxiliary series reactor in the line. When a high-voltage surge wave reaches the diverter the spark gaps break down and the disturbance is discharged to earth through the nonlinear resistor of the diverter by reason of the fact that at the high voltage involved the diverter resistance is low. As the surge voltage falls the diverter resistance automatically increases and prevents the flow of power current to earth. A diverter of this type is therefore entirely automatic in action and self extinguishing.

The internal surge impedance of a transformer winding is not a constant single-valued quantity but, on the contrary, a winding has a range of surge impedances corresponding to the frequencies of the different space harmonics of the internal voltage oscillations which connect initial hyperbolic voltage

Table 23.1

IMPEDANCES OF THREE-PHASE, 50 Hz TRANSFORMERS
H.V. OR L.V. WINDINGS DELTA OR STAR CONNECTED

Transformer kVA	Percentage impedance		
	11 kV	33 kV	66 kV
100	4·75	5·0	5·5
250	4·75	5·0	5·5
500	4·75	5·0	6·0
1000	4·75	5·0	6·0
1500	5·5	6·0	7·0
2000	6·0	6·0	7·0
3000	6·0	7·0	7·5
5000	6·0	7·0	7·5
7500	7·0	8·0	8·5
10 000	—	9·0	9·0
20 000	—	10·0	10·0
30 000	—	10·0	10·0

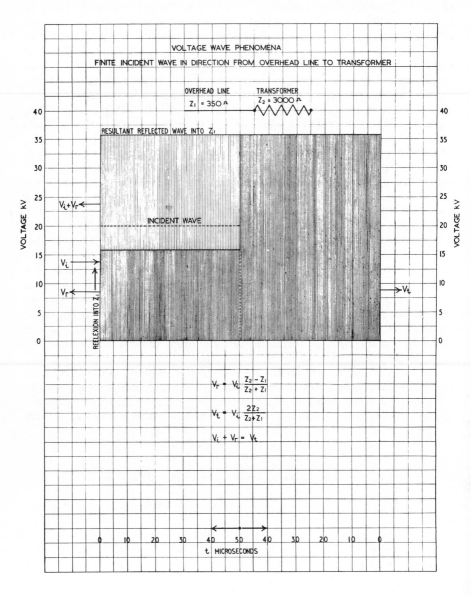

Fig. 23.2   *Diagram showing incident, transmitted and reflected voltage wave, assuming an amplitude of 20 kV for the incident wave*

distribution with final straight line conditions. The multi-harmonic surge impedances, therefore, are characteristic of what takes place in the windings, and they do not appreciably affect the terminal conditions. Moreover, the transformer terminal impedance is so high in any case, compared with the line surge impedance, that its assigned value, so long as it is of the right order, has but little influence upon the shape of the resulting surge waveform given. The wave diagrams in this Chapter show the variation of voltage and current with time at the transformer terminals and not the phenomena occurring throughout the windings subsequent to the application of surge waves to the terminals. Studies of the latter are given in Chapter 22.

Consider, first, what happens when rectangular finite voltage and current waves reach a transformer from an overhead line, there being no protective apparatus installed to intercept the disturbance. The amplitudes of the waves in the overhead line and at the transformer terminals depend upon the respective values of their surge impedance, which is given by the formula:

$$Z = \sqrt{L/C}$$

Z = surge impedance in ohms,

where L = inductance in henrys,

       C = capacitance in farads

of the circuit concerned. L and C may be taken for any convenient length of circuit.

When any travelling waves of voltage and current pass from a circuit of a certain surge impedance to a circuit of a different surge impedance, such waves in their passage to the second circuit undergo changes in amplitudes. The oncoming incident waves when reaching the transition point between the two circuits are, if the surge impedances of the two circuits are different, split up into two portions, one being transmitted into the second circuit, and the other reflected into the first. The transmitted waves always have the same sign as the incident waves, but the reflected waves may have the same or opposite sign to the incident waves depending upon the ratio of the two surge impedances. This applies both to the voltage and the current waves. If the incident waves are of finite length, the reflected waves travel back into the first circuit alone, and they are only transient waves in that circuit. If, on the other hand, the incident waves are of infinite length, the reflected waves in their passage backwards along the first circuit combine with the backs of the incident waves, so that the resultant waves in the first circuit are a combination of the two respective incident and reflected waves. Given the amplitudes of the incident voltage and current waves and the surge impedances of the two circuits, the transmitted and reflected waves may be calculated by means of the formulæ given in Table 23.2. The table gives formulæ for determining the conditions when the incident waves are finite in length, and they are based on the assumption that no distortion of the shape occurs, due to losses in the circuit. Figures 23.2 and 23.3 show the incident, transmitted, and reflected voltage and current waves respectively, assuming the incident voltage wave to have an amplitude of 20 000 V, and the incident current wave an amplitude equal to $20\,000/Z_1 = 57 \cdot 1$ A, as Z_1 is assumed to equal 350 Ω. The transmitted and reflected waves are constructed from the formulæ given on the diagrams (distortion being ignored), and it will be seen that a voltage wave arrives at the transformer terminals having an amplitude considerably higher than that of the original incident wave. It is due to this sudden increase of voltage at the transformer terminals that so many failures of the insulation on the end turns of windings have occurred in the past, as the increased voltage may be concentrated, at the first instant, across the first few turns of the winding only, though ultimately voltage is distributed evenly throughout the whole winding. The transmitted current wave is correspondingly smaller in amplitude than the incident current wave, and as such, is usually of no particular danger.

These diagrams show clearly that where the surge impedance of the second circuit is higher than that of the first, in comparison with the incident waves, the transmitted voltage wave is increased and the transmitted current wave is decreased, while the reflected waves have such signs and amplitudes as to satisfy the equations

Table 23.2

REFLECTION OF RECTANGULAR TRAVELLING WAVES
AT A TRANSITION POINT. NO PROTECTIVE DEVICE

V = voltage	Distortion due to	i = incident waves
I = current	dielectric and copper losses is ignored	r = reflected waves
		t = transmitted waves

Z_1 and Z_2 = surge impedances of circuits in Fig. 23.2.

$$\text{Current and voltage reflection factor} = \frac{Z_2 - Z_1}{Z_1 + Z_2} = n_r.$$

$$\text{Voltage transmission factor} = \frac{2Z_2}{Z_1 + Z_2} = n_{tv}.$$

$$\text{Current transmission factor} = \frac{2Z_1}{Z_1 + Z_2} = n_{tc}$$

Also
$$\frac{V_t}{V_i} = \frac{2Z_2}{Z_1 + Z_2} \quad ; \quad \frac{V_t}{V_r} = \frac{2Z_2}{Z_2 - Z_1} \quad ; \quad \frac{V_r}{V_i} = \frac{Z_2 - Z_1}{Z_2 + Z_1} \quad ;$$

$$\frac{I_t}{I_i} = \frac{2Z_1}{Z_1 + Z_2} \quad ; \quad \frac{I_t}{I_r} = \frac{2Z_1}{Z_2 - Z_1} \quad ; \quad \frac{I_r}{I_i} = \frac{Z_2 - Z_1}{Z_1 + Z_2} \quad ;$$

Then

	$Z_2 > Z_1$		$Z_2 < Z_1$	
	Voltage	*Current*	*Voltage*	*Current*
Incident wave	V_i	I_i	V_i	I_i
Transmitted wave	$n_{tv}V_i$	$n_{tc}I_i$	$n_{tv}V_i$	$n_{tc}I_i$
Reflected wave	$n_r V_i$	$-n_r I_i$	$n_r V_i$	$-n_r I_i$

For $Z_2 > Z_1$ the reflected voltage wave is positive.

For $Z_2 > Z_1$ the reflected current wave is negative.

For $Z_2 < Z_1$ the reflected voltage wave is negative.

For $Z_2 < Z_1$ the reflected current wave is positive.

Note: $\dfrac{2Z_2}{Z_1 + Z_2} - \dfrac{Z_2 - Z_1}{Z_1 + Z_2} = 1; \quad \dfrac{2Z_1}{Z_1 + Z_2} + \dfrac{Z_2 - Z_1}{Z_1 + Z_2} = 1.$

$$V_t = V_i + V_r$$
$$I_t = I_i + I_r.$$

That is, both transmitted voltage and current waves are equal to the sum of the respective incident and reflected waves.

During the time corresponding to the lengths of the incident waves, the total voltage and current in the first circuit is equal to the sum of their respective incident and reflected waves, but after that period the reflected waves alone exist in the circuit.

From the preceding formulæ and diagrams, when surges pass from one circuit to another the phenomena can be summarised as follows:

When a voltage wave passes from one circuit to another of higher surge impedance, both reflected and transmitted waves have the same sign as the incident wave, while the transmitted voltage wave is equal in amplitude to the

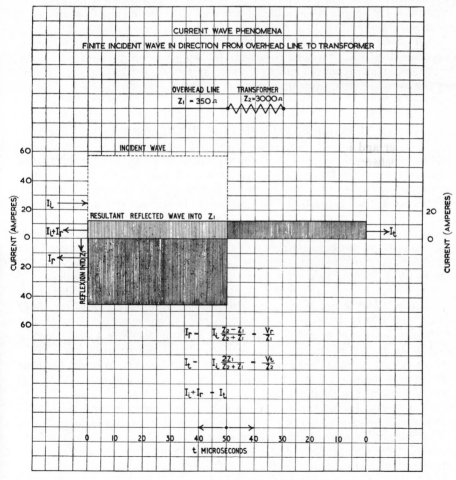

Fig. 23.3   Diagram showing incident, transmitted and reflected current wave, assuming an amplitude of 57·1 A for the incident wave

sum of the incident and reflected waves. For the same circuit conditions the transmitted current wave possesses the same sign as the incident current wave, but the reflected current wave has the opposite sign. The transmitted current wave is equal in amplitude to the sum of the incident and reflected waves.

At the transition point itself the total amplitudes of the voltage and current waves are always equal to the sum of their respective incident and reflected waves, bearing in mind, of course, the relative signs, and if the incident waves are infinite in length the same amplitudes extend throughout the whole of the circuit in which the incident waves arise. If, on the other hand, the incident waves are of finite lengths, the amplitudes of the resultant waves in the major part of the first circuit will be equal to the amplitudes of the reflected waves only.

The amplitudes of the transmitted and reflected waves are solely dependent on the ratio of the surge impedances of the two circuits.

The preceding notes have indicated that transmitted waves passing to a second circuit do so with rectangular fronts. That is, apart from any deformation of the waves which may occur due to losses in the circuits, the fronts of the transmitted waves are not modified in any way. If the second circuit is composed of inductive windings, such a rectangular fronted voltage wave represents a distinct danger to the insulation between turns and layers of the windings. It becomes desirable, therefore, to modify the shape of the wave form from the steep rectangular form to a more gradual sloping one, and this can be achieved by the use of suitable surge diverters.

SURGE PROTECTION OF TRANSFORMERS

Modern ideas on the subject of protecting transformers against the destructive effects of steep-fronted, high-voltage surges are based upon the principles governing the co-ordination of insulation strengths of the apparatus under surge conditions. The parts involved are h.v. and l.v. bushings, major insulation of h.v. and l.v. windings to earth and between windings, interphase insulation, and, finally, the interturn insulation of both windings. With the exception of the interturn insulation the various dielectric paths can be proportioned so as to *flash over* under surge conditions; the interturn insulation, however, must fail by puncture, and this occurrence must be prevented as far as possible.

The order of surge (or impulse) flashover strengths on the two sides of the transformer and in ascending values should be, bushings to earth, windings to earth, between phases and between h.v. and l.v. windings, the possibility of interturn insulation puncture being last. Bushing flashovers and internal flashovers from windings to earth, between phases and between h.v. and l.v. windings may be prevented by shunting these parts of the transformer by surge diverters or discharge gaps, and this also reduces the magnitudes of the surge waves through the windings and thus the interturn insulation stresses; the latter may also be reduced by proportioning the insulation thicknesses suitably, which, however, must be done in such a way as to avoid the possibility of occurrence of dangerous internal oscillations.

While bushings should be designed to flash over at surge voltages lower than those required to produce breakdown within the transformer, it is undesirable that frequent flashovers should occur in practice as usually the blowing of fuses, with consequent interruptions to supply, follows. The blowing of fuses is not usually occasioned by the surge current which flows to earth, for while the magnitudes of such currents may be exceedingly high their durations are, in general, far too short (being of the order of microseconds only) to affect fuses. When a surge discharges over a bushing, the surge current flows to earth via the tank and the tank earth connection, and as the impedance of this path is lower for power frequency currents than it is for impulse currents, power current follows the surge current if the system neutral is earthed, or if two bushings flash over simultaneously. Fuse rupturing, following bushing flashover, is thus due largely to the impedance characteristic of the particular discharge path to earth.

If, however, the surge wave can be discharged to earth over a path which possesses a relatively low overall impedance to the high-voltage transient, and a high impedance to the lower voltage, normal supply frequency current, then fuse blowings become eliminated or very largely reduced. For this reason, and also in order to limit the surge voltage differences which can appear across the different parts of the transformer, the major dielectric paths in the latter should be shunted by surge diverters of a design in which the discharge path possesses a resistance characteristic of the kind indicated, and protection to these paths should not be left solely to flashover of bushings.

This applies particularly to distribution transformers and other transformers located in unattended places.

Protection with surge diverters

When a travelling wave on a transmission line causes an insulator flashover, any earth fault which may be established on the system after the surge has discharged may cause the relay protective gear to operate, disconnecting the line from the supply. To avoid this outage the voltage flashover level of the line can be increased by means of larger insulators but this can only be done within limits because the higher the line insulator flashover voltage, the higher the value of the impulse wave transmitted to the transformer. Some compromise has to be made therefore between the risk of a line flashover and possible damage to equipment and thus it is necessary to know the impulse voltage strength of the transformers to be protected.

Standard impulse test level voltages have been adopted by all British transformer manufacturers and the B.S. 171 levels are given in Table 23.3.

For a 33 kV system for example, Table 23.3 shows that a transformer will withstand a positive or negative $1.2/50 \mu s$ impulse wave of 170 kV peak. Although the line insulation would permit surges of this magnitude to appear on the transformer terminals without causing damage to the transformer, it is not sound engineering practice to subject equipment repeatedly to the full test value. Hence, impulse waves that reach the transformer are usually limited to approximately 80% of the test value, namely to 135 kV peak in the example cited. If line insulators having an impulse flashover voltage of 135 kV peak are installed, the first step in co-ordinating the system insulation has been achieved.

There is still much to be considered before the co-ordination is complete, for the line insulators will have three shortcomings, namely, (i), time lag of flashover, (ii), negative polarity effect, and (iii), possible damage to the insulator itself when flashover occurs. The last two defects can be remedied by fitting the insulator with a rod gap set to flashover at the required value but this gap will also have a time lag effect. Therefore, the rod gap setting must be reduced below the 80 % flashover value to compensate for its own time lag effect or the greater liability to system outage must be accepted. The rod gap can only be considered as a partly effective form of protection and the installation of a surge diverter, although more costly than the simple rod gap, is far more effective.

A typical surge diverter consists of a series of nonlinear resistor discs separated by spark gaps. The nonlinear resistor unit is made from a material having a silicon carbide base in a clay bond. The size of grit used, method of mixing, moisture content, and firing temperature play important parts in determining the characteristics of the final product and all are very carefully controlled. The finished material is an electronic semiconductor, and its voltage/current characteristic, plotted logarithmically, is a curve with a fairly long straight portion. The equation to the straight part is given approximately by $V = kI^\beta$, where V is the voltage across the material, I the resulting current through it, k a scalar constant, and β a power index. The values of k and β are governed by the material and manufacture, so that for a given purpose these must be controlled accordingly; no one value of k or β can be used indiscriminately. For lightning protection, the ideal material will hold the voltage at a chosen maximum for all values of current, i.e., when $\beta = 0$; this cannot be realised in practice, but values of about 0·2 for β can be obtained. The value of the constant k depends not only on the material but also on the shape and size of the finished piece concerned; it is numerically equal to the voltage when unit current flows. A typical voltage/current graph of nonlinear resistor material is in Fig. 23.4, and Fig. 23.5 gives typical voltage/current curves for modern surge diverters.

For a full study of the characteristics and behaviour of the resistor material, reference should be made elsewhere, but two further features may be noted. It has a negative temperature/resistance coefficient and is therefore liable to develop thermal instability if not carefully applied. At constant voltage the

Table 23.3

System highest voltage kV, r.m.s.	Impulse voltage level		Insulation to earth
	1·2/50 full wave test kV, peak	1·2/50 chopped wave test kV, peak*	
7·2	60	These are peak values applied for chopped waves and should at least be equal to that of the specified full wave	uniform
12·5	75		
17·5	95		
24	125		
36	170		
52	250		
72·5	325		
100	380		graded
123	450		
145	550		
170	650		
245	900		
300	1050		

*The chopped wave test is a special test and is applied only if specified by the purchaser.

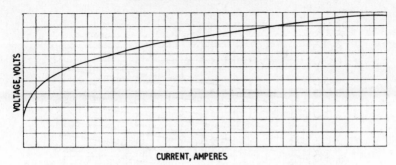

Fig.23.4   *Typical surge diverter resistor voltage/current characteristic*

current increases by about 0·6 % per degree Celsius and to maintain equilibrium the dissipation of heat must keep pace with this. Its specific heat is 0·84 watt seconds per gram per degree Celsius, between 20° and 300°C.

In a surge diverter the spark gap serves the purpose of interrupting the 'system follow' current after a surge has discharged. The multigap design is chosen because it is more sensitive to voltage variation and more effective in arc quenching. The electrodes are symmetrical, giving many of the advantages of a sphere gap, which is independent of polarity effect and quick in response to breakdown voltage. Uniform voltage grading across the gaps is attained either by direct resistance grading or by equalising the capacitance values between the gaps. Mica discs, specially selected and graded, are inserted between each pair of electrodes, and, as they are very thin and have a high dielectric constant, the capacitance is relatively high; hence the stray capacitance to earth is relatively small. This further improves the uniformity of voltage grading across the gaps. The mica used has a small loss angle, almost independent of frequency, and the change in dielectric constant when subjected to impulse voltages is therefore negligible. The thin outer edge of each mica disc points towards the gap and under impulse conditions the concentration of charge around this mica is high, so that the resulting displacement current provides electrons which irradiate the gap, thus alleviating the darkness effect and improving the consistency of flashover.

Under impulse conditions the diverter gaps are required to flash over at the lowest possible voltage and to interrupt the system follow through current at the first current zero after the surge has discharged. These two requirements are somewhat conflicting in principle since both the impulse voltage break-down and the current interrupting characteristics of the gaps increase with the number of gaps in series. Also the more nonlinear discs employed in series the smaller is the system follow current and thus fewer gaps are required to inter-rupt this current. In contrast to this principle there is the fact that the surge voltage across the diverter pile will be higher under surge conditions. Thus, some compromise must be made; if too many discs or gaps are incorporated in the diverter, adequate protection may not be provided whilst too few may result in destruction during operation under surge discharge conditions.

On a system having the neutral point solidly earthed, the power frequency voltage across the diverter will, to the first degree of approximation, not exceed line to neutral voltage. But on further consideration it will be seen that this is not always true, and that a higher voltage can appear across the diverter.

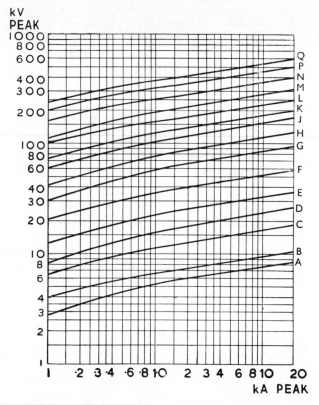

Fig. 23.5   Typical voltage/current curves ($\pm 15\%$) for surge diverters for use with neutral solidly earthed

	System voltage		System voltage		System voltage		System voltage
A	1·1 kV	F	6·6 kV	J	33 kV	N	88 kV
B	2·2 kV	F	11 kV	K	44 kV	P	110 kV
C	3·3 kV	G	16 kV	L	55 kV	Q	132 kV
D	4·4 kV	H	22 kV	M	66 kV		

Consider a three-phase system with solidly earthed neutral and diverters connected to the lines in the manner shown in Fig. 23.6. Suppose an earth fault occurs on the A phase and let us determine the resulting voltage from the B and C phases to earth, *i.e.*, across the B and C phase diverters.

The power frequency voltage of the B phase to earth is, say, V_B and by the use of symmetrical components this voltage is

$$V_B = \frac{\sqrt{3}}{2} V \left[\frac{\sqrt{3}(Z_0 + R) + j(Z_0 + 2Z_1 + 3R)}{Z_0 + Z_1 + Z_2 + 3R} \right]$$

where V is the system line to neutral voltage, Z_1, Z_2 and Z_0 are the positive, negative and zero sequence impedances of the circuit and R is the resistance of the earth fault. In the case of a power transformer, the positive and negative sequence impedances are equal, and the same applies to overhead lines. The resistance and reactance components of the zero sequence impedance, however, vary in proportion to a large extent depending on the earth resistance, etc. It is,

548

therefore, necessary to know the relative values of these before the voltage to earth of the faulty phase can be determined. As an example, suppose for a particular system the values are $Z_1 = Z_2 = 3\,\Omega$, $Z_0 = 10\,\Omega$ and $R = 0\,\Omega$. These values substituted in the above equation give the result $V_B = 1\cdot28$ V. A similar numerical value is obtained for V_C, but with the appropriate phase displacement. This means that the diverters on the B and C phases have a voltage across them 28 % in excess of the line to neutral voltage, although the system neutral is solidly earthed. Again it may be noted that it is not so much the actual values of Z_1, Z_2, etc. as the ratio of these values which determine the line to neutral voltages under fault conditions. The important relationship which determine the resultant voltages across the diverters on a system are R_0/X_1 and X_0/X_1; if these ratios are kept within the limits $R_0/X_1 < 1$ and $X_0/X_1 < 3$ the value of the maximum voltage across any diverter will not increase by more than 40 % of normal. Further, $1\cdot4$ times the line to neutral voltage is 80 % of the line to line voltage. Therefore, surge diverters for solidly earthed neutral systems are always rated at not less than 80 % of the line to line voltage. In the example given, the Z_1 and Z_0 values were assumed to be made up of $R_1 + jX_1 = 1 + j3$ for Z_1, and $R_0 + jX_0 = 3 + j9$ for Z_0 (rounded off to the nearest whole number) from which it will be seen that $R_0/X_1 = 1$ and $X_0/X_1 = 3$ in which R_0 is the zero sequence resistance and X_1 and X_0 are the positive and zero sequence reactances respectively.

In order, therefore, to provide as selective a range of protection as is economically possible, two ratings of surge diverters are made for each standard system voltage, one for solidly earthed neutral systems and one for isolated neutral systems. For any system in which the neutral point is not solidly earthed the latter type of diverter *must* be used as the use of the former type would eventually result in the destruction of the diverter.

A surge protective device should have the following qualities:

(1) rapid response to impulse overvoltages
(2) independence of wave polarity
(3) nonlinear characteristic with the lowest possible β index
(4) high thermal capacity
(5) high system follow current interrupting capacity
(6) consistent characteristics under all conditions.

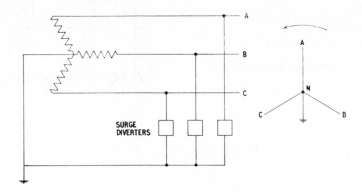

Fig. 23.6 *Surge diverters on system with solidly earthed neutral*

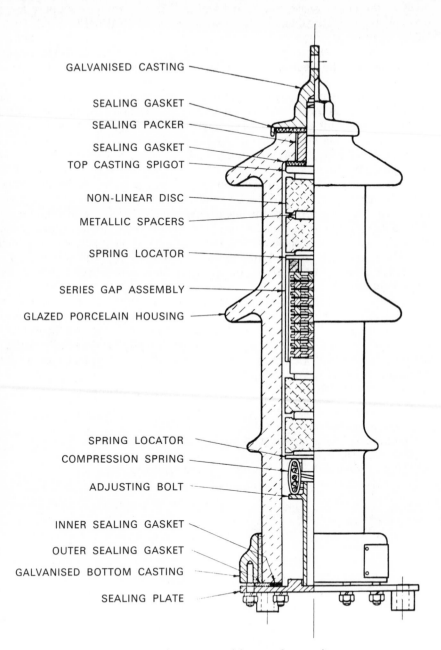

GALVANISED CASTING

SEALING GASKET

SEALING PACKER

SEALING GASKET
TOP CASTING SPIGOT

NON-LINEAR DISC

METALLIC SPACERS

SPRING LOCATOR

SERIES GAP ASSEMBLY

GLAZED PORCELAIN HOUSING

SPRING LOCATOR
COMPRESSION SPRING

ADJUSTING BOLT

INNER SEALING GASKET

OUTER SEALING GASKET
GALVANISED BOTTOM CASTING

SEALING PLATE

Fig. 23.7   Typical cross-sectional drawing of a surge diverter

The relative values of protective devices can be assessed by comparing the values given for these six items.

Figure 23.7 shows a cross-sectional drawing of a typical surge diverter.

Earthing of surge diverters

All provisions for surge protection will fail if detailed consideration is not given to earthing. It has been indicated that the surge voltage from a lightning stroke transmitted to a transformer is controlled by the voltage drop across the non-linear resistance element of the diverter. This is true only if there is no other resistance in series with the diverter, but in practice there exists the resistance of the lead from the diverter terminal to earth, the resistance of the earth plate or rods, and the resistance of the earth itself. Since the diverter surge current must also pass through these additional resistances, their voltage drop must be added to that of the diverter to obtain the total voltage transmitted to the transformer. As the current in the lightning surge may be of the order of thousands of amperes, a resistance of even a few ohms in the earth path can add considerably to the voltage passed on by the diverter.

The ideal arrangement is to run a copper conductor of adequate cross-section from the diverter earth terminal direct into the earth to a depth about 1 m, and thence parallel with the overhead lines and earth wire, terminating on the frame of the transformer. This earth conductor should be connected to a system of earth plates or rods; rods enable the same degree of earthing to be obtained at a less cost than with plates, for which the excavating cost alone is relatively heavy.

Typical diverter connections

Figure 23.8 shows schematically how the various dielectric paths of a single-phase transformer should be shunted by surge diverters in order to protect the transformer bushings and windings against surge voltages. This arrangement, it will be seen, protects the insulation to earth on both sides and the insulation between windings; discharges over any of the diverter gaps are subsequently shunted to earth. The surge stresses across all the dielectric paths are limited by the discharge values of the different diverters irrespective of the magnitudes of the original surges; the surge voltage drop over the earth connecting strip and the buried earth electrodes, which hitherto has played a prominent part in the failure of transformer insulation under surge stresses, now no longer has any effect on the distribution of surge stresses in the transformer. In other words, transformer insulation failures under surge conditions are due to high voltage differences across the parts concerned and with the arrangement shown in Fig. 23.8 these voltage differences are controlled and reduced by the setting of the shunted diverter gaps.

The theoretical scheme shown in Fig. 23.8 may be put into practical effect in several ways. Figure 23.9 shows an arrangement for a single-phase transformer with h.v. and l.v. windings completely insulated from earth. With this arrangement the surge voltage differences which can be established from the windings to the earthed tank, are limited by the surge discharge voltages of the respective

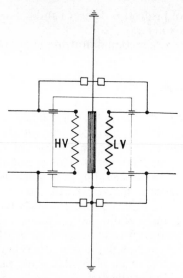

Fig. 23.8   Principle of surge protection of a single-phase transformer with two-wire secondary

diverter gaps, and these also limit the surge voltage which can appear between windings, as h.v. and l.v. diverters act together in series as a shunt to the major insulation between windings.

Figure 23.10 shows an arrangement for the protection of a single-phase transformer where the l.v. neutral conductor is earthed. In this case the h.v. surge diverters are connected to the earthed l.v. neutral conductor, so that surge disturbances become shunted to earth via the l.v. neutral earth connection as well as the diverter earth connection. A co-ordinating gap may be placed in the connection between the earth side of the diverters and the l.v. neutral conductor, this gap discharging should the surge potential of the tank reach a certain predetermined figure. With this connection it is desirable that the l.v. neutral earth circuit should have as low a resistance as possible, although experience seems to show that the connection is quite effective and safe, so far as consumers fed from the l.v. network are concerned.

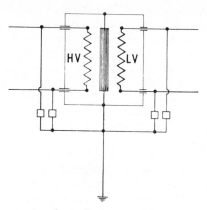

Fig. 23.9   Surge protection of a single-phase transformer with two-wire secondary

Of course, the connection shown in Fig. 23.9 may also be used if there is an earthed secondary neutral conductor.

Figures 23.11 and 23.12 show the application of the foregoing principles to the protection of a three-phase, delta/star transformer employing h.v. and l.v. diverters in the one case, and a direct connection from the earth side of the h.v. diverters to the l.v. earthed neutral in the other. These arrangements are comparable with Figs. 23.9 and 23.10 respectively. The practical results which have been achieved with these connections have shown them to be remarkably successful.

Figure 23.13 shows a typical three-phase transformer installation in a lightning area protected by valve-type surge diverters and protective discharge gap in accordance with the scheme of interconnection given in Fig. 23.12.

(c) Earth faults have different effects according to whether the neutral point of the system is earthed or isolated. In the first case, an earth fault represents a short circuit across one phase, and the same remarks regarding protection apply as outlined for the protection against short-circuit stresses. In the other

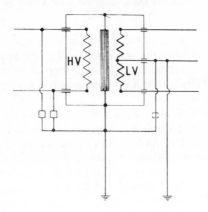

Fig. 23.10   Surge protection of a single-phase transformer with three-wire, earthed neutral secondary

case, where the neutral point is isolated, there are two conditions to consider: firstly, when the earth is a sustained one, and secondly, when it takes the form of the so called arcing ground. In the first of these two cases the voltage of the two sound lines is raised to full line voltage above earth, and after the initial shock the insulation stresses become steady, although increased by $\sqrt{3}$ above those occurring under normal service conditions. The protection for this condition is obviously earthing of the neutral point, as explained elsewhere. In the second case, where the earth fault is unstable, such as at the breakdown of an overhead line insulator, high-frequency waves are propagated along the line in both directions, and to protect the transformer against the effects of these waves some form of surge diverter gear may be installed in front of the transformer, as outlined earlier in this chapter. Alternatively, the neutral point may be earthed, thereby converting the earth fault into a short circuit across one phase. As in modern power systems it is now standard practice to earth the neutral point, the question of protection against arcing grounds or earth faults, as such, hardly needs considering.

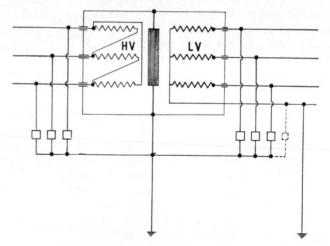

Fig. 23.11   Surge protection of a three-phase, delta/star transformer with four-wire, earthed neutral secondary

(2) Considering next the means to be adopted for protecting the system against the effects of faults arising in the transformer, the principal faults which occur are (a) breakdowns to earth either of the windings or terminal gear, and faults between phases generally on the h.v. side; and (b) short circuits between turns, chiefly of the h.v. windings.

(a) Breakdowns to earth may occur due to failure of the major insulation of transformers or of bushing insulators, these failures being due to the absence of any external surge protective apparatus or upon the failure of such apparatus to operate. When such a breakdown occurs it is essential that the transformer should be isolated from the supply with as little delay as possible, while further, it is important that the fault should not be the cause of the tripping out of circuit of any healthy apparatus.

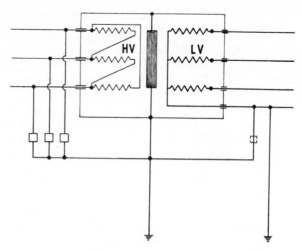

Fig. 23.12   Surge protection of a three-phase, delta/star transformer with four-wire, earthed neutral secondary

Fig. 23.13   Three-phase, 35 kVA, 11 000/400 V, 50 Hz transformer protected by surge diverters

For small transformers, plain overload and earth leakage devices will provide the necessary degree of protection to ensure the transformer being disconnected automatically from the circuit.

On larger transformers forming parts of important transmission or distribution networks, it becomes necessary to employ some form of automatic discriminative protective equipment as these remove from the circuit only faulty apparatus leaving sound apparatus intact, while the disconnection is performed in the shortest space of time and the resulting disturbance to the system is reduced to a minimum. Of the various automatic protective gear systems available, the most generally used are described here.

Comprehensive details of various forms of protective systems for generators, generator/transformer combinations, transformers, feeders and busbars are given in the J & P Switchgear Book, 7th edn.

In considering the problems of protection across a power transformer, note may first be taken of what is known as a differential rough balance scheme, as shown in Fig. 23.14. This scheme is for application where existing overcurrent and restricted earth-fault protection has become inadequate but provision of a separate differential scheme is considered unjustified. By using the overcurrent relays and current transformers, lower fault settings and faster operating times can be obtained for internal faults, with the necessary discrimination under external fault conditions.

The taps on interposing transformers are adjusted so that an inherent out-of-balance between the secondary currents of the two sets of current transformers will exist but will be insufficient to operate the overcurrent relays under normal load conditions. With an overcurrent or external fault, the out-of-balance current increases to operate the relay and choice of time and current settings of the IDMT relays will be made to permit grading with the rest of the system.

The scheme functions as a normal differential system for internal faults but is not as fast or sensitive as the more conventional schemes. It is, however, an inexpensive method of providing differential protection where IDMT relays already exist.

CIRCULATING CURRENT PROTECTION

In Fig. 23.15 is an explanatory diagram showing the principle of the circulating current system. It will be noted that current transformers (which have similar characteristics and ratio) are connected on both sides of the machine winding and a relay is connected across the pilot wires between the two current transformers. Under healthy or through fault conditions, the current distribution is as shown at (a), no current flowing in the relay winding. Should a fault occur as shown at (b), the conditions of balance are upset and current flows in the relay winding to cause operation. It will be noted that at (b) the fault is shown at a point between the two current transformers (the location of these determine the extent of the protected zone). If the fault had occurred beyond, say, the right-hand current transformer, then operation would *not* occur as the fault current would then flow through *both* current transformers thus maintaining the balance, as shown at (a). In order that the symmetry of the burden on the current transformers shall not be upset and thus cause an out-of-balance current to pass through the relay, causing operation when not intended, it is essential that the relay be connected to the pilot wires at points of equipotential. This is illustrated at (c) in Fig. 23.15, such equipotential points being those as a and b, a^1 and b^1 etc. In practice it is rarely possible to connect the relay to the actual physical mid-point in the run of the pilots and it is usual to make the connection to convenient points at the switchgear and to insert balancing resistances in the shorter length of pilot wire. The resistances should be adjustable so that accurate balance can be obtained when testing before commissioning the plant.

When circulating current protection is applied to a power transformer, some complications arise because a phase shift may be introduced which can vary with different primary/secondary connections and there will be a magnitude difference between the load current entering the primary and that leaving the secondary.

Correction for a phase shift is made by connecting the current transformers on one side of the power transformer in such a way that the resultant currents fed into the pilot cables are displaced in phase from the individual phase currents by an angle equal to the phase shift between the primary and secondary currents of the power transformer. This phase displacement of the current transformer secondary currents must also be in the same direction as that between the primary and secondary main currents.

The most familiar form of power transformer connection is that of delta/star, the phase shift between the primary and secondary sides being 30 degrees. This

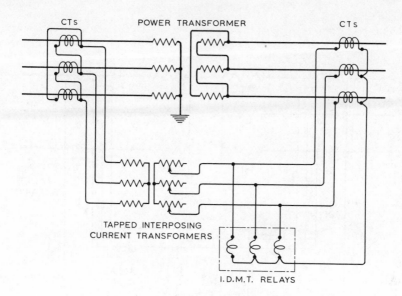

Fig. 23.14   Differential rough balance protection scheme
(GEC Measurements Ltd.)

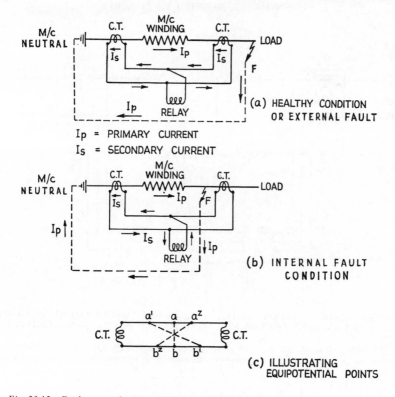

Fig. 23.15   Explanatory diagram to illustrate principle of circulating current protection

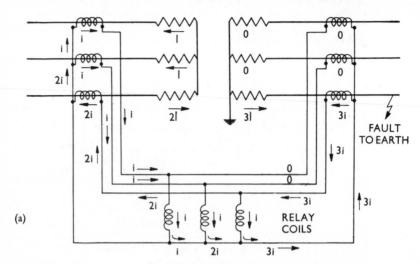

SHOWING OPERATION WHEN PROTECTIVE C.T's ARE CONNECTED IN STAR AND AN EARTH-FAULT OCCURS EXTERNAL TO THE POWER TRANSFORMER, THE RATIO OF WHICH IS ASSUMED TO BE UNITY.

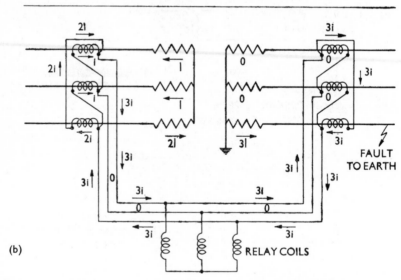

(b)

SHOWING STABILITY WHEN PROTECTIVE C.T's ARE CONNECTED IN DELTA AND AN EARTH-FAULT OCCURS EXTERNAL TO THE POWER TRANSFORMER.

Fig. 23.16   *Showing stable and unstable conditions on through earth faults, with circulating current protection applied to a star/star transformer, due to methods of connecting current transformers*

is compensated by connecting the current transformers associated with the delta winding in star and those associated with the star winding in delta.

In order that the secondary currents from the two groups of current transformers may have the same magnitude, the secondary ratings must differ, those of the star connected current transformers being 5 A and those of the delta connected group being 2·89 A, i.e., $5/\sqrt{3}$.

If the power transformer is connected delta/delta, there is no phase shift between primary and secondary line currents. Similarly, there is no phase shift in the case of star/star connected power transformers but phase correction is applied at *both* sets of current transformers, the reason being that only by this means can the protective system be stable under external earth-fault conditions. Thus, both sets of current transformers will be delta connected so that the secondary currents in the pilots from each set will be displaced in phase by 30 degrees from the line currents but both will coincide, a necessary requirement of circulating current protection. It is obvious that similarity in phase could be achieved if both sets of current transformers are connected in star, but it can be shown in this case, the protective system would be stable on through-faults between phases but *not* for earth-faults. This is demonstrated numerically in Fig. 23.16, noting that at (a) the secondary currents entering and leaving the pilots are not the same at both ends and therefore do not sum up to zero at the relays, whereas at (b) the reverse is true and no current appears in the relay coils. The 2:1:1 current distribution shown in Fig. 23.16 on the unearthed side of the transformer pertains only to such a transformer with a closed-delta tertiary winding. This winding is not shown in the diagram. Its function is to provide a short-circuit path for the flow of harmonic components in the magnetising current. The distribution applies also when the core is a three-phase type as opposed to shell-type.

As is well known, the switching in of a power transformer causes a transient surge of magnetising current to flow in the primary winding, a current which has no balancing counterpart in the secondary circuit. Because of this a 'spill' current will appear in the relay windings for the duration of the surge and will, if of sufficient magnitude, lead to isolation of the circuit. This unwanted operation can be avoided by adding time delay to the protection but, as the inrush current persists for some cycles, such delay may render protection ineffective under true fault conditions. A better solution may lie in the use of harmonic restraint, and relays of this type are shown in Figs. 21.20–21.22.

Figure 23.17 is a demonstration diagram of connections of a three-phase, delta/star connected transformer equipped with circulating current protection and shows the distribution of the short-circuit fault currents arising from a winding fault to earth on the star connected winding, when the neutral point of the latter is solidly earthed. The current phasor diagrams drawn for a one-to-one ratio, corresponding to the conditions of Fig. 23.17, are given in Fig. 23.18 in which the phasors have the following significance.

Figure 23.18 (a)

I_A, I_B, I_C are the normal balanced load currents in the primary delta connected power transformer windings.

I_R, I_Y, I_B are the normal balanced load currents in the primary main lines.

I_{AF} is the short-circuit fault current in the power transformer primary winding A_2A_1 and in the line R corresponding to the fault current I_{af}, set up in the short-circuited portion of the power transformer secondary winding a_2a_1. Its magnitude is such that the ampere-turns given by I_{AF} multiplied by the total number of turns in the primary winding A_2A_1 equal the ampere-turns given by the fault current I_{af} in the short-circuited portion of the secondary winding a_2a_1 multiplied by the number of secondary turns short circuited. The phase angle ϕ_p of I_{AF} with respect to the normal voltage across A_2A_1 is given by the expression $\cos^{-1}(R_p/Z_p)$ where R_p is the resistance of the primary winding A_2A_1 plus the resistance of the short-circuited portion of the secondary winding a_2a_1 and Z_p is the impedance of the short-circuited portion of the secondary windings a_2a_1 with respect to the whole of the primary windings A_2A_1, all quantities being

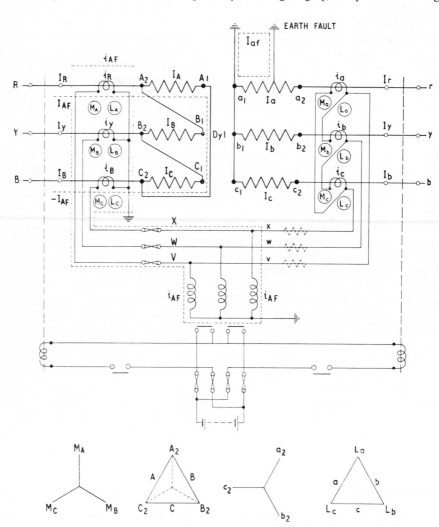

Fig. 23.17 *Circulating current protection for a three-phase delta/star connected transformer, showing operation under internal earth fault conditions*

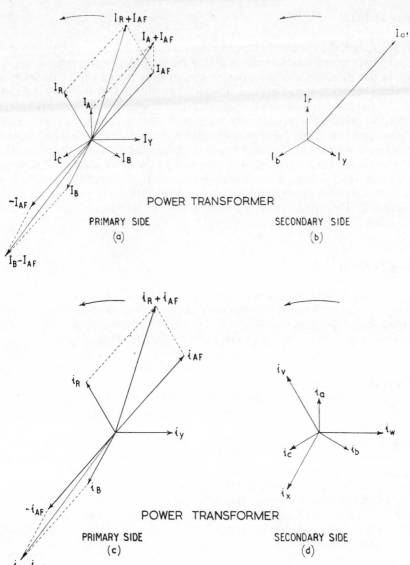

Fig. 23.18 Current phasor diagrams corresponding to the conditions of Fig. 23.17

referred to the primary side. $-I_{AF}$ is the short-circuit fault current in the line B, and is I_{AF} in the line R, but flowing in the reverse direction to I_{AF} with respect to the line R.

$I_A + I_{AF}$ is the total current in the winding A_2A_1, *i.e.*, the phasor sum of the load current and the fault current in the winding.

$I_R + I_{AF}$ is the total current in the line R, *i.e.*, the phasor sum of the load current and the fault current in the line.

$I_B - I_{AF}$ is the total current in the main line B, *i.e.*, the phasor sum of the load current and the fault current in the line.

561

Figure 23.18 (b)

I_r, I_y, I_b, are the normal balanced load currents in the secondary star connected power transformer windings and in the secondary main lines.

I_{af} is the short-circuit fault current in that part of the power transformer secondary winding a_2a_1 between the earthed neutral and the winding earth fault. Its magnitude and phase angle ϕ_s, with respect to the normal voltage across the winding a_2a_1 are determined by the impedance of the short-circuited portion of the secondary winding a_2a_1 with respect to the whole of the primary winding A_2A_1, and by the resistance R_{af} of the short-circuited portion of a_2a_1. The magnitude of I_{af} is given by the expression V_{af}/Z_{af}, where V_{af} is the normal voltage across the short-circuited portion of the winding a_2a_1 and Z_{af} is the impedance referred to earlier in terms of the secondary side of the transformer. The phase angle ϕ_s, with respect to the normal voltage across a_2a_1, is $\cos^{-1}(R_{af}/Z_{af})$.

Figure 23.18 (c)

i_R, i_Y, i_B are the normal balanced currents in the star connected secondary windings of the current transformers and in the lines connected thereto on the primary side of the power transformer. They are the currents due to the normal balanced load currents in the primary power lines R, Y, B.

i_{AF} is the fault current in the current transformer secondary winding M_AL_A and in the line V connected to it, and corresponds to the current I_{AF} in the primary power line R.

$-i_{AF}$ is the fault current transformer secondary winding M_CL_C and in the line X connected to it, and corresponds to the current $-I_{AF}$ in the primary power line B.

$i_R + i_{AF}$ is the total current in the current transformer secondary winding M_AL_A and in the line V connected to it, i.e., the phasor sum of the currents due to the load current and the fault current in the primary power line R.

$i_B - i_{AF}$ is the load current in the current transformer secondary winding M_CL_C and in the line X connected to it, i.e., the phasor sum of the current due to the load current and the fault current in the primary power line B.

The relative angular displacements between the currents of Figs. 23.18(c) are the same as those of Fig. 23.18(a).

Figure 23.18 (d)

i_a, i_b, i_c, are the normal balanced currents in the delta connected secondary windings of the current transformers on the secondary side of the power transformer. They are the currents due to the normal balanced load currents in the secondary power lines r, y, b.

i_v, i_w, i_x are the normal balanced currents in the lines to the delta connected secondary windings of the current transformers on the secondary side of the power transformer. They are the line currents corresponding to the currents in the current transformer secondary windings which are due to the normal balanced load currents in the secondary power lines r, y, b.

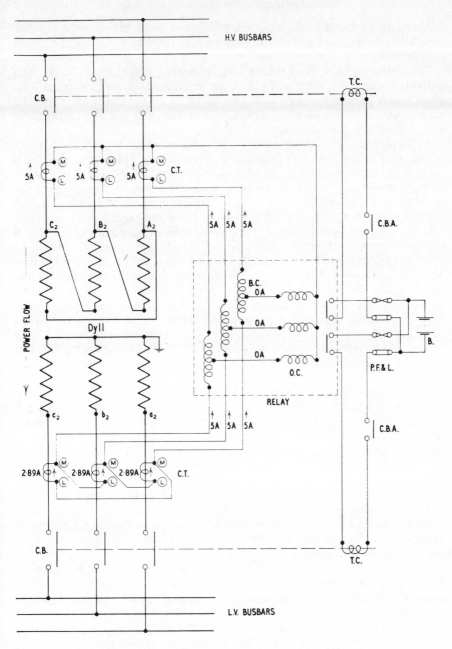

Fig. 23.19   Biased differential protection applied to a delta/star connected three-phase transformer

C.B.   Circuit breaker
C.T.   Protective current transformer
T.C.   Trip coil
C.B.A.   Circuit breaker auxiliary switch
B.C.   Bias coil
O.C.   Operating coil
P.F. & L.   Protective fuse and link
B.   Battery

This diagram bears no fault current phasors, showing that no fault currents flow through the current transformers on the secondary side of the power transformers.

The currents which flow through the protective relays are thus the fault currents i_{AF} and $-i_{AF}$ of Fig. 23.18(c), the magnitudes of which depend, for a given power transformer, upon the amount of the power transformer winding short circuited and its position with respect to the whole winding on the other side of the power transformer.

So far no mention has been made of the problem which arises when a power transformer is provided with facilities for tap changing. It has been noted that for stability under healthy or through-fault conditions, identical outputs from each group of current transformers are an essential feature of circulating current protection. It is clearly impossible for the current transformers to be matched at all tap positions unless these (the c.t.s) are also correspondingly tapped. This solution is generally impracticable if only because of the nature of the task of changing current transformer tappings each time a tap change is made on the power transformer. The latter function is often automatic so that it would then be necessary to make the tap changes on the current transformers automatic and simultaneous. Because of this and the normal inequalities which occur as between current transformers, many schemes for the protection of transformers have been devised in which steps have been taken to eliminate the difficulties and some of these schemes will be noted later. Here we can note that tap changing and current transformer inequalities can be largely avoided by using a circulating current scheme which employs a biased differential relay, indicated typically in Fig. 23.19.

In each pole of this relay, there are, in addition to the operating coil, two bias or restraining windings. Under through-fault conditions, when operation is *not* required, no current should flow through the operating coil but, because of imperfect matching of the current transformers, and the effects due to tap changing, some spill current may flow in the operating coil. This, however, will not cause operation unless the ratio of operating to bias current for which the relay is set is exceeded and the restraint or bias which is applied automatically increases as the through-fault current increases, thus enabling sensitive settings to be obtained with a high degree of stability.

To understand the operation of the bias coils, consider first the protective system under through-fault conditions (*i.e.*, a fault outside the protected zone), and then under internal fault conditions.

(a) Through-fault conditions. If a three-phase short circuit occurred on the feeder side of the system beyond the circuit-breaker the current circulating in the pilot wires would pass through the whole of the relay bias coils, and any out-of-balance current which might occur due to discrepancies in the ratios of the protective current transformers, would flow through the relay operating coil. Under these conditions the biasing torque preponderates, so preventing relay operation.

(b) Internal fault conditions. Imagine now a three-phase fault at the power transformer terminals on the star connected side and that the power flow is as shown in Fig. 23.19. Fault current flows through the three current transformers designated A on the delta connected side of the power transformer but not through the set B on the star side. Therefore, the current transformer secondary currents circulate via the pilot wires, through *one half* of the bias coils and the operating coils back to the current transformer neutral connection. Under these conditions the relay operating torque preponderates. The protective system operates correctly when the transformer is fed from either or both directions and for all types of faults.

HIGH SPEED PROTECTION OF POWER TRANSFORMERS BY BIASED DIFFERENTIAL HARMONIC RESTRAINT

The 'English Electric' Type DMH relay provides differential protection for two-winding or three-winding power transformers with a high degree of stability against through faults and is immune to the heavy magnetising current inrush that flows when a transformer is first energised. The relay is available in two forms:

(a) for use with line current transformers with ratios matched to the load current to give zero differential current under healthy conditions
(b) with tapped interposing transformers for use with standard line current transformers of any ratio.

In this relay the preponderance of second harmonic appearing in the inrush current is detected and is used to restrain its action, thus discriminating between a fault and the normal magnetising current inrush. The relay employs rectifier bridge comparators in each phase which feed their outputs through transistor amplifiers to sensitive polarised relays, resulting in:

(i) an operating current which is a function of the differential current
(ii) a restraining current, the value of which depends on the second harmonic of the differential current
(iii) a bias current which is a function of the through current and stabilises the relay against heavy through faults.

The relay is provided with an instantaneous overcurrent unit in each phase to protect against faults heavy enough to saturate the line-current transformers, under which conditions the harmonics generated would tend to restrain the main unit. These overcurrent units have a fixed setting of eight times the current-transformer secondary rating and are fed from saturable current transformers to prevent operation on peak inrush current which may momentarily exceed this value.

The operation of the main unit is briefly as follows:

Under through current conditions, current is passed by the two restraint rectifier bridges through the polarised relay in the non-operating direction. In conditions of internal fault there will be a difference between primary and secondary current, and the difference flows in the operating circuit so that the operating rectifier passes a current to the polarised relay in the operative direction. Operation depends on the relative magnitude of the total restraint and differential currents, and the ratio of these currents to cause operation is controlled by a shunt resistor across the restraint rectifiers. Under magnetising inrush conditions, the second-harmonic component is extracted by the tuned circuit and the current is passed to the relay in the non-operating condition.

In addition to the second-harmonic component, the inrush current contains a third-harmonic component, its proportion being large but less than the second.

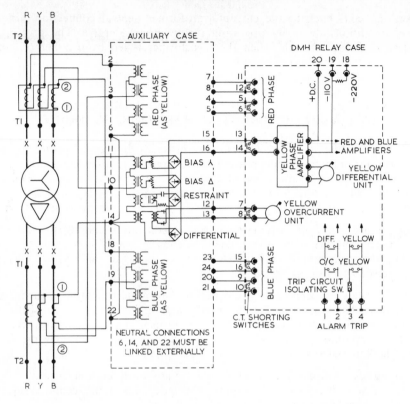

Fig. 23.20   Typical application of 'English Electric' biased differential harmonic restraint relay, Type DMH, for a three-phase two-winding transformer (GEC Measurements Ltd.)

No restraint against the third harmonic is provided as there would be danger that the relay might be delayed in operating under heavy internal fault conditions, due to the current transformer saturation producing third harmonics in the secondary waveform.

Figure 23.20 and 23.21 show typical application diagrams for three-phase two-winding, and three-phase three-winding transformers.

DUO-BIAS DIFFERENTIAL TRANSFORMER PROTECTION

Another development, basically of the conventional current-balance scheme already discussed but using a special relay compensated to override the complications associated with transformer protection is that by A. Reyrolle & Co. Ltd. This is shown in Fig. 23.22. It is a diagram of their 'Duo-Bias' relay scheme applied to a single phase, and functioning under various conditions as follows.

Under load or through-fault conditions, the current transformer secondary currents circulate through the primary winding of the bias transformer, the rectified output of which is applied to a bias winding on a transductor via a shunt resistor. Out-of-balance current flows from the centre tap on the primary

566

winding of the bias transformer, energising the transductor input winding and the harmonic-bias unit.

The input and output windings of the transductor are inductively linked but there is no inductive linking between these and the bias-windings. So long as the transformer being protected is sound the transductor bias-winding is energised by full-wave rectified current which is proportional to the load or through-fault current, and this bias current saturates the transductor. Out-of-balance currents in the transductor input winding, produced by power transformer tap changing or current transformer mismatch, superimpose an alternating m.m.f. on the d.c. bias m.m.f., as shown in Fig. 23.23, but the resulting change in working flux density is small and the output to the relay negligible.

The tappings on the shunt resistor are used for adjusting the relationship between the bias transformer primary current and the input to the transductor bias winding. This resistor also serves to suppress the ripple in the bias m.m.f. due to ripple in the bias current, because it provides a low-impedance non-inductive shunt path across the highly inductive bias winding for the a.c. content of the bias current.

If the power transformer develops a fault, the operating m.m.f. produced by the secondary fault current in the transductor input winding exceeds the bias m.m.f. resulting in a large change in working flux density which produces a correspondingly large voltage across the relay winding, and the resultant current operates the relay. Operation of the relay cannot occur unless the operating m.m.f. exceeds the bias m.m.f., and as the m.m.f. is proportional to the load or through-fault current, the required operating m.m.f. (and hence the operating current) is also proportional to the load or through-fault current.

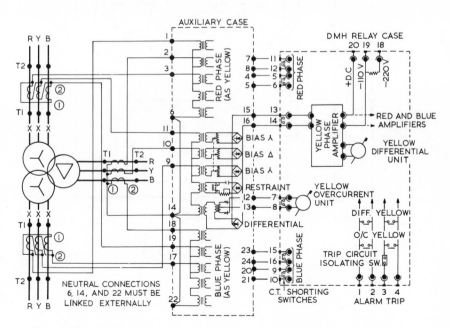

Fig. 23.21   Typical application of 'English Electric' Type DMH biased differential harmonic restraint relay for a three-phase, three-winding transformer (GEC Measurements Ltd)

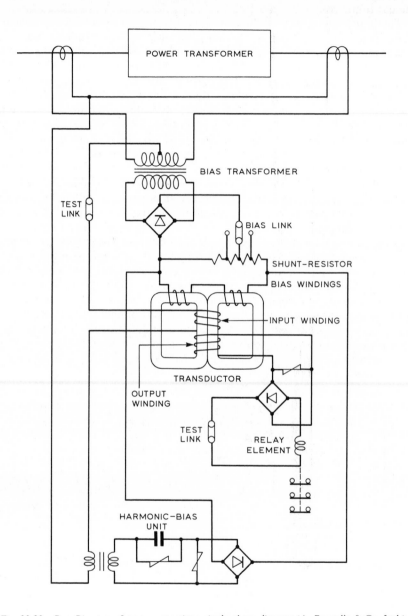

Fig. 23.22   Duo-Bias transformer protection: single-phase diagram (A. Reyrolle & Co. Ltd.)

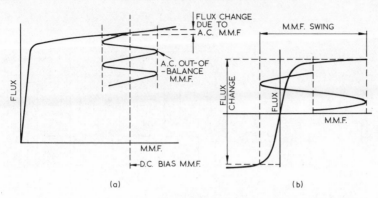

Fig. 23.23   Fluxes due to operating and biasing ampere-turns. Duo-Bias protection (A. Reyrolle & Co. Ltd.)

The harmonic bias unit shown in Fig. 23.22, is a simple tuned circuit which responds to the second-harmonic component of the magnetising current. When magnetising inrush current flows through the relay operating circuit the rectified output of the harmonic bias unit is injected into the transductor bias winding and restrains the relay.

Transformer differential relays generally have a basic setting which is the fault current required to operate them with no through current in the differential system and internal fault current fed from only one set of current transformers. In the case of the Duo-Bias relay, this is 20% of the relay rating. The actual value of the fault current at which the differential relay will operate is thus the basic setting value under no-load conditions but when load current is flowing the setting will be higher, depending upon the amount of load and the bias setting in use. With an internal earth fault in which the current is limited by a neutral-earthing resistor, the load current might well be little affected by the fault and, therefore, when considering such a condition, the effect of load current on the setting should be taken into account.

Figure 23.24 shows a diagram for a three-phase assembly of Duo-Bias relays applied to the protection of a two-winding transformer. When applied to a three-winding transformer, the relay is identical except for a change of tapping on the primary winding of the bias transformer. Further details of this type of protection are given in the J & P Switchgear Book, 7th edn.

Opposed voltage protection

The essential difference between this and the circulating current scheme is that under normal conditions no current circulates in the pilot wires, the e.m.f.s generated at either end of the pilots being balanced against each other. This is basically the well known 'opposed-voltage' scheme, a typical arrangement of which is shown in Fig. 23.25. This particular scheme is known as 'Translay' and was developed by Metropolitan-Vickers Electrical Co. Ltd.* The two diagrams illustrate the operation of the protection for through-fault conditions, and for

---

*Now part of GEC Measurements Ltd.

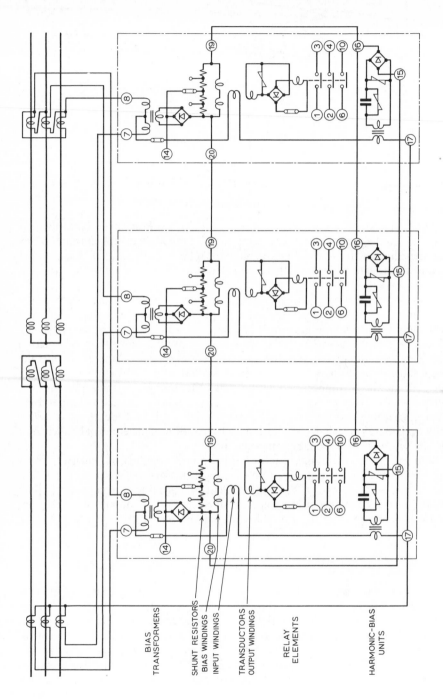

BIAS
TRANSFORMERS

SHUNT RESISTORS
BIAS WINDINGS
INPUT WINDINGS

TRANSDUCTORS
OUTPUT WINDINGS

RELAY
ELEMENTS

HARMONIC-BIAS
UNITS

Fig. 23.24 *Duo-Bias protection for a two-winding transformer* (A. Reyrolle & Co. Ltd.)

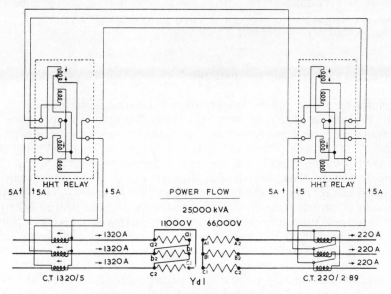

DISTRIBUTION OF RELAY CURRENTS WITH
3 PHASE THROUGH LOAD.

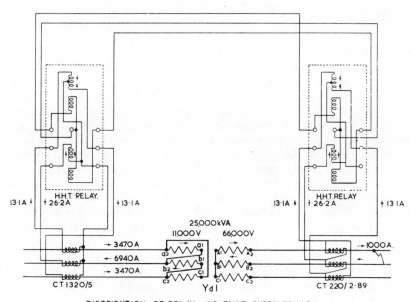

DISTRIBUTION OF RELAY AND FAULT CURRENTS WITH
SINGLE PHASE THROUGH SHORT CIRCUIT

Fig. 23.25   Translay protection applied to a transformer feeder (GEC Measurements Ltd.)

internal fault conditions. This scheme is also more fully described in the J & P Switchgear Book, which refers particularly to feeder protection, but in general applies as well to transformers.

Overcurrent and earth leakage protection

As indicated earlier, for the smaller sizes of power transformers up to, say, 1000 kVA (and in some cases larger than this), it is not always economical to fit circulating current protection, and adequate protection can be provided by means of simple overcurrent and earth fault relays, the latter preferably of the restricted form on the l.v. side. A typical diagram is shown in Fig. 23.26 where it will be seen that the h.v. side comprises three overcurrent and one earth leakage relay, whilst the l.v. arrangement is similar with the exception of a neutral current transformer if the power transformer neutral is earthed. With this type of protection no balancing of current transformers on the primary and secondary sides of the power transformer is necessary, and hence similar characteristics and definite ratios are unnecessary. Further, the earth leakage relays are instantaneous in operation, and earth fault settings as low as 20% can usually be obtained without difficulty. Line to line faults are dealt with by the overcurrent relays, which operate with a time lag and are graded with the overcurrent relays on other parts of the system. For unearthed windings (delta or star) the apparatus would consist of a three-pole overcurrent relay of the inverse, definite minimum, time lag type and a single pole instantaneous earth leakage relay with or without series resistor depending on the type of relay. This is shown at the left-hand side of Fig. 23.26 and by the full lines at the right-hand side; this is the overcurrent and plain earth leakage system of protection.

If the power transformer neutral point is earthed, as shown dotted at the right-hand side of Fig. 23.26, an additional current transformer is provided in the neutral connection with its secondary winding in parallel with the three line current transformers, and the protection is then that known as the overcurrent and restricted earth leakage system. With an external earth fault (say to the right of the current transformers on the star connected side of the power transformer), current flows in one of the line current transformers and in the neutral current transformer and the polarities are so arranged that current circulates between the two secondaries. The earth leakage relay is thus connected across equipotential points; no current flows in it, and it does not operate. With an internal earth fault, fault current flows either in the neutral current transformer only, or in opposition in the line and neutral current transformers; the relay is then energised and operates. To balance the line and neutral current transformers with external earth faults, a dummy balancing impedance equal to the impedance of one of the overcurrent elements is connected in series with the neutral current transformer as shown in Fig. 23.26 so that the burdens of the line and neutral current transformers are equalised. Figure 23.27 shows in diagrammatic form the current distribution for restricted earth fault protection for faults inside and external to the protected zone.

Dealing next with the question of protection against interturn faults within the transformer, it has already been stated that such faults are more likely to occur in the h.v. windings and therefore it is only necessary as a rule, to install protective gear on the h.v. side. When, however, the l.v. side of the transformer

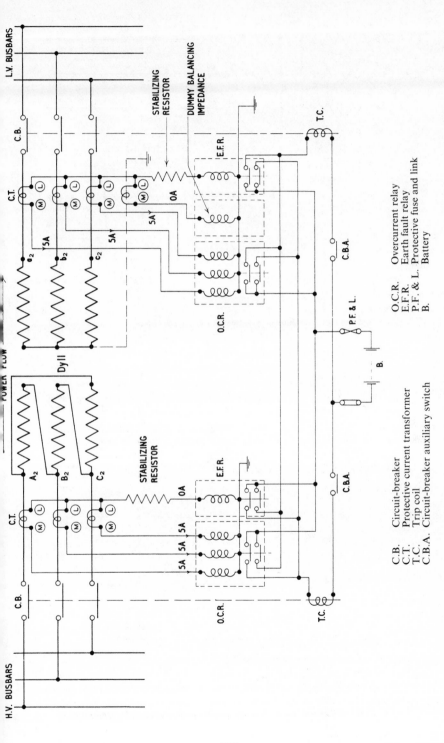

Fig. 23.26 *Overcurrent and unrestricted earth fault protection of a three-phase delta/star connected transformer*

C.B.   Circuit-breaker
C.T.   Protective current transformer
T.C.   Trip coil
C.B.A.   Circuit-breaker auxiliary switch

O.C.R.   Overcurrent relay
E.F.R.   Earth fault relay
P.F. & L.   Protective fuse and link
B.   Battery

573

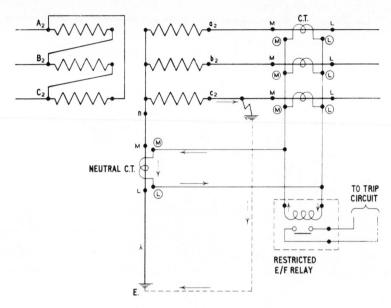

SHOWING OPERATION ON FAULT INSIDE ZONE

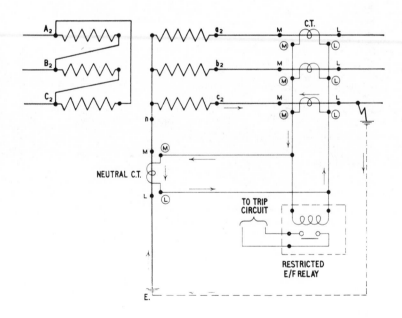

Fig. 23.27 Diagrams showing restricted earth fault protection for transformers

574

is designed for a voltage which is higher than normal, the degree of susceptibility of the windings to interturn insulation failure is comparable to that of h.v. windings, bearing in mind, of course, the influence of the type of circuit, *i.e.,* overhead lines, underground cables, or merely short connecting leads, to which the windings are connected.

The gas and oil actuated relay

The gas and oil actuated (Buchholz) relay has been used extensively in Great Britain for disconnecting a transformer from the supply upon the occurrence of an interturn fault or any other minor failure which generates gases in sufficient quantities to operate the device and to actuate the controlling circuit-breaker.

The modern transformer is a very reliable piece of electrical equipment and however infrequent breakdowns may be, they must be guarded against and all possible steps taken to maintain continuity of supply. Any means of indicating the development of a fault within the transformer, particularly in the incipient stages, may avoid major breakdowns and sudden failure of the power supply.

The gas operated relay is designed for this particular duty and depends for its operation on the fact that most internal faults within the transformer generate gases. The service record over many years shows clearly that the relay is extremely sensitive in operation and that it is possible to detect faults in their incipient stages, thus minimising damage and saving valuable time in effecting the necessary repairs. The gas operated relay can only be fitted to transformers having conservator vessels, and is installed in the pipe line between the trans-former and its conservator tank. The relay comprises an oil-tight container fitted with two internal floats which operate mercury switches connected to external alarm and tripping circuits. Normally, the device is full of oil and the floats, due to their buoyancy, rotate on their supports until they engage their respective stops. An incipient fault within the transformer generates small bubbles of gas which, in passing upwards towards the conservator, become trapped in the housing of the relay, thereby causing the oil level to fall. The upper float rotates as the oil level within the relay falls, and when sufficient oil has been displaced the mercury switch contacts close, thus completing the external alarm circuit.

In the event of a serious fault within the transformer, the gas generation is more violent and the oil displaced by the gas bubbles flows through the con-necting pipe to the conservator. This abnormal flow of oil causes the lower float to be deflected, thus actuating the contacts of the second mercury switch and completing the tripping circuit of the transformer circuit-breaker, so disconnect-ing the transformer from the supply.

Gas within the device can be collected from a small valve at the top of the relay for analysis and from the results obtained an approximate diagnosis of the trouble may be formed. Some of the faults against which the relay will give protection are:

(1) core-bolt insulation failure
(2) short-circuited core laminations
(3) bad electrical contacts
(4) local overheating

(5) loss of oil due to leakage

(6) ingress of air into the oil system.

These would normally initiate an audible or visible alarm via the upper float, whilst the following more serious faults would trip the transformer from the supply:

(a) short-circuit between phases

(b) winding earth fault

(c) winding short circuit

(d) puncture of bushings.

Typical values of the oil velocity required to operate the lower float under all surge conditions and the volume of gas required to operate the upper alarm float are given in Table 23.4.

Table 23.4

Internal diameter relay pipe, mm	Oil velocity, cm/s	gas volume, cm³
25	100	100
50	110	200
75	125	250

A view of a dismantled double float relay is shown in Fig. 23.28 and the recommended arrangement for mounting the relay is shown in Fig. 23.29. It is essential when designing the transformer tank that all gas rising from the

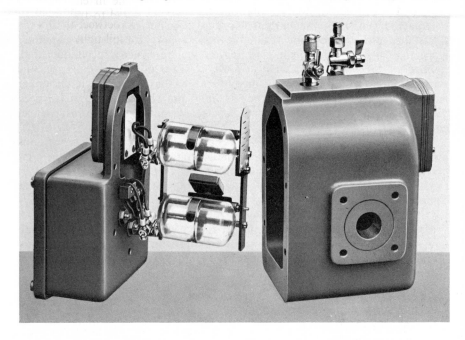

Fig. 23.28   Gas and oil actuated relay, dismantled to show position of floats

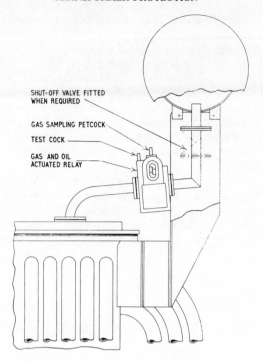

SHUT-OFF VALVE FITTED
WHEN REQUIRED

GAS SAMPLING PETCOCK

TEST COCK

GAS AND OIL
ACTUATED RELAY

Fig. 23.29   Arrangement for mounting the gas and oil actuated relay

transformer shall pass into the relay pipe and not collect in stray pockets, for otherwise an accumulation of gas would delay the operation of the alarm float. For testing purposes, a test valve is provided on the relay for connection to a source of air supply. A suitable testing equipment comprises a small air vessel with a pressure gauge and a suitable length of rubber tubing. The air chamber is filled to a pressure of approximately $4.2 \, \text{kg/cm}^2$. Slow release of the air to the relay operates the upper float whilst quick release causes the tripping float to operate.

Interturn failures

All types of coils are liable to interturn insulation failure, and the order of susceptibility may be given as crossover, continuous-disc and spiral coils. A purely interturn fault is distinguished by localised burning of the conductors of the coil affected, and often by extensive charring of the interturn insulation of the coil; distortion of the conductors is not a feature of a true interturn insulation fault. Severe coil distortion is direct and positive evidence of an external short circuit across the whole or a major portion of the winding.

It is well known among operating engineers that an initial interturn insulation failure does not draw sufficient current from the line to operate an ordinary overload circuit-breaker or even more sensitive balanced protective gear. The transformer will, in fact, only be disconnected from the line automatically when the fault has extended to such a degree as to embrace a considerable portion of the affected winding. This may take one of the forms shown in Fig. 23.30 in which the fault is confined strictly to the winding at (*a*), while at (*b*) it burns through to earth in the incipient stage of the failure.

If the fault occurs on the primary winding the short-circuited turns act as an auto transformer load on the winding, and the reactance is that between the short-circuited turns and the whole of the affected phase winding. If the fault takes place on the secondary winding the short-circuited turns act as an ordinary double winding load, and the reactance is that between the short-circuited turns and the whole of the corresponding primary phase winding.

The following example gives an idea of the relative order of magnitudes of the quantities involved.

Tests were carried out on a typical step-down 250 kVA, 50 Hz, three-phase, core-type transformer. The design data were as follows: h.v. phase voltage, 2800 V; l.v. phase voltage, 237 V; volts per turn, 7·38; turns per h.v. phase winding, 380; turns per l.v. phase winding, 32; normal impedance, 3·25%; normal reactance, 3·08%; axial length of each h.v. and l.v. phase winding, 16·4 in. The

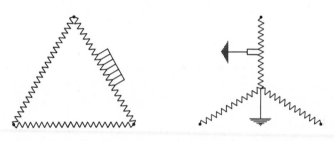

Fig. 23.30   Interturn insulation failures

(*a*) Winding insulated from earth   (*b*) Winding neutral point earthed

h.v. winding on each phase consisted of a total of 380 turns and tapping points were obtained at 28 intervals of 16 turns and two intervals of 12 turns. Both ends of each tapping point were brought out for testing so that they could be short-circuited. Impedance tests were made, first short circuiting one tapping section only at a time, starting at the top and working down the core limb, taking each consecutive interval in turn, and subsequently, short circuiting different series and parallel groups of tappings up to eight in number, at various positions throughout the entire length of the limb. This made it possible to plot impedances, primary line currents and currents in short-circuited winding sections against the relative position of the short-circuited turns in the complete winding and the number of winding sections short circuited. Tests were also made, applying voltage to the h.v. winding and to the l.v. winding, to simulate the conditions of a fault on the primary or on the secondary winding. In all cases the current in the short circuit was the normal full-load current of the h.v. winding, namely, 29·8 A.

Figures 23.31 to 23.33 inclusive give some of the results of this particular series of tests. They are fairly self-explanatory and show how the position and number of the short-circuited turns affect the primary current drawn from the

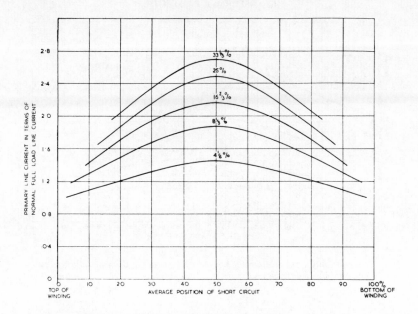

Fig. 23.31   Curves showing the relationship between primary line current and average position of short-circuited turns in a typical three-phase, core-type transformer

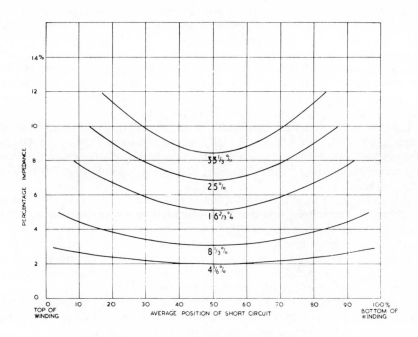

Fig. 23.32   Curves showing the relationship between the percentage impedance and average position of short-circuited turns in a typical three-phase, core-type transformer

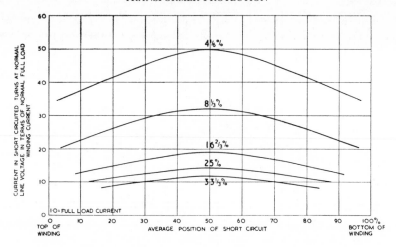

Fig. 23.33   Curves showing the relationship between the current in the short-circuited turns and the average position of short-circuited turns in a typical three-phase, core-type transformer

line. The illustrations apply to the case where the fault occurs on one phase of the primary windings, which, for this series of tests, were star connected.

It will be seen from these curves that when relatively few turns are short-circuited, on the one hand extremely large currents flow in the short-circuited turns, while *relatively* small currents are drawn from the primary lines, and at first glance these appear to be opposing facts. They are easily reconciled, however, when it is pointed out that the heavy currents in the few short-circuited turns are due to the low impedances between those turns and the primary winding, while the smallness of the current drawn from the primary lines is due to the high ratio of total primary turns to short-circuited turns. As the number of turns short-circuited increases, the impedance increases (up to a point) and the current in the short circuit decreases, while the ratio of turns cited above decreases and more current is drawn from the primary lines.

It will be noted that, bearing in mind the numbers of turns short-circuited, the impedances shown by Fig. 23.32 really are very high relative to the normal transformer impedance, and this is accounted for by the relatively high reactance produced by the dissymmetry between the primary winding and the short-circuited turns.

It is to be borne in mind that the minimum number of turns short-circuited in these tests was sixteen, and that they were all in series. In the usual interturn fault, first one turn, then a second turn, and so on are short-circuited in parallel, in which case impedances are lower than those shown in Fig. 23.32, short-circuit currents are higher, and primary line currents are lower. The usual result of this is thus: severe local burning out of the faulty turns, small primary line currents, but no untoward distortion of the windings.

The following major conclusions may be drawn from the data obtained from the tests, bearing in mind that the maximum portion of winding short-circuited was limited to one-third of the total winding on one limb.

For a given number of turns short-circuited the impedance is a minimum when the axial centre of the turns coincides with the axial centre of the winding, and the line current is then a maximum for that number of turns; the *variation* of impedance throughout the length of the winding increases with the number of turns short-circuited. Impedances increase with the number of turns short-circuited, and the increases are greatest when the short-circuited turns are at the ends of the winding. For a given number of turns short-circuited, the current in the short-circuited turns is highest when the axial centre of the turns coincides with the axial centre of the winding; the short-circuit current decreases with increasing number of turns. The primary line current increases with an increase in the number of turns short-circuited as for a given increase of the latter the impedance increase is proportionately less, so that the resulting ampere-turns in the short circuit are greater. The turns in the whole winding are constant, and therefore the line current increases proportionately to the short-circuit ampere-turns.

The characteristics disclosed by the curves apply generally to single-phase and polyphase transformers however the windings may be connected and for faults on the primary or secondary winding. Currents and impedances are of the same order of magnitude for similar interturn faults on either winding of a given transformer.

Line currents and phase voltages become unbalanced to a degree depending upon the extent of the winding fault and the transformer connections.

The curves illustrate clearly the reason why an initial breakdown of interturn insulation, involving a few turns only, fails to operate automatic protective gear, and they demonstrate that the supply can be interrupted only when sufficient turns are embraced by the fault to provide sufficient primary current to operate the protective equipment.

Chapter 24

Failures and their causes

It has been seen in the earlier chapters that a transformer consists essentially of a magnetic circuit, primary and secondary coils, terminals, cooling and insulating media and, in certain cases, auxiliary external cooling gear and tap-changing equipment, and any of these parts are liable to failure.

While the static transformer is undoubtedly one of the pieces of electrical apparatus which is least subject to breakdown, faults do occur from time to time due to various causes, but the proportion of the number of faults to the number of transformers installed is extremely low, and mechanised as distinct from electrical faults account for many failures occurring in auxiliary equipment. The faults that most frequently arise in practice may be classified broadly as follows, although to a large extent the different kinds often react upon one another:

(*a*) Failures in the magnetic circuits, *i.e.*, in the cores, yokes and adjacent clamping structure.

(*b*) Failures in the windings, *i.e.*, in the coils and minor insulation and terminal gear.

(*c*) Failures in the dielectric circuit, *i.e.*, in the oil and major insulation.

(*d*) Structural failures.

These failures may be due to (*a*) faulty manufacture, which comprises poor design, faulty material and bad workmanship; and (*b*) faulty and abnormal operation, which includes careless drying out and installation, lack of adequate supervision, and abnormal transient or sustained operating conditions. How far these various factors are responsible for transformer failures can be gathered from the following remarks.

FAILURES IN THE MAGNETIC CIRCUIT

(1) Cases have occurred from time to time in core-type transformers of a breakdown of the insulation round the bolts inserted through the cores and yokes for the purpose of clamping the laminations together. This type of fault has the effect of causing local short circuits in the laminations themselves, which produce intense local eddy currents, while if in addition two or more of the core bolts break down, heavy currents are liable to circulate in the bolts themselves,

as they form a short-circuited turn through which magnetic flux passes. This failure is most serious when one of the core bolts situated at the ends of the limb breaks down simultaneously with an adjacent bolt in the yokes, as the path between the two bolts is threaded by almost the full value of the magnetic flux when passing from the core to yoke, or vice versa. The path formed by the bolts and the thick outer clamping plates is one of low impedance, and the amount of heat generated by failures of this type is sometimes sufficient to distort the whole core to a very considerable extent. The heat generated may also char the coil insulation and cause a short circuit between turns of the adjacent winding.

It is now common practice to clamp the limb laminations of large power transformers with insulated bands. This method of construction eliminates limb bolt failures.

(2) Failures may also occur of the insulation between laminations and of the insulation between the yokes and yoke clamping plates. These failures produce large circulating eddy currents which generate considerable quantities of heat and which possibly damage the core and coil insulation. Incidentally they also increase the iron loss of the transformer.

(3) Unless very special precautions are taken to lock effectively the core-clamping bolts and the bolts tying together the core structure, vibration will be set up which will tend to weaken the core insulation and produce failures similar to those outlined above.

(4) During manufacture, the edges of the core and yoke laminations may have become burred, due to the continued use of worn tools. Unless, therefore, proper supervision is given to the cutting and punching processes, the resulting burrs will produce local short circuits in the iron laminations, and eddy currents with consequent abnormal heating will occur.

(5) It is important to make sure that in the finished transformer no metallic filings or small turnings are present between the laminations, as these are also liable to produce intense local eddy currents and excessive local heating of the cores.

(6) In certain cases, notably in high-voltage testing transformers, the top yoke may be of the butt type. If abnormal gaps are left between the cores and yoke, severe eddy currents may be set up at the butt joints, and these will result in intense heating and possibly in burning of the cores and yoke in the vicinity of the gaps. In some of the older designs this phenomenon has been intensified on account of shallow yokes, as this results in an increase of the local eddy currents, owing to the magnetic flux entering and leaving the cores at an angle other than normal to the joint. A deep yoke eliminates this by drawing the flux more vertically upwards or downwards, so tending to bring the leaving or entering angle more closely back to the normal.

(7) High core loss and consequently excessive core heating is produced when the e.m.f. waves for a given effective voltage are flat topped. This will be apparent from a consideration of the formula connecting voltage and flux, which is

$$E = 4K_f B_m A f N / 10^4 \qquad (24.1)$$

where E = the r.m.s. value of the induced voltage in the winding considered, i.e., primary or secondary

K_f = form factor of the e.m.f. wave

B_m = maximum flux density in the core in teslas

A = cross-sectional area of the core in square centimetres

f = frequency, Hz

N = number of turns in the winding considered, *i.e.*, in primary or secondary.

For a sine wave of e.m.f., $K_f = 1 \cdot 11$. For a peaked e.m.f. wave, K_f is greater than $1 \cdot 11$, and therefore B_m is lower for the same r.m.s. voltage, and the iron loss is less than if the e.m.f. wave was sine shaped. For a flat-topped e.m.f. wave, K_f is smaller than $1 \cdot 11$, and therefore B_m is higher for the same r.m.s. voltage, and the iron loss is greater than if the e.m.f. wave was sine shaped. The magnetising current in this case is also considerably larger.

(8) On rare occasions it may happen that the shops have inadvertently used thinner core laminations than instructed to use by the designer, and on account of the lower iron factor involved, which gives a smaller active cross-sectional area of core than has been allowed for and a correspondingly higher value of B_m, the use of the thinner laminations produces high iron losses and heavy magnetising currents.

(9) High flux density in the magnetic circuit often results in larger magnetising current inrushes occurring when switching a transformer into circuit on no load. While this current inrush generally dies down rapidly, large electromagnetic forces are created while the heavy current lasts and the windings are thereby strained. The phenomenon becomes more severe the nearer the transformer is located to the generating source, and repeated switching in may ultimately cause movement of the windings. The effects of switching in are dealt with more fully in the sections on switching and transient phenomena.*

(10) If transformers providing a mid-wire or neutral for use as a d.c. neutral have not their windings carefully balanced, the d.c. ampere-turns in the windings on opposite sides of the transformer neutral do not neutralise one another. The a.c. flux, therefore, instead of being symmetrically disposed on either side of the normal *zero* axis, becomes equally disposed on either side of an axis corresponding in value to the resultant d.c. flux in the core. That is, the a.c. flux becomes superposed upon the resultant d.c. flux, and the total a.c. flux waves are unsymmetrical about the *zero* axis. Therefore, the core becomes saturated during one half cycle and correspondingly less magnetised in the next half cycle, and it heats up to an extent depending upon the value of the a.c. resultant flux. This extra heating would, of course, in time adversely affect the insulation of the coils, making it brittle and likely to flake away from the conductors, while in addition the oil may become badly sludged, which would prevent dissipation of heat from the windings.

(11) Transformers for use with rotary converters are often designed for a high inherent reactance for the purpose of providing means for a.c. voltage regulation. The high reactance is obtained by the use of magnetic shunts placed in the leakage field between the primary and secondary windings, and if these shunts are designed to possess a high degree of magnetic saturation, a large amount of heat will be generated, which eventually will destroy the coil insulation and possibly cause the oil to sludge. Under certain operating conditions the high saturation in the magnetic shunts may produce commutation troubles in the rotary converter, due to the higher harmonics introduced.

*See Chapter 22.

(12) High flux density in the magnetic circuit produces higher harmonics of voltage or current of appreciable magnitudes which may have very serious effects. Generally the third harmonic only is of importance, though under very abnormal conditions harmonics above the third may create trouble. Third harmonics of magnitudes likely to be objectionable are confined to three-phase shell-type transformers and to three-phase banks of single-phase core- and shell-type transformers; they are usually negligible in three-phase core-type transformers. Their effect can still further be confined to the star/star, star/ interconnected star, or interconnected star/star connections; in the last two connections the star side only need be considered. With isolated neutrals the line to neutral voltage will contain a large third-harmonic component equal in maximum amplitude to 60 % or more of that of the fundamental. This represents an increased dielectric stress on the coil insulation which *may* shorten the life of the transformer. If the secondary neutral only of a star/star connected transformer or bank supplying a high voltage or moderately high voltage cable is earthed, the third-harmonic component of the magnetic flux becomes amplified, which further increases the third-harmonic voltage component until the transformer core becomes thoroughly saturated. The saturation, therefore, reaches a much higher value even than originally designed for, and the iron loss increases considerably. Under such conditions the transformer will reach a dangerous temperature, which will cause deterioration of the coil and core insulation, and if allowed to continue will produce sludging of the oil. This abnormal heating occurs without load on the transformer, and in practice the iron loss has been known to increase to three times the normal.

(13) If the voltage applied to the transformer *must* be increased appreciably on account of the needs of the system load, the frequency must also be increased in order to avoid high magnetic saturation of the core. An *increase* of voltage must not be accompanied by a *decrease* of frequency, as, if it is, high core saturation will occur with resultant increased iron loss and abnormal core heating. The interdependence of voltage, frequency and flux will easily be seen from the following formula:

$$B_m = (E \times 10^4)/(4K_f A f N) \tag{24.2}$$

where B_m = maximum flux density in the core in teslas

E = applied voltage of the winding considered

K_f = form factor of the e.m.f. wave

A = cross-sectional area of core in square centimetres

N = number of turns in the winding considered

f = frequency, Hz.

(14) In the older transformers ageing of the core plates may be found to have taken place. This is due to a deterioration of the material of the laminations, and the result is manifested by an increase in iron loss and rise in temperature of the transformer. This may eventually lead to a partial or complete destruction of the coil insulation and to sludging of the oil.

(15) The laminations of rectangular shell-type transformers are usually clamped between substantial top and bottom endplate structures which are tied together by vertical bolts, these bolts being in close proximity to the chamfered corners of the core. If the bolts are too near the core, flux strays from the

main magnetic circuit into the bolts and the resulting eddy currents may burn both the bolts and edges of the core plates.

FAILURES IN THE ELECTRIC CIRCUIT

(1) A short circuit between adjacent turns of a coil—usually of the high-voltage winding—may be caused by the presence of sharp edges on the copper conductors. If the transformer vibrates when on load, or if the windings are subjected to repeated electromagnetic shocks, through short circuit or switching in, these sharp edges will cut through the insulation and allow adjacent turns to make metallic contact.

(2) A short circuit between turns may result from the dislodging of one or more turns of a coil caused by a heavy external short circuit across the windings. Breakdown may not occur immediately the turns are displaced, but should the transformer vibrate while on load due to looseness of core bolts, or should it receive repeated heavy electromagnetic shocks, abrasion of the insulation between adjacent dislodged turns will most likely take place, so producing a breakdown.

(3) Occasionally the insulating covering of square wire conductors may not be wound on the copper as tightly as it should be, in which case it has a tendency to bulge over each face of the conductor. The true square configuration of the conductor is therefore masked, and in fact it may even appear to be more round than square. Consequently the coil winder has some difficulty in making certain that the conductor has not become twisted in the winding process, and cases have occurred where short circuits between turns have arisen as a result of this twisting. At some place in the coil, adjacent conductors are edge to edge or face to edge, and in time the insulation between such turns becomes chafed and breakdown ensues. The trouble is, of course, accentuated if the edges of the conductor happen to be sharp instead of rounded.

(4) Transformers for use on large systems are now usually fitted with adjustable coil clamping for the purpose of taking up any shrinkage of the insulation which may occur under service. Unless this hand adjustment of the coil supports is performed very carefully by an experienced workman, so that the correct pressure is applied to the windings, some of the conductors may become dislodged, and a short circuit between turns may occur as outlined under (2) above.

(5) Short circuits between turns are almost bound to occur sooner or later should moisture penetrate the fabric insulation of the coils. Breakdown from this cause is rendered more imminent still if the coils have been insufficiently impregnated, i.e., if the varnish has not thoroughly penetrated to the innermost layers of the coils.

(6) Drying out a transformer on site may be undertaken by an engineer not fully conversant with the operation, and due to inexperience the process may be unduly shortened. If normal voltage or a test voltage is applied while the insulation resistance of the windings is still low, the insulation between adjacent turns is very liable to fail on account of the presence of moisture vapour.

(7) Cases have occurred where thin ribbon conductors have been wound on edge, direct on to an insulating cylinder, to form spiral type coils. Such a coil is mechanically weak, and the turns are liable to lean over from the plane

normal to the cylinder surface so that the coil becomes very susceptible to damage in the event of an external short circuit occurring on the system.

(8) If a transformer is subjected to more or less rapidly fluctuating loads, the expansion and contraction of the copper conductors alternately increases and decreases the mechanical pressure on the insulation between turns. As the dielectric strength of most fabric insulation decreases with increasing mechanical pressure, the windings become more susceptible to failure should they be subjected to electrical or magnetic shocks.

(9) If individual coils, particularly of the high-voltage crossover type, are designed to have too great a radial depth in comparison with their height, so-called hot spots will arise in the interior of the coils. These will produce brittleness of the conductor insulation, and a short circuit between turns may eventually take place. This danger becomes considerably greater if the design of the oil circulation system is inadequate, *i.e.*, if the oil ducts are too narrow.

(10) Eddy currents may be produced in coil conductors—usually the low-voltage coils—where there are several wires in parallel to form each conductor. These wires are usually rectangular, and they should be wound so that the shorter side of each is presented to the flow of the leakage flux between primary and secondary windings. That is, in core-type transformers, for instance, employing concentric windings, the shorter side should be normal to the path of the leakage flux and the longer side parallel with it. If this procedure is reversed, excessive eddy currents flow in the conductors and hot spots may be produced in the windings. The various layers of subdivided conductors should be transposed during the winding process, so that each parallel portion in its path from terminal to terminal occupies each different position for an equal length. This has the effect of equalising the load taken by each individual wire. If this is not done, the different layers will not share the load equally and some will overheat, thus tending to accentuate the hot spot phenomenon previously referred to.

(11) Badly soldered joints between coils may overheat on load, and local carbonisation of the oil may occur. The heat generated at the joint will probably be transmitted to a certain contiguous length of conductor of each coil, and this may partially carbonise the insulation round the conductors and eventually result in a short circuit between turns. Such joints may eventually come apart and produce an open circuit in the winding concerned.

(12) Coil conductors may be displaced violently on the occurrence of an external short circuit, as the result of internal unbalanced electromagnetic conditions. With concentrically wound primary and secondary coils (as in the circular core type of transformer) the horizontal axes may not coincide, and therefore a vertical force acting upon the coils may be introduced in addition to the usual radial one. This vertical force has often been responsible for the distortion of end coils, particularly of the older designs of transformers in which the impedance was low. When a number of tappings are fitted, it is often very difficult to preserve electromagnetic symmetry at all ratios, and sometimes the unbalancing is unavoidable.

With sandwich wound primary and secondary coils (as in the rectangular shell type of transformer) the axes of the individual coil sides seldom coincide on account of the differences required in electrical clearances from coils to core, and of the differences in winding spaces. In this case the lack of symmetry introduces distorting forces acting radially upon the coil sides, which become pushed in towards the core and forced out away from the core, depending on the relative positions of the axes of the coil sides and upon the relative short circuit currents flowing.

(13) Short circuits between turns, breakdown of windings to earth, and puncture of insulators may take place due to the following transient phenomena:

(*a*) Concentration of voltage on the terminal end coils when switching in or when lightning surges reach the transformer. Owing to the change of surge impedance at the transition point between transformer and line, the phenomenon of reflected and transmitted voltage and current waves occurs, which may produce high voltage rises in the transformer windings. While the end coils bear the brunt of the shock the remainder of the windings is not immune against breakdown, as high voltages may still arise in any part of the windings.

(*b*) The excessive voltages set up by surges may be accentuated at open-ended tappings, at any point of change of surge impedance in the windings (such, for instance, as at the termination of conductor reinforcement where employed, or at the space between series coils), and at the neutral or mid-point. Care should be taken to insulate the conductors in these regions very thoroughly, in order to eliminate so far as possible short circuits between turns.

The disturbing cause may be either normal switching operations, lightning discharges, an arcing ground or chattering switch contacts.

(*c*) When switching out of circuit a highly inductive winding, such as a transformer primary with the secondary open-circuited, the magnetising current, and consequently the magnetic flux, tends to collapse instantaneously. While for various reasons the flux does not do this, it does sometimes drop at a much more rapid rate than that corresponding to the normal cyclic rate of change, and as a result high voltage rises are sometimes produced. The rapid cooling of the interrupting arc, particularly during the last half cycle, has been found to augment this effect.

The above various transient causes are discussed more fully in the section on transient phenomena.*

(14) Sustained heavy overloads produce high temperatures throughout the transformer. The coil insulation becomes brittle, and in time probably flakes off the conductors in places, so producing short circuits between turns. The oil deposits sludge both on the tank bottom and on the coils and core structure. This deposit exerts a blanketing effect on the coils and core, so that the excessive heating becomes cumulative in effect. Narrow oil ducts also intensify the heating. Transformers with a high ratio of copper loss to iron loss are less able to withstand overload, and are therefore more liable to fail on account of overloading.

(15) When changing voltage adjusting tappings, the terminals of which are arranged on a terminal board under oil level, care must be taken to make sure that wrong leads are not joined and part of the windings short-circuited. If they were, heavy currents would circulate in the short-circuited winding, which would produce a fault between turns unless the transformer was very speedily tripped out of circuit.

(16) Bolted joints or connections carrying current may, if not very effectively locked, become loose in service, due perhaps to some slight vibration. Such joints will heat up rapidly, and even so trivial a fault may temporarily cause a large and important transformer to be out of commission.

*See Chapter 22.

FAILURES IN THE DIELECTRIC CIRCUIT

(1) Moisture entering the oil as a result of the so-called breathing action greatly reduces its dielectric strength, so that breakdowns from coils or terminal leads to tank or core structure may take place. The greatest danger, however, is to the interturn insulation of the coils, as referred to previously.

(2) Deterioration of the oil may occur as the result of prolonged overloading of the transformer, and the action is materially assisted by the presence of bare copper and lead. Excessive oil temperature accelerates the formation of sludge, water and acids.

(3) Dielectrics having different specific permittivities are often used in series, and unless their thicknesses are properly proportioned they may be subjected to abnormally high dielectric stresses. For instance, the insulation between high-voltage and low-voltage transformer windings usually consists of paper, solid insulation and oil. Except in very high voltage transformers, the effect of the paper covering can be ignored, so that we only have to consider solid insulation with a dielectric constant of 5, and oil with a constant of 2. The total voltage across two such dielectrics in series divides up so that the voltages across equal thicknesses of each are inversely proportional to the dielectric constants, and unless the respective thicknesses are proportional so that the voltage gradient across each is within the safe working limit, first one and then the other dielectric will fail, due to corona discharge and overheating.

(4) Corona may take place from sharp conducting edges or small diameter conductors if the surface voltage gradient is high.

(5) Insulating parts, such as cylinders, tubes and terminal boards, made of compressed paper bonded with synthetic resin, may occasionally have their surfaces contaminated during the manufacturing processes, or they may hold occluded air. The former fault is responsible for surface discharge which renders the material useless, and, when occurring, generally does so on terminal boards. The latter imperfection usually results in complete puncture of the insulation concerned, as a result of ionisation of the occluded air and subsequent over-heating of the dielectric. When this type of breakdown occurs it is usually with cylinders and capacitor terminals.

(6) Earth shields placed between primary and secondary windings have been responsible for numerous breakdowns. Their presence produces a concentration of dielectric stress at their edges, which tends to strain insulation locally, while a breakdown at one point from the high-voltage windings to the shield has often resulted in almost completely destroying the high-voltage winding on the limb concerned.

(7) Narrow oil ducts are a serious menace to the serviceable life of a trans-former. Adequate cooling cannot be obtained, the coil insulation becomes brittle in time, and a fault between turns naturally follows.

(8) Occasionally metallic particles are found in the various kinds of press-boards, and these may result in a puncture of the board.

(9) Careless workmanship may result in solder splashes on the coils and on surfaces which are depended upon for creepage clearances. The possible effect of these will be obvious.

(10) During the course of time a certain amount of the oil disappears, due to evaporation and oxidation. Unless the oil is topped up to proper working level,

the transformer will overheat on account of the consequent loss of tank radiating surface and of the cooling medium. This applies particularly to transformers having tubular tanks, should the oil fall below the top of the cooling tubes.

(11) Short circuits between phases may occur if there is insufficient clearance between the phases. This may sometimes be aggravated by the insertion of a pressboard barrier between phases if the presence of the barrier upsets the distribution of dielectric stress to such an extent as to throw too high a stress across the oil spaces and across the barrier.

(12) If wooden cleats which, for instance, support the terminal leads from coils are not very thoroughly dried and impregnated the moisture present will lead to tracking and possibly to a short circuit between tapping leads.

(13) In some instances the relationship of the electrostatic capacitances between primary and secondary windings, and between individual windings and the core, may lead to the presence of high voltages in the low-voltage circuit. The solution of the problem is really one of earthing, and it has been dealt with in the section on transformer earthing.*

(14) The presence of foreign conducting particles held in suspension in the oil may cause a temporary breakdown, due to the lining up of these particles between bare parts having a difference of voltage across them. Examples of this are the flashing over under oil of terminal leads to each other and to the tank or core structure.

FAILURES DUE TO VARIOUS STRUCTURAL DEFECTS AND TO OTHER CAUSES

(1) When a transformer is provided with means for taking up shrinkage of the coil insulation, it may be necessary to take precautions against the possibility of any part or parts of the clamping structure forming a short-circuited turn. When this clamping structure consists of a steel ring placed on the tops of the windings together with adjusting screws, it is necessary to insulate these screws from the clamping ring in addition to leaving an appreciable gap in the ring itself. If this precaution is not taken the adjusting screws will complete the circuit through the yoke clamping irons and so cause a heavy short circuit current to flow. Cases have arisen where this gap has been bridged, either direct or via the yoke or clamping structure, and the steel ring melted in consequence. This possibility does not arise if the clamping rings are in separate halves.

(2) Insufficient bracing of the leads from windings to terminals may result in these becoming distorted and making contact in the event of an external short circuit taking place on the system. The danger becomes very acute on large systems on account of the large short-circuit currents which may flow.

(3) Transformer tanks sometimes give trouble on account of bad and porous welding and leaky fittings, all of which may lead to oil leakage and consequent overheating and breakdown of the transformer, if not attended to immediately. Rough handling in transit is also responsible for a large proportion of the leaky tank troubles.

(4) If, for transformers fitted with a protective gas relay, the tank is not filled

* See Chapter 20.

with oil to the correct level at the specified temperature (usually 15°C in Great Britain), or if there has been a loss of oil from the transformer tank, there is a danger that maloperation of the relay will result. In the event of a fault inside the tank this lack of protection might result in a major breakdown of the transformer.

(5) Deposits of ordinary dust, coal dust and salt spray on the surfaces of bushing insulators often cause flashover to take place. The remedy is obvious.

(6) In heavy-current, low-voltage transformers such as those used for supplying electric furnaces, a number of coils each consisting of a number of heavy conductors in parallel are also connected in parallel. It sometimes becomes a matter of difficulty to design the low-voltage winding so that all the parallel paths between terminals will have the same impedance. If they do not, however, they will not share the load equally, and parts of the winding may overheat, with the usual inevitable result.

(7) Transformers operating in parallel should preferably possess the same turns ratio, the same percentage impedances, and the same ratios of resistance to reactance voltage drops. If any of these factors are different, one transformer at least may be overloaded, and may consequently burn out. This subject has been fully treated in the section on parallel operation.*

(8) Heavy current leads between furnace transformers and the furnace should be laid out so as to keep the impedance of the circuit within desired limits. If care is not taken to achieve this, the voltage applied to the furnace will be very considerably below that required for successful operation. A case is on record where the low-voltage leads formed a loop round an iron column supporting the building, thus increasing the inductance of the low-voltage circuit and considerably reducing the voltage applied to the furnace.

(9) When housing transformers, care must be taken to provide sufficient space around them to enable the tank to dissipate the losses. If a transformer is placed too near to another unit or to the walls of the chamber, the tank will become lagged and the temperature of the transformer will increase, so endangering the coil insulation and the condition of the oil. It is essential to provide adequate ventilation for all transformers. Closed pits and small brick chambers should be avoided wherever possible. In the case of furnace transformers, effective shielding should be interposed between the furnace and the transformers.

(10) As the vapour at the top of the tank of oil immersed transformers may be of an explosive nature, a naked light should never be introduced for the purpose of examination of connections, etc. A case is on record where a transformer was wrecked and loss of human life sustained through neglect of this precaution.

(11) The temperature of artificially cooled oil immersed transformers will increase to excessive values if a stoppage of the cooling medium occurs due to some failure in the auxiliary cooling plant. If such a failure does happen, the load on the transformer should be decreased to a value that it can handle as an ordinary self-cooled transformer without attaining an excessive temperature rise.

(12) In water-cooled transformers the cooling tubes are liable to become clogged, due to deposits of lime or other matter from the water supply. If the tubes are not thoroughly cleaned out periodically, the water flow becomes

*See Chapter 18.

lessened, and the transformer will attain a temperature higher than the maximum permissible with the usual subsequent results.

(13) Cases are on record where electrolytic corrosion has occurred in transformer oil coolers, due to the use of metals in actual contact, which are situated far apart in the table of electrochemical equivalents. The remedy, of course, is to use metals in contact which are as near together in the table as possible.

(14) In artificially cooled transformers, failure may possibly occur through the leakage of water into the oil. In the case of ordinary water-cooled transformers this leakage may be due to the corrosion of the cooling tubes, and for this reason it is better to use copper rather than iron tubes, even though the initial cost may be considerably greater. A further possible cause of leakage is from the provision of joints in the cooling tube system inside the transformer tank. When external coolers are used, the risk of leakage from the cooling water to the oil is largely overcome by maintaining the oil at a much higher static pressure than the cooling water by means of a pump working against a regulating valve. In this way it is possible to ensure that any leakage which may occur will be from the oil to the water, and not vice versa.

(15) In water-cooled transformers there is a further risk of introducing moisture to the oil by condensation. When the transformer is on load, the cooling water and the pipe surrounding it are naturally at a very much lower temperature than the oil and the gases at the top of the tank. The effect of this is to cause any moisture which may be present in these gases to be deposited on the cooling tube at the point where it enters the tank. This moisture would subsequently drip into the oil, and unless means were adopted for rendering the tank airtight the process would be continuous. The risk of failure due to this phenomenon may, however, be minimised by lagging the cooling tube from the point where it enters the tank to a point below the oil level with some material which is not affected by oil and which is a non-conductor of heat.

(16) Puncture of high-voltage capacitor bushings has occurred owing to a deterioration of the paper insulation consequent upon the allowance of too high a dielectric stress per step. The dielectric loss has been so high as to generate sufficient heat to char the body of the terminal particularly near to the central conductor, and the effects have been augmented by air gaps between capacitor layers.

A study of the records of modern transformer breakdowns which have occurred over a period of years shows very conclusively that between 70 and 80% of the number of failures are finally traced to short circuits between turns, and a persual of the foregoing review will indicate that this may be caused in a number of different ways. Unfortunately, with a fault of this type it is usually very difficult to say definitely to what the failure has been due as all evidence is eliminated, as a rule, by the very nature of the breakdown. Consequently, the cause of the breakdown is often a matter for conjecture, and however unsatisfactory this may be from a point of definitely assigning the true cause, usually a very close idea of the real cause may be obtained by a careful study of the transformer itself and of the local and operating conditions. The notes in this chapter may further assist in determining the true cause.

The remaining percentage of breakdowns can often be minimised by provision of suitable transformer protective apparatus, and by regular and careful inspection of the transformers themselves and supervision of the operating conditions.

Some representative failures are illustrated by Figs. 24.1 to 24.7 inclusive.

Fig. 24.1   Interturn insulation failures of section type h.v. coils

Fig. 24.2   Failure of a steel coil-clamping ring due to short circuiting via the pressure adjusting screws and yoke clamps

Fig. 24.3 Illustration showing the effects of an external short circuit on a three-phase, core-type power transformer having an output of 500 kVA, 33 000/2300 V, 50 Hz. The transformer has crossover h.v. coils and spiral l.v. coils

Fig. 24.4 Illustration showing the effects of an external short circuit on a teaser unit of 400 kVA, 3/2 phase, 50 Hz, Scott-connected transformer group, having a ratio of 11 000/2200 V. Crossover h.v. coils. Spiral l.v. coils

Fig. 24.5   Damage at bottom of l.v. winding due to a puncture in h.v. to l.v. major insulation caused by the ingress of moisture

Fig. 24.6   Failure due to surge on h.v. side of a transformer. See Fig. 24.7

Fig. 24.7 Failure due to a surge on the h.v. side of a 5 MVA, 33/6·6 kV, 50 Hz star/delta connected transformer. Abnormally high voltages were produced across the tapping section of the windings causing an insulation breakdown between tapping risers and deformation of two tapping discs on either side of the break, due to the electromagnetic forces produced by the short circuit. A secondary effect was that of transfer of the surge to the l.v. winding causing a flashover across the l.v. risers, followed by a power frequency arc. This is clearly shown in the bottom left-hand part of Fig. 24.6 and in the enlargement of the connection Fig. 24.7

596

Chapter 25

Effects of sustained abnormal operating conditions

Occasions may arise from time to time when it is desirable to operate transformers under one or more conditions which are different from those for which the transformer was originally designed. The principal factors involved are frequency, voltage and current loading, so that we will study the general effect on an average commercial transformer of the following sustained conditions:

(1) different frequency
(2) different voltage
(3) excessive overload
(4) unbalanced loading.

(1) Dealing with variation from the originally specified frequency, it can first be stated that it is not usually possible to operate a transformer at any frequency appreciably lower than originally designed for unless the voltage and consequently the output are correspondingly reduced. The reason for this can best be appreciated from a study of the equation connecting voltage, frequency and magnetic flux

$$E \text{ or } V = B_m A f N / (22 \cdot 51 \times 10^2) \tag{25.1}$$

where
E = r.m.s. value of the voltage in the winding considered
A = cross-sectional area of the core, cm^2
B_m = magnetic flux density in the core, teslas
N = number of turns in the winding considered
f = frequency of the supply, Hz.

In industrial transformers, the flux density in the core is $1 \cdot 55 - 1 \cdot 65$ teslas, and, generally, it is not usual to allow the flux density to exceed this figure appreciably. As A and N are definitely fixed for any given design, the variables are E, B_m and f, and of these B_m cannot appreciably be increased. We are therefore left with E and f as the only permissible variables when considering using a transformer on a frequency lower than that for which it is designed. Obviously the balance of the equation must be maintained under all conditions, and therefore any reduction to the frequency f will necessitate precisely the same proportionate reduction to the voltage E if the flux density B_m is not to be exceeded and the transformer

597

core not to become overheated. As a matter of fact, the lower the frequency the higher the flux density in the core, but as this increase is relatively small over the range of the most common commercial frequencies its influence on the output is very slight, and therefore the reduction in voltage and output can be taken as being the same as the reduction in frequency.

(2) Dealing next with the question of using a transformer on voltages different from the normal rated voltage, it can be said very definitely that no attempt should ever be made to operate on voltages appreciably higher than designed for. This is inadmissible from the points of view both of electrical clearances and flux density in the core; the latter is self-evident from a study of eqn. 25.1, from which it will be seen that an increase in voltage will produce a corresponding increase in the flux density (assuming, of course, constant frequency), and this is not permissible. In passing, it is perhaps interesting to point out that some specifications state that the system frequency and voltage will not vary from the nominal figures in opposite directions, and this clause is framed in order to assure the manufacturer that the transformer will not become magnetically saturated under operating conditions. Variations to the system voltage, according to the point at which a transformer may be located is, of course, taken care of by suitable voltage adjusting tappings, and so long as the highest voltage applied or induced in the windings is not greater than the highest for which the windings are designed, no ill effects will result.

Transformers may, of course, be used on lower voltage systems than designed for, but the output would in consequence be reduced also.

(3) Overloads* are permissible or not according to the ambient temperature of the cooling air or water and the load on the transformer prior to the application of the overload. The chief point to bear in mind is that the average temperature of the windings should not exceed 95°C when measured by the resistance method, as above this average the 'hot spot' temperature becomes important. The result of excessive continuous overloads is that cellulose insulation loses its mechanical strength, and winding failure may ensue. The reader is referred to the '8°C rule' which has established the relationship between increased temperature and the reduction in strength. There is also a tendency for the oil to oxidise with excessive overloads.

(4) In considering the question of unbalanced loading it is easiest to treat the subject from the extreme standpoint of the supply to one single-phase load only, as any unbalanced three-phase load can be split up into a balanced three-phase one and one or two single-phase loads. As the conditions arising from the balanced three-phase load are those which would normally occur, it is only a question of superposing those arising from the single-phase load upon the normal conditions to obtain the sum total effects. For the purpose of this study it is only necessary to consider the more usual connections adopted for supplying three-phase and two-phase loads, as six-phase transformers which are used mainly for supplying rectifiers seldom supply unbalanced loads. The value of current distribution is based upon the assumption that the single-phase currents are not sufficient to distort the voltage phasor diagrams for the transformers or transformer banks. This assumption would approximate very closely to the truth in all cases where the primary and secondary currents in each phase are balanced. In those cases, however, where the primary current on the loaded

*See also Appendix 6.

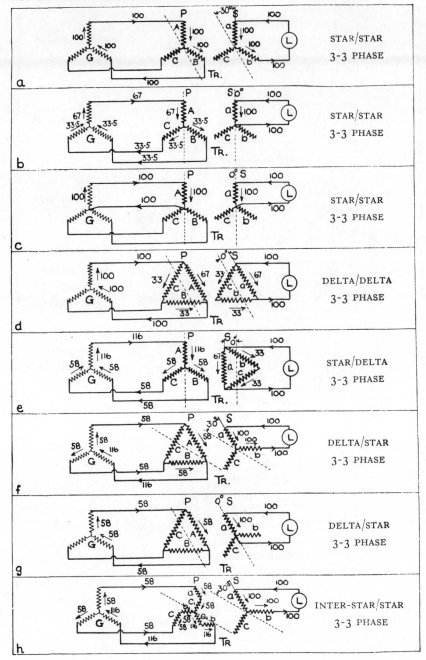

Fig. 25.1 *Current distribution due to single-phase load on polyphase transformers or transformer groups.*

Note – In all cases the dotted lines indicate the phase angle of the single-phase load currents

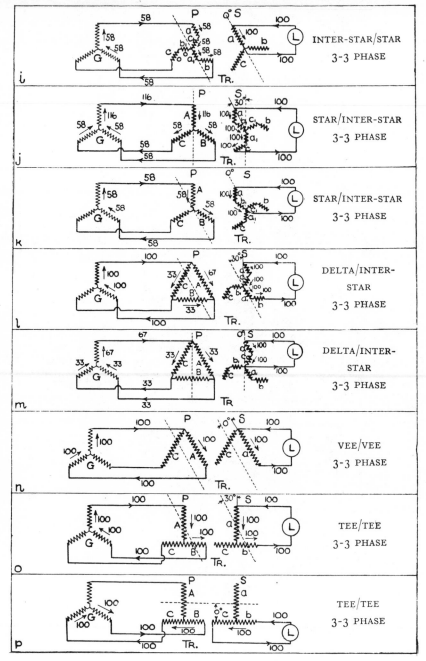

INTER-STAR/STAR
3-3 PHASE

STAR/INTER-STAR
3-3 PHASE

STAR/INTER-STAR
3-3 PHASE

DELTA/INTER-
STAR
3-3 PHASE

DELTA/INTER-
STAR
3-3 PHASE

VEE/VEE
3-3 PHASE

TEE/TEE
3-3 PHASE

TEE/TEE
3-3 PHASE

Fig. 25.2   Current distribution due to single-phase load on polyphase transformers or transformer groups

Note — In all cases the dotted lines indicate the phase angle of the single-phase load currents

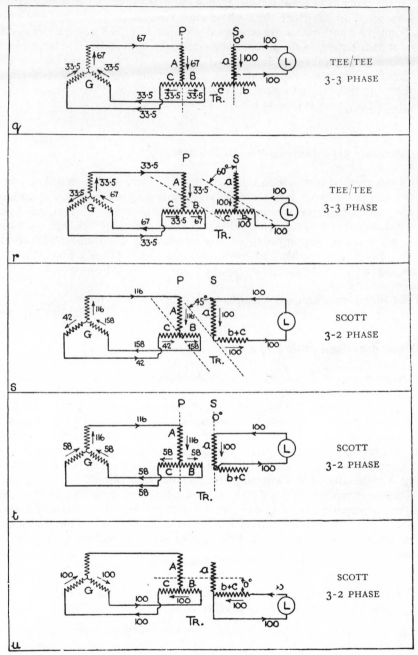

Fig. 25.3   Current distribution due to single-phase load on polyphase transformers or transformer groups.

Note – In all cases the dotted lines indicate the phase angle of the single-phase load currents

601

phase or phases has to return through phases unloaded on the secondary side, the distortion may be considerable, even with relatively small loads; this feature is very pronounced where three-phase shell-type transformers and banks of single-phase transformers are employed. Figures 25.1, 25.2 and 25.3 show the current distribution on the primary and secondary sides of three- to three-phase transformers or banks with different arrangements of single-phase loading and different transformer connections, and in Scott-connected three- to two-phase groups with different arrangements of single-phase loads. These diagrams may briefly be explained as follows:

(a) Star/star; single-phase load across two lines

With this method of single-phase loading the primary load current has a free path through the two primary windings corresponding to the loaded secondary phases, and through the two line wires to the source of supply. There is, therefore, no choking effect, and the voltage drops in the transformer windings are those due only to the normal impedance of the transformer. The transformer neutral points are relatively stable, and the voltage of the open phase is practically the same as at no load. The secondary neutral point can be earthed without affecting the conditions.

The above remarks apply equally to all types of transformers.

(b) Star/star; single-phase load from one line to neutral

With this method of single-phase loading the primary load current corresponding to the current in the loaded secondary must find a return path through the other two primary phases, and as load currents are not flowing in the secondary windings of these two phases, the load currents in the primaries act as magnetising currents to the two phases, so that their voltages considerably increase while the voltage of the loaded phase decreases. The neutral point, therefore, is considerably deflected. The current distribution shown on the primary side is approximate only, as this will vary with each individual design.

The above remarks apply strictly to three-phase shell-type transformers and to three-phase banks of single-phase transformers, but three-phase core-type transformers can, on account of their interlinked magnetic circuits, supply considerable unbalanced loads without very severe deflection of the neutral point.

(c) Star/star with generator and transformer primary neutrals joined; single-phase load from one line to neutral

In this case the connection between the generator and transformer neutral points provides the return path for the primary load current, and so far as this is concerned, the other two phases are short circuited. There is therefore no choking effect, and the voltage drops in the transformer windings are those on the one phase only, due to the normal impedance of the transformer. The transformer neutral points are relatively stable, and the voltages of the above phases are practically the same as at no load. The secondary neutral point can be earthed without affecting the conditions.

The above remarks apply equally to all types of transformers.

(*d*) Delta/delta; single-phase load across two lines

With this connection the loaded phase carries two thirds of the total load current, while the remainder flows through the other two phases, which are in series with each other and in parallel with the loaded phase. On the primary side all three windings carry load currents in the same proportion as the secondary windings, and two of the line wires only convey current to and from the generator. There is no abnormal choking effect, and the voltage drops are due to the normal impedance of the transformer only. The type of transformer does not affect the general deductions.

(*e*) Star/delta; single-phase load across two lines

On the delta side the distribution of current in the transformer windings is exactly the same as in the previous case, that is, two thirds in the loaded phase and one third in each of the other two. On the primary side the corresponding load currents are split up in the same proportions as on the secondary, and in value they are equal to the secondary currents of the different phases multiplied by 1·73 and multiplied or divided by the ratio of transformation, according to whether the transformer is a step up or step down. The primary neutral point is stable.

The above remarks apply equally to all types of transformers.

(*f*) Delta/star; single-phase load across two lines

Single-phase loading across lines of this connection gives a current distribution somewhat similar to that of (*a*), except that the currents in the two primary windings are 58 % of those occurring with the star primary, while all the three lines to the generating source carry currents in the proportions shown instead of two lines only carrying currents as in the case of the star primary. There is no choking effect, and the voltage drops in the windings are due only to the normal impedance of the transformer. The transformer secondary neutral point is relatively stable and may be earthed. The voltage of the open phase is practically the same as at no load.

The above remarks apply equally to all types of transformers.

(*g*) Delta/star; single-phase load from line to neutral

With this connection and single-phase loading the neutral, primary and secondary windings on one phase only carry load current, and on the primary side this is conveyed to and from the generating source over two of the lines only. There is no choking effect, and the voltage drops in the transformer windings are those corresponding only to the normal impedance of the transformer. The secondary neutral point is stable and may be earthed without affecting the conditions. The voltages of the open phase are practically the same as at no load. The type of transformer construction does not affect the general deductions.

(*h*) Interconnected star/star; single-phase load across two lines

With this connection and method of loading, all the primary windings take a share of the load, and although in phase C there is no current in the secondary winding, the load currents in the two halves of the primary windings of that phase flow in opposite directions, so that their magnetic effects cancel. There is no choking effect, and the voltage drops in the transformer windings are those due to the normal impedance of the transformer only. With three-phase shell-type transformers and three-phase banks of single-phase transformers the secondary neutral is not stable and should not be earthed unless the flux density is sufficiently low to permit this. With three-phase core-type transformers, however, the neutral is stable and could be earthed. The voltage of the open phase is practically that occurring at no load.

(*i*) Interconnected star/star; single-phase load from one line to neutral

With this connection and method of loading a partial choking effect occurs, due to the passage of load current in each half of the primary windings corresponding to the unloaded secondary windings. The voltage of the two phases in question, therefore, becomes increased on account of the high saturation in the cores and the voltage of the windings corresponding to the loaded phase drops. Both primary and secondary neutrals are therefore unstable and should not be earthed. The above remarks apply strictly to three-phase shell-type transformers and to three-phase banks of single-phase transformers. With three-phase core-type transformers the deflection of the neutral point is not by any means so marked, and considerable out-of-balance loads can be supplied without any very excessive deflection of the neutral points.

(*j*) Star/interconnected star; single-phase load across two lines

With this connection and method of loading the secondary windings on all three limbs carry load currents, and therefore all the primary windings carry corresponding balancing load currents. The current distribution is clearly shown on the diagram, from which it will be seen there is no choking effect, and the transformer neutral points are stable if three-phase core-type transformers are used, and so may be earthed. On the secondary side the voltage of the open phase is practically the same as at no load. The voltage drops in the transformer windings are those due only to the normal impedance of the transformer.

(*k*) Star/interconnected star; single-phase load from one line to neutral

With this method of loading there is similarly no choking effect, as the primary windings corresponding to the loaded secondaries carry balancing load currents which flow simply through two of the line wires to the generating source. The voltage drops in the transformer windings are those due only to the normal impedance of the transformer, and the voltages of the above phases are practically the same as at no load. The secondary neutral is stable and can be earthed. The primary neutral can only be earthed, however, if the transformer unit is of the three-phase core-type of construction.

(*l*) Delta/interconnected star; single-phase load across two lines

With this connection and loading the general effect is similar to the star/interconnected star connection. That is, there is no choking effect, as the primary windings corresponding to the loaded secondaries take balancing load currents, although the primary current distribution is slightly different from that occurring with a star primary. The voltage drops in the transformer windings are those due only to the normal impedance of the transformer, while the voltage of the open phase is practically the same as at no load. The secondary neutral is stable and can be earthed.

(*m*) Delta/interconnected star; single-phase load from one line to neutral

With this connection and method of loading the results are similar to those obtaining with the star/interconnected star, that is, the primary windings corresponding to the loaded secondaries carry balancing load currents so that there is no choking effect. The voltage drops in the transformer windings are those due only to the normal impedance of the transformer, while the voltages of the open phases are practically the same as at no load. The secondary neutral point is stable and can be earthed.

(*n*) Vee/vee

With this connection and method of loading there is clearly no choking effect, as this is simply a question of supplying a single-phase transformer across any two lines of a three-phase generator. The voltage drops are comparable to those normally occurring, and the voltages of the open phases are practically the same as at no load. The connection is, however, electrostatically unbalanced, and should be used only in emergency.

(*o*) Tee/tee; single-phase load across two lines, embracing the teaser and half the main windings

With this connection and method of loading there is no choking effect, as the balancing load current in the corresponding primary windings has a perfectly free path through those windings and the two line wires to the generating source. The voltage drops in the windings are those due only to the normal impedance of the transformer, and the voltages of the open phases are practically the same as at no load. The neutral points are stable and may be earthed.

(p) Tee/tee; single-phase loading across two lines, embracing the main transformer only

With this connection and method of loading the results are identical to the loading of a vee/vee transformer as at (n), that is, this is simply a case of supplying a single-phase transformer from the two lines of a three-phase generator, and consequently there is no choking effect. The voltage drops in the transformer windings are those due only to the normal impedance of the transformer, while the voltages of the open phase are practically the same as at no load. The neutral points are stable and may be earthed.

(q) Tee/tee; single-phase load from teaser line terminal to neutral

With this connection and method of loading, although there is no load current in the secondary side of the main transformer, no choking effect occurs, as the load currents in the primary side of the main transformer flow in opposite directions in the two halves, and consequently their magnetic effects annul one another. The current distribution is as shown in the diagram, and the voltage drops in the windings are those due only to the normal impedance of the transformer. The voltage of the open phase is practically the same as at no load. The neutral points can be earthed without affecting the conditions.

(r) Tee/tee; single-phase load from one main transformer line terminal to neutral

With this method of loading there is no choking effect, as the distribution of load current in the primary windings is such that the ampere-turns in the whole of the teaser and main primaries balance those of the portions of the loaded secondaries. Voltage drops are therefore those due only to the normal impedance of the windings, and the voltages of the open phases are practically the same as at no load. The transformer neutral points are relatively stable and may be earthed.

(s) Three- to two-phase Scott connection; single-phase load across two outers

With this connection and method of loading the current distribution is as shown in the diagram, from which it will be seen that the primary winding of the main transformer carries load currents of different magnitudes in the two halves. That is, one half carries the sum of the full main primary load current and half the teaser primary load current, while the other half of the main winding carries current equivalent to the difference between these two. As in both teaser and main windings the total load ampere-turns are balanced, there is no choking

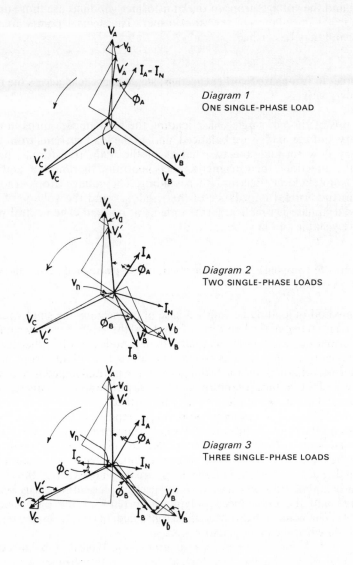

Diagram 1
ONE SINGLE-PHASE LOAD

Diagram 2
TWO SINGLE-PHASE LOADS

Diagram 3
THREE SINGLE-PHASE LOADS

V_A, V_B, V_C = *balanced no-load line to neutral voltages*
I_A, I_B, I_C = *line load currents*
I_N = *neutral load current*
V_A', V_B', V_C' = *load line to neutral voltages at receiving end*
v_a, v_b, v_c, v_n = *impedance voltage drops in each phase and in the neutral at receiving end*
ϕ_A, ϕ_B, ϕ_C = *load angles of lag at receiving end*

Fig. 25.4   Phasor diagrams showing unbalanced loadings on a delta/star, three-phase, step-down transformer

607

effect, and the voltage drops in the transformer windings are those due only to the normal impedance of the transformer. The neutral points are relatively stable and may be earthed.

(*t*) Three- to two-phase Scott connection; single-phase load across the teaser secondary only

With this method of single-phase loading the load ampere-turns in the teaser primary and secondary are balanced, and as the load currents from the teaser primary flow through the two halves of the main transformer primary in opposite directions, their magnetic effects neutralise one another, and therefore there is no choking effect in either winding. The voltage drops are those due only to the normal impedance of the windings, and the voltage of the open secondary phase is practically the same as at no load. The neutral points are relatively stable and may be earthed.

(*u*) Three- to two-phase Scott connection; single-phase load across the main transformer only

This method of loading is simply a case of connecting one single-phase transformer across two of the line wires of a normal three-phase supply, and the main transformer operates simply as such. There is no current in either teaser winding, and consequently there is no choking effect. The voltage drops in the main transformer windings are those due only to the normal impedance of the transformer, and the voltages of the open phase are practically the same as at no load. The neutral points are stable and may be earthed.

It should always be remembered that it is impossible to preserve the current balance on the primary side of a polyphase transformer or bank and in the line wires and source of supply when supplying an unbalanced polyphase load or a pure single-phase load. In most cases the voltage balance is maintained to a reasonable degree, and the voltage drops are only greater than those occurring with a balanced load on account of the greater phase differences between the voltages and the unbalanced polyphase currents or the pure single phase currents. The voltage drops become accentuated, of course, by the reactance of the circuit when the power factors are low.

Figure 25.4 shows the phasor diagrams for typical unbalanced loading conditions on a delta/star three-phase step down transformer where one, two and three separate single-phase loads are connected from lines to neutral. Voltage drops include transformer and cable or line drops. The triangles constructed on V_A', V_B', and V_C' show the resistive and reactive components of the total voltages across the respective loads. In diagram 1 the current I_N in the neutral is the same as the load current I_A; in diagram 2 the neutral current I_N is the phasor sum of the load currents I_A and I_B, while in diagram 3 I_N is the phasor sum of I_A, I_B and I_C.

Chapter 26

The influence of transformer connections upon third-harmonic voltages and currents

It is the purpose of this chapter, firstly, to enunciate briefly the fundamental principles of third-harmonic voltages and currents in symmetrical three-phase systems; secondly, to indicate their origin purely in respect of transformers; thirdly, to marshal the facts and present them in tabular form; and finally, to indicate their objectionable features.

It is not claimed that any new theories are introduced, but simply that facts, often understood in a more or less vague sort of way, are here crystallised and presented in a clear manner.

The treatment is confined to three-phase transformers with double windings, as the principles, once clearly understood, are so easily applicable to two-phase and six-phase transformers and to polyphase auto transformers as not to require further elaboration.

PRINCIPLES OF THIRD HARMONICS IN SYMMETRICAL THREE-PHASE SYSTEMS

There are two general forms of connections of three-phase systems to consider: (*a*) star connection and (*b*) delta connection.

(*a*) Star

In any star connected system of conductors it is a basic law that the instantaneous sum of the currents flowing to and from the common junction or star point is zero.

In a symmetrical three-phase, three-wire star connected system, the currents and voltages of each phase at fundamental frequency are spaced 120° apart. At any instant the instantaneous current in the most heavily loaded phase is equal and opposite in direction to the sum of the currents in the other two phases, and at fundamental frequency this balance is maintained throughout the cycle. At third-harmonic frequency, however, currents flowing in each phase would be $3 \times 120 = 360°$ apart, or, in other words, in phase with one another

609

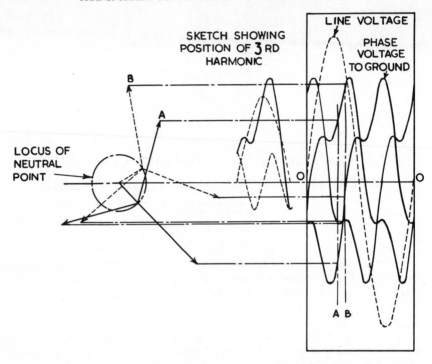

Fig. 26.1   Phenomenon of 'oscillating neutral' in a symmetrical three-phase, three-wire, star connected system with unearthed neutral

and flowing in the same relative direction in the phases at the same instant. The sum of the currents in the star connection would therefore not be zero, and consequently in a symmetrical three-phase, three-wire star connected system third-harmonic currents cannot exist.

If, however, a connection is taken from the neutral point in such a manner that it completes the circuit of each phase independently (though through a common connection), a current at three times the fundamental frequency can circulate through each phase winding and through the lines and the fourth lead from the neutral point. The fourth lead acting as a drain for third-harmonic currents preserves the current balance of the system; it has, of course, no effect on the currents at fundamental frequency, as these are already balanced.

Third-harmonic voltages, on the other hand, can exist in each phase of a symmetrical three-phase, three-wire star connected system, that is, from each line to ground,* but they cannot appear in the voltages between lines. The third-harmonic voltages in each phase are in phase with one another, so that there is one third-harmonic phasor only, and the neutral point of the star is located at the end of this phasor. The potential of the neutral point is consequently not zero, but oscillates round the zero point at triple frequency and third-harmonic voltage. Figure 26.1 illustrates this phenomenon and also shows how the third-harmonic voltages to ground cancel out, so far as the voltages

*With unearthed neutral, or from each line to neutral with earthed neutral.

between lines are concerned, leaving the line terminal voltages free from their influence.

When a connection is taken from the neutral point in such a manner as to allow third-harmonic currents to flow, the third-harmonic voltages to neutral are expended in forcing the currents round the circuits. It will be seen subsequently that according to the characteristics of the circuit in which these currents flow, the third-harmonic voltages may be suppressed totally or only partially.

(b) Delta

In any delta connected system of conductors it is a basic law that the resultant fundamental voltage round a delta is zero. That is, the addition of the voltage phasors at fundamental frequency which are spaced $360°/m$ (where m = number of phases) apart forms a regular closed polygon.

In a symmetrical three-phase delta connected system third-harmonic voltages tending to occur in each phase would be spaced $360°$ apart, and so would be in phase with each other and act in the closed delta circuit as a single-phase voltage of third-harmonic frequency. Such a voltage could not actually exist in a closed delta system, so that third-harmonic currents circulate round the delta without appearing in the lines and the third-harmonic voltages are suppressed.

In discussing the third-harmonic aspect of various combinations of star and delta connections for three-phase transformer operation, we therefore have the following bases to work upon:

(1) With a three-wire star connection, third-harmonic voltages may exist between lines and neutral or ground, but not between lines.

(2) With a three-wire connection, third-harmonic currents cannot exist.

(3) With a four-wire star connection, third-harmonic voltages from lines to neutral or ground are suppressed partially or completely according to the impedance of the third-harmonic circuit.

(4) With a four-wire star connection, third-harmonic currents may flow through the phases and through the line wires and fourth lead from the neutral.

(5) With a three-wire delta connection, third-harmonic voltages in the phases and hence between the lines are suppressed.

(6) With a three-wire delta connection, third-harmonic current may flow round the closed delta, but not in the lines.

ORIGIN OF THIRD-HARMONIC VOLTAGES AND CURRENTS IN TRANSFORMERS

It should be understood that this discussion is quite distinct and apart from higher harmonic functions of the source of supply, and it is limited to those which are inherent in the magnetic and electric circuits of the transformer. The two circuits being closely interlinked, it is a natural sequel that the higher harmonic phenomena occurring in both should be interdependent.

It is well known that there are two characteristics in the behaviour of sheet steel transformer laminations when under the influence of an alternating electromagnetic field, which produce an appreciable distortion in the waveform (from the standard sine wave) of certain alternating functions. These functions are no-load current, flux and induced voltages, any distortion of which is due to the varying permeability of the core steel plates and to cyclic magnetic hysteresis. For the purpose of this chapter, the range of the phenomena involved is more briefly and cogently explained by means of diagrams with short explanations than by lengthy dissertation and tedious mathematical equations. Figures 26.2 to 26.8 inclusive, together with the following remarks, aim at attaining this

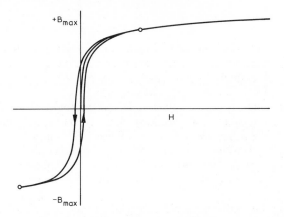

Fig. 26.2   Typical B/H curve and hysteresis loop for cold rolled steel

end. Figure 26.2 shows a typical B/H curve with hysteresis loop for silicon steel transformer core steel; the hysteresis loop illustrates the general shapes that would occur in practice.

Figure 26.3 shows the waveform relation between the no-load current, flux and induced e.m.f. when the e.m.f. is a sine wave and when hysteresis is absent. From a study of these curves it will be seen that the current is a true magnetising current, being in phase with the flux, its peaked form showing the presence of a prominent third harmonic. It will also be noted that this wave is symmetrical about the horizontal axis, and each half wave about a vertical axis. The flux must, of course, be sinusoidal on account of the assumption of a sine wave induced e.m.f.

Figure 26.4 is similar to Fig. 26.3, with the exception that hysteresis is taken into account. In this case the current is not a true magnetising current on account of the hysteresis component which is introduced, which makes the no-load current lead the flux by a certain angle θ, the hysteretic angle of advance. This figure also shows that for the same maximum flux the maximum values of the true magnetising and no-load current are the same, but that when taking hysteresis into account the no-load current becomes unsymmetrical about a vertical axis passing through its peak. It will, however, be seen by comparing Figs. 26.3 and 26.4 that the third-harmonic component is contained almost entirely in the true magnetising current, and very little, if any, in the current component due to hysteresis, thus indicating that third-harmonic currents are produced as a result of the varying permeability of the core steel, and only in a very minor degree by magnetic hysteresis.

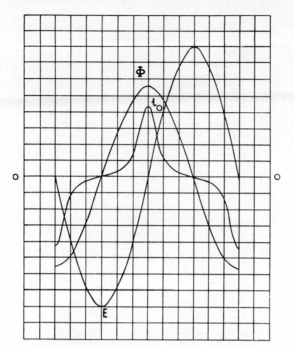

Fig. 26.3   No-load current, flux, and induced voltage waves, with a sine wave of applied voltage; hysteresis effects excluded.

$$i_0 = 100 \sin \theta - 54 \cdot 7 \sin 3\theta + 31 \cdot 5 \sin 5\theta + \ldots$$

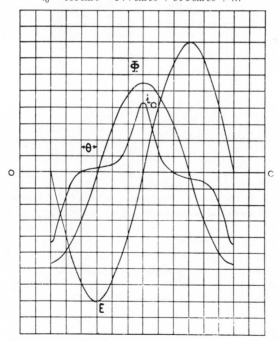

Fig. 26.4   No-load current, flux and induced voltage waves, with a sine wave of applied voltage; hysteresis effects included

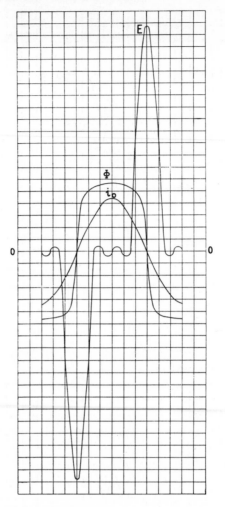

Fig. 26.5   No-load current, flux and induced voltage waves, with a sine wave of no-load current; hysteresis effects excluded.

$$\Phi_m = 100 \sin\theta + 22{\cdot}9 \sin 3\theta + 5{\cdot}65 \sin 5\theta + \ldots$$
$$E = 100 \cos\theta + 69{\cdot}0 \cos 3\theta + 28{\cdot}4 \cos 5\theta + \ldots$$

Figure 26.5 shows the waveform relation between the no-load current, flux and induced e.m.f. when the current is a sine wave and when hysteresis is absent. As in the case of Fig. 26.3, the current is a true magnetising current and in phase with the flux. The flux wave is flat topped, which indicates the presence of a third harmonic in phase with the fundamental, the harmonic having a negative maximum coincident with the positive maximum of the fundamental, and so producing a flat-topped resultant wave. It will be noticed that the flux wave is symmetrical about the horizontal axis, and each half wave about a vertical axis. The induced e.m.f. is, of course, affected by the departure of the shape of the flux wave from the sine, a flat-topped flux wave producing a highly peaked wave of induced e.m.f. (as shown in the figure), in which also appears a prominent third harmonic. In the case of the voltage wave the third harmonic is in opposition

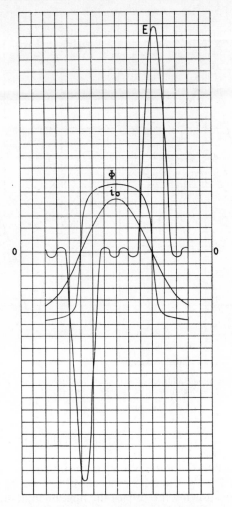

Fig. 26.6 No-load current, flux and induced voltage waves with a sine wave of no-load current, hysteresis effects included

$$\Phi_m = 100 \sin \theta + 22 \cdot 9 \sin 3\theta + 5 \cdot 65 \sin 5\theta + \ldots$$
$$E = 100 \cos \theta + 69 \cdot 0 \cos 3\theta + 28 \cdot 4 \cos 5\theta + \ldots$$

to the fundamental, the positive maximum of fundamental and harmonic waves occurring at the same instant, so that the resultant voltage wave becomes peaked.

Figure 26.6 is similar to Fig. 26.5 with the exception that hysteresis is taken into account. In this case the no-load current leads the flux, thereby producing the hysteretic angle of advance θ as in the case of Fig. 26.4. The flux wave is somewhat more flat-topped, and while still symmetrical about the horizontal axis, each half wave is unsymmetrical about a vertical axis passing through its peak.

The induced voltage waves of Figs. 26.5 and 26.6 do not take into account harmonics above the fifth, and this accounts for the ripples on the zero axis.

Hysteresis does not alter the maximum value of the flux wave, though it

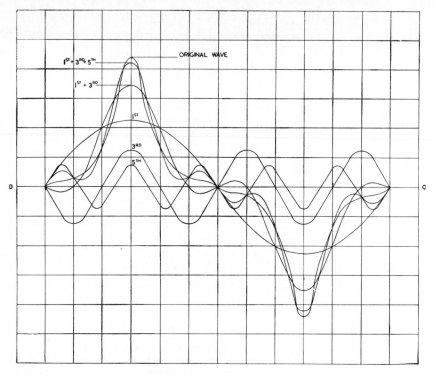

Fig. 26.7   *Harmonic analysis of peaked no-load current wave of Fig. 26.3.*
$$i_0 = 100 \sin \theta - 54{\cdot}7 \sin 3\theta + 31{\cdot}5 \sin 5\theta + \ldots$$

increases its dissymmetry; the wider the hysteresis loop the greater the dissymmetry of the flux wave.

Figure 26.7 and 26.8 show the analysis up to the fifth harmonic of the magnetising current wave, i_0, Fig. 26.3, and the induced voltage wave E, Fig. 26.5; in each case waves are given showing the sum of the fundamental and third harmonic, and indicating the degree of the error involved in ignoring harmonics beyond the third. In order to obtain some idea at a glance of the approximate phase of the third harmonic relative to the fundamental in a composite wave, Fig. 26.9 shows the shape of the resultant waves obtained when combining the fundamental and third harmonic alone with different positions of the harmonic.

From the foregoing discussion on the origin of third harmonics the following conclusions are to be drawn:

(1) With a sine wave of flux, and consequently induced voltages, the no-load current contains a prominent third harmonic which produces a peakiness in the wave. The third harmonic is introduced mainly into the true magnetising current component through the variation in the permeability of the sheet steel and only in a very small degree into the hysteresis component of the current by the cyclic hysteresis.

(2) With a sine wave of no-load current the flux and consequently the induced voltages contain prominent third harmonics which produce a flat-topped flux wave and peaked induced voltage waves.

616

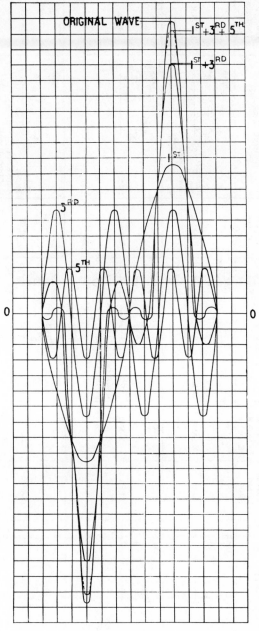

Fig. 26.8   Harmonic analysis of peaked induced voltage wave of Fig. 26.5.
$$E = 100 \cos \theta + 69 \cos 3\theta + 28{\cdot}4 \cos 5\theta + \ldots$$

Table 26.1   WAVE SHAPE RELATIONSHIP BETWEEN FLUX, INDUCED VOLTAGE AND NO-LOAD CURRENT

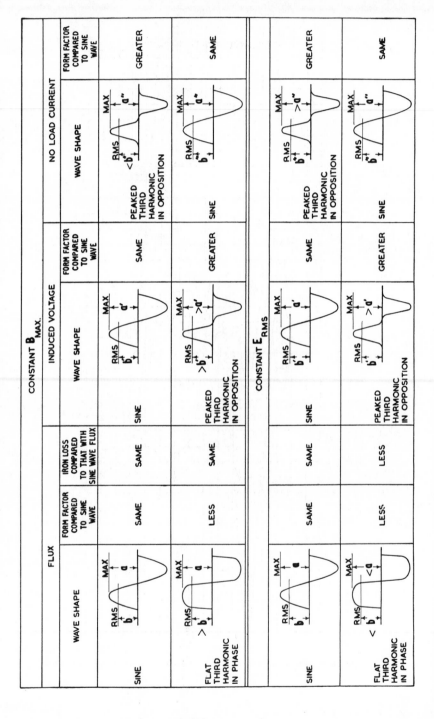

Table 26.1 summarises the above conclusions (a) for constant B_{max}. (b) for constant E_{rms}.

Table 26.2 is a summary of the conditions obtaining with different connections of double wound three-phase transformers. It has already been seen under the heading of 'origin' that third harmonics must occur in either the magnetising current or the voltage of each phase, and it should now be borne in mind that as a power transformer is connected across the supply voltage, and not in series with the line, the magnetising current will, wherever possible, take the shape required by the particular connections and condition of service. In consequence of this, where it is not possible for the third-harmonic component of the magnetising current to circulate in the primary, it will be transferred to the secondary by magnetic induction if the secondary connection is such as to allow circulation of the third harmonic to take place.

When both primary and secondary connections are such as to allow the circulation of the third-harmonic component of the magnetising current, the component will split up and circulate round each winding in inverse proportion to the impedance of the two windings.

This is the case whether the closed circuit is formed due to the connection chosen, such as with the delta connection, or due to the conditions of service, such as the star connection with a fourth lead either to the generator neutral on the primary side or to the load on the secondary side.

With reference to Table 26.2 the following remarks will make the whole question clear and indicate the broad principles determining the existence of third harmonics in no-load currents, flux and induced voltages.

(1) Star/star, both neutrals isolated

As there are no fourth leads to provide a drain for the third-harmonic component of the magnetising current, the flux and consequently the voltage from line to ground must contain third harmonics which appear on both primary and secondary sides. The line voltages and the line currents are not affected. The magnetising current is, of course, sinusoidal.

(2) Star/star, primary and generator neutrals connected together, secondary neutral isolated

The connection between the transformer primary and generator neutrals provides a return circuit for the third-harmonic components of the no-load currents, so that these currents in the transformer windings will contain third harmonics. There will be no circulation of magnetising current on the secondary side on account of the isolated neutral point. The sinusoidal components of the no-load currents do not flow through the connection between neutrals, as they are already balanced in the three phases. As the third-harmonic currents must flow through the lines and generator windings then, due to this additional impedance, the currents will be less than if they simply had to circulate round, say, a closed delta secondary only. Consequently, the m.m.f. per phase of the transformers is less than that required to eliminate the third-harmonic voltages to neutral; that is, part of the e.m.f.s neutralising the third-harmonic voltages

619

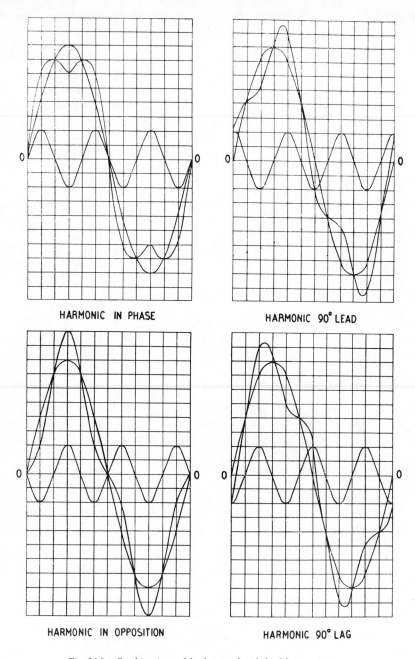

HARMONIC IN PHASE

HARMONIC 90° LEAD

HARMONIC IN OPPOSITION

HARMONIC 90° LAG

Fig. 26.9   Combinations of fundamental and third-harmonic waves

Table 26.2
STATEMENT ON THE INFLUENCE OF THREE-PHASE TRANSFORMER CONNECTIONS ON THIRD HARMONICS

Third harmonics in double-wound, single-phase, core- and shell-type transformers and in three-phase shell-type transformers for three-phase service. Sine wave of applied voltage

No.	Connections	Primary					Secondary			
		Currents		Voltages		Flux	Currents		Voltages	
		No-load	Line	Line	Phase		No-load	Line	Line	Phase
1	Star I.N./Star I.N.	sine	sine	sine	contains 3rd H.(P.)	contains 3rd H.(F.T.)	—	sine	sine	contains 3rd H.(P.)
2	Star N. to G./Star I.N.	*contains 3rd H.(P.)	*contains 3rd H.(P.)	sine	*contains 3rd H.(P.)	*contains 3rd H.(F.T.)	—	sine	sine	*contains 3rd H.(P.)
3	Star I.N./Star, 4-Wire	sine	sine	sine	*contains 3rd H.(P.)	*contains 3rd H.(F.T.)	*contains 3rd H.(P.)	*contains 3rd H.(P.)	sine	*contains 3rd H.(P.)
4	Star I.N. Tertiary Delta/Star I.N.	sine in star 3rd H. in Delta (P.)	sine	sine	sine	sine	contains 3rd H.(P.)	sine	sine	sine
5	Star I.N./Delta	sine	sine	sine	sine	sine	contains 3rd H.(P.)	sine	sine	sine
6	Star N. to G./Delta	*contains 3rd H.(P.)	*contains 3rd H.(P.)	sine	sine	sine	*contains 3rd H.(P.)	sine	sine	sine
7	Star I.N./Interconnected Star I.N.	sine	sine	sine	contains 3rd H.(P.)	contains 3rd H.(F.T.)	—	sine	sine	sine
8	Star I.N./Interconnected Star, 4-Wire	sine	sine	sine	contains 3rd H.(P.)	contains 3rd H.(F.T.)	—	sine	sine	sine
9	Delta/Star I.N.	contains 3rd H.(P.)	sine	sine	sine	sine	—	sine	sine	sine
10	Delta/Star, 4-Wire	contains 3rd H.(P.)	sine	sine	sine	sine	contains 3rd H.(P.)	contains 3rd H.(P.)	sine	sine
11	Delta/Delta	contains 3rd H.(P.)	sine	sine	sine	sine	contains 3rd H.(P.)	sine	sine	sine

* In all these cases the third-harmonic component is less than it otherwise would be if (a) the circulating third-harmonic current flowed through a closed delta winding only, or (b) the neutral point was isolated. I.N. means 'isolated neutral'. N. to G. means 'transformer primary neutral connected to generator neutral'. (P.) means 'peaked wave'. (F.T.) means 'flat top wave'.

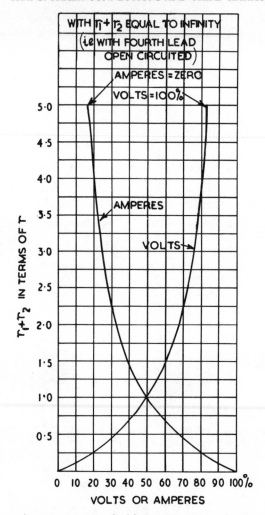

Fig. 26.10   Curves showing percentage third-harmonic current and voltage in star/star connected banks of single-phase transformers with fourth lead on primary side

act across the transformers and part across the lines and generator windings. Therefore, a portion of the original third-harmonic voltage still appears between lines and ground on both primary and secondary sides, while in addition a third-harmonic component of the no-load current circulates in the primary windings and in the lines between transformer and generator.

Figure 26.10 shows the proportion of third-harmonic current and voltage components occurring with this connection with different ratios of the impedance of lines plus generator windings to transformer windings, and it is to be noted that as this ratio approaches infinity (*i.e.*, disconnected neutrals) the current approaches zero and the voltage its maximum value.

The flux wave, therefore, contains a small third-harmonic component corresponding to the resultant third-harmonic voltage component, which later appears in the voltage from lines to ground on both primary and secondary sides. The voltage waves between lines on both sides of the transformer are sinusoidal. The no-load current flowing in the primary winding contains a third-harmonic component smaller than that which normally occurs in accordance with the ratio of the external to internal impedance. The line currents on the secondary side of the transformer remain sinusoidal.

(3) Star/star, primary neutral isolated, fourth lead on the secondary side

In this case the fourth lead on the secondary side operates exactly in the same way for the third-harmonic components of the no-load current as does the fourth lead on the primary side in case (2). The factors governing the magnitude of the third-harmonic current and voltage components are the same as given under the previous heading, so that with this connection a small third harmonic occurs in the flux wave corresponding to the resultant third-harmonic voltage component from line to ground which appears on both the primary and secondary sides. A third-harmonic component of the no-load current circulates in the transformer secondary windings and through the line wires and neutral connection. The no-load current in the primary side is sinusoidal. All other functions are sinusoidal, as stated in Table 26.2.

(4) Star, tertiary delta/star, primary and secondary neutrals both isolated

In this case the tertiary delta provides a closed path for circulation of the third-harmonic component of the no-load current, so that the current in the star primary is sinusoidal and the third-harmonic current appears in the tertiary delta only. The flux is a sine wave, and so also are all the voltages and line currents.

(5) Star/delta, primary neutral isolated

The delta connected secondary provides a closed path, and allows circulation of the third-harmonic component of the no-load current in a similar manner to the tertiary delta in case (4). On the primary side, therefore, the no-load current is sinusoidal, and as the flux wave is also sinusoidal the remaining functions will have this same characteristic.

(6) Star/delta, primary and generator neutrals connected together

In this case a closed path is provided on both primary and secondary sides for the circulation of the third-harmonic component of the no-load current so that this component will split up, part circulating through each winding in inverse proportion to the impedance of the two circuits. On account of the additional impedance introduced on the primary side by the lines and generator windings, the greater portion of the third-harmonic component of the no-load current will circulate in the delta secondary, the remainder flowing through the primary

623

windings, lines and generator windings. The flux wave will be sinusoidal so that the voltages on both sides will be free from third harmonics.

(7) Star/interconnected star, primary and secondary neutrals both isolated

As there is no closed circuit on either primary or secondary sides which will allow the third-harmonic component of the no-load current to circulate, a third-harmonic component will be contained in the flux wave which on the primary star connected side will produce a third harmonic in the voltage wave from line to ground, though not appearing in the voltage between lines. On the secondary side the voltage across the whole of the winding on each transformer or limb if connected in series would also contain a third-harmonic component, but by interconnecting the windings of the *different* transformers or limbs the third-harmonic voltage components of the halves of the windings which are now in series, and electrically constitute each phase, oppose one another and so cancel out. On the secondary side, therefore, neither third-harmonic components of voltage from line to neutral nor of current exist. The secondary line voltage will also be sinusoidal. It is to be noted, however, that should a connection be taken from the point where the half of the windings of one transformer is joined in series to the half of the windings of another transformer, a third-harmonic voltage component will be contained in the voltage from this point either to neutral or to the line. The no-load current flowing in the primary side and also the primary and secondary load currents will be sinusoidal.

(8) Star/interconnected star, primary neutral isolated, fourth lead on secondary side

It was shown under the previous heading that no third-harmonic component of the voltage from line to neutral existed on the secondary side, and therefore the addition of a fourth lead on this side will not result in a circulation of third-harmonic current component, as there is no voltage component to maintain such a flow. The conditions, therefore, will be exactly the same as with the previous connection.

(9) Delta/star, secondary neutral isolated

The conditions with this connection are identical with those obtaining for the star/delta under (5), with the exception that the whole of the no-load current is supplied on the primary side as the third-harmonic component circulates in the delta winding. All other functions are sinusoidal.

(10) Delta/star, fourth lead on secondary side

From the point of view of third harmonics this case is the reverse of that outlined under (6). A closed path is similarly provided on both primary and secondary sides for the circulation of the third-harmonic component of the no-load current, so that this component will be split up, part circulating through each winding in inverse proportion to the impedance of the two circuits. In this case additional impedance is introduced on the secondary side by the fourth lead and the load, so that the greater portion of the third-harmonic component of the no-load current will circulate in the delta primary, the remainder flowing through the secondary windings, lines and load. The flux wave will be sinusoidal, so that the voltages on both sides will be free from third harmonics.

(11) Delta/delta

In this case the closed delta windings on each side of the transformer offer a path of minimum impedance for the flow of the third-harmonic component of the no-load current which will be split up, part circulating through each winding in inverse proportion to the impedance of the two circuits. The flux wave will be sinusoidal so that the voltages on both sides will be free from third harmonics.

It should be noted that it is not permissible to join together the generator and transformer primary neutrals (either via earth or direct) *if the transformer primary is connected in the interconnected star manner*. If the neutrals are joined a third-harmonic current may flow in the fourth lead resultant from third-harmonic voltages in the source of supply, and due to the splitting and interconnection of the transformer primary windings, the third harmonic exciting ampere-turns, and consequently the induced back e.m.f.s in the halves of the windings on each limb of the transformer neutralise each other, thereby providing a low impedance path (virtually a short circuit so far as transformer reactance is concerned) for the flow of third-harmonic currents. Such currents would only be limited in magnitude by the impedance of the generator winding and by the resistance of the connecting cables and transformer windings, and they may reach such large values in practice as seriously to overload the transformer and generator windings. Figure 26.11 shows the paths of the third-harmonic currents, and the opposition of third-harmonic exciting ampere-turns in the halves of the windings on each limb of the transformer will be seen.

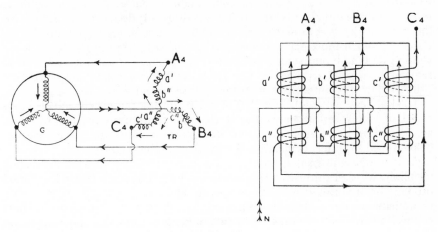

Fig. 26.11   Paths of third-harmonic currents

The final conclusions arrived at under headings (1) to (10) apply strictly to single-phase transformers of both shell and core types, and to three-phase shell-type transformers, as with each of these types of construction the magnetic circuit of each phase is complete in itself, and no interchange of m.m.f. and flux occurs between the phases. With three-phase core-type transformers of the usual construction in which there are three limbs the case is somewhat different. With this type of transformer, any third-harmonic fluxes tending to flow would have the same instantaneous direction up or down each limb, and in order to preserve the magnetic balance of the core the third-harmonic flux in each limb must find its return path outside the core through the oil (or air) and tank. Such a path is of very high reluctance, and consequently the third harmonics in the flux and induced voltage waves are very small.

OBJECTIONABLE FEATURES OF THIRD HARMONICS

These come under two headings: and may be summarised as follows:

Due to third-harmonic currents
(*a*) Overheating of transformer windings and of load.
(*b*) Telephone and discriminative protective gear magnetic disturbances.
(*c*) Increased iron loss in transformers.

Due to third-harmonic voltages
(*d*) Increased transformer insulation stresses.
(*e*) Electrostatic charging of adjacent lines and telephone cables.
(*f*)Possible resonance at third-harmonic frequency of transformer windings and line capacitance.

These disadvantages may briefly be referred to as follows:
(*a*) In practice, overheating of the transformer windings and load due to the circulation of third-harmonic current very rarely occurs, as care is taken to design the transformer so that the flux density in the core is not so high as to increase the third-harmonic component of the no-load current unduly. Apart from the question of design, a transformer might, of course, have a higher voltage impressed upon it than that for which it was originally designed, but in this case the increased heating from the iron loss due to the resulting higher flux density would be much more serious than the increased heating in the transformer windings due to larger values of the third-harmonic circulating current. These remarks hold good, irrespective of whether the transformer windings are delta connected or star connected with a fourth lead system.

The only case where the circulation of the third-harmonic currents is likely to become really serious in practice is where the transformer primary windings are connected in interconnected star, the generator and transformer neutrals being joined together, as mentioned previously in this chapter.

(*b*) It is well known that harmonic currents circulating in lines paralleling telephone wires or through the earth where a telephone earth return is adopted produce disturbances in the telephone circuit. This is only of practical importance in transmission or distribution lines of some length (as distinct from short connections to load), and then as a rule only occurs with the star connection using a fourth lead, which may be one of the cable cores or the earth.

Similar interference may take place in the pilot cores of discriminative protective gear systems, and unless special precautions are taken relays may operate incorrectly.

The remedy is, of course, obvious, and consists either of using a delta connected transformer winding or omitting the fourth lead and earthing at one point of the circuit only.

(c) In the case of a three-phase bank of single-phase transformers using a star/star connection, it has been proved experimentally that a fourth lead connection on the primary side between the transformer bank and generator neutrals (which allows the circulation of third-harmonic currents) results in an increase of the iron loss of the transformers of a figure of the order of 20% higher than that given with the neutrals disconnected. This figure is, of course, variable according to the design of the transformers and the impedance of the primary circuit. The conditions are similar for three-phase shell-type transformers.

Under certain conditions, the third-harmonic component of the phase voltage of star/star connected three-phase and shell-type transformers or banks of single-phase transformers may be amplified by the line capacitances. This occurs when the h.v. neutral is earthed, so that third-harmonic currents may flow through the transformer windings, returning through the ground and the capacitances of the line wires to ground. The amplification occurs only when the capacitance of the circuit is small as compared to its inductance, in which case the third-harmonic currents will lead the third-harmonic voltages almost by 90°, and they will be in phase with the third-harmonic component of the magnetic fluxes in the transformer cores. The third harmonic component of the fluxes therefore becomes intensified, which in turn produces an increase in the third-harmonic voltages, and a still further increase of the third-harmonic capacitance currents. This process continues until the transformer cores becomes saturated, at which stage it will be found the induced voltages are considerably higher and more peaked than the normal voltages, and the iron loss of the transformer is correspondingly greater. In practice, the iron loss has been found to reach three times the normal iron loss of the transformer, and apparatus has failed in consequence.

This phenomenon does not occur with three-phase core-type transformers on account of the absence of third harmonics.

(d) It has been pointed out previously that with the three wire star connection and isolated neutral a voltage occurs from the neutral point to ground having a frequency of three times the fundamental, so that while measurements between the lines and from lines to neutral indicate no abnormality, a measurement from neutral to ground with a sufficiently low reading voltmeter would indicate the magnitude of the third harmonic. In practice, with single-phase transformers the third-harmonic voltages may reach a magnitude of 60% of the fundamental, and this is a measure of the additional stress on the transformer windings to ground. While due to the larger margin of safety it may not be of great importance in the case of ordinary power distribution transformers, it will have considerable influence on the reliability of transformers as higher voltages are approached.

(e) Due also to the conditions outlined in (d), star connected banks of single-phase transformers connected to an overhead line or underground cable, and operated with an earthed or unearthed neutral, may result in an electrostatic charging at third-harmonic frequency of adjacent power and telephone cables.

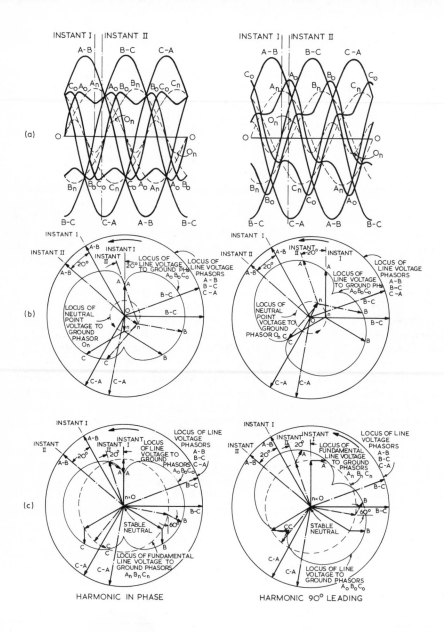

Fig. 26.12   Rectangular and polar co-ordinate forms of fundamental and third-harmonic

628

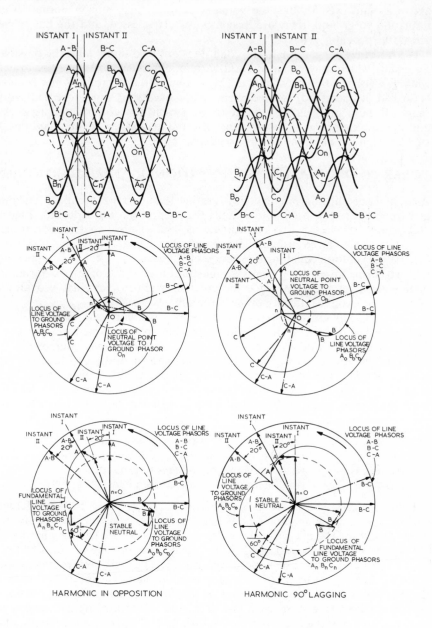

voltages in a symmetrical three-phase, three-wire, star connected system

This produces abnormal induced voltages to ground if the adjacent circuits are not earthed, the whole of the circuit being raised to an indefinite potential above ground even though the voltages between lines remain normal. The insulation to ground, therefore, becomes unduly stressed, and the life of the apparatus probably reduced.

(f) A further danger due to the conditions outlined under (e) is the possible resonance which may occur at third-harmonic frequency of the transformer windings with the line capacitance. This can happen if the transformer neutral is earthed or unearthed, and the phenomenon occurs perhaps more frequently than is usually appreciated, but due to the present-day complicated networks and the resulting large damping constants, the magnitude of the quantities is such as often to render the disturbances innocuous.

POLAR CO-ORDINATE REPRESENTATION OF THIRD HARMONICS

As in practice third-harmonic voltages rather than currents are more likely to give trouble it will be profitable to study the phenomena a little further. Figure 26.12 shows the rectangular co-ordinate method of wave construction translated into the polar co-ordinate form in which the angle represents time, one complete revolution equalling one period of $1/f$ seconds. Taken in conjunction, the two forms give a very complete illustration of the progressive variation of the quantities involved, while the polar diagram in particular provides a clear insight into the influence of the harmonic.

Figure 26.12(a) is in part a reproduction of Fig. 26.9, which shows the resultant voltage waves to ground in a symmetrical three-phase, three-wire, star connected system, with different positions of the third-harmonic component. In addition, the voltage waves between lines are shown, and these illustrate the well known fact that no matter what the relative position of harmonic and fundamental may be, the third-harmonic voltages in the phases cancel out and leave the line voltages free from their influence. In Fig. 26.12 the symbols have the following significance:

An, Bn, Cn	= fundamental sine waves of phase voltages.
On	= third harmonic sine wave voltage component.
Ao, Bo, Co	= resultant waves of voltages to ground.
A-B, B-C, C-A	= sine waves of voltages between lines.

Figure 26.12(b) shows the polar diagrams corresponding to the different rectangular co-ordinate diagrams of Fig. 26.12(a) for a system having an unearthed neutral. The diagrams show clearly the locus of the voltage phasors of the neutral point to ground, the lines to ground, and the line to line. The first and third of these, it will be noticed, are true circles having their centres at the point O, thus indicating that the voltage from the neutral point to ground and also the voltages between lines follow the sine law. The non-sinusoidal characteristic of the voltages from each line to ground is indicated by the expansion and contraction of the curve representing the loci of the line voltage to ground phasors, and the four cases given, corresponding to different positions of the harmonic, indicate that the locus curves are identical in shape, varying only in their relative position. That is, their maximum and minimum values occur at different angles or instants of time, this alone accounting for the different shapes of the resultant waves Ao, Bo, and Co in Fig. 26.12. The angular distances

between the maximum and minimum values of the loci curves of the line voltages to ground phasors remain constant. To show just how the radial phasors themselves are affected the conditions at two definite instants are given, the instants being 20° of the fundamental apart; their corresponding position is also shown in the diagrams of Fig. 26.12(a). In all cases the vertical components of the radial phasors will be found to be equal to the heights of the corresponding waves above or below the zero line of Fig. 26.12(a).

Figure 26.12(c) shows the polar diagrams for the same instances as given in Figs. 26.12(a) and 26.12(b), but for the case where the neutral point of the system is earthed. These diagrams show that earthing the neutral in no way affects the loci of the voltage phasors of the lines to ground or the line to line, nor does it affect the minimum and maximum values of the former, or the instants at which they attain these values. As the neutral point is earthed, the third-harmonic voltages appear at the line ends of the phasors representing the phase voltages, with the result, as shown, that for the cases of unearthed and earthed neutrals the voltage stresses on the system so far as third harmonics alone are concerned remain the same.

FURTHER NOTES ON THIRD HARMONICS WITH THE STAR/STAR CONNECTION

It is generally appreciated that three-phase shell-type transformers and three-phase groups of single-phase transformers should not have their windings connected star/star on account of the third-harmonic voltages which may be generated in the transformers at the normal flux densities usually employed. It is, however, not so equally well known that under certain operating conditions the star/star connection of the type of transformers referred to above may produce serious overheating in the iron circuit in addition to augmented stresses in the dielectrics. The conditions referred to are when the secondary neutral of the transformer or group is earthed, the connecting lines having certain relative values of electrostatic capacitance.

Consider a three-phase step-up group of single-phase transformers having their windings star/star connected, each transformer of such a group having a flux density in the core of approximately 1·55 teslas.

With isolated neutrals on both sides, no third-harmonic currents can flow, and consequently the magnetic fluxes and induced voltages would contain large third-harmonic components, the flux waves being flat-topped and the induced voltage waves peaked. The magnetising current waves would be sinusoidal. At the flux density stated, the flux waves would have a third-harmonic component approximately equal in amplitude to 20% of that of the fundamental, and the resulting induced voltage waves would have third-harmonic components of amplitudes of approximately 60% of that of their fundamentals. With isolated neutrals the third-harmonic components of the voltage waves would be measurable from each neutral to earth by an electrostatic voltmeter. Their *effects* would be manifested when measuring the voltages from each line terminal to the neutral point by an ordinary moving iron or similar voltmeter. There would be no trace of them when measuring between line terminals on account of their opposition in the two windings which are in series between any two line terminals so far as third harmonics are concerned.

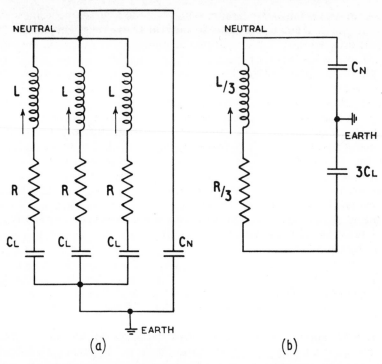

Fig. 26.13   Third-harmonic distribution of inductance, resistance and capacitance in an unearthed neutral three-phase circuit consisting of the secondaries of a three-phase group of single-phase transformers supplying an open-ended transmission line

With isolated neutrals the only drawback to the third-harmonic voltage components is the increased dielectric stress in the transformer insulation.

It should be borne in mind that so far as third harmonics of either voltage or current are concerned the transformer windings of each phase are really in parallel and the harmonics in each winding have the same time phase position. When such transformers are connected to transmission or distribution lines on open circuit, the parts which are effective so far as third harmonics are concerned can be represented as shown in Fig. 26.13(a) where we have three circuits in parallel, each consisting of one limb of the transformer with the capacitance to earth of the corresponding line, this parallel circuit being in series with the capacitance between earth and the neutral point of the transformer. By replacing the three parallel circuits by a simple equivalent circuit consisting of a resistance, inductance and capacitance Fig. 26.13(a) can be simplified to that shown in Fig. 26.13(b). The inductance is simply that of the three phases of the transformer in parallel, and the voltage across these is simply the third-harmonic voltage generated in each secondary phase of the transformer. As the third-harmonic voltages are generated in the transformer windings on account of the varying permeability of the magnetic cores, the inductance shown in Fig. 26.13(b) can be looked upon as being a triple frequency generator supplying a voltage equal to the third harmonic voltage of each phase across the two capacitors in series. The capacitor $3C_L$ is equal to three times the capacitance to earth of each line while the capacitor C_N represents the capacitance from the neutral point

632

to earth. By comparison the latter capacitor is infinitely small, so that as a voltage applied across series capacitors divides up in inverse proportion to their capacitances, practically the whole of the third-harmonic voltage appears across the capacitor formed between the transformer neutral point and earth. This explains why, in star/star connected banks having isolated neutrals, the third-harmonic voltage can be measured from the neutral point to earth by means of an electrostatic voltmeter.

Now consider the conditions when the secondary neutral point is earthed, the secondary windings being connected to a transmission or distribution line on open circuit. This line, whether overhead or underground, will have certain values of capacitance from each wire to earth, and so far as third harmonics are concerned the circuit is as shown diagrammatically in Fig. 26.14(a). It will be seen that the only difference between this figure and Fig. 26.13(a) is that the capacitor C_N between the neutral point and earth has been short circuited. The effect of doing this may, under certain conditions, produce very undesirable results. The compound circuit shown in Fig. 26.14(a) may be replaced by that shown in Fig. 26.14(b), where resistance, inductance and capacitance are respectively the single equivalents of the three shown in parallel in Fig. 26.14(a), and from this diagram it will be seen that all the third-harmonic voltage is concentrated from each line to earth. Under this condition the third-harmonic component cannot be measured direct, but its effects are manifested when measuring from each line terminal to ground by an ordinary moving iron or similar instrument.

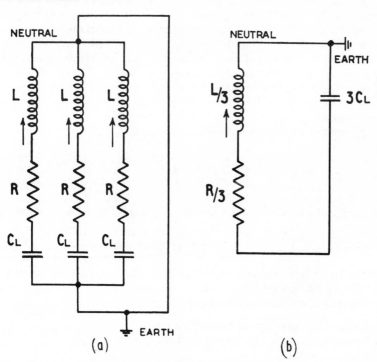

Fig. 26.14   *Third-harmonic distribution of inductance, resistance and capacitance in an earthed neutral three-phase circuit consisting of the secondaries of a three-phase group of single-phase transformers supplying an open-ended transmission line*

633

The chief difference between the conditions illustrated in Figs. 26.13(a) and 26.14(a) is that where in the first case no appreciable third-harmonic current could flow on account of the small capacitance between the neutral point and earth, in the second case triple-frequency currents can flow through the transformer windings completing their circuit through the capacitances formed between the lines and earth. We thus see that the conditions are apparently favourable for the elimination of the third-harmonic voltages induced in the transformer windings on account of the varying permeability of the magnetic cores.

This, however, is not all the story, for in order that the third-harmonic voltages induced in the transformer windings shall be eliminated, the third-harmonic currents must have a certain phase relationship with regard to the fundamental sine waves of magnetising currents which flow in the primary windings. In practice the third-harmonic currents flowing in such a circuit as shown in Fig. 26.14(a) may or may not have the desired phase relationship, for the following reasons.

The circuit shown in Fig. 26.14(b) is a simple series circuit of inductance L, resistance R and capacitance C, the impedance of which is given by the equation,

$$Z = \sqrt{\left\{R^2 + \left(2\pi f L - \frac{1}{2\pi f C}\right)^2\right\}}$$

The resistance R is the combined resistance of the three circuits in parallel shown in Fig. 26.14(a), namely, the transformer windings which are earthed, the lines, and the earth. The capacitance C is the combined capacitance of the three lines to earth in parallel, as the capacitance of the transformer windings to earth is so small that it can be ignored. The inductance L is the combined inductance of the three transformer windings in parallel which are earthed, the inductance between the lines and earth being ignored on account of their being very small. The inductance of the transformer windings, of course, corresponds to open-circuit conditions, as the triple-frequency currents are confined to the secondary windings only, on account of the connections adopted.

For a circuit of this description the power factor is given by the expression,

$$\cos \phi = \frac{R}{Z} = \frac{R}{\sqrt{\left\{R^2 + \left(2\pi f L - \frac{1}{2\pi f C}\right)^2\right\}}}$$

and the angle of lead or lag of the current with respect to the applied voltage is,

$$\phi = \tan^{-1}\left(\frac{2\pi f L - 1/2\pi f C}{R}\right) \tag{26.1}$$

If the value of $2\pi f L$ is greater than that of $1/2\pi f C$ the angle ϕ is lagging, and if smaller the angle ϕ is leading.

There are three extreme conditions to consider:

(1) when C is very large compared with L
(2) when L is very large compared with C
(3) when L and C are equal.

If C is large compared with L the impedance of the combined circuit is relatively low, so that the line capacitances to earth form, more or less, a short

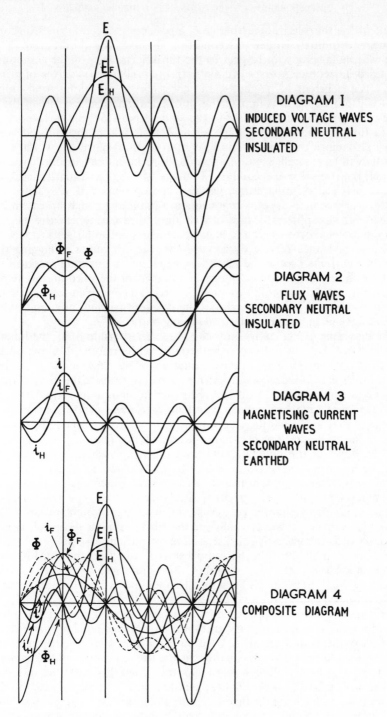

DIAGRAM I
INDUCED VOLTAGE WAVES
SECONDARY NEUTRAL
INSULATED

DIAGRAM 2
FLUX WAVES
SECONDARY NEUTRAL
INSULATED

DIAGRAM 3
MAGNETISING CURRENT
WAVES
SECONDARY NEUTRAL
EARTHED

DIAGRAM 4
COMPOSITE DIAGRAM

Fig. 26.15   Induced voltage, flux and magnetising current waves in a three-phase, star/star connected group of single-phase transformers with secondary neutral solidly earthed and supplying an open-ended line such that $C_L > L_L$

635

circuit to the third-harmonic voltage components induced in the transformer secondary windings. Under this condition the resulting third-harmonic currents will be lagging with respect to the third-harmonic voltage components. The third-harmonic currents will act with the fundamental waves of primary magnetising current to magnetise the core, and the resulting total ampere-turns will more or less eliminate the third-harmonic components of the flux waves, bringing the latter nearer to the sine shape. This will correspondingly reduce the third-harmonic voltage components, making the induced voltage waves also more sinusoidal. The reduction in third-harmonic voltage components will have a reflex action upon the third-harmonic currents circulating through the transformer secondary windings and the line capacitances, and a balance between third-harmonic voltages and currents will be reached when the third-harmonic voltage components are reduced to such an extent as to cause no further appreciable flow of secondary third-harmonic currents.

In the extreme case where the line capacitances are so large as to make the capacitive reactance practically zero, almost the full values of lagging third-harmonic currents flow in the secondary windings to eliminate practically the whole of the third-harmonic voltages, so that from the third-harmonic point of view this condition would be equivalent to delta connected secondary transformer windings. Figure 26.15 shows the different current, flux, and induced voltage wave phenomena involved, assuming that $C_L > L_L$.

The diagrams of Fig. 26.15 show the phase relationship of all the functions involved, but they do not show the actual third-harmonic flux and voltage-reduction phenomena. The composite diagram of Fig. 26.15 shows very clearly that the third-harmonic secondary current is in opposition to the third-harmonic flux component, and the result is a reduction in amplitude of the latter. As a consequence the induced voltage waves become more nearly sinusoidal, and ultimately they approach the true sine wave to an extent depending upon the value of the capacitance reactance of the secondary circuit.

When, however, the inductance of the transformer windings is high compared with the line capacitances to ground, the third-harmonic components of the voltage waves become intensified. In this case the inductive reactance is very high compared with the capacitive reactance, so that the third-harmonic voltage components impressed across the line capacitances produce third-harmonic secondary currents which lead the third-harmonic secondary voltages. The angle of lead is given by eqn. 26.1 and in the extreme case where the capacitance is very small the third-harmonic current will lead the third-harmonic voltage almost by 90°. The resulting third-harmonic ampere-turns of the secondary winding act together with the fundamental exciting ampere-turns in the primary, and as the two currents are in phase with one another their effect is the same as that produced by a primary exciting current equal to the sum of the fundamental primary and third-harmonic secondary currents. The sum of two such currents in phase is a dimpled current wave, and compared with the fundamental sine wave of exciting current the r.m.s. value of the composite current wave is higher, though more important than this is the fact that such a current wave produces a very flat topped flux wave. In other words, the third-harmonic components of the flux waves are intensified, and on this account the third-harmonic voltage components of the induced voltage waves are also intensified. Higher third-harmonic voltage waves react upon the secondary circuit to produce larger third-harmonic currents, which in turn increase the

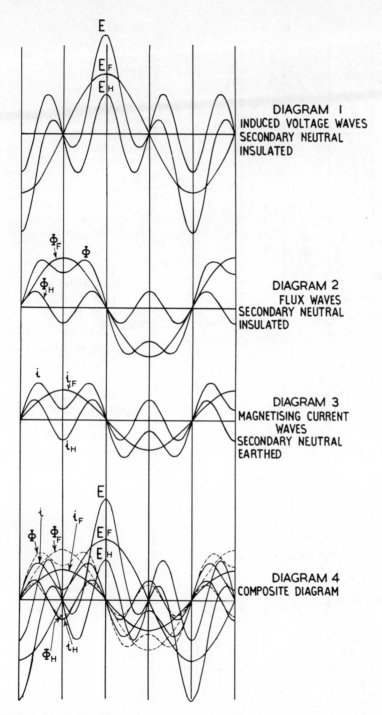

DIAGRAM 1
INDUCED VOLTAGE WAVES
SECONDARY NEUTRAL
INSULATED

DIAGRAM 2
FLUX WAVES
SECONDARY NEUTRAL
INSULATED

DIAGRAM 3
MAGNETISING CURRENT
WAVES
SECONDARY NEUTRAL
EARTHED

DIAGRAM 4
COMPOSITE DIAGRAM

Fig. 26.16   Induced voltage, flux and magnetising current waves in a three-phase, star/star connected group of single-phase transformers with secondary neutral solidly earthed and supplying an open-ended line such that $L_L > C_L$

637

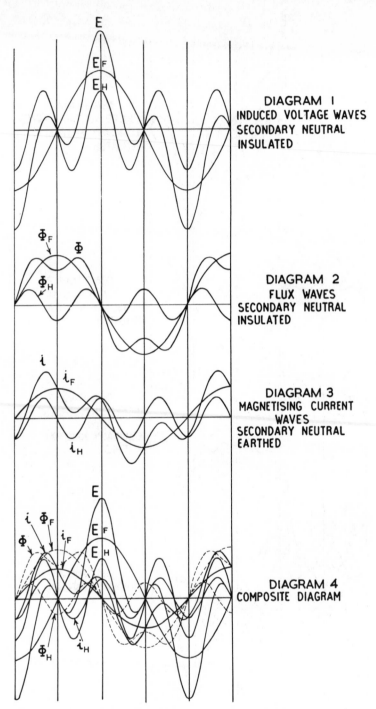

DIAGRAM 1
INDUCED VOLTAGE WAVES
SECONDARY NEUTRAL
INSULATED

DIAGRAM 2
FLUX WAVES
SECONDARY NEUTRAL
INSULATED

DIAGRAM 3
MAGNETISING CURRENT
WAVES
SECONDARY NEUTRAL
EARTHED

DIAGRAM 4
COMPOSITE DIAGRAM

Fig. 26.17   Induced voltage, flux and magnetising current waves in a three-phase, star/star connected group of single-phase transformers with secondary neutral solidly earthed and supplying an open-ended line such that $C_C = L_L$

638

third-harmonic flux waves, and again the third-harmonic voltage waves. This process of intensification continues until a further increase of magnetising current produces no appreciable increase of third-harmonic flux, so that the ultimate induced voltages become limited only by the saturation characteristics of the magnetic circuit. It should be noted that the third-harmonic currents circulate in the secondary windings only, as the connections on the primary side do not permit the transfer of such currents.

Figure 26.16 shows the phase relationship of the different current, flux and induced voltage waves involved, assuming the third-harmonic currents lead the third-harmonic voltage components by 90°. The diagrams of this figure do not show the actual amplification phenomena. The composite diagram shows very clearly that the third-harmonic secondary current is in phase with the third-harmonic flux component, and the result is an amplification of the latter. Therefore, the induced voltage waves become more highly peaked and ultimately reach exceedingly high values, producing excessive dielectric stresses, high iron losses, and severe overheating.

Cases have occurred in practice of transformer failures due to this third-harmonic effect, and one case is known where, on no-load, the transformer oil reached a temperature rise of 53°C in six hours, the temperature still rising after that time at the rate of 3°C per hour.

In the resonant condition where the capacitive and inductive reactances are equal, the flow of third-harmonic current is limited only by the resistance of the secondary circuit. The third-harmonic currents would be in phase with the third-harmonic voltage components, and being of extremely high values they would produce exceedingly high voltages from each line to earth and across the transformer windings. The transformer core would reach even a higher degree of saturation than that indicated in the previous case, and the transformers would be subjected to excessive dielectric and thermal stresses. Figure 26.17 shows the wave phenomena apart from the amplification due to resonance.

The resonant condition fortunately, however, is one that may be seldom met with, but the other two cases are likely to occur on any system employing star/star connected transformers with earthed secondary neutral, and unless some provision can be made for allowing the circulation of third-harmonic currents under permissible conditions three-phase shell-type transformers or three-phase groups of single-phase transformers should not so be connected.

With three-phase core-type transformers there is still theoretically the same disadvantage, but as in such transformers the third-harmonic voltage components do not exceed about 5% of the fundamental, the dangers are proportionately reduced. However, at high transmission voltages even a 5% third-harmonic voltage component may be serious in star/star connected three-phase core-type transformers when the neutral point is earthed, and it is therefore best to avoid this connection entirely if neutral points have to be earthed.

Precisely the same reasoning applies to three-phase transformers or groups having interconnected star/star windings if it is desired to earth the neutral point on the star connected side. With this connection the third-harmonic voltages are eliminated by opposition on the interconnected star side only, but they are present on the star connected side in just the same way as if the windings were star/star connected, their average magnitudes being of the order of 5% for three-phase core-type transformers and 50 to 60% in three-phase shell-type and three-phase groups of single-phase transformers.

Chapter 27

Electromagnetic forces in transformer windings*

A power transformer must be constructed so that it is able to withstand the mechanical stresses caused by external system faults, and B.S.171, Section 8, specifies the duration of the short-circuit fault current. Although the provisions of Section 8 do not necessarily correspond to service conditions they form a satisfactory basis of mechanical design.

To the designer the problem has two aspects, (i) the calculation or measurement of the electromagnetic forces and (ii) the mechanical design of the transformer windings to withstand these forces. The latter requires a knowledge of the properties of the materials used in the construction, which are copper and insulating materials. Stresses produced by the radial components of the electromagnetic forces, *i.e.* forces acting at right angles to the axis of the limb, are quite different from those produced by the components acting parallel to the axis and it is convenient to consider the two components separately.

Short-circuit currents

It is usual to determine the electromagnetic forces acting in a power transformer when a current corresponding to the first peak of a three-phase short circuit flows in the windings, it being assumed that the transformer is connected to a supply system of infinite fault capacity. It is considered that an asymmetry factor of 1·8 as defined in B.S.171 represents the worst conditions likely to occur in practice.

On this basis the first peak of the short-circuit current of a three-phase transformer has a maximum value of

$$\hat{I}_{sc} = \frac{1 \cdot 8 \sqrt{2}\,\mathrm{MVA}\,10^6}{\sqrt{3}\,V e_z} \quad \text{amperes} \qquad (27.1)$$

where, V = line terminal voltage

e_z = fractional (per unit) impedance voltage.

---

*Based upon ERA Report Ref. Q/T134, 'The Measurement and Calculation of Axial Electromagnetic Forces in Concentric Transformer Windings', by M. Waters, B.Sc., F.I.E.E., and a paper with the same title published in the Proceedings of the Institution of Electrical Engineers, Vol. 10, Part II, No. 79, February, 1954.

The limb current I_{max} corresponding to this value is used in force calculations.

The impedance voltage e_z is dependent upon the tapping position, and to calculate the forces accurately it is necessary to use the value of impedance corresponding to the tapping position being considered. For normal tapping arrangements the change in the percentage impedance due to tappings is of the order of 10 %, and if this is neglected the force may be in error by an amount up to plus or minus 20 %.

For preliminary calculations, or if a margin of safety is required, the minimum percentage impedance which may be obtained on any tapping should be used, and in the case of tapping arrangements shown in column one of Table 27.1 this corresponds to the tapping giving the best balance of ampere-turns along the length of the limb. However, in large transformers, where a good ampere-turn balance is essential to keep the forces within practical limits, the change in percentage impedance is small and can usually be neglected.

When calculating forces the magnetising current of the transformer is neglected, and the primary and secondary windings are assumed to have equal and opposite ampere-turns. All forces are proportional to the square of the ampere-turns, with any given arrangement of windings.

Mechanical strength

It has been suggested by other authors that the mechanical strength of a power transformer should be defined as the ratio of the r.m.s. value of the symmetrical short-circuit current to the rated full-load current. The corresponding stresses which the transformer must withstand are based upon the peak value of the short-circuit current assuming an asymmetry referred to earlier. A transformer designed to withstand the current given by eqn. 27.1 would thus have a strength of i/e_z.

It will be appreciated that the strength of a transformer for a single fault may be considerably greater than that for a series of faults, since weakening of the windings may be progressive. Moreover, a transformer will have a mechanical strength equal only to the strength of the weakest component in a complex structure. Progressive weakening also implies a short-circuit 'life' in addition to a short-circuit strength. The problem of relating system conditions to short-circuit strength is a complex one and insufficient is yet known about it for definite conclusions to be drawn.

RADIAL ELECTROMAGNETIC FORCES

These forces are relatively easy to calculate since the axial field producing them is accurately represented by the simple two dimensional picture used for reactance calculations. They produce a hoop stress in the outer winding, and a compressive stress in the inner winding.

The mean hoop stress in the copper of the outer winding at the peak of the first half wave of short-circuit current, assuming an asymmetry factor of 1·8 and an infinite system capacity, is given by

Table 27.1

Arrangement of tappings	Residual ampere-turn diagram	P_A, tonnes	$\Lambda\left(\dfrac{Window\ height}{Core\ circle}\right) = 4.2$	$\Lambda\left(\dfrac{Window\ height}{Core\ circle}\right) = 2.3$
A		 $\dfrac{2\pi a (NI_{max})^2 \Lambda}{10^{11}}$	5·5	6·4
B		 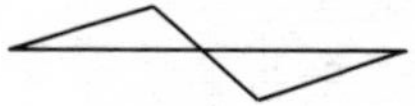 $\dfrac{\pi a (NI_{max})^2 \Lambda}{2 \times 10^{11}}$	5·8	6·6
C		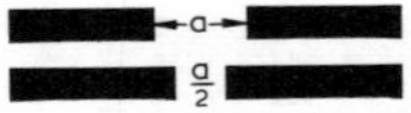 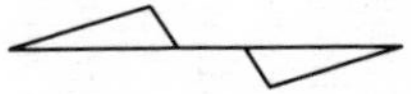 $\dfrac{\pi a (NI_{max})^2 \Lambda}{4(1-\frac{1}{2}a) \times 10^{11}}$	5·8	6·6
D		 $\dfrac{\pi a (NI_{max})^2 \Lambda}{8 \times 10^{11}}$	6·0	6·8
E		$\dfrac{\pi a (NI_{max})^2 \Lambda}{16(1-\frac{1}{2}a) \times 10^{11}}$	6·0	6·8

642

$$\hat{\sigma}_{mean} = \frac{0.031 \, W_{cu}}{h e_z^{\,2}} \text{ tonne/cm}^2 \text{ (peak)} \tag{27.2}$$

where $\qquad W_{cu} = I^2 R_{dc}$ loss in the winding in kW at rated full load and at 75°C.

h = axial height of the windings in centimetres.

Normally this stress increases with the kVA per limb but it is important only for ratings above about 10 MVA per limb. Fully annealed copper has a very low mechanical strength and a great deal of the strength of a copper conductor depends upon the cold working it receives fortuitously after annealing, due to coiling, wrapping, etc. It has been suggested that 0.54 tonne/cm² represents the maximum permissible stress in the copper, if undue permanent set in the outer winding is to be avoided. For very large transformers, some increase in strength may be obtained by lightly cold working the copper or by some form of mechanical restraint. Ordinary high-conductivity copper when lightly cold worked softens very slowly at transformer temperatures and retains adequate strength during the life of an oil-filled transformer.

The radial electromagnetic force is greatest for the inner conductor and decreases linearly to zero for the outermost conductor. The internal stress relationship in a disc coil is such that considerable levelling up takes place and it is usually considered that the mean stress as given in eqn. 27.2 may be used in calculations.

The same assumption is often made for multilayer windings, when the construction is such that the spacing blocks between layers are able to transmit the pressure effectively from one layer to the next. If this is not so then the stress in the layer next to the duct is twice the mean value.

Inner windings tend to become crushed against the core, and it is common practice to support the winding from the core and to treat the winding as a continuous beam with equidistant supports, ignoring the slight increase of strength due to curvature. The mean radial load per cm length of conductor of a disc coil is

$$W = \frac{0.31 \hat{\sigma} A_c}{D_w} \text{ tonne/cm length}$$

or alternatively,

$$W = \frac{5.1 \, U \times 1}{e_z f d_1 \, \pi D_m N} \text{ tonne/cm length} \tag{27.3}$$

where A_c = cross-sectional area of the conductor upon which the force is required, cm²

D_w = mean diameter of winding, cm

U = rated kVA per limb

f = frequency, Hz

$\hat{\sigma}$ = peak value of mean hoop stress, tonne/cm², from eqn. 27.2

d_1 = equivalent duct width, cm

D_m = mean diameter of transformer, (i.e. of h.v. and l.v. windings), cm

N = number of turns in the winding.

Equation 27.3 gives the total load per centimetre length upon a turn or conductor occupying the full radial thickness of the winding. In a multilayer winding with k layers the value for the layer next to the duct would be $(2k-1)/k$ times this value, for the second layer $(2k-3)/k$, and so on.

Where the stresses cannot be transferred directly to the core, the winding itself must be strong enough to withstand the external pressure. Some work has been carried out on this problem, but no method of calculation proved by tests has yet emerged. It has been proposed, however, to treat the inner winding as a cylinder under external pressure, and although not yet firmly established by tests, this method shows promise of being useful to designers.

AXIAL ELECTROMAGNETIC FORCES

Forces in the axial direction can cause failure by producing collapse of the winding, fracture of the end rings or clamping system, bending of the conductors between spacers, or by compressing the insulation to such an extent that slackness occurs leading to displacement of spacers and subsequent failure.

Measurement of axial forces

A simple method is available, developed by the Electrical Research Association, for measuring the total axial force upon the whole or part of a concentric winding. This method does not indicate how the force is distributed round the circumference of the winding but this is only a minor disadvantage.

If the axial flux linked with each coil of a disc winding at a given current is plotted against the axial position, the curve represents, to a scale which can be calculated, the axial compression curve of the winding. From such a curve the total axial force upon the whole or any part of a winding may be read off directly.

The flux density of the radial component of leakage field is proportional to the rate of change of axial flux with distance along the winding. The curve of axial flux plotted against distance thus represents the integration of the radial flux density and gives the compression curve of the winding if the points of zero compression are marked.

The voltage per turn is a measure of the axial flux, and in practice the voltage of each disc coil is measured, and the voltage per turn plotted against the mid-point of the coil on a diagram with the winding length as abscissa. The method can only be applied to a continuous disc winding by piercing the insulation at each crossover.

The test is most conveniently carried out with the transformer short circuited as for the copper-loss test.

The scale of force at 50 Hz is given by

$$1 \text{ volt (r.m.s.)} = \frac{\text{r.m.s. ampere-turns per centimetre}}{15,750} \text{ tonne (peak).}$$

To convert the measured voltages to forces under short-circuit conditions the values must be multiplied by $(1 \cdot 8 \, I_{sc}/I_t)^2$ where I_{sc} is the symmetrical short-circuit current and I_t the current at which the test is carried out.

To obtain the compression curve it is necessary to know the points of zero compression, and these have to be determined by inspection. This is not difficult since each arrangement of windings produces zero points in well defined positions.

A simple mutual inductance potentiometer can be used instead of a voltmeter, and a circuit of this type is described in E.R.A. Report, Ref. Q/T 113, the balance being independent of current and frequency.

Figure 27.1 shows typical axial compression curves obtained on a transformer having untapped windings of equal heights. There are no forces tending to separate the turns in the axial direction. The ordinates represent the forces between coils at all points, due to the current in the windings. Since the slope of the curve represents the force developed per coil it will be seen that only in the end coils are there any appreciable forces. The dotted curve, which is the sum of the axial compression forces for the inner and outer curves, has a maximum value given by

$$P_c = \frac{5 \cdot 1 U}{e_z f h} \text{ tonnes} \qquad (27 \cdot 4)$$

in terms of U the rated kVA per limb and h the axial height of the windings in centimetres. This is the force at the peak of the first half cycle of fault current, assuming an asymmetry factor of 1·8.

The results shown in Fig. 27.1 and other similar figures appearing later in this chapter were obtained on a three-phase transformer constructed so that the voltage across each disc coil in both inner and outer coil stacks could easily be measured. To ensure very accurate ampere-turn balance along the whole length of the windings, primary and secondary windings consisted of disc coils identical in all respects except diameter, and spacing sectors common to both windings were used so that each disc coil was in exactly the same axial position as the corresponding coil in the other winding.

It would be noted that the forces in a transformer winding depend only upon its proportions and on the total ampere-turns, and not upon its physical size. Thus, model transformers are suitable for investigating forces, and for

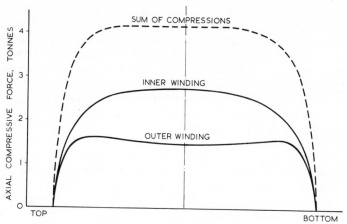

Fig. 27.1   Axial compression curves for untapped transformer windings

large units where calculation is difficult it may be more economical to construct a model and measure the forces than to carry out elaborate calculations.

The voltage per turn method has proved very useful in detecting small accidental axial displacements of two windings from their normal position.

Calculation of axial electromagnetic forces

The problem of calculating the magnitude of the radial leakage field and hence the axial forces of transformer windings has received considerable attention and precise solutions have been determined by various authors. These methods are complex and a computer is necessary if results are to be obtained quickly and economically. The residual ampere-turn method gives reliable results, and attempts to produce closer approximations add greatly to the complexity without a corresponding gain in accuracy. This method does not give the force on individual coils, but a number of simple formulae of reasonable accuracy are available for this purpose.

Residual ampere-turn method

The axial forces are calculated by assuming the winding is divided into two groups, each having balanced ampere-turns. Radial ampere-turns are assumed to produce a radial flux which causes the axial forces between windings.

The radial ampere-turns at any point in the winding are calculated by taking the algebraic sum of the ampere-turns of the primary and secondary windings between that point and either end of the windings. A curve plotted for all points is a residual or unbalanced ampere-turn diagram from which the method derives its name. It is clear that for untapped windings of equal length and without displacement there are no residual ampere-turns or forces between windings. Nevertheless, although there is no axial thrust between windings, internal compressive forces and forces on the end coils are present. A simple expedient enables the compressive forces present when the ampere-turns are balanced to be taken into account with sufficient accuracy for most design purposes.

The method of determining the distribution of radial ampere-turns is illustrated in Fig. 27.2 for the simple case of a concentric winding having a fraction a of the total length tapped out at the end of the outer winding. The two components I and II of Fig. 27.2(b) are both balanced ampere-turn groups which, when superimposed, produce the given ampere-turn arrangement. The diagram showing the radial ampere-turns plotted against distance along the winding is a triangle, as shown in Fig. 27.2(c), having a maximum value of $a(NI_{max})$, where (NI_{max}) represents the ampere-turns of either the primary or secondary winding.

To determine the axial forces, it is necessary to find the radial flux produced by the radial ampere-turns or, in other words, to know the effective length of path for the radial flux for all points along the winding. The assumption is made that this length is constant and does not vary with axial position in the winding. Tests show that this approximation is reasonably accurate, and that the flux does, in fact, follow a triangular distribution curve of a shape similar to the residual ampere-turn curve.

The calculation of the axial thrust in the case shown in Fig. 27.2 can now be

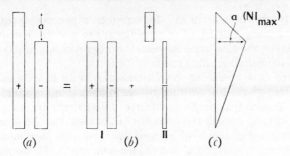

Fig. 27.2 *Determination of residual ampere-turn diagram for winding tapped at one end*

made as follows. If ℓ_{eff} is the effective length of path for the radial flux, and since the mean value of the radial ampere-turns is $\frac{1}{2}a(NI_{max})$, the mean radial flux density at the mean diameter of the transformer is

$$B_r = \frac{4\pi}{10^5} \frac{a(NI_{max})}{2\ell_{eff}} \text{ teslas}$$

and the axial force on either winding of NI_{max} ampere-turns is

$$P_A = \frac{4\pi}{10^7} \frac{a(NI_{max})^2}{2\ell_{eff}} \pi D_m \text{ newtons}$$

The last equation may be rewritten as

$$P_A = \frac{2\pi a(NI_{max})^2}{10^{11}} \frac{\pi D_m}{\ell_{eff}} \text{ tonnes} \qquad (27.5)$$

The second factor of this expression, $\pi D_m / \ell_{eff}$, is the permeance per unit axial length of the limb for the radial flux, referred to the mean diameter of the transformer. It is independent of the physical size of the transformer and depends only upon the configuration of the core and windings. Forces are greatest in the middle limb of a three-phase transformer, and therefore the middle limb only need be considered. A review of the various factors involved indicates that the forces are similar in a single-phase transformer wound on two limbs. Thus if eqn. 27.5 is written as

$$P_A = \frac{2\pi a(NI_{max})^2}{10^{11}} \Lambda \text{ tonnes} \qquad (27.6)$$

where $\Lambda = \pi D_m / \ell_{eff}$ and is the permeance per unit axial length of limb, it gives the force for all transformers having the same proportions whatever their physical size. Since the ampere-turns can be determined without difficulty, in order to cover all cases it is necessary to study only how the constant Λ varies with the proportions of the core, arrangement of tappings, dimensions of the winding duct and proximity of tank.

Reducing the duct width increases the axial forces slightly, and this effect is greater with tapping arrangements which give low values of residual ampere-turns. However, for the range of duct widths used in practice the effect is small.

Where the equivalent duct width is abnormally low, say less than 8% of the mean diameter, forces calculated using the values given in Table 27.1 should be increased by approximately 20 % for tappings at two points equidistant from the middle and ends, and 10 % for tappings at the middle. The axial forces are also influenced by the clearance between the inner winding and core. The closer the core is to the windings, the greater is the force.

The effect of tank proximity is to increase Λ in all cases, and for the outer limbs of a three-phase transformer by an appreciable amount; however, the middle limb remains practically unaffected unless the tank sides are very close to it. As would be expected, the presence of the tank has the greatest effect for tappings at one end of the winding, and the least with tappings at two points equidistant from the middle and ends of the winding. As far as limited tests can show, the presence of the tank never increases the forces in the outer limbs to values greater than those in the middle limb, and has no appreciable effect upon the middle limb with practical tapping arrangements. The only case in which the tank would have appreciable effect is in that of a single-phase transformer wound on one limb, and in this case the value of Λ would again not exceed that for the middle phase of a three-phase transformer.

The location of the tappings is the predominating influence on the axial forces since it controls the residual ampere-turn diagram. Forces due to arrangement E in Table 27.1 are only about one thirty-second of those due to arrangement A. The value of Λ is only slightly affected by the arrangement of tappings so that practically the whole of the reduction to be expected from a better arrangement of tappings can be realised. It varies slightly with the ratio of limb length to core circle diameter, and also if the limbs are more widely spaced.

In Table 27.1 values of Λ are given for the various tapping arrangements and for two values of the ratio, window height/core circle diameter. The formula for calculating the axial force on the portion of either winding under each triangle of the residual ampere-turn diagram is given in each case. The values of Λ apply to the middle limb with three-phase excitation, and for the tapping sections in the outer winding.

AXIAL FORCES FOR VARIOUS TAPPING ARRANGEMENTS

Additional axial forces due to tappings can be avoided by arranging the tappings in a separate coil so that each tapping section occupies the full winding height. Under these conditions there are no ampere-turns acting radially and

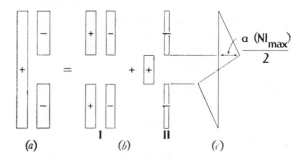

Fig. 27.3   Determination of residual ampere-turn diagram for winding tapped at middle

the forces are the same as for untapped windings of equal length. Another method is to arrange the untapped winding in a number of parallel sections in such a way that there is a redistribution of ampere-turns when the tapping position is changed and complete balance of ampere-turns is retained.

(i) Transformer with tappings at the middle of the outer winding

To calculate the radial field the windings are divided into two components as shown in Fig. 27.3. Winding group II produces a radial field diagram as shown in (c). The two halves of the outer winding are subjected to forces in opposite directions towards the yokes while there is an axial compression of similar magnitude at the middle of the inner winding.

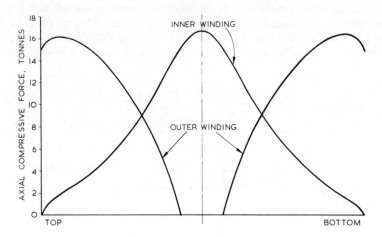

Fig. 27.4   Axial compression curves for $13\frac{1}{3}\%$ tapped out of the middle of the outer winding

Measured curves are given in Fig. 27.4 for the case of $13\frac{1}{3}\%$ tapped out of the middle of the outer winding. The maximum compression in the outer winding is only slightly greater than the end thrust, and it occurs at four to five coils from the ends. The maximum compression in the inner winding is at the middle.

Axial end thrust

The axial end thrust is given by

$$P_A = \frac{\pi a (NI_{max})^2 \Lambda}{2 \times 10^{11}} \text{ tonnes} \tag{27.7}$$

Maximum compression

If P_c is the sum of both compressions as given by eqn. 27.6 and it is assumed that two thirds of this is the inner winding, then the maximum compression in the inner winding is given by

$$P_{max} = \frac{2}{3} \times \frac{5 \cdot 1 U}{e_z f h} + \frac{\pi a (N I_{max})^2 \Lambda}{2 \times 10^{11}} \text{ tonnes} \qquad (27.8)$$

The maximum compression in the outer winding is slightly less than this.

Figure 27.5 shows curves of maximum compression in the inner and outer windings, and of end thrust plotted against the fraction of winding tapped out for the same transformer. Equation 27.7 represents the line through the origin.

Most highly stressed turn or coil

The largest electromagnetic force is exerted upon the coils immediately adjacent to the tapped out portion of a winding and it is in these coils that the maximum bending stresses occur when sector spacers are used. The force upon a coil or turn in the outer winding immediately adjacent to the gap is given theoretically by

$$P_A = 0 \cdot 773 \, q P_r \log_{10}\left(\frac{2a'}{w} + 1\right) \text{tonnes} \qquad (27.9)$$

where, P_r = total radial bursting force of transformer, tonnes

q = fraction of total ampere-turns in a coil or winding

w = axial length of coil including insulation, centimetres

a' = axial length of winding tapped out.

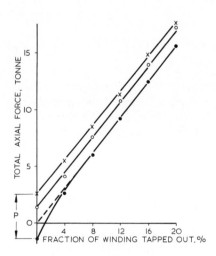

Fig. 27.5   Curves of end thrust and maximum compression for tappings at the middle of the outer winding

● ● ● ●   end thrust
○ ○ ○ ○   maximum compression in outer winding
×   ×   ×   ×   maximum compression in inner winding
P = sum of compressions in both windings

There is reasonable agreement between calculated and measured forces; the calculated values are 10 to 20% high, no doubt owing to the assumption that the windings have zero radial thickness.

The coils in the inner winding exactly opposite to the most highly stressed coils in the outer winding have forces acting upon them of a similar, but rather lower, magnitude.

(ii) Tappings at the middle of the outer winding but with thinning of the inner winding

The forces in the previous arrangement may be halved by thinning down the ampere-turns per unit length to half the normal value in the portion of the un-tapped winding opposite the tappings. Alternatively, a gap may be left in the untapped winding of half the length of the maximum gap in the tapped winding. With these arrangements there is an axial end thrust from the untapped winding when all the tapped winding is in circuit, and an end thrust of similar magnitude in the tapped winding when all the tappings are out of circuit. In the mid position there are no appreciable additional forces compared with untapped windings.

(a) Axial end thrust

When all tappings are in circuit the end thrust of the untapped winding may be calculated by means of eqn. 27.7, substituting for a the fractional length of the gap in the untapped winding. When all tappings are out of circuit the end thrust is given by

$$P_A = \frac{\pi a (N I_{max})^2 \Lambda}{4(1 - \frac{1}{2}a)10^{11}} \text{ tonnes} \tag{27.10}$$

where a, the fraction of the axial length tapped out, is partially compensated by a length $\frac{1}{2}a$ omitted from the untapped winding. The constant Λ has the same value as in eqn. 27.7. The forces are similar when the ampere-turns are thinned down instead of a definite gap being used.

(b) Maximum compression

In either of the two preceding cases the maximum compression exceeds the end thrust by an amount rather less than the force given by eqn. 27.4.

(c) Most highly stressed coil or turn

When all tappings are in circuit, the force upon the coil or turn adjacent to the compensating gap in the untapped winding may be calculated by applying eqn. 27.9; in such a case a would be the length of the gap expressed as a fraction. It should be noted, however, that since thinning or provision of a compensating gap is usually carried out on the inner winding, the presence of the core increases the force slightly. Hence this equation is likely to give results a few per cent low

651

in this case. On the other hand, when thinning is used, the force upon the coil adjacent to the thinned-out portion of winding is rather less than given by eqn. 27.9.

(iii) Two tapping points midway between the middle and ends of the outer winding

(a) Without thinning of the untapped winding

A typical example of the compression in the inner and outer windings is given in Fig. 27.6 for the case of approximately 13 % tapped out of the outer winding, half being at each of two points midway between the middle and ends of the winding.

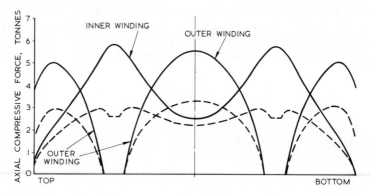

Fig. 27.6   Axial compression curve for tappings at two points in the outer winding.
Dotted curves show forces with thinning of the inner winding opposite each tapping point

There are three points of maximum compression in the outer winding, the middle one being the largest. In the inner winding there are two equal maxima opposite the gaps in the outer winding.

The axial force upon each quarter of either winding due to the tappings is given by

$$P_A = \frac{\pi a (N I_{max})^2 \Lambda}{8 \times 10^{11}} \text{ tonnes} \tag{27.11}$$

where a is the total fraction of axial length tapped out, and the constant Λ has the value given in Table 27.1.

This force acts towards the yokes in the two end sections of the outer winding, so that eqn. 27.7 gives the axial end thrust for the larger values of a. The curve of end thrust plotted against the fraction tapped out can be estimated without difficulty since it deviates only slightly from the straight line of eqn. 27.11.

The forces with this arrangement of tappings are only about one-sixteenth of the forces due to tappings at one end of the winding, and they are of the same order as the forces in the untapped winding.

The most highly stressed coils are those adjacent to the tapping points, and the forces may be calculated by means of eqn. 27.9 substituting $\frac{1}{2}a$ for a.

(b) *With thinning of the untapped winding*

This practice represents the optimum method of reducing forces when a section is tapped out of a winding, and the dotted curves in Fig. 27.6 show the forces obtained when the inner winding is thinned opposite each of the two gaps in the outer winding to an extent of 50% of the total tapping range. The force upon each quarter of either winding is

$$P_A = \frac{\pi a (N I_{max})^2 \Lambda}{16 \times 10^{11}} \text{ tonnes} \tag{27.12}$$

when all tappings are in circuit, and

$$P_A = \frac{\pi a (N I_{max})^2 \Lambda}{16(1 - \frac{1}{2}a)10^{11}} \text{ tonnes} \tag{27.13}$$

when all tappings are out of circuit. In these equations Λ has the value given in Table 27.1, and a represents the total fraction tapped out.

The forces upon the coils immediately adjacent to the gaps may be calculated as described in eqn. 27.9, since these forces are determined by the lengths of the gaps and not by their positions in the winding.

Chapter 28

Transformer noise

Noise inevitably emitted by transformers in operation sometimes gives rise to complaints which are difficult to resolve for various reasons. The two main difficulties, however, are due to the fact that distribution transformers are normally located closer to houses or offices than are other types of equipment and that, since they operate throughout 24 hours of every day, the noise continues during the night, when it is most noticeable.

In approaching the noise problem it is therefore essential to consider not only its engineering aspects, but to bear in mind that noise is a subjective phenomenon involving the vagaries of human nature.

THE SUBJECTIVE NATURE OF NOISE

The subjective nature of noise is underlined by the standard definition* which states it is 'sound which is undesired by the recipient'. It is thus easy to see how people participating in a party can enjoy it, while neighbours wishing to sleep find it disturbing. It also shows why some sounds, such as the dripping of a tap, can be classed as a noise, especially since intermittent sounds are usually more annoying than continuous sounds.

Fortunately, transformer noise is not only continuous, but is also largely confined to the medium range of audio frequencies, which are least disagreeable to the human ear. The absence of inherently objectionable features means that the annoyance value of transformer noise is roughly proportional to its apparent loudness. A good starting point for tackling the problem is therefore to determine the apparent loudness of the noise emitted by transformers of different types and sizes.

Methods of measuring noise

The measurement of noise is by no means as simple as the measurement of physical or electrical quantities. Loudness, like annoyance, is a subjective sensation dependent to a large extent on the characteristics of the human ear.

*See B.S. 661 'Glossary of acoustical terms'.

It must therefore be dealt with on a statistical basis and investigators in this field have shown that the loudness figure allocated to a given sound by a panel of average observers is a reasonably well defined function of its sound pressure and frequency. Since sound pressure and frequency are the objective characteristics measured by a sound level meter or a sound level indicator, it is thus possible to obtain a rating proportional to the loudness of a sound from the appropriate meter readings. A sound level meter is illustrated in Fig. 28.1, whilst Fig. 28.2 shows a typical sound level indicator.

Fig. 28.1 Dawe Type 1400 sound level meter (Dawe Instruments Ltd.)

Fig. 28.2 Dawe Type 1408 sound level indicator (Dawe Instruments Ltd.)

To enable meter readings to be correlated with loudness values, a quantitative picture of the response of the human ear to different sounds must be available. Work carried out at the National Physical Laboratory by Robinson and Dadson forms the basis of the standardised loudness curves* reproduced in Fig. 28.3. They show how the sensitivity of hearing of the average human observer varies with variations in the frequency of the sound and the sound pressure. Sensitivity decreases towards the low and high limits of the audio frequency range, so that sounds falling outside the band from approximately 16 Hz to 16 kHz are inaudible to most human observers.

The microphone of sound measuring instruments is in effect a transducer for measuring sound pressures which are normally expressed in dyne/cm^2. Since the sensitivity of the human ear falls off in a roughly logarithmic fashion with increasing sound pressure, it is usual to calibrate instruments for measuring

*See B.S. 3383.

sound levels on a logarithmic scale, graduated in decibels, or dB. This scale uses as a base an r.m.s. pressure level of 0·0002 dyne/cm², which is approximately the threshold of hearing of an acute ear at 1000 Hz. Thus, a noise having an r.m.s. pressure level of d dynes/cm² would be said to have a sound pressure level (usually abbreviated to sound level) of $20\log_{10}d/0\cdot0002$ dB. The decibel scale is used for the ordinates of Fig. 28.3, each 20 dB rise in sound level representing a tenfold increase in sound pressure.

The curves of Fig. 28.3 represent equal loudness contours for a pure note under free field conditions. They show that the human ear will ascribe equal loudness to pure notes of 77 dB at 30 Hz, 51 dB at 100 Hz, 40 dB at 1000 Hz, 34 dB at 3000 Hz, 40 dB at 6000 Hz and 47 dB at 10000 Hz. By definition the loudness levels in phons of all points on this equal loudness contour are equated

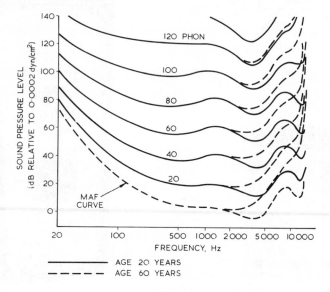

Fig. 28.3   *Equal loudness curves* (Robinson and Dadson)

to the dB rating of the 1000 Hz tone (that is to say, to 40 dB), therefore the loudness level in phons of all 1000 Hz sounds is numerically equal to their dB rating. The loudness level of all other pure notes is numerically equal to the decibel rating of the 1000 Hz note appearing to be equally loud. Since the loudness level in phons is numerically equal to the equivalent decibel rating at 1000 Hz, it is usual to mark the scales of meters for measuring sound level in decibels to imply the method by which the measurement has been made. The dB graduation can be seen clearly on the meters shown in Figs. 28.1 and 28.2.

Determining loudness

The equal loudness curves show how the sensitivity of the ear varies with frequency but do not indicate how the ear responds to changes in sound pressure level. For this purpose the sone scale of loudness has been standardised;*

again from the results of subjective tests with many listeners. The reference point of this scale is taken arbitrarily as a loudness of 1 sone for a level of 40 phons, that is, 40 dB at 1000 Hz. It has been shown, moreover, that each rise or fall of 10 phons in the loudness level appears to the ear as a doubling or halving respectively of the loudness. This provides the following convenient relationship between apparent loudness (A sones) and loudness level (P phons):

$$A = 10^{0.03\,(P-40)} \text{ sones} \qquad (28.1)$$

By substitution in this expression or in Fig. 28.4 it is possible to determine the effect on an average observer of any change in reading of a sound level meter.

The sone scale is linear, so that a noise having a loudness of $2A$ sones sounds twice as loud as a noise of A sones.

Reference to eqn. 28.1 shows that the noise emitted by two similar sources does not sound twice as loud as the noise emitted by each source separately. The sound level is increased only by 3 dB and the apparent loudness by about one quarter.

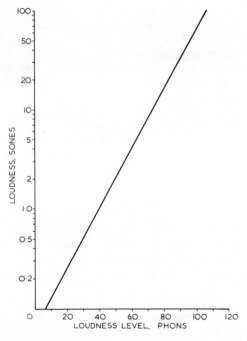

Fig. 28.4   Relation between loudness and loudness level

Sound measuring instruments

The equal loudness contours shown in Fig. 28.3 form the basis of the so called weighting networks incorporated into modern instruments for measuring

*See B.S.3045. The relation between the sone scale of loudness and the phon scale of loudness level.

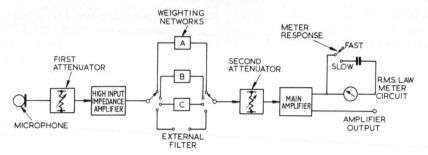

Fig. 28.5   Block diagram of the circuit of the Dawe type 1400 sound level meter

sound level. In Fig. 28.5 the construction of a sound level meter is shown diagrammatically. Historically the weighting networks *A*, *B* and *C* were intended to simulate the response of the ear at low, medium and high sound levels respectively. However, extensive tests have shown that in many cases sound level *A* is found to correlate best with subjective noise ratings and is thus called for in many specifications. The *C* weighting is used for frequency analysis purposes and where the physical sound pressure level is to be measured. In the sound level meter illustrated in Fig. 28.1, any network can be brought into use simply by pressing the appropriate push-button control.

The microphone incorporated into a sound level meter is non-directional and the dynamic response of the meter closely follows that of the human ear. The signal from the microphone is transmitted through the first attenuator to the measuring circuit. The attenuators are inserted to give the meter a range (24–140 dB) which is approaching that of the human ear, which has the remarkable ability of accommodating a range of pressures of a million to one between the thresholds of hearing and of pain. With the correct attenuator in circuit, the reading on the meter added to the value of the attenuators in use gives the sound level rating of the noise.

A sound level meter or indicator sums up a given noise in terms of a single decibel rating. Although convenient, such a rating yields little information as to the character of the noise, as it represents only its magnitude. To determine the character of a noise, a frequency spectrum must be prepared by means of an audio frequency analyser an example of which is shown in Fig. 28.6. Such an analyser is essentially a variable filter which suppresses sound at all frequencies outside the desired band.

Analysers are of two main types. The first uses a filter of narrow bandwidth to give a continuous spectrum and any marked amount of noise at a particular frequency is clearly demonstrated by a sharp rise in the meter reading. The second type of analyser incorporates a filter of wider bandwidth to sum the noise over a certain range of frequencies. The waveband normally chosen for the latter type of analyser is one octave, with midband frequencies of 125 Hz, 250 Hz, 500 Hz, 1000 Hz, 2000 Hz and 4000 Hz. The 125 Hz band, for example, spans the octave 90–180 Hz, while the 4000 Hz band covers 2800–5600 Hz. Frequencies above and below this range (*i.e.*, 31·5–90 Hz and 5600–8000 Hz) are covered by low- and high-pass sections respectively. Prior to 1964, octave bands with different frequency bands were used (75–150 Hz, 150–300 Hz etc.).

In practice, an octave band analyser is normally used to obtain quick results

658

during initial surveys. It can then be replaced by a tunable analyser to plot a detailed frequency spectrum over any octave shown to be unduly noisy.

SOUND LEVEL MEASUREMENTS OF TRANSFORMERS

Noise measurements are made with a sound level meter having characteristics complying with B.S.3489, and it is an advantage if the instrument employed is provided with means for checking the overall acoustic calibration, for example, by means of a falling ball acoustic calibrator. Measurements are taken at no-load and all readings are normally recorded using the *A* weighting.

The transformer is excited on its principal tapping at rated voltage and frequency, but preliminary check tests are usually made to see if there is any significant variation of noise level between the different positions of the tapping switch or tap linking device.

Measurements are taken with the microphone at a height above ground level corresponding to one half of the height of the tank, and at points around the tank perimeter spaced at no less than 60 cm nor more than 90 cm apart. The microphone is placed at a distance of approximately 30 cm from an imaginary taut string line passed around all major projections from the tank walls.

For the purpose of noise tests, major projections are defined as:

(1) For plain tanks: any stiffening member permanently affixed to the tank
(2) For tubular tanks: the outer contour of the cooling tubes.

Where cable boxes, bushing turrets or tap changing compartments project below the tank cover, the string contour is increased locally in their vicinity, but it is not usual to include any projections beyond the contour so determined if such are due to valves, jacking and transport lugs, or any projection above the tank cover height.

The background noise level is also measured and the maximum value in the vicinity of the transformer is recorded. It should be at least 7 dB and preferably 10 dB below the 'average surface noise level' of the transformer. This level is

Fig. 28.6   Dawe type 1461AF analyser (Dawe Instruments Ltd.)

defined as the arithmetic average of all the readings obtained around the transformer. It is desirable that there are no reflecting surfaces within 3 m of the transformer during the measurement of transformer noise levels.

INTERPRETATION OF TRANSFORMER NOISE

A typical analysis of transformer noise is reproduced in Fig. 28.7, which can be considered as a composite graph of a large number of readings. In this diagram, the ordinates indicate the magnitude of the various individual constituents of the noise whose frequencies represent the abscissae. The most striking point is the strength of the component at 100 Hz or twice the normal operating frequency of the transformer. Consideration of magnetostrictive strain in the transformer core reveals that magnetostriction can be expected to produce a longitudinal vibration in the laminations at just this measured frequency. Unfortunately, the magnetostrictive strain is not truly sinusoidal in character, which leads to the introduction of the harmonics seen in Fig. 28.7. Deviation from a 'square-

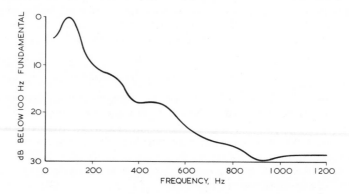

Fig. 28.7   *Typical analysis of noise emitted by transformers*

law' magnetostrictive characteristic would result in even harmonics (at 200, 400, 600 Hz, etc.), while the different values of magnetostrictive strain for increasing and decreasing flux densities—a pseudo hysteresis effect—lead to the introduction of odd harmonics (at 300, 500, 700 Hz, etc.).

Reference to Fig. 28.3 indicates that the sensitivity of the ear to noise increases rapidly at frequencies above 100 Hz. On the 40 phon contour, it requires an increase of 21 dB in intensity to make a sound at 100 Hz appear as loud as one at 1000 Hz. The harmonics in a transformer noise may thus have a substantial effect on an observer even though their level is 10 dB or more lower than that of the 100 Hz fundamental.

Although longitudinal vibration is the natural consequence of magnetostriction, the need to restrain the laminations by clamping also leads to transverse vibrations, this effect being illustrated in Fig. 28.8. Measurements taken on this effect suggest that transverse vibrations contribute roughly as much sound energy to the total noise as do the longitudinal vibrations. As already pointed out, two similar sources sound about 23% louder than one. By the same token, complete elimination of the transverse vibration would reduce the loudness of the transformer noise by only about a fifth. Although valuable, this reduction, even if technically and economically possible, is insignificant compared with the halving of the loudness which can be achieved by a reduction of 10 dB in the noise level of both longitudinal and transverse vibrations.

Figure 28.7 covers typical transformers incorporating cold-rolled laminated cores operating at flux densities between 1·5 and 1·6 teslas. Even variations of 10 % in flux density have been shown to produce changes of noise level of the order of only about 2 dB, although the character of the noise may vary appreciably. From this it will be apparent that it is most uneconomic to obtain a reduction in noise level by the employment of low flux densities. This is perhaps demonstrated best by reference to experience with cold-rolled steel. To make the optimum use of these newer materials, it is necessary to operate them at higher flux densities of the order of 1·65 teslas. While this higher flux density tends to lead to a higher noise level for a given size of core, current results suggest that the difference is quite small for a given transformer rating, due to the smaller core made possible by the use of cold-rolled material. Considerable work is being undertaken to obtain even quieter operation by suitable treatment of the raw material and by particularly careful assembly of the finished core laminations.

Turning to other possible sources of noise emitted by a transformer, the forces present between the individual conductors in the winding when the transformer is loaded must be considered. These forces are however of a sinusoidal nature so that any vibration consists of a fundamental at 100 Hz with negligible harmonics. The fundamental is thus effectively dwarfed by the much greater 100 Hz fundamental generated by the core, while there are no harmonics to add to the annoyance value. Acoustic measurements confirm this conclusion by showing that the noise level on most transformers increases by not more than 2 dB (15 % rise in loudness) from no load to full load. Any variation is in fact attributable more to changes in flux density than to variations in the forces in the windings.

These comments on transformer noise assume the absence of resonance in any part of the unit. Normally the minimum natural frequency of the core and windings lies in the region of 1000 Hz. Figure 28.7 indicates that the exciting forces are very low at this or higher frequencies. Accordingly, it can confidently be expected that the unfortunate effects associated with resonance will be avoided. The natural frequency of the tank or fittings being lower, resonance of these is much more likely to occur, since the vibrations of the core can be transmitted by the oil to the tank. If any part of the structure has a natural frequency at or near 100, 200, 300, 400 Hz, etc., the result will be an amplification of noise at that particular frequency. Naturally, this point is closely watched by reputable manufacturers and the stiffness of any offending part is duly adjusted before manufacture commences.

NOISE REDUCTION ON SITE

Control of the noise emitted by a transformer rests almost entirely with the manufacturer, who will endeavour to make his product as quiet as practicable. A certain amount of noise is, however, inevitable and, if it proves offensive, must

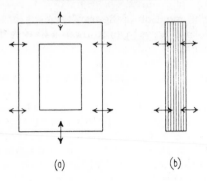

(a)   (b)

Fig. 28.8   Vibrations due to magnetostriction: (a) longitudinal and (b) transverse

be dealt with by the purchaser who can do much to ensure acceptance of the transformer long before it is delivered. Typical average sound levels of a range of transformers are given in Fig. 28.9. They should of course be compared with any test figures for the actual transformer to be installed, as soon as any figures become available. The levels quoted in Fig. 28.9 will, however, provide a reasonable basis for preparatory action. The reduction of noise level with distance must be allowed for. Doubling the distance from a point source of noise means that a given amount of sound must be spread over four times the area. From this cause alone, doubling of distance results in a 6 dB fall in sound level. In practice, scattering combined with the absorption by the air itself ensures that the noise reduction is greater, particularly at the higher frequencies. Figure 28.10 shows measured values of attenuation with distance for typical transformer ratings. Assuming that there is no screening between the transformer and a given building these curves enable the noise level outside the building to be computed.

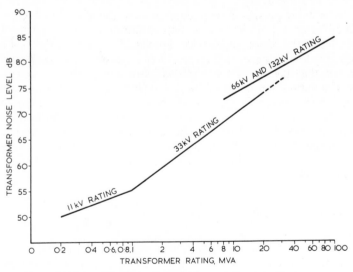

Fig. 28.9   Typical transformer average surface noise levels

Normally it is not necessary to reduce transformer noise in the vicinity of residential buildings to such a level that it is inaudible. Experience suggests that the transformer noise will be acceptable if it is not audible inside a bedroom of the nearest house at night time when a small window of the room is open. Under these conditions the transformer noise level outside the house can be considered as the permitted maximum transformer noise at this position. Provided the sound level meter reading measured outside the house is not more than 2 dB above the bedroom background level, both being measured at the *A* weighting, the acceptable noise level inside will not be exceeded. From Table 28.1, which also gives the calculated equivalent 'phon' values of the transformer noise as obtained from the typical composition and Churcher-King equal loudness contours and tone summation curve, it will be seen that three types of background level have been given, and these are considered to be representative of conditions existing at night in neighbourhoods of the kinds referred to.

Using the values from Table 28.1 as a basis, it is possible to determine whether the noise level within nearby bedrooms will be acceptable if the transformer is

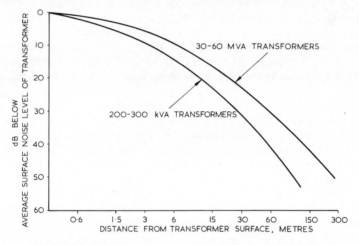

Fig. 28.10   Curves showing measured attenuation of transformer noise with distance

erected at various alternative locations. One or other location may well ensure that no householder is subjected to an unduly high noise level. Failing this, the investigation may still show the minimum attenuation necessary to bring the noise level down to an acceptable level. The most appropriate method of achieving this object can then be selected and work put in hand immediately, so that the site is ready when the transformer is delivered.

Provided the noise level resulting from transformer operation is below that given in Table 28.1, conditions should be satisfactory and no corrective action is necessary. In fact, with well designed transformers, acoustic conditions will normally be satisfactory under urban conditions at all points beyond 15 metres from the transformer for a rating of 200 kVA and 25 metres for a rating of 500 kVA. Assuming that bedroom windows do not face directly on the transformer, it is possible to decrease these distances by about two-thirds.

In urban areas, it is normally impracticable to site transformers more than 100 metres from the nearest dwelling. In this case, transformers with ratings in excess of about 15 MVA will probably need to be provided with some form of attenuation giving a noise reduction of between 10 and 25 dB.

The most obvious method of attenuation is by the provision of a suitable barrier between the transformer and the listener. The simplest form of barrier is a screening wall, the effectiveness of which will vary with height and density as well as with the frequency of the noise. The attenuation of a 100 Hz noise by a 6 metre wall will not normally exceed 10 dB outside the immediate 'shadow' cast by the wall.

Such attenuation just reaches the bottom of the range cited but some slight further attenuation can be achieved by judicious use of absorbent material. This treatment may result in an attenuation as small as 2 dB and will seldom give a figure in excess of 6 dB. While absorbent material may give some relief on existing installations or may make a single wall shielding a transformer in one direction more effective, it will not usually provide a complete solution where an untreated screen wall is itself unsatisfactory. In difficult cases, resort must be had to more effective means of silencing, such as total enclosure. Enclosing structures should naturally have the minimum number of openings for ventilation and doors, which sometimes may raise problems of cooling. Where large transformers have separate radiator banks, difficulties of this kind are minimised since with care it is possible to locate the radiators outside the enclosure without appreciably affecting the noise emitted. In any case, openings in walls of the structure should be located on the side remote from nearby dwellings.

Provided openings are located at the optimum position, the attenuation given by any reasonable structure will normally be above that necessary to give tolerable conditions in nearby houses. An important factor in this connection is the relatively great attenuation of the harmonic content of the noise, which has been shown earlier in this chapter to have a nuisance value out of all proportion to its magnitude.

Once such a structure has been erected, adequate maintenance is of the utmost importance if the initial attenuation is to be maintained. Any openings and doors should be checked frequently, as gaps which may develop to a sufficient size would permit considerable escape of sound energy.

It is often advisable to compare a frequency spectrum of the noise emitted by the transformer to be installed against the spectrum of the background noise at the proposed location. In the values quoted earlier, typical frequency spectra have been assumed. Any marked deviations of the actual noises from the character attributed to them can lead to a considerable reduction in the masking power of the background noise. For this reason alone, a frequency analysis on site is valuable, even if it is compared only with an average spectrum for transformer noise, such as that given in Fig. 28.7.

Topographical features of the site should be exploited to the full in order to reduce noise. Where possible the transformer should be located in the prevailing down wind direction from houses. Existing walls and mounds should, if possible, be kept between dwellings and the transformer. Natural hollows can sometimes be used to increase the effective height of screen walls, as can artificially constructed pits. Cultivated shrubs and trees form only an ineffective barrier to noise as sound attenuation is largely determined by the mass of the barrier. In some cases where the smaller ratings of distribution transformers are installed, the psychological effect, however, may be sufficient to avoid a complaint,

Table 28.1*

ACCEPTABLE MAXIMUM NOISE LEVELS OUTSIDE DWELLINGS

Locality	Bedroom† background noise level in dB	Sound† level in dB	Transformer noise Calculated equivalent phon (or dB at 1000 Hz) value
Rural	20	22	30
Urban-residential	26	28	40
Industrial	35	37	50

†A weighting.

simply because the transformer becomes hidden by the trees and is therefore not visible.

In conclusion, a word should be said about two technical ideas which have been suggested as a solution to the noise problem. The first attempts to interrupt the transmission of vibration from the core to the tank of the transformer by inserting a resilient gas filled barrier in the oil close to the wall. The core is further isolated from the tank by placing it on anti-vibration mountings. Although attractive in theory, such a design introduces considerable practical difficulties and has not yet been exploited on a commercial basis.

The second suggestion is as revolutionary in character. It makes use of what may be termed 'anti-sound': a loudspeaker is provided to broadcast a noise similar to that emitted by the transformer, but 180° out of phase. This difference of phase should have the effect that the compressions of the broadcast noise cancel out the rarefactions of the transformer noise, and vice versa, so that an observer should hear nothing. Unfortunately, tests have tended to show that, while it is possible to generate a limited pool of silence in this way, this benefit is only achieved at the cost of worsening conditions generally. For the present, therefore, users will have to continue to rely on the well proved methods of noise attenuation, involving distance, screening walls and enclosures. It is considered that these methods are the most economic way of dealing with the problem.

*Table 28.1 and the first paragraph of page 663 are taken substantially from C.I.G.R.E. paper no. 108, 1956, 'Transformer noise limitation,' by C. M. Brownsey, I. Glover and G. B. Harper.

Appendix 1

Transformer equivalent circuit

The calculations of a combined electrical system or circuit comprising transformers, transmission and distribution lines are often simplified by the use of the equivalent circuit diagram. The characteristics of a loaded transformer also can often be indicated more clearly by the same means. Figure A1.1 shows the more general form of diagram of connections, and Fig. A1.2 the corresponding phasor diagram for a loaded transformer.

V_1	primary terminal voltage
E_1	primary induced e.m.f.
E_2	secondary induced e.m.f.
V_2	secondary terminal voltage
I_2	secondary load current
I_2'	load component of total primary current
I_1	total primary current (including I_0 and I_2')
I_0	primary no-load current
I_m	primary magnetising current
I_c	primary core loss current
R_1	primary resistance
X_1	primary leakage reactance
Z_1	primary impedance
R_2	secondary resistance
X_2	secondary leakage reactance
Z_2	secondary impedance
$I_1 R_1$	primary resistance voltage drop
$I_1 X_1$	primary reactance voltage drop
$I_1 Z_1$	primary impedance voltage drop
$I_2 R_2$	secondary resistance voltage drop
$I_2 X_2$	secondary reactance voltage drop
$I_2 Z_2$	secondary impedance voltage drop
$R\ell$	secondary load resistance
$X\ell$	secondary load reactance
$Z\ell$	secondary load impedance
N_1	primary turns
N_2	secondary turns
$\cos\phi_2$	secondary load power factor

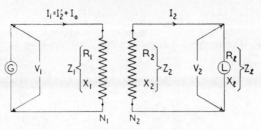

Fig. A1.1 Circuit diagram

Figure A1.3 shows the equivalent circuit diagram corresponding to Fig. A1.1 and this applies to step-up and step-down transformers. This diagram enables the primary voltage V_1 necessary to maintain a given load voltage V_2 to be determined. Those characteristics of Fig. A1.3 which apply to the secondary circuit are shown as referred to the primary circuit by the turns ratio n. The admittance Y_0 (no-load current divided by the primary induced e.m.f.) is simply such as to represent the no-load characteristics of the transformer; that is, the resistance branch takes a current equal to the core-loss current, and the reactance branch takes a current equal to the true magnetising current. This method of treatment takes account of the efficiency of the transformer, and the copper losses appear as voltage drops. The phasor diagram corresponding to Fig. A1.3 is shown in Fig. A1.4. E_1 is the e.m.f. across the admittance Y_0, and $V_1 - E_1$ is the voltage drop (which is not measurable) that is assumed to occur in the primary circuit if half the transformer reactance is allotted to the primary side of the transformer.

The required voltage V_1 can be calculated simply by multiplying the equivalent load current in the circuit by the total equivalent impedance of the circuit. In calculating the equivalent impedance, the individual equivalent ohmic

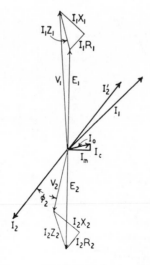

Fig. A1.2 Phasor diagram of loaded transformers. Assumed turns ratio 1:1

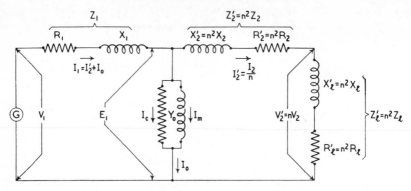

Fig. A1.3   Equivalent circuit diagram for determining the primary voltage V_1

resistances may be added arithmetically as also may be the equivalent ohmic reactances; including, of course, the equivalent load resistance and reactance; the total equivalent impedance is then the phasor sum of the total equivalent resistances and reactances. This is illustrated in Fig. A1.5. (This method neglects the very small phase displacements that exist between the individual ohmic impedances but the approximation is normally justified.)

Figure A1.6 is similar to Fig. A1.3 except that the notation is framed so as to enable the secondary load voltage V_2 at a given primary voltage V_1 to be determined. That is, the primary characteristics are referred to the secondary circuit by the turns ratio n.

The phasor diagram corresponding to Fig. A1.6 is shown in Fig. A1.7 and the simplified phasor diagram for calculating the required voltage V_2 is shown in Fig. A1.8. In constructing the latter diagram the very small phase displacements that exist between the individual ohmic impedances have again been neglected.

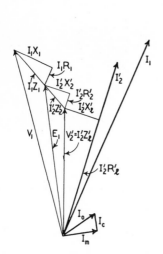

Fig. A1.4   Phasor diagram of equivalent circuit shown in Fig. A1.3

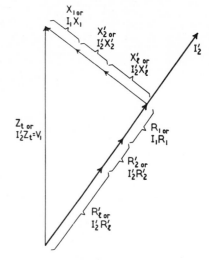

Fig. A1.5   Resultant phasor diagram corresponding to Fig. A1.4

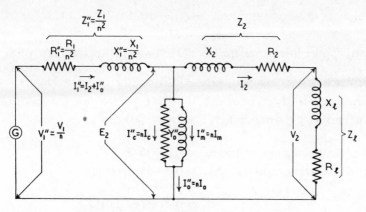

Fig. A1.6   Equivalent circuit diagram for determining the secondary load voltage V_2

Example on Fig. A1.3

Consider the case of a 200 kVA, 11 000/415 V, three-phase, delta/star, 50 Hz transformer.

Tested phase voltage ratio n $\quad = 11\,000\sqrt{3}/415 = 45{\cdot}9$

$\qquad$ and therefore $n^2 \quad = 45{\cdot}9^2 = 2110$

Tested core loss per phase $\quad = 270$ watts

I_c per phase $\quad = 270/11\,000 = 0{\cdot}025$ A

Tested no-load current per phase, $I_0 = 0{\cdot}212$ A

I_n per phase $\qquad\qquad = \sqrt{(0{\cdot}212^2 - 0{\cdot}025^2)}$

$\qquad\qquad = 0{\cdot}211$ A

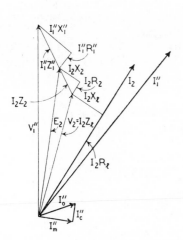

Fig. A1.7   Phasor diagram of equivalent circuit shown in Fig. A1.6

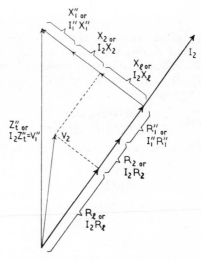

Fig. A1.8   Resultant phasor diagram corresponding to Fig. A1.7

Primary no-load power factor, $\cos \phi_0 = 0.025/0.212 = 0.118$

and therefore $\phi_0 \quad = 83.3°$

Tested copper loss per phase $\quad = 1130$ watts $\Big\rbrace$ Approximately 5%

increase in losses on

Calculated copper loss per phase $= 1080$ watts $\Big\rbrace$ calculated figure

Calculated primary resistance $\quad = 16.6\ \Omega$

Actual primary resistance, $R_1 \quad = 16.6 \times 1.05$

$= 17.4\ \Omega$

Calculated secondary resistance $\quad = 0.00606\ \Omega$

Actual secondary resistance, $R_2 \quad = 0.00606 \times 1.05\ \Omega$

$= 0.00637\ \Omega$

$R_2' = 0.00637 \times 2110\ \Omega$

$= 13.4\ \Omega$

Tested h.v. impedance per phase $= 524$ V at 6.06 A per phase

therefore $\quad Z_e' = 524/6.06 = 86.3\ \Omega$

and $\qquad X_e' = \sqrt{\{(Z_e')^2 - (R_1 + R_2')^2\}}$

$= \sqrt{\{86.3^2 - (17.4 + 13.4)^2\}}$

$= 80.6\ \Omega$

Assuming the reactive voltage drop to be equal in the primary and secondary winding, gives $X_1 = 40.3\ \Omega$, and $X_2' = 40.3\ \Omega$

The phase constants of the transformer derived from test figures are thus

n	$= 45.9$	ϕ_0	$= 83.2°$
n^2	$= 2110$	R_1	$= 17.4\ \Omega$
I_c	$= 0.025$ A	R_2'	$= 13.4\ \Omega$
I_o	$= 0.212$ A	X_1	$= 40.3\ \Omega$
I_m	$= 0.211$ A	X_2'	$= 40.3\ \Omega$

(a) With an applied primary terminal voltage V_1 of 11 000 V it is required to find the secondary terminal voltage V_2 when the transformer supplies a secondary load of 200 kVA at a power factor $\cos \phi_2 = 0.8$ lagging.

All calculations are made on a phase to neutral basis.

It is necessary to start by assuming a value of V_2, say 288 V, (i.e., assuming a 5% drop).

$I_2 = 200\,000/(\sqrt{3} \times 228 \times \sqrt{3}) = 292$ A

$I_2' = 292/45.9 = 6.36$ A

$R\ell = (V_2 \cos \phi_2)/I_2 = 228 \times 0.8/292 = 0.624\ \Omega$

$R\ell' = 0.624 \times 2110$

$= 1310\ \Omega$

$X_1 = (V_2 \sin \phi_2)/I_2$

$= 228 \times 0.6/292$

$= 0.468\ \Omega$

$X\ell' = 0.468 \times 2110$

$= 987\ \Omega$

Resistance drop to SS $= I_2' (R\ell' + R_2') = 6\cdot36(1310 + 13\cdot4) = 8420$ V

Reactance drop to SS $= I_2' (X\ell' + X_2') = 6\cdot36(987 + 40\cdot3) = 6540$ V

$$\phi_{ss} = \tan^{-1} 6540/8420 = \tan^{-1} 0\cdot776 = 37\cdot8°$$

$$\begin{aligned} I_1 &= \sqrt{\{[I_2' + I_0 \cos(\phi_0 - \phi_{ss})]^2 + [I_0 \sin(\phi_0 - \phi_{ss})]^2\}} \\ &= \sqrt{\{[6\cdot36 + 0\cdot212\cos(83\cdot2° - 37\cdot8°)]^2 + [0\cdot212\sin(83\cdot2° - 37\cdot8°)]^2\}} \\ &= 6\cdot51 \text{ A} \end{aligned}$$

primary line current $= 6\cdot51 \times \sqrt{3} = 11\cdot3$ A

$$\begin{aligned} \theta_1 &= \tan^{-1} \frac{I_0 \sin(\phi_0 - \phi_{ss})}{I_2' + I_0 \cos(\phi_0 - \phi_{ss})} \\ &= \tan^{-1} \frac{0\cdot212\sin(83\cdot2° - 37\cdot8°)}{6\cdot36 + 0\cdot212\cos(83\cdot2° - 37\cdot8°)} \\ &= \tan^{-1} 0\cdot0232 = 1\cdot3° \end{aligned}$$

Resistance drop to PP $= 8420 + I_1 R_1 = 8420 + 6\cdot51 \times 17\cdot4 = 8530$ V

Reactance drop to PP $= 6540 + I_1 X_1 = 6540 + 6\cdot51 \times 40\cdot3 = 6800$ V

$$V_1 = \sqrt{(8530^2 + 6800^2)} = 10\,910 \text{ V}$$

$$\text{Percentage regulation} = \frac{V_1 - V_2'}{V_1} \times 100$$

$$= \frac{10\,910 - 228 \times 45\cdot9}{10\,910} \times 100 = 4\cdot2\%$$

Thus, with $V_1 = 11\,000$ V

we have $V_2 = 11\,000 \dfrac{100 - 4\cdot2}{100} \times \dfrac{1}{45\cdot9}$

i.e., $V_2 = 230$ V.

Secondary terminal line voltage $= 230\sqrt{3} = 398$ V

$(\phi_1 - \theta_1) = \tan^{-1} 6800/8530 = \tan^{-1} 0\cdot798 = 38\cdot6°$

$\phi_1 = 38\cdot6° + 1\cdot3° = 39\cdot9°$

$\cos\phi_1 = 0\cdot767$

Power input $= V_1 I_1 \cos\phi_1 = 10\,910 \times 6\cdot51 \times 0\cdot767 = 54\,500$ watts

Power output $= V_2 I_2 \cos\phi_2 = 228 \times 292 \times 0\cdot8 = 53\,300$ watts

therefore percentage efficiency $= (53\,300/54\,500)\,100 = 97\cdot80\%$.

With $V_1 = 11\,000$ V and $V_2 = 230$ V,

the corrected value of $I_2 = 200\,000/(\sqrt{3} \times 230 \times \sqrt{3}) = 290$ A

and, the corrected value of $I_2' = 290/45\cdot9 = 6\cdot32$ A.

The corrected value of $R\ell' = 230 \times 0\cdot8 \times 2110/290 = 1340$ Ω

and, the corrected value of $X\ell' = 230 \times 0\cdot6 \times 2110/290 = 1000$ Ω.

(b) Results can also be obtained by means of the symbolic method which allows for the very small phase displacements between the individual ohmic impedances. Consider the general form of Fig. A1.9 as shown in Fig. A1.11:

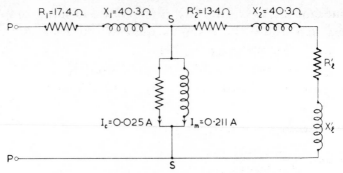

$R_1 = 17.4\,\Omega$ $X_1 = 40.3\,\Omega$ $R'_2 = 13.4\,\Omega$ $X'_2 = 40.3\,\Omega$

$I_c = 0.025\,A$ $I_m = 0.211\,A$

Fig. A1.9   Equivalent circuit diagram for the given example

It can be derived that

$$V'_2 = \frac{V_1 Z\ell' Z_0}{(Z\ell' + Z'_2)(Z_0 + Z_1) + Z_1 Z_0} \tag{A1.1}$$

$$I_1 = \frac{V_1(Z_0 + Z'_2 + Z\ell')}{(Z\ell' + Z'_2)(Z_0 + Z_1) + Z_1 Z_0} \tag{A1.2}$$

$$I'_2 = \frac{V_1 Z_0}{(Z\ell' + Z'_2)(Z_0 + Z_1) + Z_1 Z_0} \tag{A1.3}$$

In order to determine Z_0 it is first necessary to calculate E_1.

$$
\begin{aligned}
E_1 &= V_2 + I'_2 Z'_2 \\
&= 230 \times 45.9\,(0.8 + j0.6) + 6.32\,(13.4 + j40.3) \\
&= 8530 + j6590 = 10\,800\ \text{V} \\
Z_0 &= E_1/I_0 = E_1/(I_c - jI_m) \\
&= \frac{E_1 I_c}{I_c^2 + I_m^2} + j\frac{E_1 I_m}{I_c^2 + I_m^2} \\
&= \frac{10\,800 \times 0.025}{0.025^2 + 0.211^2} + j\frac{10\,800 \times 0.211}{0.025^2 + 0.211^2} \\
&= 5990 + j50\,500\ \Omega
\end{aligned}
$$

Thus,

$$
\begin{aligned}
Z_1 &= 17.4 + j40.3\ \Omega \\
Z'_2 &= 13.4 + j40.3\ \Omega \\
Z_0 &= 5990 + j50\,500\ \Omega \\
Z\ell' &= 1340 + j1000\ \Omega
\end{aligned}
$$

and $(Z\ell' + Z'_2)(Z_0 + Z_1) + Z_1 Z_0 = (-46.3 + j75.6)10^6$.

Thus, substituting these values in eqn. A1.1,

$$V'_2 = \frac{11\,000\,(1340 + j1000)(5990 + j50\,500)}{(-46.3 + j75.6)10^6}$$

672

$$= 10\,540\ \overline{\diagdown 1{\cdot}6^\circ}\ V$$

$$V_2 = 10\,540/45{\cdot}9 = 230\ V$$

and the secondary terminal line voltage $= 230\sqrt{3} = 398\ V$.
Substituting in eqn. A1.2,

$$I_1 = \frac{11\,000(5990 + j50\,500 + 13{\cdot}4 + j40{\cdot}3 + 1340 + j1000)}{(-46{\cdot}3 + j75{\cdot}6)10^6}$$

$$= 6{\cdot}45\ \overline{\diagdown 39{\cdot}6^\circ}\ A$$

Primary line current $= 6{\cdot}45\sqrt{3}\ \overline{\diagdown 39{\cdot}6^\circ} = 11{\cdot}2\ A$

$$\cos\phi_1 = \cos 39{\cdot}6^\circ$$
$$= 0{\cdot}771$$

Power input $= V_1 I_1 \cos\phi_1$
$$= 11\,000 \times 6{\cdot}45 \times 0{\cdot}771$$
$$= 54\,600\ \text{watts}$$

Substituting in eqn. A1.3,

$$I_2' = \frac{11\,000(5990 + j50\,500)}{(-46{\cdot}3 + j75{\cdot}6)10^6} = 6{\cdot}31\ \overline{\diagdown 38{\cdot}3^\circ}\ A$$

and $I_2 = 6{\cdot}31 \times 45{\cdot}9\ \overline{\diagdown 38{\cdot}3^\circ}\ A = 290\ \overline{\diagdown 38{\cdot}3^\circ}\ A$

Power input $= V_2 I_2 \cos\phi_2 = 230 \times 290 \times 0{\cdot}8 = 53\,300\ \text{watts}$

Percentage efficiency $= (53\,300/54\,600)\,100 = 97{\cdot}62\,\%$.

Thus it will be seen that the results obtained by the two methods of calculation show close agreement.

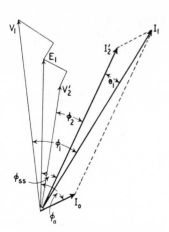

Fig. A1.10   Phasor diagram

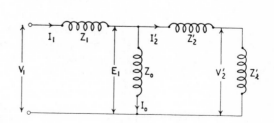

Fig. A1.11   The general form of Fig. A1.9

Appendix 2

Geometry of the transformer phasor diagram

The following diagrams and equations apply equally well to single-phase transformers and to the individual phases of polyphase transformers.

Specification data

V_1 primary terminal voltage
V_2 secondary terminal voltage
I_2' load component of total primary current
I_1 total primary current
I_2 secondary load current
$n(= N_1/N_2 = E_1/E_2)$ turns ratio
$\cos\phi_2$ secondary load power factor

Test data

R_1 primary resistance
R_2 secondary resistance
$I_2'Z_e'$ total equivalent impedance voltage drop (referred to the primary side)
$I_2 Z_e''$ total equivalent impedance voltage drop (referred to the secondary side)
I_0 primary no-load current
I_0'' secondary no-load current
P_f iron loss

For the phasor diagram see Fig. A2.1. Voltage drops due to no-load current are ignored.

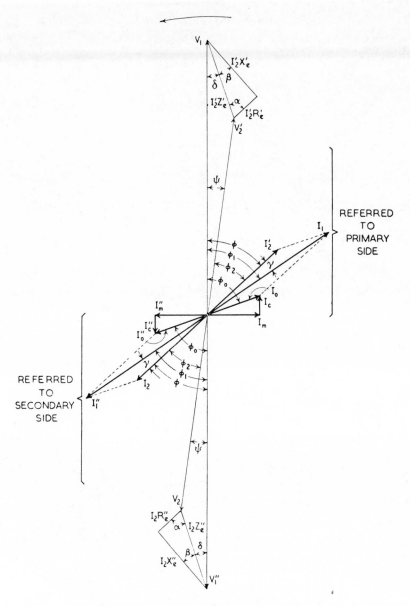

Fig. A2.1   Phasor diagram for no-load and for load conditions

675

Table A2.1

Characteristics referred to	Primary side	Secondary side
No-load (Fig. A2.1)		
(1) Primary terminal voltage	V_1	$V_1'' = V_1/n$
(2) Secondary terminal voltage	$V_2' = nV_2$	V_2
(3) Primary no-load current	I_0	$I_0'' = nI_0$
(4) Primary core loss current	$I_c = P_I/V_1$	$I_c'' = nI_c = P_I/V_1''$
(5) Primary magnetising current	$I_m = \sqrt{I_0^2 - I_c^2}$	$I_m'' = nI_m = \sqrt{(I_0'')^2 - (I_c'')^2}$
(6) Primary no-load power factor	$\cos\phi_0 = I_c/I_0$	$\cos\phi_0 = I_c''/I_0'' = I_c/I_0$
On-load (Fig. A2.1)		
(7) Secondary load current	$I_2' = I_2/n$	I_2
(8) Secondary load power factor	$\cos\phi_2$	$\cos\phi_2$
(9) Load component of total primary current	I_2'	$I_2' = nI_2'$
(10) Total primary current	$I_1 = \sqrt{\{(I_2'\cos\phi + I_c)^2 + (I_2'\sin\phi + I_m)^2\}}$	$I_1'' = nI_1 = \sqrt{\{(I_2\cos\phi + I_c'')^2 + (I_2\sin\phi + I_m'')^2\}}$
(11) Primary total load power factor	$\cos\phi_1 = \cos\left[\tan^{-1}\dfrac{I_2'\sin\left\{\sin^{-1}\left(\dfrac{I_2'Z_e'}{V_1}\cos\beta + \phi_2\right) + \phi_2\right\} + I_m}{I_2'\cos\left\{\sin^{-1}\left(\dfrac{I_2'Z_e'}{V_1}\cos\beta + \phi_2\right) + \phi_2\right\} + I_c}\right]$	$\cos\phi_1 = \cos\left[\tan^{-1}\dfrac{I_2\sin\left\{\sin^{-1}\left(\dfrac{I_2Z_e''}{V_1''}\cos\beta + \phi_2\right) + \phi_2\right\} + I_m''}{I_2\cos\left\{\sin^{-1}\left(\dfrac{I_2Z_e''}{V_1''}\cos\beta + \phi_2\right) + \phi_2\right\} + I_c''}\right]$
(12) Primary resistance	R_1	$R_1'' = R_1/n^2$
(13) Secondary resistance	$R_2' = n^2R_2$	R_2
(14) Total equivalent resistance	$R_e' = R_1 + R_2' = n^2 R_e''$	$R_e'' = R_1'' + R_2 = R_e'/n^2$
(15) Total equivalent resistance voltage drop	$I_2'R_e' = nI_2R_e''$	$I_2R_e'' = I_2'R_e'/n$
(16) Total equivalent reactance voltage drop	$I_2'X_e' = nI_2X_e''$	$I_2X_e'' = I_2'X_e'/n$
(17) Total equivalent impedance voltage drop	$I_2'Z_e' = nI_2Z_e''$	$I_2Z_e'' = I_2'X_e'/n$

Table A2.1 (continued)

Characteristics referred to	Primary side	Secondary side
(18) Voltage regulation	$\|V_1 - V_2'\| = I_2'R_e'\cos\phi_2 + I_2'X_e'\sin\phi_2 + \dfrac{(I_2'X_e'\cos\phi_2 - I_2'R_e'\sin\phi_2)^2}{200}$	$\|V_1'' = V_2\| = (\|V_1 - V_2\|)/n = I_2R_e''\cos\phi_2 + I_2X_e''\sin\phi_2 + \dfrac{(I_2X_e''\cos\phi_2 - I_2R_e''\sin\phi_2)^2}{200}$
(19) α	$\cos^{-1}(I_2'R_e'/I_2'Z_e')$ or $\sin^{-1}(I_2'X_e'/I_2'Z_e')$	$\cos^{-1}(I_2R_e''/I_2Z_e'')$ or $\sin^{-1}(I_2X_e''/I_2Z_e'')$
(20) β	$\cos^{-1}(I_2'X_e'/I_2'Z_e')$ or $\sin^{-1}(I_2'R_e'/I_2'Z_e')$	$\cos^{-1}(I_2X_e''/I_2Z_e'')$ or $\sin^{-1}(I_2R_e''/I_2Z_e'')$
(21) γ	$2\left(\cos^{-1}\sqrt{\dfrac{S'(S'-I_0)}{I_1I_2'}}\right)$ where, $S' = \frac{1}{2}(I_1+I_2'+I_0)$	$2\left(\cos^{-1}\sqrt{\dfrac{S''(S''-I_0'')}{I_1'I_2}}\right)$ where, $S'' = \frac{1}{2}(I_1'+I_2+I_0'')$
(22) δ	$2\left(\cos^{-1}\sqrt{\dfrac{U'(U''-V_2')}{V_1'I_2'Z_e'}}\right)$ where, $U' = \frac{1}{2}(V_1+V_2'+I_2'Z_e')$	$2\left(\cos^{-1}\sqrt{\dfrac{U''(U''-V_2)}{V_1''I_2Z_e''}}\right)$ where, $U'' = \frac{1}{2}(V_1''+V_2+I_2Z_e'')$
(23) ψ	$2\left(\cos^{-1}\sqrt{\dfrac{U'(U''-I_2'Z_e')}{V_1'V_2'}}\right)$ where, $U' = \frac{1}{2}(V_1+V_2'+I_2'Z_e')$	$2\left(\cos^{-1}\sqrt{\dfrac{U''(U''-I_2Z_e'')}{V_1''V_2}}\right)$ where, $U'' = \frac{1}{2}(V_1''+V_2+I_2Z_e'')$
(24) ϕ	$\phi_2 + \psi$	$\phi_2 + \psi$
On Short circuit (Fig. A2.2)		
(25) Total equivalent resistance voltage drop	$I_2'R_e'(S.C.) = I_2'R_e V_1/I_2'Z_e'$	$I_2R_e''(S.C.) = I_2R_e''V_1''/I_2Z_e''$
(26) Total equivalent reactance voltage drop	$I_2'X_e'(S.C.) = I_2'X_e'V_1/I_2'Z_e'$	$I_2X_e''(S.C.) = I_2X_e''V_1''/I_2Z_e''$
(27) Total equivalent impedance voltage drop	$I_2'Z_e'(S.C.) = V_1$	$I_2Z_e''(S.C.) = V_1''$
(28) Short circuit current	$I_2'(S.C.) = I_2'V_1/I_2'Z_e'$ (no-load current ignored)	$I_2(S.C.) = I_2V_1''/I_2Z_e''$
(29) α	$\cos^{-1}(I_2'R_e'(S.C.)/I_2'Z_e'(S.C.))$ or $\sin^{-1}(I_2'X_e'(S.C.)/I_2'Z_e'(S.C.))$	$\cos^{-1}(I_2R_e''(S.C.)/I_2Z_e''(S.C.))$ or $\sin^{-1}(I_2X_e''(S.C.)/I_2Z_e''(S.C.))$
(30) β	$\cos^{-1}(I_2'X_e'(S.C.)/I_2'Z_e'(S.C.))$ or $\sin^{-1}(I_2'R_e'(S.C.)/I_2'Z_e'(S.C.))$	$\cos^{-1}(I_2X_e''(S.C.)/I_2Z_e''(S.C.))$ or $\sin^{-1}(I_2R_e''(S.C.)/I_2Z_e''(S.C.))$

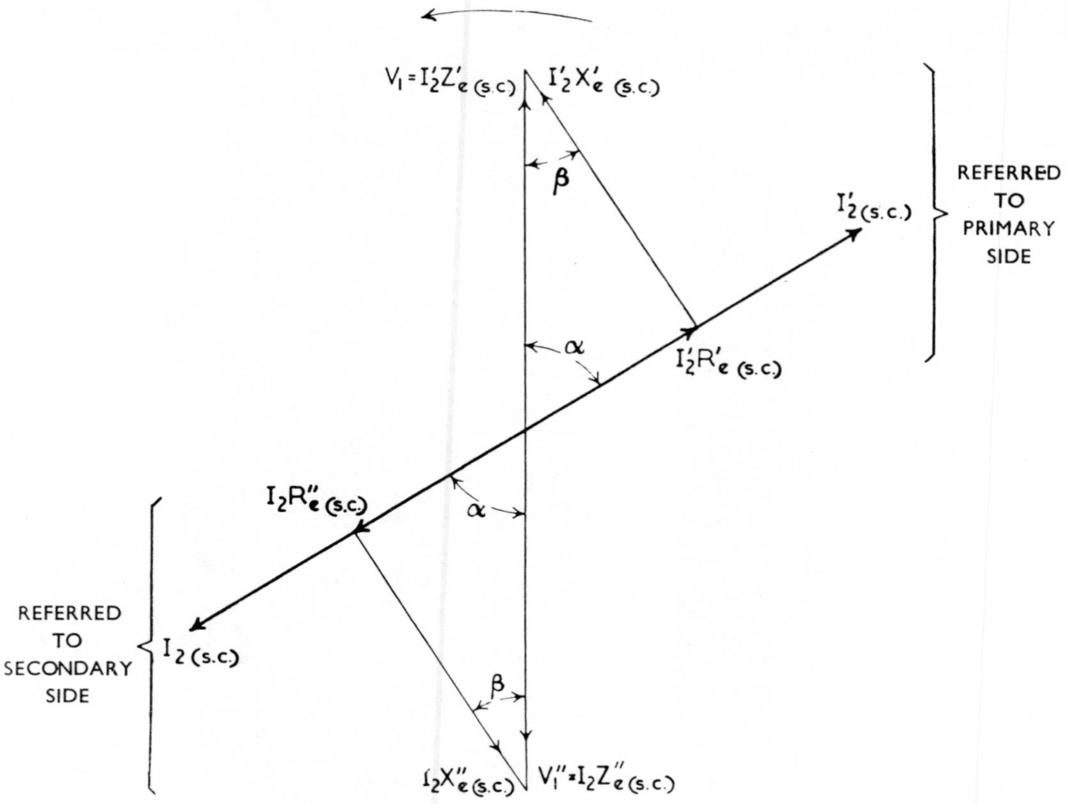

Fig. A2.2   Phasor diagram for short-circuit conditions

678

Appendix 3

The transformer circle diagram

By means of the circle diagram the loci of the ends of the phasors representing terminal voltages and currents in single and polyphase transformers may be located at all power factors and all loads.

The amounts of, and the phasor relations between, the primary and secondary voltages and currents under any conditions of load and power factor may be determined and the regulation may be obtained graphically.

Referring all quantities to the secondary side, and working on a 'per phase' basis:

Let N_1 and N_2 be the number of primary and secondary turns respectively.

V_1'' primary terminal voltage, reversed in time phase and multiplied by the ratio N_2/N_1

V_2 secondary terminal voltage

I_1'' total primary full-load current, reversed in time phase and multiplied by the ratio N_1/N_2

I_2 secondary full-load current

I_0'' primary no-load current, reversed in time phase and multiplied by the ratio N_1/N_2

I_c'' primary core loss current, reversed in time phase and multiplied by the ratio N_1/N_2

I_m'' primary magnetising current, reversed in time phase and multiplied by the ratio N_1/N_2

$\cos \phi_0$ primary no-load power factor

$\cos \phi_1$ primary total load power factor

$\cos \phi_2$ secondary load power factor

R_e'' total equivalent resistance

X_e'' total equivalent reactance

Z_e'' total equivalent impedance

Note: If it is desired to refer quantities to the primary side, the secondary current phasors must be reversed in time phase and multiplied by the ratio N_2/N_1, and the secondary voltage phasors must be reversed in time phase and multiplied by the ratio N_1/N_2. In addition, the values of resistance, reactance and impedance as referred to the secondary side must be multiplied by the ratio $(N_1/N_2)^2$ in order to transfer them to the primary side. In calculating the resistance, reactance and impedance voltage drops, the effect of the no-load current has been ignored.

The following quantities can be obtained from design calculations as well as from the test results:

Total iron loss, total copper loss, percentage reactance, and percentage magnetising current.

Then, $$I''_m = \frac{\text{percentage magnetising current}}{100} I_2 \text{ amperes}$$

$$I''_c = \frac{\text{total watts iron loss}}{\text{number of phases } V''_1} \text{amperes}$$

$$I''_0 = \sqrt{\{(I''_m)^2 + (I''_c)^2\}}$$

$$X''_c = \frac{\text{percentage reactance}}{100} \frac{V''_1}{I''_1} \text{ ohms}$$

$$R''_e = \frac{\text{total watts copper loss}}{\text{number of phases } I_2{}^2} \text{ ohms}$$

$$Z = \sqrt{(X^2 + R^2)} \text{ohms}$$

$$Z''_e = \sqrt{\{(X''_e)^2 + (R''_e)^2\}} \text{ ohms}$$

CONSTRUCTION OF CIRCLE DIAGRAMS (see Figs. A3.1, A3.2)

First draw the phasor $OA = V''_1$. With centre A and radius $= I''_2 Z''_e$ describe the circle BCD. This circle is the locus of the end of the secondary load terminal voltage phasor V_2 for various values of $\cos \phi_2$.

Draw the radius CA of the circle BCD such that $\widehat{OAC} = \beta$ where $\cos \beta = X''_2/$

Z''_e. Draw YY', the right bisector of OA. Now draw AQ so that $\widehat{CAQ} = \phi_2$, the phase angle of the secondary load current, and let AQ cut YY' in Q.

(*Note:* If ϕ_2 is lagging, draw $\widehat{CAQ}$ clockwise; if ϕ_2 is leading, draw $\widehat{CAQ}$ counter-clockwise.) With centre Q and radius QA describe an arc cutting the first circle BCD in B. Join OB, AB. Then $OB = V_2$.

Draw $OO' = I''_m$ at right angles to OA, and $O'O'' = I''_c$ parallel to OA. Then $OO'' = I''_0$. With centres O and O'' and radii $= I_2$, draw two circles FGH and JKL as shown in Figs. A3.1, A3.2.

Draw the radius OM of the circle FGH such that $\widehat{BOM} = \phi_2$; then $OM = I_2$. Draw MN parallel to OO'', cutting the circle JKL in N. Join ON. Then $ON = I''_1$

and $\widehat{AON} = \phi_1 = $ primary input current phase angle, *i.e.*, primary input power factor $= \cos \phi_1$.

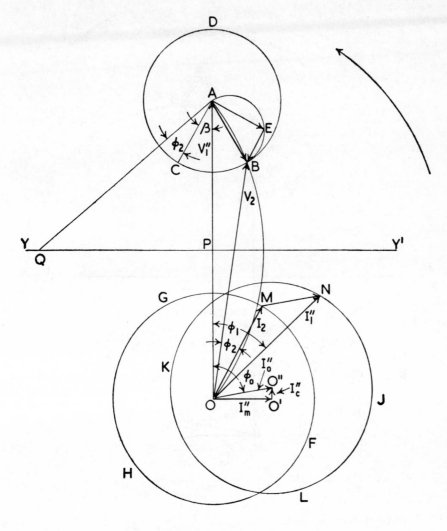

Fig. A3.1   Circle diagram—lagging power factor load

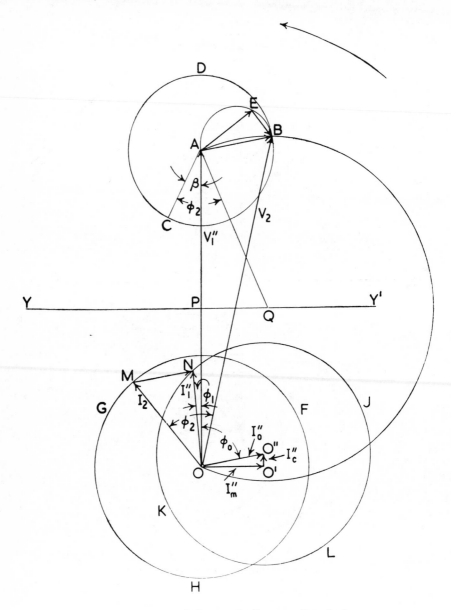

Fig. A3.2   Circle diagram—leading power factor load

Also the phase angle of the primary no-load current $= \widehat{AOO''} = \phi_0$, *i.e.*, the primary no-load power factor $= \cos \phi_0$.

It is evident that, $\qquad\qquad \cos \phi_0 = I_c''/I_0''$

and it may be shown that,

$$\cos \phi_1 = \cos \left[\tan^{-1} \frac{I_2 \sin \left\{ \sin^{-1} \left(\dfrac{I_2 Z_e''}{V_1''} \overline{\cos \beta + \phi_2} \right) + \phi_2 \right\} + I_m''}{I_2 \cos \left\{ \sin^{-1} \left(\dfrac{I_2 Z_e''}{V_1''} \overline{\cos \beta + \phi_2} \right) + \phi_2 \right\} + I_c''} \right]$$

On AB as diameter, describe a semicircle ABE. Draw BE parallel to OM, cutting the semicircle ABE in E. Join AE.

Then, BE $= I_2 R_e'' =$ total resistance voltage drop per phase referred to the secondary side,

and, $\quad$ AE $= I_2 X_e'' =$ reactance voltage drop per phase referred to the secondary side.

The percentage regulation is given by the expression,

$$\text{percentage regulation} = \frac{OA - OB}{OA} \times 100.$$

Appendix 4

Transformer regulation

The standard formula for determining the percentage regulation of a transformer at full load and at a power factor $\cos \phi$ is,*

$$V_X \sin \phi_2 + V_R \cos \phi_2 + \frac{(V_X \cos \phi_2 - V_R \sin \phi_2)^2}{200} \qquad (A4.1)$$

where V_X = percentage reactance voltage at full load

V_R = percentage resistance voltage at full load

ϕ = angle of lag of the full-load current.

This formula is correct for the determination of the regulation at any load differing from full load, and it is only necessary to divide V_X and V_R wherever they appear in the formula by the factor given by dividing the full-load current by the current corresponding to the particular load at which the regulation is desired. In most practical cases the load current flowing through a transformer has lagging power factor so that no doubts can arise with regard to the correct signs to be used, for these are exactly as given in the above general equation.

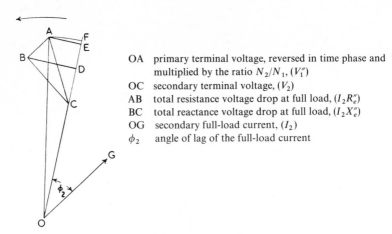

OA   primary terminal voltage, reversed in time phase and multiplied by the ratio N_2/N_1, (V_1'')

OC   secondary terminal voltage, (V_2)

AB   total resistance voltage drop at full load, $(I_2 R_e'')$

BC   total reactance voltage drop at full load, $(I_2 X_e'')$

OG   secondary full-load current, (I_2)

ϕ_2 angle of lag of the full-load current

Fig. A4.1   Regulation diagram—lagging power factor load

*For impedances above 20% refer to B.S.171.

From time to time, however, it is necessary to calculate the regulation for currents at leading power factors and it is, therefore, interesting to consider whether the standard formula given above applies in such cases.

Like many other problems of this kind, the solution can be obtained from the geometry of the figure, and the following investigation has been conducted upon this basis. Referring all quantities to the secondary side, and working on a per phase basis.

Drop perpendiculars from A and B to OC produced, meeting OC produced at E and D respectively. With radius OA and centre O draw an arc AF to meet OC produced at F.

$$\text{Power factor} = \cos\phi_2 = \cos\widehat{COG}$$

Since AB is parallel to OG, AB must make an angle ϕ_2 with OC produced and BC must make an angle $90° - \phi_2$ with OC.

$$\text{Percentage regulation} = 100\,\frac{OA - OC}{OA}$$

$$= 100\,\frac{OF - OC}{OA}$$

$$= 100\,\frac{CF}{OA}$$

$$= 100\left(\frac{CD + DE + EF}{OA}\right)$$

$$= 100\left(\frac{BC\sin\phi_2 + AB\cos\phi_2}{OA} + \frac{EF}{OA}\right)$$

$$= V_X\sin\phi_2 + V_R\cos\phi_2 + 100\,\frac{EF}{OA}$$

where V_X and V_R are the percentage reactance and resistance voltage drops at full load respectively.

In order to evaluate EF it must be remembered that OF is the radius of a circle and that AE is a perpendicular to it from a point on the circumference, and that, therefore,

$$\frac{EF}{AE} = \frac{AE}{OE + OF}$$

$$EF = \frac{AE^2}{OE + OF}$$

Now although EF may be appreciable as compared with CF, it is negligible compared with so large a quantity as OE + OF, and therefore it is permissible to write 2OF for the latter.

Thus
$$EF = \frac{AE^2}{2OF} = \frac{AE^2}{2OA}$$

therefore
$$100\,\frac{EF}{OA} = 100\,\frac{AE^2}{2OA^2}$$

$$= 100\frac{(\text{BC}\cos\phi_2 - \text{AB}\sin\phi_2)^2}{2\text{OA}^2}$$

$$= \frac{(V_X\cos\phi_2 - V_R\sin\phi_2)^2}{200}$$

percentage regulation $= V_X\sin\phi_2 + V_R\cos\phi_2 + \dfrac{(V_X\cos\phi_2 - V_R\sin\phi_2)^2}{200}$

In Fig. A4.2 the power factor of the load

$$= \cos\theta_2 \text{ leading}$$
$$= \cos(360° - \theta_2) \text{ lagging}$$
$$= \cos\phi_2 \text{ lagging}$$

AB and BC make angles θ_2 and $90° - \theta_2$ respectively with OC. Therefore,

percentage regulation $= 100\dfrac{(\text{OA} - \text{OC})}{\text{OA}}$

$$= 100\frac{(\text{OF} - \text{OC})}{\text{OA}}$$

$$= 100\frac{(-\text{CF})}{\text{OA}}$$

$$= 100\frac{(-\text{CD} + \text{DE} + \text{EF})}{\text{OA}}$$

$$= -V_X\sin\theta_2 + V_R\cos\theta_2 + 100\frac{\text{AE}^2}{2\text{OA}^2}$$

$$= -V_X\sin\theta_2 + V_R\cos\theta_2 + \frac{(V_X\cos\theta_2 + V_R\sin\theta_2)^2}{200}$$

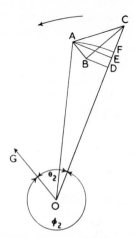

Fig. A4.2   Regulation diagram—leading power factor load

Now, $\qquad \theta_2 = 360° - \phi_2$

and therefore, $\quad \sin \theta_2 = -\sin \phi_2$

$$\cos \theta_2 = \cos \phi_2$$

and percentage regulation

$$= V_X \sin \phi_2 + V_R \cos \phi_2 + \frac{(V_X \cos \phi_2 - V_R \sin \phi_2)^2}{200} \qquad \text{(A4.2)}$$

It will be seen from eqns. A4.1 and A4.2 that the final regulation formula is the same for lagging and leading power factors provided the angle ϕ_2 is the true angle of lag measured clockwise from the position of the secondary terminal voltage phasor. Thus, in the case of a lagging power factor $\cos \phi_2$ the angle ϕ_2 is substituted directly into the regulation formula, but in the case of a leading power factor $\cos \theta_2$ the angle to be substituted is not θ_2 but $\phi_2 = (360° - \theta_2)$, and the following relationship must be observed:

$$\cos \phi_2 = \cos(360° - \theta_2) = \cos \theta_2$$
$$\text{and} \qquad \sin \phi_2 = \sin(360° - \theta_2 = -\sin \theta_2$$

If the percentage regulation comes out negative, it indicates that the load has produced a rise in voltage.

Appendix 5

Symmetrical components in unbalanced three-phase systems

Modern technique in the calculation of system fault conditions demands a knowledge of the theory of symmetrical components and the phase sequence characteristics of the individual parts of the system. It would be out of place here to deal with symmetrical components as extensively as the subject demands. As transformers are involved in system fault calculations, a very brief study of the application of the theory to the phasor analysis of unbalanced three-phase systems may quite properly be given, and such is therefore presented in what follows. A consideration of phase sequence characteristics of transformers subsequently appears.

When a short-circuit fault occurs in a three-phase network, currents and voltages in the three phases become unequal in magnitude and unbalanced in their phase displacements, so that the phasors representing them are no longer equal and spaced 120° apart.

It is possible to analyse any given system of three-phase unbalanced phasors into three other balanced phasor systems which are called positive, negative and zero phase sequence phasors respectively.

The positive phase sequence system is that in which the phase (or line) voltages and/or currents reach their maxima in the same order as do those of the normal supply.

It is conventionally assumed that *all* phasors rotate in a counter-clockwise direction, and the positive phase sequence system is that in which the phase maxima occur in the order ABC.

Conversely, the negative phase sequence system is that in which the phasors, while rotating in the same direction as the positive phase sequence phasors, namely, counter-clockwise, reach their maxima in the order ACB.

The zero phase sequence system is a single-phase phasor, and it represents the residual voltage or current which is present in a three-phase circuit under fault conditions when a fourth wire is present either as a direct metallic connection or as a double earth on the system.

The positive phase sequence systems of voltages and currents are those which correspond to the normal load conditions.

The negative phase sequence systems of voltages and currents are those which are set up in the circuit by the fault, and their magnitudes are a direct measure of the superposed fault conditions between phases. The individual voltages and currents of this system are confined to the three lines.

APPENDIX 5

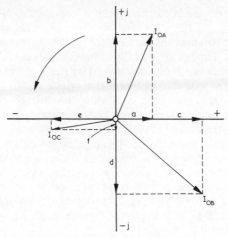

Fig. A5.1   Typical three-phase unbalanced phasor system

The zero phase sequence systems of voltages and currents are also set up in the circuit by the fault, and their magnitudes are a direct measure of the super-posed fault conditions *to earth*. The voltages and currents of this system embrace the fourth wire (or ground) in addition to the three line wires.

A balanced three-phase system, which corresponds to normal balanced load conditions, contains a positive phase sequence system only.

An unbalanced three-phase system, in which the phasor sum is zero, contains both positive and negative phase sequence systems, but no zero phase sequence phasors. In practice this corresponds to the case of a short circuit between two line wires. The positive phase sequence system is that part of the total unbalanced phasor system which corresponds to the normal loading condition. The negative phase sequence system is that which is introduced by the particular fault conditions.

An unbalanced three-phase system, in which the phasor sum has some definite magnitude, contains positive, negative and zero phase sequence systems.

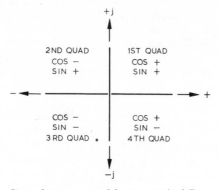

Fig. A5.2   Signs of trigonometrical functions in the different quadrants

689

In practice this corresponds to the case of a line earth fault on a three-phase circuit having an earthed neutral. The positive phase sequence system is that part of the total unbalanced phasor system which corresponds to the normal loading conditions. The negative phase sequence system is that which is introduced by the particular fault conditions and which is confined to the three line wires. The zero phase sequence system is that which is introduced as a residual component by the particular fault condition, the voltages appearing between the lines and earth, while equal and co-phasal currents flow in the line wires, giving a resultant through the ground of three times their individual magnitudes.

Figure A5.1 shows a typical three-phase unbalanced phasor system which may be of current or voltage. The treatment is unaffected by the character of the quantities and it is assumed, in what follows, that we are dealing with current phasors. Counter-clockwise phasor rotation is taken to be positive, and the usual convention of positive and negative rectangular co-ordinates is adopted.

Let it be assumed the rectangular components of the unbalanced phasor system of Fig. A5.1 are:

$$\text{phase A}: a+jb$$
$$\text{phase B}: c+jd$$
$$\text{phase C}: e+jf$$

These, of course, are the general expressions and do not indicate the relative positions of the phasors to each other. This would be given by inserting before each symbol letter the actual components sign shown by the diagram, thus:

$$\text{phase A}: a+jb$$
$$\text{phase B}: c-jd$$
$$\text{phase C}: -e-jf$$

Let it be assumed, further, that the rectangular components of the, as yet undetermined, positive phase sequence phasor system are:

$$\text{phase A}: \quad m+jn \tag{A5.1}$$

$$\text{phase B}: \quad o+jp = (m+jn)\left(-\tfrac{1}{2}-j\frac{\sqrt{3}}{2}\right) \tag{A5.2}$$

$$\text{phase C}: \quad q+jr = (m+jn)\left(-\tfrac{1}{2}+j\frac{\sqrt{3}}{2}\right) \tag{A5.3}$$

and of the negative phase sequence system:

$$\text{phase A}: \quad s+jt \tag{A5.4}$$

$$\text{phase B}: \quad u+jv = (s+jt)\left(-\tfrac{1}{2}+j\frac{\sqrt{3}}{2}\right) \tag{A5.5}$$

$$\text{phase C}: \quad w+jx = (s+jt)\left(-\tfrac{1}{2}-j\frac{\sqrt{3}}{2}\right) \tag{A5.6}$$

while the zero phase sequence system is:

$$\left.\begin{array}{l}\text{phase A}\\ \text{phase B}\\ \text{phase C}\end{array}\right\} y+jz \tag{A5.7}$$

The terms

$$\left(-\tfrac{1}{2}-j\frac{\sqrt{3}}{2}\right) \quad \text{and} \quad \left(-\tfrac{1}{2}+j\frac{\sqrt{3}}{2}\right)$$

are operators and correspond with the clockwise or counter-clockwise rotation of the phasor representing phase A to the positions occupied by the phasors representing phases B and C, thus allowing the latter to be expressed in terms of the former.

The general operator is the expression $\cos\alpha + j\sin\alpha$, where α is the angle through which the original phasor is turned; the precise sign to be inserted before $\cos\alpha$ and $j\sin\alpha$ depends upon the quadrant into which the phasor is turned, as shown by Fig. A5.2. The operator for rotating through any angle, counter-clockwise or clockwise, can be obtained from this expression in the

Table A5.1

OPERATIONS FOR 30° INCREMENTS FOR COUNTER-CLOCKWISE
AND CLOCKWISE TURNING OF PHASORS

Angle, degrees	Direction of turning					
	Counter-clockwise*			Clockwise		
	Cos	Sin	Operator	Cos	Sin	Operator
0	1	0	$1+j0$	1	0	$1+j0$
30	$\sqrt{3}/2$	$\tfrac{1}{2}$	$\sqrt{3}/2+j\tfrac{1}{2}$	$\sqrt{3}/2$	$-\tfrac{1}{2}$	$\sqrt{3}/2-j\tfrac{1}{2}$
60	$\tfrac{1}{2}$	$\sqrt{3}/2$	$\tfrac{1}{2}+j\sqrt{3}/2$	$\tfrac{1}{2}$	$-\sqrt{3}/2$	$\tfrac{1}{2}-j\sqrt{3}/2$
90	0	1·0	$0+j1$	0	-1	$0-j1$
120	$-\tfrac{1}{2}$	$\sqrt{3}/2$	$-\tfrac{1}{2}+j\sqrt{3}/2$	$-\tfrac{1}{2}$	$-\sqrt{3}/2$	$-\tfrac{1}{2}-j\sqrt{3}/2$
150	$-\sqrt{3}/2$	$\tfrac{1}{2}$	$-\sqrt{3}/2+j\tfrac{1}{2}$	$-\sqrt{3}/2$	$-\tfrac{1}{2}$	$-\sqrt{3}/2-j\tfrac{1}{2}$
180	-1	0	$-1+j0$	-1	0	$-1+j0$
210	$-\sqrt{3}/2$	$-\tfrac{1}{2}$	$-\sqrt{3}/2-j\tfrac{1}{2}$	$-\sqrt{3}/2$	$\tfrac{1}{2}$	$-\sqrt{3}/2+j\tfrac{1}{2}$
240	$-\tfrac{1}{2}$	$-\sqrt{3}/2$	$-\tfrac{1}{2}-j\sqrt{3}/2$	$-\tfrac{1}{2}$	$\sqrt{3}/2$	$-\tfrac{1}{2}+j\sqrt{3}/2$
270	0	-1	$0-j1$	0	1	$0+j1$
300	$\tfrac{1}{2}$	$-\sqrt{3}/2$	$\tfrac{1}{2}-j\sqrt{3}/2$	$\tfrac{1}{2}$	$\sqrt{3}/2$	$\tfrac{1}{2}+j\sqrt{3}/2$
330	$\sqrt{3}/2$	$-\tfrac{1}{2}$	$\sqrt{3}/2-j\tfrac{1}{2}$	$\sqrt{3}/2$	$\tfrac{1}{2}$	$\sqrt{3}/2+j\tfrac{1}{2}$
360	1	0	$1+j0$	1	0	$1+j0$

manner shown by Table A5.1. Thus a combination of the 90° j operator with ordinary trigonometrical functions gives an operator for any angle of rotation.

The study will be facilitated by introducing at this stage the graphical methods of separating out the positive, negative and zero phase sequence phasor systems. These are shown in Figs. A5.3, A5.4 and A5.5 respectively. In Fig. A5.3 the positive phase sequence phasor for each phase is derived as follows: For phase A, add to the phasor I_{OA} the phasor I_{OB} rotated, in the positive counter-clockwise direction, through 120°, as shown by I'_{OB} and to I'_{OB} add the phasor I_{OC} rotated in the same direction through 240° as shown by I'_{OC}. Join the extremity of I'_{OC} to the star point O and trisect the line so obtained. This gives the phasor I_{OA+} which is the positive phase sequence phasor for phase A. For phase B, add to I_{OB} the phasor I_{OC} rotated positively through 120° as shown by I'_{OC}, and to

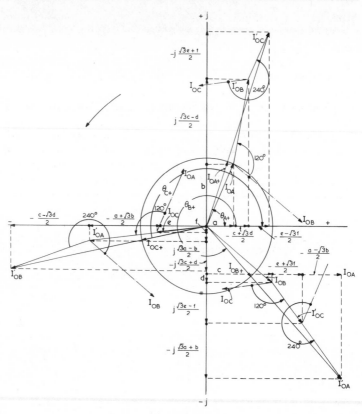

Fig. A5.3   *Derivation of positive phase sequence phasor system*

I'_{OC} add the phasor I_{OA} rotated positively through 240° as shown by I'_{OA}. Join the extremity of I'_{OA} to the star point O and trisect as before to obtain I_{OB+}, the positive phase sequence phasor for phase B. For phase C, add to I_{OC} the phasor I_{OA} rotated positively through 120° as shown by I'_{OA}, and to I'_{OA} add the phasor I_{OB} rotated positively through 240° as shown by I'_{OB}. Join the extremity of I'_{OB} to the star point O and trisect as before to obtain I_{OC+}, the positive phase sequence phasor for phase C.

In Fig. A5.4 the negative phase sequence phasor for each phase is obtained in exactly the same way as the positive phase sequence phasors of Fig. A5.3, except that the rotations through 120° and 240° are affected in the negative clockwise direction.

Having obtained the positive or negative phase sequence component for phase A, the corresponding components for phases B and C can be obtained without repeating the graphical performance for those phases but simply by drawing the phasors $I_{OB\pm}$ and $I_{OC\pm}$ equal in length to $I_{OA\pm}$ and spaced 120° and 240° therefrom in the sequences shown in Figs. A5.3 and A5.4. The graphical construction is shown for all phases, however, in order to clarify the derivation of the final equations for the phase sequence components.

In Fig. A5.5 the single-phase zero phase sequence phasor, which is the same for all three phases, is obtained by adding I_{OA}, I_{OB} and I_{OC} together without any rotation, joining the extremity of I_{OC} to the star point O and trisecting the line so obtained.

It has already been stated that the equation of a phasor turned through 120° in a clockwise direction is the equation to the phasor in the original position multiplied by,

$$-\tfrac{1}{2}-j\frac{\sqrt{3}}{2}, \text{ (eqn. A5.2)};$$

for clockwise turning through 240° the multiplier is

$$-\tfrac{1}{2}+j\frac{\sqrt{3}}{2}, \text{ (eqn. A5.3)};$$

for counter-clockwise turning through 120° the multiplier is

$$-\tfrac{1}{2}+j\frac{\sqrt{3}}{2}, \text{ (eqn. A5.5)};$$

for counter-clockwise turning through 240° the multiplier is

$$-\tfrac{1}{2}-j\frac{\sqrt{3}}{2}, \text{ (eqn. A5.6)}.$$

We are thus equipped for expressing mathematically the phasor rotations through 120° and 240° in both directions shown in Figs. A5.3 and A5.4. The relevant equations are as follows:

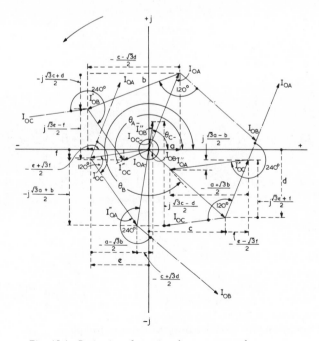

Fig. A5.4 *Derivation of negative phase sequence phasor system*

693

POSITIVE PHASE SEQUENCE

Phase A

$$I_{OA} = a + jb \tag{A5.8}$$

$$I'_{OB} = (c+jd)\left(-\tfrac{1}{2}+j\frac{\sqrt{3}}{2}\right) = -\left(\frac{c+\sqrt{3}d}{2}\right)+j\left(\frac{\sqrt{3}c-d}{2}\right) \tag{A5.9}$$

$$I'_{OC} = (e+jf)\left(-\tfrac{1}{2}-j\frac{\sqrt{3}}{2}\right) = -\left(\frac{e-\sqrt{3}f}{2}\right)-j\left(\frac{\sqrt{3}e+f}{2}\right) \tag{A5.10}$$

$$I_{OA}+I'_{OB}+I'_{OC} = (a+jb)+\left\{-\left(\frac{c+\sqrt{3}d}{2}\right)+j\left(\frac{\sqrt{3}c+d}{2}\right)\right\}$$

$$+\left\{-\left(\frac{e-\sqrt{3}f}{2}\right)-j\left(\frac{\sqrt{3}e+f}{2}\right)\right\}$$

$$= \left\{a-\frac{c+\sqrt{3}d}{2}-\frac{e-\sqrt{3}f}{2}\right\}+j\left\{b+\frac{\sqrt{3}c-d}{2}-\frac{\sqrt{3}e+f}{2}\right\} \tag{A5.11}$$

Simplifying eqn. A5.11 and dividing by 3 gives the positive phase sequence component for phase A thus:

$$I_{OA+} = \left(\frac{a}{3}-\frac{c+e}{6}+\frac{f-d}{2\sqrt{3}}\right)+j\left(\frac{b}{3}-\frac{d+f}{6}+\frac{c-e}{2\sqrt{3}}\right) \tag{A5.12}$$

Putting eqn. A5.12 as in A5.1,

$$I_{OA+} = m+jn$$

then
$$\theta_{A+} = \tan^{-1} n/m$$

Phase B

$$I_{OB} = c + jd$$

$$I'_{OC} = (e+jf)\left(-\tfrac{1}{2}+j\frac{\sqrt{3}}{2}\right) = -\left(\frac{e+\sqrt{3}f}{2}\right)+j\left(\frac{\sqrt{3}e-f}{2}\right)$$

$$I'_{OA} = (a+jb)\left(-\tfrac{1}{2}-j\frac{\sqrt{3}}{2}\right) = -\left(\frac{a-\sqrt{3}b}{2}\right)-j\left(\frac{\sqrt{3}a+b}{2}\right)$$

$$I_{OB}+I'_{OC}+I'_{OA} = (c+jd)+\left\{-\left(\frac{e+\sqrt{3}f}{2}\right)+j\left(\frac{\sqrt{3}e-f}{2}\right)\right\}$$

$$+\left\{-\left(\frac{a-\sqrt{3}b}{2}\right)-j\left(\frac{\sqrt{3}a+b}{2}\right)\right\}$$

$$= \left\{c-\frac{e+\sqrt{3}f}{2}-\frac{a-\sqrt{3}b}{2}\right\}+j\left\{d+\frac{\sqrt{3}e-f}{2}-\frac{\sqrt{3}a+b}{2}\right\} \tag{A5.13}$$

Simplifying eqn. A5.13 and dividing by 3 gives the positive phase sequence component for phase B, thus,

$$I_{OB+} = \left(\frac{c}{3} - \frac{e+a}{6} + \frac{b-f}{2\sqrt{3}}\right) + j\left(\frac{d}{3} - \frac{f+b}{6} + \frac{e-a}{2\sqrt{3}}\right) \qquad \text{(A5.14)}$$

Putting eqn. A5.14 as in eqn. A5.2,

$$I_{OB+} = o + jp$$

then
$$\theta_{B+} = \tan^{-1} p/o$$

Phase C

$$I_{OC} = e + jf$$

$$I'_{OA} = (a+jb)\left(-\tfrac{1}{2} + j\frac{\sqrt{3}}{2}\right) = -\left(\frac{a+\sqrt{3}b}{2}\right) + j\left(\frac{\sqrt{3}a-b}{2}\right)$$

$$I'_{OB} = (c+jd)\left(-\tfrac{1}{2} - j\frac{\sqrt{3}}{2}\right) = -\left(\frac{c-\sqrt{3}d}{2}\right) - j\left(\frac{\sqrt{3}c+d}{2}\right)$$

$$I_{OC} + I'_{OA} + I'_{OB} = (e+jf) + \left\{-\left(\frac{a+\sqrt{3}b}{2}\right) + j\left(\frac{\sqrt{3}a-b}{2}\right)\right\}$$

$$+ \left\{-\left(\frac{c-\sqrt{3}d}{2}\right) - j\left(\frac{\sqrt{3}c+d}{2}\right)\right\}$$

$$= \left\{e - \frac{a+\sqrt{3}b}{2} - \frac{c-\sqrt{3}d}{2}\right\} + j\left\{f + \frac{\sqrt{3}a-b}{2} - \frac{\sqrt{3}c+d}{2}\right\} \qquad \text{(A5.15)}$$

Simplifying eqn. A5.15 and dividing by 3 gives the positive phase sequence component for phase C, thus,

$$I_{OC} = \left(\frac{e}{3} - \frac{a+c}{6} + \frac{d-b}{2\sqrt{3}}\right) + j\left(\frac{f}{3} - \frac{b+d}{6} + \frac{a-c}{2\sqrt{3}}\right) \qquad \text{(A5.16)}$$

Putting eqn. A5.16 as in A5.3,

$$I_{OC+} = q + jr$$

then
$$\theta_{C+} = \tan^{-1} r/q$$

NEGATIVE PHASE SEQUENCE

Phase A

$$I_{OA} = a + jb$$

$$I''_{OB} = (c+jd)\left(-\tfrac{1}{2} - j\frac{\sqrt{3}}{2}\right) = -\left(\frac{c-\sqrt{3}d}{2}\right) - j\left(\frac{\sqrt{3}c+d}{2}\right)$$

$$I''_{OC} = (e+jf)\left(-\tfrac{1}{2} + j\frac{\sqrt{3}}{2}\right) = -\left(\frac{e+\sqrt{3}f}{2}\right) + j\left(\frac{\sqrt{3}e-f}{2}\right)$$

$$I_{OA} + I''_{OB} + I''_{OC} = (a+jb) + \left\{ -\left(\frac{c-\sqrt{3}d}{2}\right) - j\left(\frac{\sqrt{3}c+d}{2}\right) \right\}$$

$$+ \left\{ -\left(\frac{e+\sqrt{3}f}{2}\right) + j\left(\frac{\sqrt{3}e-f}{2}\right) \right\}$$

$$= \left\{ a - \frac{c-\sqrt{3}d}{2} - \frac{e+\sqrt{3}f}{2} \right\} + j\left\{ b - \frac{\sqrt{3}c+d}{2} + \frac{\sqrt{3}e-f}{2} \right\} \qquad (A5.17)$$

Simplifying eqn. A5.17 and dividing by 3 gives the negative phase sequence component for phase A, thus,

$$I_{OA-} = \left(\frac{a}{3} - \frac{c+e}{6} + \frac{d-f}{2\sqrt{3}}\right) + j\left(\frac{b}{3} - \frac{d+f}{6} + \frac{e-c}{2\sqrt{3}}\right) \qquad (A5.18)$$

Putting eqn. A5.18 as in A5.4,

$$I_{OA-} = s + jt$$

then
$$\theta_{A-} = \tan^{-1} t/s$$

Phase B

$$I_{OB} = c + jd$$

$$I''_{OC} = (e+jf)\left(-\tfrac{1}{2} - j\frac{\sqrt{3}}{2}\right) = -\left(\frac{e-\sqrt{3}f}{2}\right) - j\left(\frac{\sqrt{3}e+f}{2}\right)$$

$$I''_{OA} = (a+jb)\left(-\tfrac{1}{2} + j\frac{\sqrt{3}}{2}\right) = -\left(\frac{a+\sqrt{3}b}{2}\right) + j\left(\frac{\sqrt{3}a-b}{2}\right)$$

$$I_{OB} + I''_{OC} + I''_{OA} = (c+jd) + \left\{ -\left(\frac{e-\sqrt{3}f}{2}\right) - j\left(\frac{\sqrt{3}e+f}{2}\right) \right\}$$

$$+ \left\{ -\left(\frac{a+\sqrt{3}b}{2}\right) + j\left(\frac{\sqrt{3}a-b}{2}\right) \right\}$$

$$= \left\{ c - \frac{e-\sqrt{3}f}{2} - \frac{a+\sqrt{3}b}{2} \right\} + j\left\{ d - \frac{\sqrt{3}e+f}{2} + \frac{\sqrt{3}a-b}{2} \right\} \qquad (A5.19)$$

Simplifying eqn. A5.19 and dividing by 3 gives the negative phase sequence component for phase B, thus,

$$I_{OB-} = \left(\frac{c}{3} - \frac{e+a}{6} + \frac{f-b}{2\sqrt{3}}\right) + j\left(\frac{d}{3} - \frac{f+b}{6} + \frac{a-e}{2\sqrt{3}}\right) \qquad (A5.20)$$

Putting eqn. A5.20 as in A5.5,

$$I_{OB-} = u + jv$$

then
$$\theta_{B-} = \tan^{-1} v/u$$

Phase C

$$I_{OC} = e + jf$$

$$I''_{OA} = (a+jb)\left(-\tfrac{1}{2}-j\frac{\sqrt{3}}{2}\right) = -\left(\frac{a-\sqrt{3}b}{2}\right)-j\left(\frac{\sqrt{3}a+b}{2}\right)$$

$$I''_{OB} = (c+jd)\left(-\tfrac{1}{2}+j\frac{\sqrt{3}}{2}\right) = -\left(\frac{c+\sqrt{3}d}{2}\right)+j\left(\frac{\sqrt{3}c-d}{2}\right)$$

$$I_{OC}+I''_{OA}+I''_{OB} = (e+jf)+\left\{-\left(\frac{a-\sqrt{3}b}{2}\right)-j\left(\frac{\sqrt{3}a+b}{2}\right)\right\}$$

$$+\left\{-\left(\frac{c+\sqrt{3}d}{2}\right)+j\left(\frac{\sqrt{3}c-d}{2}\right)\right\}$$

$$= \left\{e-\frac{a-\sqrt{3}b}{2}-\frac{c+\sqrt{3}d}{2}\right\}+j\left\{f-\frac{\sqrt{3}a+b}{2}+\frac{\sqrt{3}c-d}{2}\right\} \tag{A5.21}$$

Simplifying eqn. A5.21 and dividing by 3 gives the negative phase sequence component for phase C, thus,

$$I_{OC-} = \left(\frac{e}{3}-\frac{a+c}{6}+\frac{b-d}{2\sqrt{3}}\right)+j\left(\frac{f}{3}-\frac{b+d}{6}\quad\frac{c-a}{2\sqrt{3}}\right) \tag{A5.22}$$

Putting eqn. A5.22 as in A5.6,

$$I_{OC-} = w + jx$$

then

$$\theta_{C-} = \tan^{-1} x/w.$$

In practice it is not necessary to calculate out the positive and negative phase sequence components for all three phases as all positives are equal and all negatives are equal.

ZERO PHASE SEQUENCE

Phases A, B and C

As this is a single-phase phasor, common to all three phases, one calculation only is involved.

$$I_{OA} = a + jb$$

$$I_{OB} = c + jd$$

$$I_{OC} = e + jf$$

$$I_{OA}+I_{OB}+I_{OC} = (a+jb)+(c+jd)+(e+jf)$$

$$= (a+c+e)+j(b+d+f) \tag{A5.23}$$

Dividing eqn. A5.23 by 3 gives the zero phase sequence component for each phase, thus,

$$I_{\substack{OAO\\OBO\\OCO}} = \left(\frac{a+c+e}{3}\right)+j\left(\frac{b+d+f}{3}\right) \tag{A5.24}$$

Putting eqn. A5.24 as in A5.7,

$$I_{\substack{OAO\\OBO\\OCO}} = y + jz$$

then

$$\theta_{\substack{AO\\BO\\CO}} = \tan^{-1} z/y$$

When zero phase sequence currents exist in an unbalanced three-phase system the current in each line wire is that given by eqn. A5.24. The current in the return circuit, that is, the ground or a fourth wire, is the sum of the currents in the three lines. Zero phase sequence voltages are simply three voltages in parallel between each line and the return.

The magnitudes of the phase sequence components obtained by means of eqns. A5.12, A5.14, A5.16, A5.18, A5.20, A5.22 and A5.24 are not affected in any way if the non-standard convention of *clockwise* direction of phasor rotation is adopted.

The whole of the foregoing treatment gives the *general* formulæ for calculating the different quantities, and in evaluating them care must be taken to insert the actual co-ordinate sign before each component value according to the quadrant in which each unbalanced phasor lies. Similarly, in evaluating the angular displacements of the phase sequence components from the assumed reference phasor position, *i.e.*, the $+x$ axis of Figs. A5.3, A5.4 and A5.5, due account must be taken of the quadrant in which the component is found. In the first quadrant the total angle is that given directly by the $\tan^{-1}$ value; in the second quadrant the total angle is $180°$ minus the $\tan^{-1}$ value; in the third quadrant the angle is $180°$ plus the $\tan^{-1}$ value, and in the fourth quadrant the angle is $360°$ minus the $\tan^{-1}$ value.

An interesting example of the application of symmetrical components is afforded by the phasor analysis of the conditions which arise when a three-phase star/star core-type transformer, having a three-wire primary and a four-wire

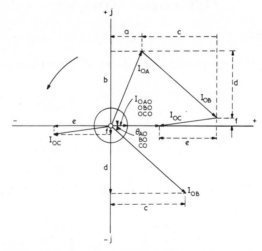

Fig. A5.5   Derivation of zero phase sequence phasor system

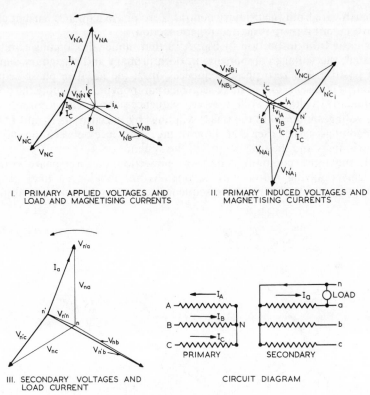

I. PRIMARY APPLIED VOLTAGES AND
LOAD AND MAGNETISING CURRENTS

II. PRIMARY INDUCED VOLTAGES AND
MAGNETISING CURRENTS

III. SECONDARY VOLTAGES AND
LOAD CURRENT

CIRCUIT DIAGRAM

Fig. A5.6   Phasor diagrams showing the voltage and current conditions in a three-phase, core-type, star/star connected transformer with three-wire primary and four-wire secondary when supplying a single-phase load from one line to neutral

secondary, supplies an unbalanced load. Taking the extreme case of a single load on one phase from line to neutral, the phasor quantities are illustrated typically by Fig. A5.6, in which diagram I shows all primary currents and applied voltages, diagram II primary magnetising currents and induced voltages, and diagram III secondary current and induced voltages. Loss currents are neglected, and it is assumed that primary and secondary coils are wound in opposite directions. The secondary load has a unity power factor. The primary current I_A is 67% of the secondary current I_a, while I_B and I_C are each 50% of I_A.

The current I_a in the loaded secondary phase winding is a true zero phase sequence current, having its return path through the neutral conductor. There is, however, no zero phase sequence current on the primary side of the transformer, as is shown by the phasor analysis of Fig. A5.7, in which diagram I shows the summation of the load and magnetising currents in the primary phase windings, diagram II the resulting positive phase sequence currents, and diagram III the negative phase sequence currents; diagram IV gives the construction for the zero phase sequence current, which, it will be noted, is *nil*. The reason for this is that zero phase sequence current in the secondary winding becomes converted to zero phase sequence voltage in the primary windings by the choking effect of the two unloaded primary windings, resulting from the absence of a fourth wire from the primary neutral. Zero phase sequence current cannot flow in a circuit if there is no neutral connection.

It is clear from inspection of Fig. A5.6 that, under the loading conditions illustrated, the voltage components in both primary and secondary windings are of positive and zero phase sequences only; the negative phase sequence component is absent. The positive sequence components are simply the no-load line to neutral voltages, while the zero sequence components are represented by the voltages induced in the phase windings by the currents I_B and I_C in the unloaded phases; the latter is also the voltage difference between the star point of the windings and the true neutral of the system.

With the more common delta/star connection of three-phase core-type transformers having a four-wire secondary, an unbalanced load produces positive, negative and zero phase sequence secondary line currents, but only

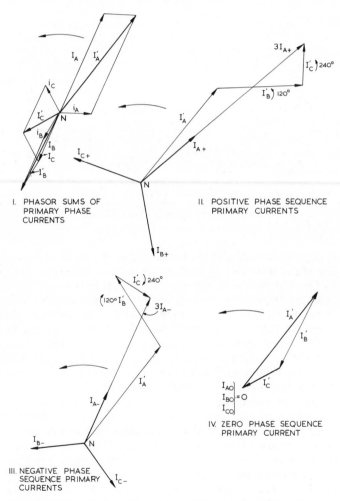

I. PHASOR SUMS OF PRIMARY PHASE CURRENTS

II. POSITIVE PHASE SEQUENCE PRIMARY CURRENTS

III. NEGATIVE PHASE SEQUENCE PRIMARY CURRENTS

IV. ZERO PHASE SEQUENCE PRIMARY CURRENT

Fig. A5.7   *Phasor diagrams showing total primary currents and their phase sequence analysis corresponding to Fig. A5.6*

positive and negative phase sequence primary line currents. Zero phase sequence currents flow round the primary delta winding, however.

These conditions are shown in phasor form by the diagrams of Fig. A5.8 on the assumption of a one to one ratio of phase windings.

From diagram I it is seen there are zero phase sequence currents in the secondary star windings and in the lines connected thereto; there are also zero phase sequence currents in the primary delta windings but not in the primary lines. Diagrams II and III show that there are positive and negative phase sequence currents in the primary and secondary windings and lines; in the respective windings the corresponding primary and secondary currents are in phase, but corresponding line currents are displaced by 30°; corresponding winding currents are equal, but line currents are in the ratio of 1·73 to 1. Thus zero phase sequence currents flow in the secondary lines but not in the primary lines; positive and negative phase sequence currents flow in both primary and secondary lines.

In transformers, positive and negative phase sequence impedances* are the normal load leakage impedances of the transformer; they are series impedances in the equivalent network diagrams. In those cases where zero phase sequence currents can flow in *both* primary and secondary *lines* the zero phase sequence impedance* per phase also is the normal load leakage impedance of the phase windings, assuming symmetry of the phases; it is also a series impedance. Where zero phase sequence currents cannot flow in the *lines on both sides* the series zero phase sequence impedance is open circuited and is thus equivalent to infinity in its relation to the series network.

Thus the series zero phase sequence impedance per phase is the same as the positive or negative phase sequence impedance per phase, or, alternatively, it is equivalent to infinity, due to an insulated star neutral or a delta connection, or to an interconnected star winding on the other side.

As, under certain conditions, a star or interconnected star winding may present an impedance to earth to the flow of zero phase sequence current if the neutral is earthed, a zero phase sequence impedance shunted to earth from the star end of the series impedance branch of the equivalent circuit diagram may quite properly be included. This must not be confused with shunted exciting admittance, which is neglected. If zero phase sequence current flows *only* to earth on the star side the shunted impedance is shown as connected directly to earth; if the current flows from one winding to the other over the normal load leakage series impedance of the transformer the shunt impedance to earth is shown open circuited. With star connected windings this shunt zero phase sequence impedance is, in general, considerably higher than the normal load leakage impedance, being of the average order of 50%; for interconnected star windings it is much lower than the normal load leakage impedance being of the same order as that of an interconnected star neutral earthing transformer.

In the case of an insulated neutral star winding, zero phase sequence currents cannot flow either from the lines or in the transformer winding, so that in the zero phase sequence network a star connected winding with an insulated neutral is denoted by open links at the star end of the series impedance and at the earth end of the star side shunt impedance. When a star connected winding is earthed or a fourth wire is used, zero phase sequence currents can flow *through* the

---

* Strictly speaking, the impedances to positive, negative and zero phase sequence currents respectively.

particular winding and the external circuit connected thereto, but they cannot *circulate* in the transformer winding itself, so that in the equivalent network diagram an earthed neutral, or four-wire, star connected winding is denoted by closed links at the star end of the series impedance and at the earth end of the star side shunt impedance branches of the network.

When a delta connected winding is used, zero phase sequence currents cannot flow from the delta to the connected lines, or vice versa, but they can circulate in the delta winding without flowing through the external circuit. A delta winding thus represents a closed path with respect to the transformer but an open circuit with respect to that side of the equivalent network to which the delta winding is connected. The zero phase sequence connections for the equivalent circuit of a delta winding thus are represented by an open link between the delta end of the series impedance branch of the network and the lines, and a direct shunt connection to earth of the delta end of the equivalent series impedance.

The interconnected star neutral earthing transformer has an open circuit for applied positive or negative phase sequence voltage. For zero phase sequence, however, the currents in all the lines have the same value, so that the zero phase sequence impedance per phase is the normal load leakage impedance between the two winding halves on the same limb. The zero phase sequence connection is, therefore, a simple shunt impedance to earth.

In single-phase transformers, positive, negative and zero phase sequence impedances are the same when the circuit conditions are such as to permit the flow of zero phase sequence currents in the lines on both sides.

It is important to distinguish the difference between zero phase sequence series and shunt impedances, and these are summarised in Fig. A5.9 for the different three-phase transformer connections and conditions of earthing.

The true shunt impedances Z_{AN} and Z_{BN}, shown in Fig. A5.9, are effective only for the star/star, star/interconnected star, and interconnected star/star connections, and then only when the neutral point is earthed on *one* side with the star/star connection, but on one or both sides with the other two connections. For transformers connected star/star with one neutral earthed, the shunt impedance on the earthed side is, on the average, of the order of 50% for three-phase core-type transformers, while for three-phase shell-type and three-phase groups of single-phase transformers it is of the average order of 400%. For star/interconnected star and interconnected star/star transformers with the star neutral earthed, the shunt impedance on the star side is of the average order of 50% for three-phase core-type transformers and 400% for three-phase shell-type and three-phase groups of single-phase transformers; where the interconnected star winding is earthed the shunt impedance on the interconnected star side is much lower than the normal series impedance of the transformer for all types of transformers, being the impedance between winding halves on the same limb of the core. In those cases where *both* primary and secondary neutrals of star/interconnected star and interconnected star/star windings are earthed the respective shunt impedances are of the same orders of magnitudes as given above.

Where both neutrals of star/star connected windings are earthed, as in diagram IV of Fig. A5.9, the shunt impedance in the equivalent circuit is a small exciting impedance which, being neglected, is shown open circuited. A delta winding in conjunction with an earthed neutral star winding also results in the true shunt

702

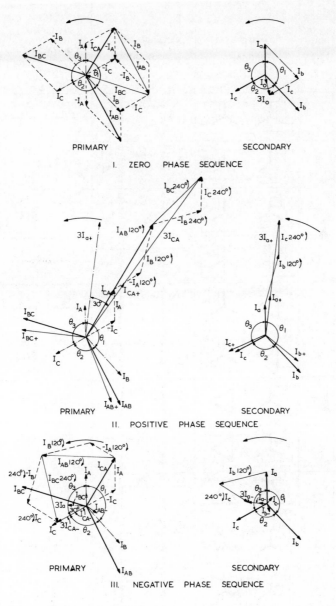

PRIMARY   **SECONDARY**

I.   ZERO   PHASE   SEQUENCE

PRIMARY   **SECONDARY**

II.   POSITIVE   PHASE   SEQUENCE

PRIMARY   **SECONDARY**

III.   NEGATIVE   PHASE   SEQUENCE

Fig. A5.8   Phasor diagrams showing phase sequence currents in a three-phase delta/star connected transformer with three-wire primary and four-wire secondary supplying unbalanced loads to neutral

703

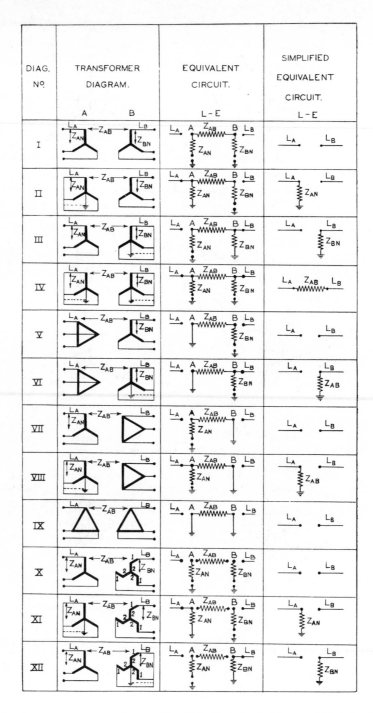

Fig. A5.9 Zero phase sequence equivalent circuits and

704

DIAG. No.	TRANSFORMER DIAGRAM.		EQUIVALENT CIRCUIT.	SIMPLIFIED EQUIVALENT CIRCUIT.
	A	B	L–E	L–E
XIII				
XIV				
XV				
XVI				
XVII				
XVIII				
XIX				
XX				
XXI				
XXII				
XXIII				
XXIV				

impedances for two-winding and three-winding transformers

impedance of the latter being open circuited in the equivalent circuit diagram.

Where an interconnected star connection is used on one side, either winding is non-inductive to the other to zero phase sequence currents, so that Z_{AN} and Z_{BN}, as the case may be, is the leakage impedance between the winding halves on the same core limb for zero phase sequence currents in the interconnected star winding or the self-inductive impedance to earth in the star connected winding. The series impedance Z_{AB} is thus infinity even when both neutrals are earthed, as in diagrams XIII and XVII of Fig. A5.9.

For an interconnected star neutral earthing transformer the shunt impedance to earth, Z_{12}, is the leakage impedance between the winding halves on the same core limb, as is diagram XVIII.

For three-winding transformers the following expressions show the relationship of the impedances between the different windings, assuming, in each case, that the third winding is open circuited:

$$Z_A = \tfrac{1}{2}(Z_{AT} + Z_{AB} - Z_{TB}) \qquad Z_{AT} = Z_A + Z_T$$
$$Z_T = \tfrac{1}{2}(Z_{AT} + Z_{TB} - Z_{AB}) \qquad Z_{TB} = Z_T + Z_B$$
$$Z_B = \tfrac{1}{2}(Z_{BA} + Z_{TB} - Z_{AT}) \qquad Z_{BA} = Z_B + Z_A$$

Where an impedance having an assigned value is open circuited, its circuit value thereby becomes converted to infinity.

The shunt impedances to which an approximate average value of 400 % has been assigned are based upon average normal load leakage impedances of 5 % and average normal magnetising currents of 5 %. The actual value of shunt impedance varies with the size and design of transformer, and as a percentage it is,

$$\text{(short circuit kVA} \div \text{magnetising kVA)} \times 100$$

The shunt impedances to which an approximate average value of 50 % has been given are based upon tests carried out on three-phase core-type transformers.

Where, in the foregoing remarks, reference is made to the equality of the zero phase sequence impedance and the normal impedance, the qualifying statement that this is not exact for three-phase core-type transformers should be borne in mind. The normal load impedance is due to three-phase currents in the phase windings, while the series zero phase sequence impedance is due to single-phase winding currents; the impedance due to the latter is thus affected by the interlinking of the magnetic circuits of the three phases, but it does not differ very considerably from the normal load leakage impedance. If exact figures are required they should be obtained from the manufacturer.

All the references to three-phase core-type transformers apply to the three-limb core construction. For the five-limb core-type the equality of zero phase sequence impedances and normal load leakage impedances, where applicable, are exact.

It should be remembered that zero phase sequence currents in transformer windings depend not only upon the connections of the windings and the earthing of the winding neutral points but also upon the external circuit conditions, particularly as regards earthing. Thus in any complete zero phase sequence network involving transformers the effective zero phase sequence transformer impedances given in Fig. A5.9 may be modified by the external circuit conditions. The impedances shown, therefore, in Fig. A5.9, and also the orders of magnitudes given in the foregoing remarks, assume that the external circuit conditions are such as regards arrangement and earthing as to permit zero phase sequence currents to flow in the earthed transformer windings.

In brief summary the position is that series positive and negative phase sequence impedances of a transformer are the normal load leakage impedance; the shunt positive phase sequence impedance due to normal no-load magnetising impedance is ignored, as usually it does not enter into short circuit calculations. The zero phase sequence impedance of a transformer with an earthed neutral and no electrical connection between windings (either direct or via earth) constitutes a shunt impedance. Shunt zero phase sequence impedances are those over which zero phase sequence currents flow from the lines to neutral, while zero phase sequence impedances are those over which zero phase sequence currents flow from one transformer winding to the other.

The zero phase sequence impedances depend upon the connections of the transformer windings and also upon the earthing conditions of the windings and of the rest of the circuit.

Appendix 6

Loading of oil immersed transformers*

The working life of a transformer is dependent on the life of its insulation, and the rate at which this deteriorates increases with increasing temperature which, in turn, depends upon the overload cycle. An assessment of the overload which can be carried by a given transformer, has in the past, been based on two factors, (i) the temperature of the cooling medium and (ii) the prior loading condition. This basis has been modified and a separate B.S. Code of Practice C.P.1010: 1959 supersedes the earlier recommendations. In this Code a new consideration is introduced, in which four separate 'daily loading conditions' are related to maximum winding temperature with a view to obtaining the best utilisation of life expectancy, whilst at the same time due consideration is given to the increased rate of insulation deterioration at higher temperatures. This is applicable to oil immersed transformers with Class A insulation manufactured to B.S.171.

While it is not possible to present accurate data for all conditions of use or variations in transformer design, it is intended that the Code will enable the user to determine the range of loading conditions which are permissible, consistent with an average life expectancy.

Care should be taken, when increasing the load on a transformer, that any associated cables and switchgear are adequately rated for such increases and that any transformer ancillary equipment, e.g., tap changers, bushings, etc., do not impose any limitation. The voltage regulation will increase when the load on the transformer is increased, but it is possible to calculate the value from the regulation formula for a given load and power factor. (See Chapter 1.)

As in past editions of B.S.171, the thermal time constant of the transformer is employed, together with other factors, in order to determine the permissible loading and the following definition applies:

Thermal time constant

The time which would be required for a transformer loaded at rated kVA and

*This Appendix is based upon C.P.1010 (1959), 'Guide to Loading of Transformers' by permission of The British Standards Institution, Sales Branch, Newton House, 101–113 Pentonville Road, London W1Y 4AA, from whom copies of the complete Standard may be obtained.

in a cooling medium at constant temperature, to reach the temperature rise attainable under steady state conditions if the initial rate of temperature rise, when no heat was being dissipated, were maintained.

Typical values of thermal time constants in hours for ONAN transformers are given in Table A6.1.

Table A6.1 (1T)*
THERMAL TIME CONSTANT IN HOURS FOR ONAN TRANSFORMERS

Rated kVA	System highest voltage (kV on h.v. side)			
	12·5 and below	37	73	123 and above
0– 50	3·5	5·0	—	—
51– 250	3·0	4·0	5·0	—
251– 1000	2·75	3·5	4·0	5·0
1001–10 000	2·5	2·75	3·5	4·0
Above 10 000	2·5	2·75	3·0	3·5

An estimate of the loading of an oil immersed transformer is based upon the assumed conditions that:

  (i) the transformer has an average ratio of copper to iron losses, (consistent with the method of cooling). This ratio is not critical and a fairly wide departure from average ratios will not materially affect the figures.

 (ii) the mineral oil used complies with B.S.148 and is maintained in accordance with C.P.1009: 1959

(iii) the temperature of the cooling medium does not increase by more than 10°C during the loading period

(iv) transformers complying with B.S.171 are suitable for operation up to a height of 1000 metres above sea level as regards kVA loading.

To obtain the appropriate loading of a transformer it is assumed that the daily (24 h) operating conditions are correlated to the maximum temperature of the windings. The 24 hour period is subdivided into 'heavy' and 'light' loading cycles. Light loading conditions are those which produce a temperature in any part of the windings not exceeding 80°C, at which temperature the rate of using the transformer life is negligible. If the winding temperatures specified are not exceeded for the relevant operating condition, each of the four daily operating conditions shown in Table A6.2 is considered to represent an equivalent use in potential life of the transformer.

The designations R, L, M and S in the last column of Table A6.2 correspond to the four loading conditions shown, and the permissible transformer loading is obtained from the corresponding Tables A6.3 to A6.6 in this appendix.

Table A6.3 (1R) indicates the maximum recurrent constant load which can be applied daily for the whole of the heavy loading period for different temperatures of cooling media. If the anticipated peak loading is not in excess of the values corresponding to the loading period and the temperature of cooling air, then no reference to other tables is necessary.

*Table 1T, C.P.1010: 1959.

Table A6.2*
BASIS FOR RECOMMENDATIONS CONTAINED IN C.P.1010

Daily operating conditions hours of loading		Maximum temperature of windings °C		Loading	
Heavy	Light	Heavy load period	Light load period	Table	C.P.1010 designation
24	0	95	—	A6.3	1R
16	8	100	80	A6.4	1L
8	16	105	80	A6.5	1M
3	21	115	80	A6.6	1S

The following example illustrates the use of Table A6.3 (1R). A 500 kVA, ONAN transformer is expected to be used as follows:

Heavy load for 8 hours—9 a.m. to 5 p.m.
Light load for 16 hours—for the remainder of each day.

From 11 a.m. until 5 p.m. the average cooling air temperature is expected to be 25°C, and for the remainder of the 24 hour period, 10°C. It is required to determine the load which the transformer could carry in each period.

Considering each period of *steady* air temperature and load, it is seen from

Table A6.3 (1R)
RECURRENT DAILY LOADINGS FOR THE OPERATING CONDITIONS STATED IN THE TABLE, EXPRESSED AS A PERCENTAGE OF RATED KVA

1	2	3	4	5	6	7	8
	All day	Long period		Medium period		Short period	
Cooling air temperature °C	24 hours heavy load	16 hours heavy load	Remaining 8 hours light load	8 hours heavy load	Remaining 16 hours light load	3 hours heavy load	Remaining 21 hours light load
0	120	125	105	130	105	150	105
5	115	120	100	125	100	145	100
10	110	115	94	120	94	140	94
15	105	110	88	115	88	135	88
20	100	105	82	110	82	130	82
25	94	100	76	105	76	125	76
30	88	94	70	100	70	120	70
35	82	88	64	94	64	115	64
40	76	82	57	88	57	110	57
45	70	76	49	82	49	105	49
50	64	70	40	76	40	100	40

*Schedule 2, C.P.1010: 1959.

Table A6.3 (1R) that the loading corresponds to the medium period, and that columns 5 and 6 will give the loading as a percentage of the transformer rated kVA.

Hence, from 9 a.m. to 11 a.m. at 10°C air, the permissible loading =
120% load, *i.e.*, 600 kVA

from 11 a.m. to 5 p.m. at 25°C air, the permissible loading =
105% load, *i.e.*, 525 kVA

from 5 p.m. to 9 a.m. at 10°C air, the permissible loading =
94% load, *i.e.*, 470 kVA.

It will be noted that in this case the thermal time constant is not required.

In cases where the peak loading exceeds the values in Table A6.3 (1R) reference is necessary to Tables A6.4, A6.5 or A6.6 (1L, 1M or 1S of C.P.1010) which are used for assessing the time for which a peak load of a particular magnitude can be carried, within the heavy loading period. These tables take into account not only the temperature of the cooling air but also that of the top oil in the transformer at the time when the peak load starts, and they are applicable to peak loads imposed within:

(*a*) the long period of heavy loading, Table A6.4 (1L)
(*b*) the medium period of heavy loading, Table A6.5 (1M)
(*c*) the short period of heavy loading, Table A6.6 (1S).

From Table A6.4, A6.5 or A6.6 (1L, 1M or 1S of C.P.1010) as appropriate to the loading condition under investigation, the value in hours against the known temperature conditions of cooling air and top oil is obtained. The thermal time constant applicable to the transformer from Table A6.1 is then multiplied by the time in hours obtained from Table A6.4, 5 or 6, and the product is the period for which the peak load may be maintained. At the end of this period the load should be reduced for the remainder of the loading period to a value not exceeding the appropriate loading in columns 3, 5 or 7 of Table A6.3 (1R) and thereafter the load should further be reduced for the remainder of the 24 hours to the appropriate loading in columns 4, 6 or 8 of Table A6.3 (1R).

The example now given illustrates the use of the peak loading tables.

A 500 kVA, 11 000/400 V, ONAN cooled transformer is working on a load cycle of 8 hours heavy load followed by 16 hours at light load. It is anticipated that a peak load of 130% will be required during the heavy load period. The average temperature of the cooling air during the heavy loading will be 10°C and the transformer oil at the start of the peak load will be at approximately 60°C. The problem is to determine for how long the peak load can be maintained and, in addition, the permissible loads that the transformer can carry for the remainder of the day.

Reference to Table A6.3 (1R) shows that for this condition the maximum loading during the heavy loading period is 120%. Since 130% is required, it is necessary to consult Table A6.5 (1M) in order to ascertain the peak load within the medium period loading. It is seen that a load of 130% corresponds to a value of 0·57 hours per hour of thermal time constant. The product of this value and the appropriate time constant (from Table A6.1) gives the duration of the peak load,

i.e., $0.57 \times 2.75 = 1.5$ hours (approximately).

711

During the remaining 6·5 hours of the heavy load period at 10°C air temperature, 120% load may be maintained; for the subsequent light loading period of 16 hours a load of 94% may be maintained.

The permissible loading schedule is therefore as follows:

> Peak load of 130% for 1·5 hours; *i.e.*, 650 kVA
> Heavy load of 120% for 6·5 hours; *i.e.*, 600 kVA
> Light load of 94% for 16 hours; *i.e.*, 470 kVA.

Table A6.4 (1L)
DURATION OF PEAK LOAD WITHIN LONG PERIOD LOADING

Cooling air temperature °C	Top oil temperature at start of required peak loading °C	Load as a percentage of rated kVA							
		150	140	130	120	110	100	90	80
		Hours per hour of thermal time constant							
0	0	1·07	1·51	2·53					
	10	0·94	1·37	2·37					
	20	0·79	1·20	2·19					
	30	0·62	1·00	1·95					
	40	0·41	0·75	1·65					
	50	0·14	0·42	1·22					
10	10	0·77	1·04	1·49	2·56				
	20	0·64	0·90	1·33	2·38				
	30	0·49	0·73	1·14	2·18				
	40	0·31	0·53	0·91	1·89				
	50	0·10	0·28	0·61	1·52				
	60	—	—	0·18	0·91				
20	20	0·53	0·72	0·99	1·43	2·50			
	30	0·40	0·58	0·83	1·25	2·30			
	40	0·25	0·41	0·64	1·04	2·05			
	50	0·08	0·21	0·41	0·77	1·72			
	60	—	—	0·11	0·40	1·23			
30	30	0·35	0·48	0·66	0·92	1·34	2·30		
	40	0·22	0·34	0·50	0·74	1·14	2·08		
	50	0·07	0·17	0·31	0·53	0·90	1·79		
	60	—	—	0·08	0·25	0·57	1·38		
	70	—	—	—	—	0·08	0·69		
40	40	0·19	0·28	0·41	0·58	0·82	1·20	2·02	
	50	0·06	0·14	0·25	0·40	0·62	0·98	1·77	
	60	—	—	0·06	0·19	0·38	0·69	1·42	
	70	—	—	—	—	0·05	0·29	0·90	
50	50	0·05	0·12	0·21	0·33	0·48	0·69	1·03	1·67
	60	—	—	0·05	0·15	0·28	0·47	0·78	1·38
	70	—	—	—	—	0·04	0·18	0·44	0·97

Table A6.5 (1M)
DURATION OF PEAK LOAD WITHIN MEDIUM PERIOD LOADING

0	0	1·27	1·87						
	10	1·14	1·73						
	20	0·99	1·57						
	30	0·82	1·37						
	40	0·60	1·11						
	50	0·33	0·78						

10	10	0·91	1·25	1·88				
	20	0·78	1·11	1·72				
	30	0·63	0·94	1·53				
	40	0·46	0·74	1·30				
	50	0·25	0·49	1·00				
	60	—	0·16	0·57				
20	20	0·65	0·87	1·21	1·84			
	30	0·52	0·73	1·05	1·67			
	40	0·37	0·56	0·86	1·45			
	50	0·19	0·36	0·63	1·18			
	60	—	0·11	0·33	0·81			
30	30	0·44	0·59	0·81	1·14	1·76		
	40	0·31	0·45	0·65	0·96	1·56		
	50	0·16	0·29	0·46	0·75	1·32		
	60	—	0·09	0·24	0·48	0·99		
	70	—	—	—	0·11	0·50		
40	40	0·27	0·38	0·53	0·73	1·05	1·61	
	50	0·14	0·24	0·37	0·56	0·85	1·39	
	60	—	0·07	0·18	0·35	0·60	1·10	
	70	—	—	—	0·07	0·28	0·76	
50	50	0·12	0·20	0·31	0·45	0·64	0·92	1·41
	60	—	0·06	0·15	0·27	0·44	0·70	1·16
	70	—	—	—	0·06	0·19	0·41	0·82

Table A6.6 (1S)
DURATION OF PEAK LOAD WITHIN SHORT PERIOD LOADING

10	10	1·27			
	20	1·14			
	30	0·99			
	40	0·82			
	50	0·60			
	60	0·33			
20	20	0·91	1·25		
	30	0·78	1·11		
	40	0·63	0·94		
	50	0·46	0·74		
	60	0·25	0·49		
30	30	0·65	0·87	1·21	
	40	0·52	0·73	1·05	
	50	0·37	0·56	0·86	
	60	0·19	0·36	0·63	
	70	—	0·11	0·33	
40	40	0·44	0·59	0·81	1·14
	50	0·31	0·45	0·65	0·96
	60	0·16	0·29	0·46	0·75
	70	—	0·09	0·24	0·48
50	50	0·27	0·38	0·53	0·73	1·05
	60	0·14	0·24	0·37	0·56	0·85
	70	—	0·07	0·18	0·35	0·60

PROCEDURE: Against actual cooling air temperature and top oil temperature read off the factor in the table and multiply by the thermal time constant.

Data and examples given in this appendix are confined to ONAN cooled transformers. For units designed for other cooling methods, reference should be made to the B.S. Code of Practice, C.P.1010: 1959.

713

A symmetrical component study of earth faults in transformers in parallel

The behaviour of transformers in parallel under earth fault conditions is governed largely by the neutral point earthing of the circuit in which the fault occurs. The current and voltage distributions may be determined in a direct and simple manner by the application of symmetrical components, and the present study demonstrates the procedure and shows the influence of the neutral point earth circuits.

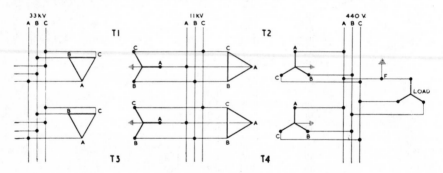

Fig. A7.1   Layout of system

For the examples is chosen a typical three-phase, 50 Hz, duplicate transformer grid substation of the smaller type, from the secondary busbars of which are fed two duplicate step-down transformers supplying a factory load, as shown in Fig. A7.1. The specifications of these transformers are as follows:

Grid transformers, each:
  5000 kVA
  33 000 delta to 11 000 star volts
  87·5 to 262·5 line amperes
  6·7% reactance
  Primary resistance per phase winding, 2·3 Ω
  Secondary resistance per phase winding 0·07 Ω.

Consumer's transformers, each:
  1500 kVA
  11 000 delta to 440 star volts
  78·8 to 1970 line amperes
  4·5% reactance
  Primary resistance per phase winding, 1·05 Ω
  Secondary resistance per phase winding 0·0005 Ω.

The following constants will be used in the investigation:

Grid transformers:
      Primary line to neutral voltage = 19 050
      Secondary line to neutral voltage = 6350
      Secondary line to neutral reactance voltage = 6·7% of 6350 = 425·5

Consumer's transformers:
      Primary line to neutral voltage = 6350
      Secondary line to neutral voltage = 254
      Secondary line to neutral reactance voltage = 4·5% of 254 = 11·42

Three cases are investigated, namely:

(1) both consumer's transformers in commission and both secondary neutrals earthed solidly
(2) as (1), but the neutral of one transformer only earthed
(3) one consumer's transformer only in commission, its neutral being earthed solidly.

In all cases the neutrals of both the grid transformers are earthed solidly.

A dead earth fault (*i.e.*, zero resistance) is assumed to occur on one of the secondary busbars of the consumer's transformers or on the l.v. distributor to the consumer's premises sufficiently near to the transformers for the distributor impedance to the fault to be neglected.

It is first assumed that the neutral earths of the consumer's transformers have zero resistance, and subsequently it is shown how the fault currents and voltages are modified by earth resistance.

It is further assumed that the applied voltages at the primary terminals of the grid transformers remain balanced under fault conditions.

As the fault is assumed to occur on the l.v. secondary side of the consumer's transformers, the constants of all transformers are referred to the 440 V circuit. This gives fault currents and voltages in terms of that circuit, and subsequently they are converted to equivalent 11 and 33 kV values in order to show the true magnitudes in all the other windings.

Resistances and reactances are expressed in ohms, and this avoids the use of an arbitrary kVA base. All constants and values are per phase, that is, line to neutral, and where delta windings are involved, their equivalent line to neutral resistances are determined first, before conversion to a different voltage base, for the sake of uniformity of treatment.

The transformer constants, referred to the 440 V side, are then as follows:

Each grid transformer:
(*a*) Equivalent primary resistance, line to neutral, by the usual delta star conversion formulæ $= \dfrac{2\cdot3^2}{2\cdot3\times3} = 0\cdot767$ Ω

Equivalent primary resistance in terms of secondary line to neutral

$$\text{voltage} = 0.767\left(\frac{6350}{19\,050}\right)^2 = 0.0853\,\Omega$$

Total resistance to neutral referred to 11 kV side $= 0.0853 + 0.07$
$= 0.1553\,\Omega$

The total equivalent resistance to neutral referred to the 440 V side is then

$$0.1553\left(\frac{254}{6350}\right)^2 = 0.0002485\,\Omega$$

which will be rounded up to $0.00025\,\Omega$.

(b) The ohmic reactance, line to neutral, referred to the 11 kV secondary side is $425.5/262.5 = 1.62\,\Omega$, and referred to the 440 V side it is

$$1.62\left(\frac{254}{6350}\right)^2 = 0.00259\,\Omega$$

which will be rounded up to $0.0026\,\Omega$.

The impedance of each grid transformer, line to neutral, referred to the 440 V circuit is thus

$$Z = R + jX = 0.00025 + j0.0026 = 0.00261\,\Omega.$$

Each consumer's transformer:

(c) Equivalent primary resistance, line to neutral, by delta star conversion

$$= \frac{1.05^2}{1.05 \times 3} = 0.35\,\Omega$$

Equivalent primary resistance in terms of secondary line to neutral

$$\text{voltage} = 0.35\left(\frac{254}{6350}\right)^2 = 0.00056\,\Omega$$

Total equivalent resistance to neutral referred to 440 V side

$$= 0.00056 + 0.0005 = 0.00106\,\Omega.$$

(d) The ohmic reactance, line to neutral, referred to the 440 V secondary side is $11.42/1970 = 0.0058\,\Omega$.

The impedance of each consumer's transformer, line to neutral, referred to the 440 V circuit is thus

$$Z = R + jX = 0.00106 + j0.0058 = 0.0059\,\Omega.$$

As, with the grid transformers, the equivalent reactance at 254 V is only 0.04% less than the total impedance, while with the consumer's transformers the equivalent reactance at 254 V is only 1.7% less than the total impedance, the error involved in assuming impedances to be in phase, and treating them as reactances, is negligible. This course will be adopted, therefore, as it saves a good deal of labour in the subsequent calculations.

In this study the phases and lines on the 440 V side are lettered A, B and C, rotation being in the order named. The earth fault is assumed to occur on line A, and the normal voltage to neutral of this line is taken as the reference phasor. In accordance with established procedure the voltage phasor V_A is regarded as lying on the $+Y$ axis of the usual X, Y system of co-ordinates, in order to clarify the presentation of the current terms. The ultimate results are not affected thereby.

In the study of short-circuit currents by symmetrical components the phase sequence networks are derived to embrace the entire circuit from the source of supply *up to the point of the system fault* but not beyond it. If the network is fed from more than one source, the circuits between the fault and all the sources are included. Similarly, if an earth fault occurs at some point along one of a pair of paralleled transmission lines fed from one end, the whole of the lines (and transformers, if any) up to the point at the receiving end where they are paralleled, are included in the phase sequence diagrams, but nothing beyond that point affects the problem. That is, so far as all three *phase sequence networks* are concerned, it is assumed all three lines or busbars of the faulty circuit are connected together and to earth for the study of earth fault short-circuit currents and voltages, and thus only those parts of the actual network which can supply such a three-phase short circuit to earth can be included in the phase sequence networks.

PART I: CURRENTS

The general theorems controlling the currents and voltages in this study are as follows:

The total earth fault current is,

$$I_F = 3V/Z \tag{A7.1}$$

where, V is the normal line to neutral voltage of the system on the voltage base adopted.

Z is the sum of the impedances of the zero, positive and negative phase sequence networks, so that,

$$Z = Z_0 + Z_1 + Z_2 \tag{A7.2}$$

As we are dealing only with static plant, the positive and negative sequence impedances are equal and the same as the normal circuit impedances. The zero sequence impedances depend upon the normal and fault earthing conditions, but for delta/star transformers they have the same values as the normal load impedances, or alternatively infinity, according to whether or not the earthing conditions permit the flow of zero phase sequence currents. This will become clear from the subsequent diagrams.

The zero, positive and negative sequence currents *in the fault* are equal and each one-third of I_F, so that,

$$I_{F0} = I_{F1} = I_{F2} = I_F/3 \tag{A7.3}$$

These are also the total currents in the respective phase sequence networks, and they each divide into the branches of the networks in inverse proportion to the branch sequence impedances. The total sequence network currents are then,

$$I_0 = I_{F0}, I_1 = I_{F1}, \text{ and } I_2 = I_{F2} \tag{A7.4}$$

respectively.

In the faulty phase A the total fault current is the sum of the three sequence total currents, I_0, I_1 and I_2, so that,

$$I_A = I_0 + I_1 + I_2 = I_F \tag{A7.5}$$

and it divides up into the branches of the faulty phase in inverse proportion to the branch sequence impedances. Alternatively, the total fault current in each branch of the faulty phase is the sum of the sequence currents in the corresponding branches of the three sequence networks.

The total fault currents in the other two phases B and C are given by the expressions,

$$\left.\begin{aligned} I_B &= I_0 + h^2 I_1 + h I_2 \\ I_C &= I_0 + h I_1 + h^2 I_2 \end{aligned}\right\} \tag{A7.6}$$

in which the phasor operators h and h^2 are,

$$h = -\tfrac{1}{2} + j\frac{\sqrt{3}}{2} = -0\cdot5 + j0\cdot866$$

$$h^2 = -\tfrac{1}{2} - j\frac{\sqrt{3}}{2} = -0\cdot5 - j0\cdot866$$

so that,

$$\begin{aligned} I_B &= I_0 + I_1(-0\cdot5 - j0\cdot866) + I_2(-0\cdot5 + j0\cdot866) \\ &= I_0 - 0\cdot5(I_1 + I_2) - j0\cdot866(I_1 - I_2) \\ I_C &= I_0 + I_1(-0\cdot5 + j0\cdot866) + I_2(0\cdot5 - j0\cdot866) \\ &= I_0 - 0\cdot5(I_1 + I_2) + j0\cdot866(I_1 - I_2) \end{aligned} \tag{A7.7}$$

These fault currents divide up into the branches of their respective phases in inverse proportion to the branch impedances. Alternatively, the total fault current in each branch of the two sound phases is the sum of the sequence currents in the corresponding branches of the three sequence networks.

The final short-circuit fault currents derived in this way are all in terms of the 440 V star circuit, and they must be converted to delta and star currents at 11 kV, and at 33 kV where applicable. The conversion factors to be used are as follows:

$$\text{Star current at 440 V to delta current at 11 kV} = \frac{254}{11\,000} = 0\cdot0231$$

$$\text{Star current at 440 V to star current at 33 kV} = \frac{440}{33\,000} = 0\cdot01333.$$

The proper application of these conversion factors to the final short-circuit currents in the complete system, derived on $440/\sqrt{3}$ V base, gives the true fault currents in the respective paths.

Case 1

The complete circuit is as shown in Fig. A7.1, and all transformer neutrals are earthed. The complete sequence networks are shown in Fig. A7.2, those parts which do not enter into the final sequence diagrams as carrying any of the sequence current concerned being left open circuited. The simplified sequence networks are given in Fig. A7.3, and the component impedance values are inserted.

From Fig. A7.3 the sequence impedances are,

$$Z_0 = 0.0059/2 = 0.00295 \,\Omega$$
$$Z_1 = (0.00261 + 0.0059)/2 = 0.004255 \,\Omega$$
$$Z_2 = 0.004255 \,\Omega$$

From eqn. A7.2 the total impedance of the entire circuit of Fig. A7.3 is,

$$Z = Z_0 + Z_1 + Z_2 = 0.00295 + 0.004255 + 0.004255 = 0.01146 \,\Omega$$

From eqn. A7.1 the total current in the fault is,

$$I_F = 3V/Z = (3 \times 254)/0.01146 = 66\,498 \, A.$$

From eqns. A7.3 and A7.4 the total current in each of the sequence networks is,

$$I_0 = I_1 = I_2 = 66\,498/3 = 22\,166 \, A$$

and these currents divide equally into the two branches of each sequence diagram, as shown in Fig. A7.3, since the impedances of the parallel connected branches are equal.

The total current in the faulty phase throughout the circuit is, as given by eqn. A7.5, the sum of the various sequence currents, so that, remembering the 440 V base and equivalent star network throughout, we have,

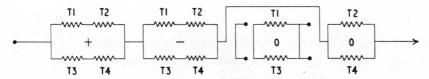

Fig. A7.2   Complete phase sequence networks.
Case 1.   E.R.* = 0

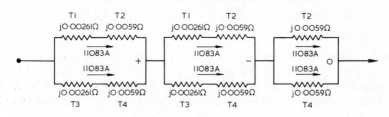

Fig. A7.3   Simplified phase sequence networks with sequence impedances and currents.
Case 1.   E.R. = 0

*In this appendix E.R. denotes 'earth resistance'.

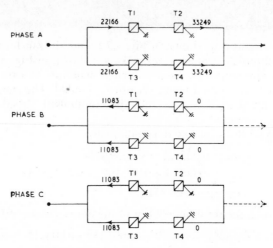

Fig. A7.4   *Phase currents on 440 V star base.*
Case 1.   E.R. = 0

I_A in primary line of T1 and T3 = 11 083 + 11 083 = 22 166 A

I_A in secondary line of T2 and T4 = 22 166 + 11 083 = 33 249 A

In phases B and C the total currents are, from eqns. A7.7,

I_B in primary line of T1 and T3
   $= 0 - 0 \cdot 5(11\,083 + 11\,083) - j0 \cdot 866(11\,083 - 11\,083) = -11\,083$ A

I_B in secondary line of T2 and T4 = 11 083 − 11 083 = 0 A

I_C in primary line of T1 and T3
   $= 0 - 0 \cdot 5(11\,083 + 11\,083) + j0 \cdot 866(11\,083 - 11\,083) = -11\,083$ A

I_C in secondary line of T2 and T4 = 11 083 − 11 083 = 0 A.

These currents are shown in Fig. A7.4.

Applying the delta star current conversion factors, we have,

I_A in primary line of T1 and T3 = 22 166 × (−0·0133)* = −295 A

I_A in secondary line of T2 and T4 = 33 249 A

I_B in primary line of T1 and T3 = 11 083 × (−0·0133)* = 147·5 A

I_B in secondary line of T2 and T4 = 0

I_C in primary line of T1 and T3 = −11 083 × (−0·0133)* = 147·5 A

I_C in secondary line of T2 and T4 = 0

In the case of the secondary circuits of transformers T1 and T3 and of the primary circuits of T2 and T4, phase displacements with respect to the 440 V star base are involved, and account must be taken of these. From Fig. A7.5 the following equations apply for conversion of delta to star currents and vice versa.

___

*The minus sign is introduced here to take account of the reversal of line currents brought about by the cascade delta star transformations in each of the parallel circuits between the 33 kV and 440 V busbars (see Fig. A7.5).

$$I'_a = n(I''_b - I''_c) = I'_B - I'_C$$
$$I'_b = n(I''_c - I''_a) = I'_C - I'_A \qquad\qquad \text{(A7.8)}$$
$$I'_c = n(I''_a - I''_b) = I'_A - I'_B$$

where n is the turns per phase ratio of transformation in whichever transformation direction is being considered.

From eqn. A7.8, the actual currents in the secondary circuits of T1 and T3 and in the primary circuits of T2 and T4 are,

I_A in primary line of T2 $= 0 \cdot 0231(0-0) = 0$

I_A in primary line of T4 $= 0$

I_B in primary line of T2 $= 0 \cdot 0231(0-33\,249) = -768$ A

I_B in primary line of T4 $= -768$ A

I_C in primary line of T2 $= 0 \cdot 0231(33\,249-0) = 768$ A

I_C in primary line of T4 $= 768$ A

I_A in secondary line of T1 $= 0$

I_A in secondary line of T3 $= 0$

I_B in secondary line of T1 $= 768$ A

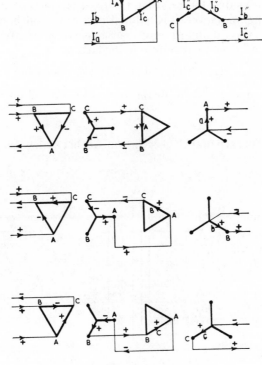

Fig. A7.5   Star delta and delta star current conversions

721

I_B in secondary line of T3 $= -768$ A

I_C in secondary line of T1 $= 768$ A

I_C in secondary line of T3 $= 768$ A.

The final true current distribution throughout the circuit is then as shown in Fig. A7.6.

In practice, currents of the magnitudes derived in the foregoing would not be attained due to earth resistance values. Suppose the total resistance R in the earth circuit between the neutral point of *each* transformer T2 and T4 and the fault to be $0.25 \, \Omega$. The only diagrams affected by these additions are the zero sequence networks in Figs. A7.2 and A7.3, in which a resistance equal to $3R = 0.75 \, \Omega$ is inserted in each branch of the network in series with the zero sequence impedances of T2 and T4 respectively.

Then the total zero sequence impedance of T2 and T4 in parallel including the earth resistances is,

$$Z_0 = R + jX = (0.75106 + j0.0058)/2 = 0.37553 + j0.0029 = 0.376 \, \Omega$$

and, $Z_1 = 0.004255 \, \Omega$ as before

$Z_2 = 0.004255 \, \Omega$ as before.

From eqn. A7.2,

$$Z = Z_0 + Z_1 + Z_2 = 0.37553 + j(0.0029 + 0.004255 + 0.004255)$$
$$= 0.37553 + j0.01141 = 0.376 \, \Omega.$$

It will be seen that the total fault current is now controlled by the earth resistances, and, moreover, it is reduced to a value which, compared with the previous one makes it almost unrecognisable as a short-circuit current.

From eqn. A7.1 the total current in the fault is, then,

$$I_F = 3V/Z = (3 \times 254)/0.376 = 2028 \text{ A}.$$

From eqns. A7.3 and A7.4 the total current in each of the sequence networks is,

$$I_0 = I_1 = I_2 = 2028/3 = 676 \text{ A}.$$

The current in each branch of each sequence network is one half the foregoing, that is, 338 A.

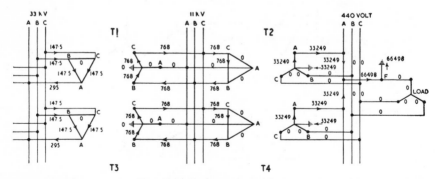

Fig. A7.6   Fault current distribution.
Case 1.   E.R. $= 0$

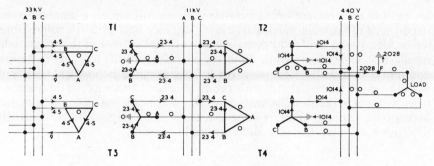

Fig. A7.7 Fault current distribution.
Case 1.   E.R. = 0·25 Ω

By the same procedure as already indicated and combining the different steps, the final fault currents throughout the entire circuit prior to star delta conversions are:

I_A in primary line of T1 and T3 = (338 + 338)(−0·0133) = −9 A

I_A in secondary line of T2 and T4 = 676 + 338 = 1014 A

I_B in primary line of T1 and T3
= [0 − 0·5(338 + 338) − j0·866(338 − 338)](−0·0133) = 4·5 A

I_B in secondary line of T2 and T4 = 338 − 338 = 0

I_C in primary line of T1 and T3
= [0 − 0·5(338 + 338) + j0·866(338 − 338)](−0·0133) = 4·5 A

I_C in secondary line of T2 and T4 = 338 − 338 = 0

Applying the star delta conversions of eqns. A7.8, the actual currents in the secondary circuits of T1 and T3 and in the primary circuits of T2 and T4 are

I_A in primary line of T2 = 0·0231(0 − 0) = 0

I_A in primary line of T4 = 0

I_B in primary line of T2 = 0·0231(0 − 1014) = −23·4 A

I_B in primary line of T4 = −23·4 A

I_C in primary line of T2 = 0·0231(1014 − 0) = 23·4 A

I_C in primary line of T4 = 23·4 A

I_A in secondary line of T1 = 0

I_A in secondary line of T3 = 0

I_B in secondary line of T1 = −23·4 A

I_B in secondary line of T3 = −23·4 A

I_C in secondary line of T1 = 23·4 A

I_C in secondary line of T3 = 23·4 A.

The final true current distribution throughout the circuit is shown in Fig. A7.7, and it will be seen that, compared with Fig. A7.6, the proportionality between *all* currents is the ratio of the total currents in the fault.

It will be noticed that in this example the fault current in the earthed secondary line A of each consumer's transformers T2 and T4 is only about 50% of the

723

normal full-load current of the transformer, and this serves to emphasise the importance of securing particularly low resistance neutral earth connections for l.v. circuits. At light load periods a short-circuit current of this magnitude may be inadequate (apart from 'doubling' effects) to operate line overload trips, and unless sufficiently sensitive earth leakage tripping is provided, the fault would be maintained with disastrous consequences. In h.v. circuits the absolute *ohmic* value of the earth resistance connection may be higher, although the same limitation is imposed upon its value in relation to the impedance of the circuit to which it is connected, as is necessary in l.v. circuits.

Case 2

The complete circuit is the same as Fig. A7.1, except that the secondary neutral of the consumer's transformer T4 is insulated from earth. The phase sequence networks are shown in Figs. A7.8 and A7.9, and it will be seen that these differ from Figs. A7.2 and A7.3 only by reason of the elimination of the impedance of transformer T4 from the zero sequence network.

For this case the sequence impedances are, then,

$$Z_0 = 0.0059 \, \Omega$$
$$Z_1 = 0.004255 \, \Omega$$
$$Z_2 = 0.004255 \, \Omega$$

From eqn. A7.2 the total impedance of the entire circuit of Fig. A7.9 is,

$$Z = Z_0 + Z_1 + Z_2 = 0.0059 + 0.004255 + 0.004255 = 0.01441 \, \Omega$$

From eqn. A7.1 the total current in the fault is,

$$I_F = 3V/Z = (3 \times 254)/0.01441 = 52\,896 \text{ A}$$

From eqns. A7.3 and A7.4 the total current in each of the sequence networks is,

$$I_0 = I_1 = I_2 = 52\,896/3 = 17\,632 \text{ A}$$

The current in each branch of the positive and negative sequence networks is one half of this, or 8816 A, as shown in Fig. A7.9.

By the same procedure as adopted in case 1 the line currents on the 440 V base and equivalent star network throughout are,

by eqn. A7.5,

I_A in primary line of T1 and T3 = $8816 + 8816 = 17\,632$ A
I_A in secondary line of T2 = $17\,632 + 17\,632 + 17\,632 = 35\,264$ A
I_A in secondary line of T4 = $8816 + 8816 = 17\,632$ A

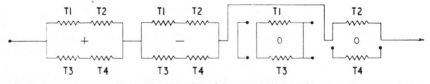

Fig. A7.8   Complete phase sequence networks
Case 2.   E.R. = 0

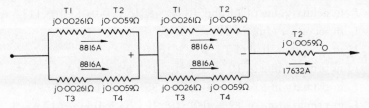

Fig. A7.9 *Simplified phase sequence networks with sequence impedances and currents.*
Case 2. E.R. = 0

by eqn. A7.7,

I_B in primary line of T1 and T3

$$= 0 - 0.5(8816 + 8816) = j0.866(8816 - 8816) = -8816 \text{ A}$$

I_B in secondary line of T2 $= 17\,632 - 8816 = 8816 \text{ A}$

I_B in secondary line of T4 $= -8816 \text{ A and similarly}$

I_C in primary line of T1 and T3

$$= 0 - 0.5(8816 + 8816) + j0.866(8816 - 8816) = 8816 \text{ A}$$

I_C in secondary line of T2 $= 17\,632 - 8816 = 8816 \text{ A}$

I_C in secondary line of T4 $= -8816 \text{ A}.$

These currents are shown in Fig. A7.10.

Applying the delta star conversion factors previously given together with the provisions of eqns. A7.8, the actual currents throughout the entire circuit are,

I_A in primary line of T1 and T3 $= 17\,632 \times (-0.0133) = -235 \text{ A}$

I_A in secondary line of T2 $= 35\,264 \text{ A}$

I_A in secondary line of T4 $= 17\,632 \text{ A}$

I_B in primary line of T1 and T3 $= -8816 \times (-0.0133) = 117.5 \text{ A}$

I_B in secondary line of T2 $= 8816 \text{ A}$

I_B in secondary line of T4 $= -8816 \text{ A}$

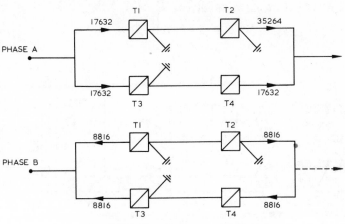

Fig. A7.10 *Phase currents on 440 V star base.*
Case 2. E.R. = 0

I_C in primary line of T1 and T3 $= -8816 \times (-0 \cdot 0133) = 117 \cdot 5$ A

I_C in secondary line of T2 $= 8816$ A

I_C in secondary line of T4 $= -8816$ A

I_A in primary line of T2 $= 0 \cdot 0231(8816 - 8816) = 0$ A

I_A in primary line of T4 $= 0 \cdot 0231(-8816 + 8816) = 0$ A

I_B in primary line of T2 $= 0 \cdot 0231(8816 - 35\,264) = -612$ A

I_B in primary line of T4 $= 0 \cdot 0231(-8816 - 17\,632) = -612$ A

I_C in primary line of T2 $= 0 \cdot 0231(35\,264 - 8816) = 612$ A

I_C in primary line of T4 $= 0 \cdot 0231(17\,632 + 8816) = 612$ A

I_A in secondary line of T1 $= 0$

I_A in secondary line of T3 $= 0$

I_B in secondary line of T1 $= -612$ A

I_B in secondary line of T3 $= -612$ A

I_C in secondary line of T1 $= 612$ A

I_C in secondary line of T3 $= 612$ A

These currents are shown in Fig. A7.11.

If, now, it is assumed the resistance R of the earth circuit between the neutral point of T2 and the fault is $0 \cdot 25 \ \Omega$, the sequence impedances are as given in Figs. A7.8 and A7.9, except that a resistance of $3R = 0 \cdot 75 \ \Omega$ is inserted in series in the zero sequence network.

Then, $\qquad Z_0 = R + jX = 0 \cdot 75106 + j0 \cdot 0058 = 0 \cdot 752 \ \Omega$

$\qquad\qquad Z_1 = 0 \cdot 004255 \ \Omega$ as before

$\qquad\qquad Z_2 = 0 \cdot 004255 \ \Omega$ as before.

From eqn. A7.2,

$$Z = Z_0 + Z_1 + Z_2 = 0 \cdot 75106 + j(0 \cdot 0058 + 0 \cdot 004255 + 0 \cdot 004255)$$
$$= 0 \cdot 75106 + j0 \cdot 01431 = 0 \cdot 752 \ \Omega.$$

From eqn. A7.1 the total fault current is,

$$I_F = 3V/Z = (3 \times 254)/0 \cdot 752 = 1014 \ \text{A}.$$

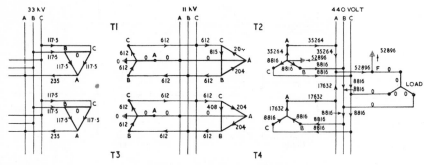

Fig. A7.11   Fault current distribution.
Case 2.   E.R. = 0

This current, it will be seen, is one half that obtaining in case 1, where the same resistance was assumed to be present in the neutral earth circuit.

The values of the true currents throughout the entire circuit are then those given in Fig. A7.11 reduced in direct proportion to the respective total currents in the earth fault, that is, multiplied by the ratio $1014/52\,896 = 0.0192$.

It will be noticed how much less likely is the transformer with insulated neutral to be tripped out of circuit should an earth fault occur when there is appreciable resistance in the earth fault circuit.

Case 3

In this case the system is the same as in Fig. A7.1, except that transformer T4 is disconnected from the busbars on both sides. The simplified phase sequence networks are shown in Fig. A7.12, together with their impedance values. The total sequence impedances are then,

$$Z_0 = 0.0059\,\Omega$$
$$Z_1 = (0.00261/2) + 0.0059 = 0.0072\,\Omega$$
$$Z_2 = 0.0072\,\Omega.$$

The total impedance of the entire circuit of Fig. A7.12 is,

$$Z = Z_0 + Z_1 + Z_2 = 0.0059 + 0.0072 + 0.0072 = 0.0203\,\Omega.$$

The total fault current is,

$$I_F = 3V/Z = (3 \times 54)/0.0203 = 37\,500\,\text{A}.$$

The total current in each sequence network is,

$$I_0 = I_1 = I_2 = 37\,500/3 = 12\,500\,\text{A}.$$

and the distribution is as in Fig. A7.12.

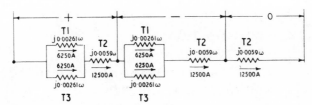

Fig. A7.12 Simplified phase sequence networks with sequence impedances and currents. Case 3. E.R. = 0

The total current in the faulty phase, on the 440 V base and equivalent star network throughout is,

I_A in primary line of T1 and T3 $= 6250 + 6250 = 12\,500\,\text{A}$

I_A in secondary line of T2 $= 12\,500 + 12\,500 + 12\,500 = 37\,500\,\text{A}$

I_B in primary line of T1 and T3
$$= 0 - 0.5(6250 + 6250) - j0.866(6250 - 6250) = -6250\,\text{A}$$

I_B in secondary line of T2
$$= 12\,500 - 0.5(12\,500 + 12\,500) - j0.866(12\,500 - 12\,500) = 0$$

I_C in primary line of T1 and T3
$$= 0 - 0.5(6250 + 6250) + j0.866(6250 - 6250) = 6250 \text{ A}$$
I_C in secondary line of T2 $= 12\,500 - 12\,500 = 0$

Applying the star delta current conversion factors,

I_A in primary line of T1 and T3 $= 12\,500 \times (-0.0133) = -166 \text{ A}$

I_A in secondary line of T2 $= 37\,500 \text{ A}$

I_B in primary line of T1 and T3 $= -6250 \times (-0.0133) = 83 \text{ A}$

I_B in secondary line of T2 $= 0$

I_C in primary line of T1 and T3 $= -6250 \times (-0.0133) = 83 \text{ A}$

I_C in secondary line of T2 $= 0$

Applying eqns. A7.8, the final currents in the primary lines of T2 and the secondary lines of T1 and T3 are,

I_A in primary line of T2 $= 0.0231(0 - 0) = 0$

I_B in primary line of T2 $= 0.0231(0 - 37\,500) = -866 \text{ A}$

I_C in primary line of T2 $= 0.0231(37\,500 - 0) = 866 \text{ A}$

I_A in secondary line of T1 $= 0$

I_A in secondary line of T3 $= 0$

I_B in secondary line of T1 $= 866/2 = -433 \text{ A}$

I_B in secondary line of T3 $= -866/2 = -433 \text{ A}$

I_C in secondary line of T1 $= 866/2 = 433 \text{ A}$

I_C in secondary line of T3 $= 866/2 = 433 \text{ A}$.

The final current distribution is shown in Fig. A7.13.

If a resistance of $0.25\,\Omega$ be assumed in the neutral earth circuit of consumer's transformer T2,

$$Z_0 = R + jX = 0.75106 + j0.0058 = 0.752\,\Omega$$
$$Z_1 = 0.0072\,\Omega \text{ as before}$$
$$Z_2 = 0.0072\,\Omega \text{ as before}$$

and,
$$Z = Z_0 + Z_1 + Z_2 = 0.75106 + j(0.0058 + 0.0072 + 0.0072)$$
$$= 0.75106 + j0.0202 = 0.752\,\Omega.$$

The total current in the fault is,

$$I_F = 3V/Z = (3 \times 254)/0.752 = 1014 \text{ A}$$

being one half the current flowing in the fault when T1 and T2 are both in commission and with assumed earth circuit resistances of $0.25\,\Omega$ each. This, of course, is because of the predominating effect of the earth resistances upon the fault current magnitudes.

The qualitative current distribution is the same as in Fig. A7.13 and quantitatively the currents are those of Fig. A7.13 multiplied by the ratio of the total earth fault currents in the two cases, namely, $1014/37\,500 = 0.02705$.

In the foregoing treatments the fault currents throughout the circuits are $\pm j$ with respect to the voltage to neutral reference phasor of phase A on the

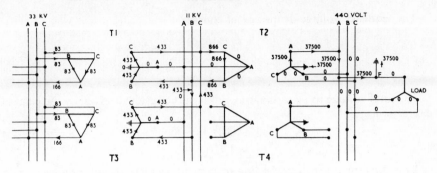

Fig. A7.13   Fault current distribution.
Case 3.   E.R. = 0

secondary side of T2 and T4, when the earth resistances are zero. This is due to the transformer resistances being negligible compared with the reactances. When resistances are included in the earth circuits, however, the fault currents practically are in phase or in opposition with the voltage reference phasor V_A (or V_a), due to the overwhelming effect of the earth resistances, the transformer reactances being negligible by comparison.

PART 2: VOLTAGES

The voltages under fault conditions, at any point in the system, may be obtained in the following way. The phase sequence voltages are first calculated at the 440 V and at the 11 kV busbars on the 440 V base. The voltages from all three lines to neutral are then determined at the same points, still on the 440 V base. The values so derived at the 440 V busbars, *i.e.*, on the secondary sides of transformers T2 and T4, are the actual ones. The true 11 kV delta primary voltages of T2 and T4 and star secondary voltages of T1 and T3 are obtained by applying star delta conversion factors and phase transformation terms. The 33 kV delta primary voltages of T1 and T3 have been postulated as remaining balanced under fault conditions.

The general equations used to determine the circuit voltages are then as follows:

For each phase sequence, the sequence component of line to neutral voltage at any point in the system is equal to the generated line to neutral voltage V_G (which in this case is the voltage at the 33 kV busbars) minus the sequence voltage drop v at that point. Thus, in general, we have, for the sequence line to neutral voltages at any point in a given system,

positive sequence component, $\qquad V_1 = V_G - v_1$

negative sequence component, $\qquad V_2 = 0 \ -v_2 \qquad$ (A7.9)

zero sequence component, $\qquad V_0 = 0 \ -v_0$

The generated voltage V_G is of positive sequence only, so that for negative and zero sequences $V_G = 0$.

729

The sequence voltage drops are, of course,

$$v_1 = I_1 Z_1$$
$$v_2 = I_2 Z_2 \qquad \text{(A7.10)}$$
$$v_0 = I_0 Z_0$$

The total voltage to neutral at any point in the system is given by the general expression,

$$V = V_0 + V_1 + V_2$$
$$= V_G - (v_0 + v_1 + v_2)$$

In a three-phase system the line to neutral voltages at any point are then,

$$V_A = V_0 + V_1 + V_2$$
$$V_B = V_0 + h^2 V_1 + h V_2$$
$$V_C = V_0 + h V_1 + h^2 V_2$$

when applying the usual phasor operators h and h^2, given on page 718.

For use in the calculations, the equations above are rewritten,

$$V_A = V_0 + V_1 + V_2$$
$$V_B = V_0 + V_1(-0.5 - j0.866) + V_2(-0.5 + j0.866)$$
$$= V_0 - 0.5(V_1 + V_2) - j0.866(V_1 - V_2) \qquad \text{(A7.11)}$$
$$V_C = V_0 + V_2(-0.5 + j0.866) + V_2(-0.5 - j0.866)$$
$$= V_0 - 0.5(V_1 + V_2) + j0.866(V_1 - V_2)$$

being comparable with the current eqns. A7.5, A7.6 and A7.7.

Voltage conversion factors are as follows:

Star voltage of 254 V to delta voltage of 11 kV = 0.0231.
Star voltage of 254 V to star voltage of 6.35 kV = 0.01333.

Figure A7.14 shows the conversions adopted for star delta and delta star voltage conversions. The corresponding equations are

$$V_A' = -V_a''/n = V_b' - V_c'*$$
$$V_B' = -V_b''/n = V_c' - V_a' \qquad \text{(A7.12)}$$
$$V_c' = -V_c''/n = V_a' - V_b'$$

$$V_a' = \tfrac{1}{3}V_A' + \tfrac{2}{3}V_C'$$
$$V_b' = \tfrac{1}{3}V_B' + \tfrac{2}{3}V_A' \qquad \text{(A7.13)}$$
$$V_c' = \tfrac{1}{3}V_C' + \tfrac{2}{3}V_B'$$

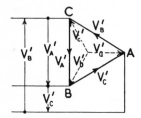 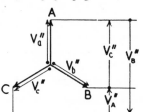

Fig. A7.14 Star delta and delta star voltage conversions

*The minus signs are introduced into the central terms of these equations to denote opposition of primary and secondary phase winding voltage phasors.

730

$$V_A'' = V_b'' - V_c''$$
$$V_B'' = V_c'' - V_a'' \qquad\qquad (A7.14)$$
$$V_C'' = V_a'' = V_b''$$

Additions and subtractions are, of course, carried out vectorially. The factor n is the turns per phase ratio of transformation in whichever transformation direction is being considered.

The detailed calculations can now be proceeded with by application of the foregoing equations.

Case 1. Circuit diagrams (Figs. A7.1 to A7.6)

Voltages at 440 V busbars

Phase sequence voltages, from eqns. A7.9 and A7.10 and Fig. A7.3:

Positive sequence voltage is,

$$V_1 = V_G - I_1 Z_1$$
$$= j254 - 11\,083(j0 \cdot 00261 + j0 \cdot 0059)$$
$$= j254 - j94 \cdot 3 = j159 \cdot 7 \text{ V.}$$

Negative sequence voltage is,

$$V_2 = V_G - I_2 Z_2$$
$$= 0 - 11\,083(j0 \cdot 00261 + j0 \cdot 0059)$$
$$= 0 - j94 \cdot 3 = -j94 \cdot 3 \text{ V.}$$

Zero sequence voltage is,

$$V_0 = V_G - I_0 Z_0$$
$$= 0 - (11\,083 \times j0 \cdot 0059) = -j65 \cdot 4 \text{ V.}$$

Line to neutral voltages, from eqns. A7.11:

Line A to neutral,

$$V_A = V_a'' = V_0 + V_1 + V_2$$
$$= -j65 \cdot 4 + j159 \cdot 7 - j94 \cdot 3 = 0.$$

Line B to neutral,

$$V_B = V_b'' = V_0 - 0 \cdot 5(V_1 + V_2) - j0 \cdot 866(V_1 - V_2)$$
$$= -j65 \cdot 4 - 0 \cdot 5(j159 \cdot 7 - j94 \cdot 3) - j0 \cdot 866(j159 \cdot 7 + j94 \cdot 3)$$
$$= -j65 \cdot 4 - j32 \cdot 7 + 220$$
$$= -j98 \cdot 1 + 220 = 241 \text{ V.}$$

Line C to neutral,

$$V_C = V_c'' = V_0 - 0 \cdot 5(V_1 + V_2) + j0 \cdot 866(V_1 - V_2)$$
$$= -j65 \cdot 4 - 0 \cdot 5(j159 \cdot 7 - j94 \cdot 3) + j0 \cdot 866(j159 \cdot 7 + j94 \cdot 3)$$
$$= -j65 \cdot 4 - j32 \cdot 7 - 220$$
$$= -j98 \cdot 1 - 220 = 241 \text{ V.}$$

Line to line voltages, from eqns. A7.14:

$$V_A'' = V_b'' - V_c''$$
$$= -j98{\cdot}1 + 220 + j98{\cdot}1 + 220 = 440 \text{ V}.$$
$$V_B'' = V_c'' - V_a''$$
$$= -j98{\cdot}1 - 220 - 0$$
$$= -j98{\cdot}1 - 220 = 241 \text{ V}.$$
$$V_C'' = V_a'' - V_b''$$
$$= 0 + j98{\cdot}1 - 220$$
$$= j98{\cdot}1 - 220 = 241 \text{ V}.$$

Voltages at 11 kV busbars

Phase sequence voltages, from eqns. A7.9 and A7.10 and Fig. A7.3:

Positive sequence voltage is,

$$V_1 = V_G - I_1 Z_1$$
$$= j254 - (11\,083 \times j0{\cdot}00261)$$
$$= j254 - j28{\cdot}9 = j225{\cdot}1 \text{ V}.$$

Negative sequence voltage is,

$$V_2 = V_G - I_2 Z_2$$
$$= 0 - (11\,083 \times j0{\cdot}00261)$$
$$= -j28{\cdot}9 \text{ V}.$$

Zero sequence voltage is $V_0 = 0$.

Line to neutral voltages, from eqns. A7.11:

Line A to neutral,

$$V_A = V_a'' = V_0 + V_1 + V_2$$
$$= 0 + j225{\cdot}1 - j28{\cdot}9 = j196{\cdot}2 \text{ V}.$$

Line B to neutral,

$$V_B = V_b'' = V_0 - 0{\cdot}5(V_1 + V_2) - j0{\cdot}866(V_1 - V_2)$$
$$= 0 - 0{\cdot}5(j225{\cdot}1 - j28{\cdot}9) - j0{\cdot}866(j225{\cdot}1 + j28{\cdot}9)$$
$$= -j98{\cdot}1 + 220 = 241 \text{ V}.$$

Line C to neutral,

$$V_C = V_c'' = V_0 - 0{\cdot}5(V_1 + V_2) + j0{\cdot}866(V_1 - V_2)$$
$$= 0 - 0{\cdot}5(j225{\cdot}1 - j28{\cdot}9) + j0{\cdot}866(j225{\cdot}1 + j28{\cdot}9)$$
$$= -j98{\cdot}1 - 220 = 241 \text{ V}.$$

Converting these line to neutral voltages on a 440 V star base to true 11 kV delta values, from Fig. A7.14 and eqns. A7.12:

$$V_A' = -V_a''/n$$
$$= -j196{\cdot}2/0{\cdot}0231 = -j8500 \text{ V}$$

$$V_B' = -V_b''/n$$
$$= (j98.1 - 220)/0.0231$$
$$= j4250 - 9530 = 10\,420 \text{ V}$$
$$V_C' = -V_c''/n$$
$$= (j98.1 + 220)/0.0231$$
$$= j4250 + 9530 = 10\,420 \text{ V}.$$

The corresponding true line to neutral voltages on the 11 kV side of T1 and T3 are obtained from eqns. A7.13, thus,

$$V_a' = \tfrac{1}{3}V_A' + \tfrac{2}{3}V_C'$$
$$= \tfrac{1}{3}(-j8500) + \tfrac{2}{3}(j4250 + 9530)$$
$$= -j2833 + j2833 + 6353$$
$$= 6353 \text{ V}$$
$$V_b' = \tfrac{1}{3}V_B' + \tfrac{2}{3}V_A'$$
$$= \tfrac{1}{3}(j4250 - 9530) + \tfrac{2}{3}(-j8500)$$
$$= j1417 - 3177 - j5667$$
$$= -j4250 - 3177 = 5300 \text{ V}$$
$$V_c' = \tfrac{1}{3}V_C' + \tfrac{2}{3}V_B'$$
$$= \tfrac{1}{3}(j4250 + 9530) + \tfrac{2}{3}(j4250 - 9530)$$
$$= j1417 + 3177 + j2833 - 6354$$
$$= j4250 - 3177 = 5300 \text{ V}.$$

The voltages of T1 and T3 are the same and those of T2 and T4 are identical. All the voltages throughout the circuits are summarised as follows:

Transformers T1 and T3:

All primary line voltages, 33 kV.
Secondary line to neutral voltage,

$$V_a' = 6353 \text{ V}$$
$$V_b' = j4250 - 3177 = 5300 \text{ V}$$
$$V_c' = j4250 - 3177 = 5300 \text{ V}.$$

Transformers T2 and T4:

Primary line voltages,

$$V_A' = -j8500 \text{ V}$$
$$V_B' = j4250 - 9530 = 10\,420 \text{ V}$$
$$V_C' = j4250 + 9530 = 10\,420 \text{ V}.$$

Secondary line to neutral voltages,

$$V_a'' = 0$$
$$V_b'' = -j98.1 + 220 = 241 \text{ V}$$
$$V_c'' = -j98.1 - 220 = 241 \text{ V}.$$

11 kV   440 V

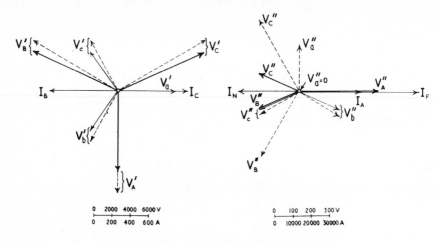

Fig. A7.15   Currents and voltages.
Case 1.   E.R. = 0 Broken lines show no-load voltages

Secondary line voltages,

$$V_A'' = +440 \text{ V}$$
$$V_B'' = -j98 \cdot 1 - 220 = 241 \text{ V}$$
$$V_C'' = j98 \cdot 1 - 220 = 241 \text{ V}.$$

These are shown in phasor form in Fig. A7.15.

The voltages throughout the circuit are very different from the foregoing if neutral earth resistances be included. Taking $0 \cdot 25 \, \Omega$ in the neutral earth circuits of transformers T2 and T4 as before, and by the same procedure as just given, the circuit voltages are as follows:

Voltages at 440 V busbars

Phase sequence voltages, from eqns. A7.9 and A7.10 and Fig. A7.3, bearing in mind the inclusion of a series resistance $R_0 = 3R = 0 \cdot 75 \, \Omega$ in each branch of the zero sequence network:

Positive sequence voltage is,

$$V_1 = V_G - I_1 Z_1$$
$$= j254 - j338 * (j0 \cdot 00261 + j0 \cdot 0059)$$
$$= j254 + 2 \cdot 9 = j254 \text{ V}.$$

Negative sequence voltage is,

$$V_2 = V_G - I_2 Z_2$$
$$= 0 - j338(j0 \cdot 00261 + j0 \cdot 0059)$$
$$= 0 + 2 \cdot 9 = 2 \cdot 9 \text{ V}.$$

*The j operator is attached to the current throughout as it is virtually in phase with V_G.

734

Zero sequence voltage is,

$$V_0 = V_G - I_0 Z_0$$
$$= 0 - j338(0\cdot75106 + j0\cdot0059)$$
$$= 0 - j254 + 2 = -j254 \text{ V.}$$

Line to neutral voltages, from eqns. A7.11:

Line A to neutral,

$$V_A = V_a'' = V_0 + V_1 + V_2$$
$$= -j254 + 2 + j254 + 2\cdot9 + 2\cdot9 = 7\cdot8 \text{ V.}$$

Line B to neutral,

$$V_B = V_b'' = V_0 - 0\cdot5(V_1 + V_2) - j0\cdot866(V_1 - V_2)$$
$$= -j254 + 2 - 0\cdot5(j254 + 2\cdot9 + 2\cdot9)$$
$$- j0\cdot866(j254 + 2\cdot9 - 2\cdot9)$$
$$= -j254 + 2 - j127 - 1\cdot45 - 1\cdot45 + 220 - j2\cdot5 + j2\cdot5$$
$$= -j381 + 219\cdot1 = 440 \text{ V.}$$

Line C to neutral,

$$V_C = V_c'' = V_0 - 0\cdot5(V_1 + V_2) + j0\cdot866(V_1 - V_2)$$
$$= -j254 + 2 - 0\cdot5(j254 + 2\cdot9 + 2\cdot9)$$
$$+ j0\cdot866(j254 + 2\cdot9 - 2\cdot9)$$
$$= -j254 + 2 - j127 - 1\cdot45 - 1\cdot45 - 220 + j2\cdot5 - j2\cdot5$$
$$= -j381 - 220\cdot9 = 440 \text{ V.}$$

Line to line voltages, from eqns. A7.14:

$$V_A'' = V_b'' - V_c''$$
$$= -j381 + 219\cdot1 + j381 + 220\cdot9 = 440 \text{ V}$$

$$V_B'' = V_c'' - V_a''$$
$$= -j381 - 220\cdot9 - 7\cdot8$$
$$= -j381 - 228\cdot7 = 443 \text{ V}$$

$$V_C'' = V_a'' - V_b''$$
$$= 7\cdot8 + j381 - 219\cdot1$$
$$= j381 - 211\cdot3 = 435 \text{ V.}$$

The voltage drop across the neutral earth resistance is, of course,

$$V_R = I_0 R_0 = I_0 3R = j338 \times 0\cdot75 = j254 \text{ V.}$$

Voltages at 11 kV busbars

Phase sequence voltages, from eqns. A7.9 and A7.10 and Fig. A7.3:

Positive sequence voltage is,

$$V_1 = V_G - I_1 Z_1$$

$$= j254 - (j338 \times j0 \cdot 00261)$$
$$= j254 + 0 \cdot 9 = 254 \text{ V.}$$

Negative sequence voltage is,

$$V_2 = V_G - I_2 Z_2$$
$$= 0 - (j338 \times j0 \cdot 00261) = +0 \cdot 9 \text{ V.}$$

Zero sequence voltage is $V_0 = 0$.

Line to neutral voltages, from eqns. A7.11:

Line A to neutral,

$$V_A = V_a'' = V_0 + V_1 + V_2$$
$$= 0 + j254 + 0 \cdot 9 + 0 \cdot 9 = j254 + 1 \cdot 8 = 254 \text{ V}$$

$$V_B = V_b'' = V_0 - 0 \cdot 5(V_1 + V_2) - j0 \cdot 866(V_1 - V_2)$$
$$= 0 - 0 \cdot 5(j254 + 0 \cdot 9 + 0 \cdot 9) - j0 \cdot 866(j254 + 0 \cdot 9 - 0 \cdot 9)$$
$$= -j127 - 0 \cdot 45 - 0 \cdot 45 + 220 - j0 \cdot 8 + j0 \cdot 8$$
$$= -j127 + 219 \cdot 1 = 253 \text{ V}$$

$$V_C = V_c'' = V_0 - 0 \cdot 5(V_1 + V_2) + j0 \cdot 866(V_1 - V_2)$$
$$= 0 - 0 \cdot 5(j254 + 0 \cdot 9 + 0 \cdot 9) + j0 \cdot 866(j254 + 0 \cdot 9 - 0 \cdot 9)$$
$$= -j127 - 0 \cdot 45 - 0 \cdot 45 - 220 + j0 \cdot 8 - j0 \cdot 8$$
$$= -j127 - 220 \cdot 9 = 254 \text{ V.}$$

True 11 kV delta voltages are, from eqns. A7.13:

$$V_A' = -V_a''/n$$
$$= (-j254 - 1 \cdot 8)/0 \cdot 0231$$
$$= -j11 \, 000 - 77 = 11 \, 000 \text{ V}$$

$$V_B' = -V_b''/n$$
$$= (j127 - 219 \cdot 1)/0 \cdot 0231$$
$$= j5500 - 9500 = 10 \, 750 \text{ V}$$

$$V_C' = -V_c''/n$$
$$= (j127 + 220 \cdot 9)/0 \cdot 0231$$
$$= j5500 + 9560 = 11 \, 020 \text{ V.}$$

The corresponding true line to neutral voltages on the 11 kV side of T1 and T3 are, from eqns. A7.14:

$$V_a' = \tfrac{1}{3}V_A' + \tfrac{2}{3}V_C'$$
$$= \tfrac{1}{3}(-j11 \, 000 - 77) + \tfrac{2}{3}(j5500 + 9560)$$
$$= -j3667 - 26 + j3667 + 6373 = 6347 \text{ V}$$

$$V_b' = \tfrac{1}{3}V_B' + \tfrac{2}{3}V_A'$$
$$= \tfrac{1}{3}(j5500 - 9500) + \tfrac{2}{3}(-j11 \, 000 - 77)$$
$$= j1833 - 3167 - j7333 - 51$$
$$= -j5500 - 3218 = 6380 \text{ V}$$

$$V_c' = \tfrac{1}{3}V_C' + \tfrac{2}{3}V_B'$$
$$= \tfrac{1}{3}(j5500 + 9560) + \tfrac{2}{3}(j5500 - 9550)$$
$$= j1833 + 3187 + j3667 - 6367$$
$$= j5500 - 3180 = 6350 \text{ V}.$$

The voltages of T1 and T3 are the same, and those of T2 and T4 are identical. All voltages throughout the circuits are summarised as follows:

Transformers T1 and T3:

All primary line voltages, 33 kV.
Secondary line to neutral voltages,

$$V_a' = 6347 \text{ V}$$
$$V_b' = -j5500 - 3218 = 6380 \text{ V}$$
$$V_c' = j5500 - 3180 = 6350 \text{ V}$$

Transformers T2 and T4:

Primary line voltages,

$$V_A' = -j11\,000 - 77 = 11\,000 \text{ V}$$
$$V_B' = j5500 - 9500 = 10\,750 \text{ V}$$
$$V_C' = j5500 + 9560 = 11\,020 \text{ V}.$$

Secondary line to neutral voltages,

$$V_a'' = 7 \cdot 8 \text{ V}$$
$$V_b'' = j381 + 219 \cdot 1 = 440 \text{ V}$$
$$V_c'' = -j381 - 220 \cdot 9 = 440 \text{ V}.$$

11 kV   440 V

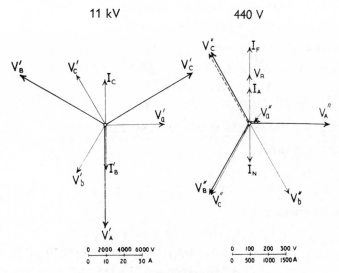

Fig. A7.16 Currents and voltages.
Case 1. E.R. $= 0 \cdot 25 \ \Omega$. Broken lines show no-load voltages

Secondary line voltages,

$$V_A'' = 440 \text{ V}$$
$$V_B'' = -j381 - 228\cdot7 = 443 \text{ V}$$
$$V_C'' = j381 - 211\cdot3 = 435 \text{ V}.$$

Voltage drop across earth resistance, $V_R = j254$ V.

Figure A7.16 shows in phasor form the voltages at the 440 V and the 11 kV busbars. The current phasors qualitatively are the same as those of Fig. A7.15, but rotated through 90°, and quantitatively are of the magnitudes shown in Fig. A7.7. Small apparent discrepancies in the co-ordinate values of the different voltages are due to slide rule working and to minor adjustments made in rounding off certain figures.

Case 2. Circuit diagrams (Figs. A7.1 and A7.7 to A7.10)

Voltages at 440 V busbars

Phase sequence voltages, from eqns. A7.9 and A7.10 and Fig. A7.8:

Positive sequence voltage is,

$$V_1 = V_G - I_1 Z_1$$
$$= j254 - 8816(j0\cdot00261 + j0\cdot0059)$$
$$= j254 - j75 = j179 \text{ V}.$$

Negative sequence voltage is,

$$V_2 = V_G - I_2 Z_2$$
$$= 0 - 8816(j0\cdot00261 + j0\cdot0059)$$
$$= 0 - j75 = -j75 \text{ V}.$$

Zero sequence voltage is,

$$V_0 = V_G - I_0 Z_0$$
$$= 0 - (17\,632 \times j0\cdot0059)$$
$$= -j104 = -j104 \text{ V}.$$

Transformer T2, line to neutral voltages, from eqns. A7.11:

Line A to neutral,

$$V_A = V_a'' = V_0 + V_1 + V_2$$
$$= -j104 + j179 - j75 = 0$$

Line B to neutral,

$$V_B = V_b'' = V_0 - 0\cdot5(V_1 + V_2) - j0\cdot866(V_1 - V_2)$$
$$= -j104 - 0\cdot5(j179 - j75) - j0\cdot866(j179 + j75)$$
$$= -j104 - j89\cdot5 + j37\cdot5 + 155 + 65$$
$$= -j156 + 220 = 270 \text{ V}.$$

Line C to neutral,

$$V_C = V_c'' = V_0 - 0.5(V_1 + V_2) + j0.866(V_1 - V_2)$$
$$= -j104 - 0.5(j179 - j75) + j0.866(j179 + j75)$$
$$= -j104 - j89.5 + j37.5 - 155 - 65$$
$$= -j156 - 220 = 270 \text{ V.}$$

Line to line voltages, from eqns. A7.14:

$$V_A'' = V_b'' - V_c''$$
$$= -j156 + 220 + j156 + 220 = 440 \text{ V.}$$

$$V_B'' = V_c'' - V_a''$$
$$= -j156 - 220 - 0 = 270 \text{ V.}$$

$$V_C'' = V_a'' - V_b''$$
$$= 0 + j156 - 220 = 270 \text{ V.}$$

Transformer T4, line to neutral voltages, from eqns. A7.11:

Line A to neutral,

$$V_A = V_a'' = V_0 + V_1 + V_2$$
$$= 0 + j179 - j75 = j104 \text{ V.}$$

Line B to neutral,

$$V_B = V_b'' = V_0 - 0.5(V_1 + V_2) - j0.866(V_1 - V_2)$$
$$= 0 - 0.5(j179 - j75) - j0.866(j178 + j75)$$
$$= -j89.5 + j37.5 + 155 + 65$$
$$= -j52 + 220 = 226 \text{ V.}$$

Line C to neutral,

$$V_C = V_c'' = V_0 - 0.5(V_1 + V_2) + j0.866(V_1 - V_2)$$
$$= 0 - 0.5(j179 - j75) + j0.866(j179 + j75)$$
$$= -j89.5 + j37.5 - 155 - 65$$
$$= -j52 - 220 = 226 \text{ V.}$$

Line to line voltages, from eqns. A7.14:

$$V_A'' = V_b'' - V_c''$$
$$= -j52 + 220 + j52 + 220 = 440 \text{ V}$$

$$V_B'' = V_c'' - V_a''$$
$$= -j52 - 220 - j104$$
$$= -j156 - 220 = 270 \text{ V.}$$

$$V_C'' = V_a'' - V_b''$$
$$= j104 + j52 - 220$$
$$= j156 - 220 = 270 \text{ V.}$$

Voltages at 11 kV busbars

Phase sequence voltages, from eqns. A7.9 and A7.10 and Fig. A7.8:

Positive sequence voltage is,

$$V_1 = V_G - I_1 Z_1$$
$$= j254 - (8816 \times j0.00261)$$
$$= j254 - j23 = j231 \text{ V}.$$

Negative sequence voltage is,

$$V_2 = V_G - I_2 Z_2 = 0 - (8816 \times j0.00261) = -j23 \text{ V}.$$

Zero sequence voltage is $V_0 = 0$.

Line to neutral voltages, from eqns. A7.11:

Line A to neutral,

$$V_A = V_a'' = V_0 + V_1 + V_2$$
$$= 0 + j231 - j23 = j208 \text{ V}.$$

Line B to neutral,

$$V_B = V_b'' = V_0 - 0.5(V_1 + V_2) - j0.866(V_1 - V_2)$$
$$= 0 - 0.5(j231 - j23) - j0.866(j231 + j23)$$
$$= -j115.5 + j11.5 + 200 + 20$$
$$= -j104 + 220 = 243 \text{ V}.$$

Line C to neutral,

$$V_C = V_c'' = V_0 - 0.5(V_1 + V_2) + j0.866(V_1 - V_2)$$
$$= 0 - 0.5(j231 - j23) + j0.866(j231 + j23)$$
$$= -j115.5 + j11.5 - 200 - 20$$
$$= -j104 - 220 = 243 \text{ V}.$$

Converting these voltages to true 11 kV delta values, from Fig. A7.14 and eqns. A7.12:

$$V_A' = -V_a''/n$$
$$= -j208/0.0231 = -j9000 = 9000 \text{ V}$$
$$V_B' = -V_b''/n$$
$$= (j104 - 220)/0.0231$$
$$= j4500 - 9520 = 10\,500 \text{ V}$$
$$V_C' = -V_c''/n$$
$$= (j104 + 220)/0.0231$$
$$= j4500 + 9520 = 10\,500 \text{ V}.$$

The corresponding line to neutral voltages on the 11 kV side of T1 and T3 are, from eqns. A7.13:

740

$$V'_a = \tfrac{1}{3}V'_A + \tfrac{2}{3}V'_C$$
$$= \tfrac{1}{3}(-j9000) + \tfrac{2}{3}(j4500 + 9520)$$
$$= -j3000 + j3000 + 6347 = 6347 \text{ V}$$

$$V'_b = \tfrac{1}{3}V'_B + \tfrac{2}{3}V'_A$$
$$= \tfrac{1}{3}(j4500 - 9520) + \tfrac{2}{3}(-j9000)$$
$$= j1500 - 3173 - j6000$$
$$= -j4500 - 3173 = 5500 \text{ V}$$

$$V'_c = \tfrac{1}{3}V'_C + \tfrac{2}{3}V'_B$$
$$= \tfrac{1}{3}(j4500 + 9520) + \tfrac{2}{3}(j4500 - 9520)$$
$$= j1500 + 3173 + j3000 - 6347$$
$$= j4500 - 3174 = 5500 \text{ V}.$$

All the voltages throughout the circuits are summarised as follows:

Transformers T1 and T3:

All primary line voltages, 33 kV.
Secondary line to neutral voltages,

$$V'_a = 6347 \text{ V}$$
$$V'_b = -j4500 - 3173 = 5500 \text{ V}$$
$$V'_c = j4500 - 3174 = 5500 \text{ V}.$$

Secondary line voltages,

$$V'_A = -j9000 = 9000 \text{ V}$$
$$V'_B = j4500 - 9520 = 10\,500 \text{ V}$$
$$V'_C = j4500 + 9520 = 10\,500 \text{ V}.$$

Transformers T2 and T4:

Primary line voltages,

$$V'_A = -j9000 = 9000 \text{ V}$$
$$V'_B = j4500 - 9520 = 10\,500 \text{ V}$$
$$V'_C = j4500 + 9520 = 10\,500 \text{ V}.$$

Secondary line voltages,

$$V''_A = +440 \text{ V}$$
$$V''_B = -j156 - 220 = 270 \text{ V}$$
$$V''_C = j156 - 220 = 270 \text{ V}.$$

Transformer T2:

Secondary line to neutral voltages,

$$V''_a = 0 \text{ V}$$
$$V''_b = -j156 + 220 = 270 \text{ V}$$
$$V''_c = -j156 - 220 = 270 \text{ V}.$$

Transformer T4:

Secondary line to neutral voltages,

$$V_a'' = j104 = 104 \text{ V}$$
$$V_b'' = -j52 + 220 = 226 \text{ V}$$
$$V_c'' = -j52 - 220 = 226 \text{ V}.$$

Neutral to earth voltage, $V_{NE} = j104 = 104$ V.

These voltages are shown in phasor form in Fig. A7.17.

If a resistance of $0.25 \, \Omega$ is now included in the earth fault current circuit, the sequence impedances are as given in Fig. A7.8, except that the resistance $R_0 = 3R = 0.75$ is inserted in series in the zero sequence network; the overall zero sequence impedance is then,

$$Z_0 = 0.75106 + j0.0058 = 0.752 \, \Omega$$

as given before.

The total current in each sequence network is 338 A, and one half this in each of the parallel branches of the positive and negative sequence networks. The voltages through the circuit are then as follows:

Voltages at 440 V busbars

Phase sequence voltages, from eqns. A7.9 and A7.10 and Fig. A7.8, bearing in mind the inclusion of the series resistance $R_0 = 0.75 \, \Omega$ in the zero sequence network:

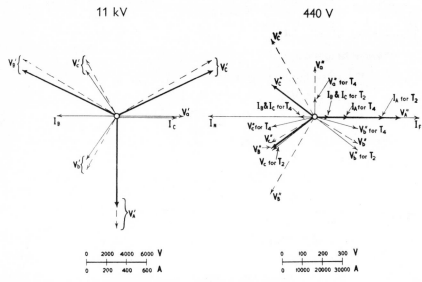

Fig. A7.17   Currents and voltages.
Case 2.   E.R. = 0. Broken lines show no-load voltages

Positive sequence voltage is,

$$V_1 = V_G - I_1 Z_1$$
$$= j254 - j169(j0\cdot00261 + j0\cdot0059)$$
$$= j254 + 1\cdot44 = 254 \text{ V.}$$

Negative sequence voltage is,

$$V_2 = V_G - I_2 Z_2$$
$$= 0 - j169(j0\cdot00261 + j0\cdot0059)$$
$$= +1\cdot44 = 1\cdot44 \text{ V.}$$

Zero sequence voltage is,

$$V_0 = V_G - I_0 Z_0$$
$$= 0 - j338(0\cdot75106 + j0\cdot0058)$$
$$= -j254 + 2 = 254 \text{ V.}$$

Transformer T2, line to neutral voltages, from eqns. A7.11:

Line A to neutral,

$$V_A = V_a'' = V_0 + V_1 + V_2$$
$$= -j254 + 2 + j254 + 1\cdot44 + 1\cdot44 = 4\cdot88 \text{ V.}$$

Line B to neutral,

$$V_B = V_b'' = V_0 - 0\cdot5(V_1 + V_2) - j0\cdot866(V_1 - V_2)$$
$$= -j254 + 2 - 0\cdot5(j254 + 1\cdot44 + 1\cdot44)$$
$$- j0\cdot866(j254 + 1\cdot44 - 1\cdot44)$$
$$= -j254 + 2 - j127 - 0\cdot72 + 220 - j1\cdot25 + j1\cdot25$$
$$= -j381 + 220\cdot55 = 440 \text{ V.}$$

Line C to neutral,

$$V_C = V_c'' = V_0 - 0\cdot5(V_1 + V_2) + j0\cdot866(V_1 - V_2)$$
$$= -j254 + 2 - 0\cdot5(j254 + 1\cdot44 + 1\cdot44)$$
$$+ j0\cdot866(j254 + 1\cdot44 - 1\cdot44)$$
$$= -j254 + 2 - j127 - 0\cdot72 - 0\cdot72 - 220 + j1\cdot25 - j1\cdot25$$
$$= -j381 - 219\cdot45 = 440 \text{ V.}$$

Line to line voltages, from eqns. A7.14:

$$V_A'' = V_b'' - V_c''$$
$$= -j381 + 220\cdot55 + j381 + 219\cdot45$$
$$= 440 \text{ V}$$

$$V_B'' = V_c'' - V_a''$$
$$= -j381 - 219\cdot45 - 4\cdot88$$
$$= -j381 - 224\cdot35 = 443 \text{ V}$$

$$V_C'' = V_a'' - V_b''$$
$$= 4 \cdot 9 + j381 - 220 \cdot 55$$
$$= j381 - 215 \cdot 65 = 438 \text{ V}.$$

Transformer T4, line to neutral voltages, from eqns. A7.11:

Line A to neutral,

$$V_A = V_a'' = V_0 + V_1 + V_2$$
$$= 0 + j254 + 1 \cdot 44 + 1 \cdot 44$$
$$= j254 + 2 \cdot 88 = 254 \text{ V}.$$

Line B to neutral,

$$V_B = V_b'' = V_0 - 0 \cdot 5(V_1 + V_2) - j0 \cdot 866(V_1 - V_2)$$
$$= 0 - 0 \cdot 5(j254 + 1 \cdot 44 + 1 \cdot 44) - j0 \cdot 866(j254 + 1 \cdot 44 - 1 \cdot 44)$$
$$= -j127 - 0 \cdot 72 - 0 \cdot 72 + 220 - j1 \cdot 25 + j1 \cdot 25$$
$$= -j127 + 218 \cdot 55 = 252 \text{ V}.$$

Line C to neutral,

$$V_C = V_c'' = V_0 - 0 \cdot 5(V_1 + V_2) + j0 \cdot 866(V_1 - V_2)$$
$$= 0 - 0 \cdot 5(j254 + 1 \cdot 44 + 1 \cdot 44) + j0 \cdot 866(j254 + 1 \cdot 44 - 1 \cdot 44)$$
$$= -j127 - 0 \cdot 72 - 0 \cdot 72 - 220 + j1 \cdot 25 - j1 \cdot 25$$
$$= -j127 - 221 \cdot 45 = 255 \text{ V}.$$

Line to line voltages, from eqns. A7.14:

$$V_A'' = V_b'' - V_c''$$
$$= -j127 + 218 \cdot 55 + j127 + 221 \cdot 45 = 440 \text{ V}$$

$$V_B'' = V_c'' - V_a''$$
$$= -j127 - 221 \cdot 45 - j254 - 2 \cdot 88$$
$$= -j381 - 224 \cdot 35 = 443 \text{ V}$$

$$V_C'' = V_a'' - V_b''$$
$$= j254 + 2 \cdot 88 + j127 - 218 \cdot 55$$
$$= j381 - 215 \cdot 65 = 438 \text{ V}.$$

The voltage drop across the neutral earth resistance is,

$$V_R = I_0 R_0 = I_0 \times 0 \cdot 3R = j338 \times 0 \cdot 75 = j254 \text{ V}.$$

Voltages at 11 kV busbars

Phase sequence voltages from eqns. A7.9 and A7.10 and Fig. A7.8:

Positive sequence voltage is,

$$V_1 = V_G - I_1 Z_1$$
$$= j254 - (j169 \times j0 \cdot 00261)$$
$$= j254 + 0 \cdot 44 = 254 \text{ V}.$$

Negative sequence voltage is,

$$V_2 = V_G - I_2 Z_2$$
$$= 0 - (j169 \times j0 \cdot 00261)$$
$$= +0 \cdot 44 = 0 \cdot 44 \text{ V.}$$

Zero sequence voltage is $V_0 = 0$.

Line to neutral voltages, from eqns. A7.11:

Line A to neutral,

$$V_A = V_a'' = V_0 + V_1 + V_2$$
$$= 0 + j254 + 0 \cdot 44 + 0 \cdot 44$$
$$= j254 + 0 \cdot 88 = 254 \text{ V.}$$

Line B to neutral,

$$V_B = V_b'' = V_0 - 0 \cdot 5(V_1 + V_2) - j0 \cdot 866(V_1 - V_2)$$
$$= 0 - 0 \cdot 5(j254 + 0 \cdot 44 + 0 \cdot 44) - j0 \cdot 866(j254 + 0 \cdot 44 - 0 \cdot 44)$$
$$= -127 - 0 \cdot 22 - 0 \cdot 22 + 220 - j0 \cdot 38 + j0 \cdot 38$$
$$= -j127 + 219 \cdot 55 = 254 \text{ V.}$$

Line C to neutral,

$$V_C = V_c'' = V_0 - 0 \cdot 5(V_1 + V_2) + j0 \cdot 866(V_1 - V_2)$$
$$= 0 - 0 \cdot 5(j254 + 0 \cdot 44 + 0 \cdot 44) + j0 \cdot 866(j254 + 0 \cdot 44 - 0 \cdot 44)$$
$$= -j127 - 0 \cdot 22 - 0 \cdot 22 - 220 + j0 \cdot 38 - j0 \cdot 38$$
$$= -j127 - 220 \cdot 45 = 254 \text{ V.}$$

Converting these voltages to true 11 kV delta values, from Fig. A7.14 and eqns. A7.12:

$$V_A' = -V_a''/n$$
$$= (-j254 - 0 \cdot 88)/0 \cdot 0231$$
$$= -j11\,000 - 39 = 11\,000 \text{ V}$$

$$V_B' = V_b''/n$$
$$= (j127 - 219 \cdot 55)/0 \cdot 0231$$
$$= j5500 - 9500 = 10\,750 \text{ V}$$

$$V_C' = -V_c''/n$$
$$= (j127 + 220 \cdot 45)/0 \cdot 0231$$
$$= j5500 + 9550 = 11\,020 \text{ V.}$$

The corresponding line to neutral voltages on the 11 kV side of T1 and T3 are, from eqns. A7.13:

$$V_a' = \tfrac{1}{3}V_A' + \tfrac{2}{3}V_C'$$
$$= \tfrac{1}{3}(-j11\,000 - 39) + \tfrac{2}{3}(j5500 + 9550)$$
$$= -j3667 - 13 + j3667 + 6367 = 6354 \text{ V.}$$

$$V_b' = \tfrac{1}{3}V_B' + \tfrac{2}{3}V_A'$$
$$= \tfrac{1}{3}(j5500 - 9500) + \tfrac{2}{3}(-j11\,000 - 39)$$
$$= j1833 - 3167 - j7333 - 26$$
$$= -j5500 - 3193 = 6360 \text{ V}$$

$$V_c' = \tfrac{1}{3}V_C' + \tfrac{2}{3}V_B'$$
$$= \tfrac{1}{3}(j5500 + 9550) + \tfrac{2}{3}(j5500 - 9500)$$
$$= j1833 + 3183 + j3667 - 6333$$
$$= j5500 - 3150 = 6340 \text{ V}.$$

All the voltages throughout the circuits are summarised as follows:
Transformer T1 and T3:

All primary line voltages, 33 kV.
Secondary line to neutral voltages,

$$V_a' = +6354 = 6354 \text{ V}$$
$$V_b' = -j5500 - 3193 = 6360 \text{ V}$$
$$V_c' = j5500 - 3150 = 6340 \text{ V}.$$

Secondary line voltages,

$$V_A' = -j11\,000 - 39 = 11\,000 \text{ V}$$
$$V_B' = j5500 - 9500 = 10\,750 \text{ V}$$
$$V_C' = j5500 + 9550 = 11\,020 \text{ V}.$$

Transformers T2 and T4:

Primary line voltages,

$$V_A' = -j11\,000 - 39 = 11\,000 \text{ V}$$
$$V_B' = j5500 - 9500 = 10\,750 \text{ V}$$
$$V_C' = j5500 + 9550 = 11\,020 \text{ V}.$$

Secondary line voltages,

$$V_A'' = +440 = 440 \text{ V}$$
$$V_B'' = -j381 - 224 \cdot 35 = 443 \text{ V}$$
$$V_C'' = j381 - 215 \cdot 65 = 438 \text{ V}.$$

Transformer T2:

Secondary line to neutral voltages,

$$V_a'' = +4 \cdot 9 = 4 \cdot 9 \text{ V}$$
$$V_b'' = -j381 + 220 \cdot 55 = 440 \text{ V}$$
$$V_c'' = -j381 - 219 \cdot 45 = 440 \text{ V}.$$

Transformer T4:

Secondary line to neutral voltages,

$$V_a'' = j254 + 2 \cdot 9 = 254 \text{ V}$$
$$V_b'' = -j127 + 218 \cdot 55 = 252 \text{ V}$$

11 kV   440 V

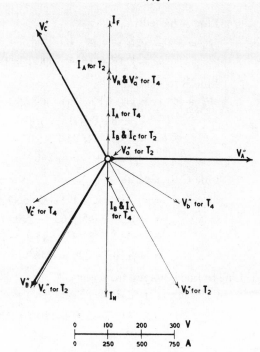

ALL H.V. VOLTAGES
PRACTICALLY
NO-LOAD
VOLTAGES

Fig. A7.18   Currents and voltages.
Case 2.   E.R. = 0·25 Ω

$$V_c'' = -j127 - 221·45 = 255 \text{ V}.$$

Voltage drop across earth resistance, $V_R = j254 = 254$ V.
These voltages are shown in phasor form in Fig. A7.18.

Case 3.   Circuit diagrams (Figs. A7.1, A7.12 and A7.13)

Voltages at 440 V busbars

Phase sequence voltages, from eqns. A7.9 and A7.10 and Fig. A7.12:

Positive sequence voltage is,

$$V_2 = V_G - I_2 Z_2 = 0 - [(6250 \times j0·00261) + (12\ 500 \times j0·0059)]$$
$$= j254 - j16·3 - j73·8 = j163·9 \text{ V}.$$

Negative sequence voltage is,

$$V_2 = V_G - I_2 Z_2 = 0 - [(6250 \times j0·00261) + (12\ 500 \times j0·0059)]$$
$$= 0 - j16·3 - j73·8 = -j90·1 \text{ V}.$$

Zero sequence voltage is,

$$V_0 = V_G - I_0 Z_0 = 0 - (12\ 500 \times j0·0059)$$
$$= 0 - j73·8 = -j73·8 \text{ V}.$$

747

Line to neutral voltages, from eqns. A7.11:

Line A to neutral,

$$V_A = V_a'' = V_0 + V_1 + V_2$$
$$= -j73\cdot8 + j163\cdot9 - j90\cdot1 = 0 \text{ V.}$$

Line B to neutral,

$$V_B = V_b'' = V_0 - 0\cdot5(V_1 + V_2) - j0\cdot866(V_1 - V_2)$$
$$= -j73\cdot8 - 0\cdot5(j163\cdot9 - j90\cdot1) - j0\cdot866(j163\cdot9 + j90\cdot1)$$
$$= -j73\cdot8 - j81\cdot95 + j45\cdot05 + 142 + 78$$
$$= -j110\cdot7 + 220 = 246 \text{ V.}$$

Line C to neutral,

$$V_C = V_c'' = V_0 - 0\cdot5(V_1 + V_2) + j0\cdot866(V_1 - V_2)$$
$$= -j73\cdot8 - 0\cdot5(j163\cdot9 - j90\cdot1) + j0\cdot866(j163\cdot9 + j90\cdot1)$$
$$= -j73\cdot8 - j81\cdot95 + j45\cdot05 - 142 - 78$$
$$= -j110\cdot7 - 220 = 246 \text{ V.}$$

Line to line voltages, from eqns. A7.14:

$$V_A'' = V_b'' - V_c''$$
$$= -j110\cdot7 + 220 + j110\cdot7 + 220$$
$$= 440\text{V}$$

$$V_B'' = V_c'' - V_a''$$
$$= -j110\cdot7 - 220 - 0$$
$$= -j110\cdot7 - 220 = 246 \text{ V}$$

$$V_C'' = V_a'' - V_b''$$
$$= 0 + j110\cdot7 - 220$$
$$= j110\cdot7 - 220 = 246 \text{ V.}$$

Voltages at 11 kV busbars

Phase sequence voltages, from eqns. A7.9 and A7.10 and Fig. A7.12:

Positive sequence voltage is,

$$V_1 = V_G - I_1 Z_1$$
$$= j254 - (6250 \times j0\cdot00261)$$
$$= j254 - j16\cdot3 = j237\cdot7 \text{ V.}$$

Negative sequence voltage is,

$$V_2 = V_G - I_2 Z_2$$
$$= 0 - (6250 \times j0\cdot00261)$$
$$= 0 - j16\cdot3 = -j16\cdot3 \text{ V.}$$

Zero sequence voltage is $V_0 = 0$.

Line to neutral voltages, from eqns. A7.11:

Line A to neutral,

$$V_A = V_a'' = V_0 + V_1 + V_2$$
$$= 0 + j237 \cdot 7 - j16 \cdot 3 = j221 \cdot 4 \text{ V}.$$

Line B to neutral,

$$V_B = V_b'' = V_0 - 0 \cdot 5(V_1 + V_2) - j0 \cdot 866(V_1 - V_2)$$
$$= 0 - 0 \cdot 5(j237 \cdot 7 - j16 \cdot 3) - j0 \cdot 866(j237 \cdot 7 + j16 \cdot 3)$$
$$= -j118 \cdot 85 + j8 \cdot 15 + 205 \cdot 9 + 14 \cdot 1$$
$$= -j110 \cdot 7 + 220 = 246 \text{ V}.$$

Line C to neutral,

$$V_C = V_c'' = V_0 - 0 \cdot 5(V_1 + V_2) + j0 \cdot 866(V_1 - V_2)$$
$$= 0 - 0 \cdot 5(j237 \cdot 7 - j16 \cdot 3) + j0 \cdot 866(j237 \cdot 7 + j16 \cdot 3)$$
$$= -j118 \cdot 85 + j8 \cdot 15 - 205 \cdot 9 - 14 \cdot 1$$
$$= -j110 \cdot 7 - 220 = 246 \text{ V}.$$

Converting these to true 11 kV delta values, from Fig. A7.14 and eqns. A7.12:

$$V_A' = -V_a''/n$$
$$= -j221 \cdot 4/0 \cdot 0231 = -j9600 = 9600 \text{ V}$$

$$V_B' = -V_b''/n$$
$$= (j110 \cdot 7 - 220)/0 \cdot 0231$$
$$= j4800 - 9530 = 10\,650 \text{ V}$$

$$V_C' = -V_c''/n$$
$$= (j110 \cdot 7 + 220)/0 \cdot 0231$$
$$= j4800 + 9530 = 10\,650 \text{ V}.$$

The corresponding true line to neutral voltages on the 11 kV side of T1 and T3 are, from eqns. A7.13:

$$V_a' = \tfrac{1}{3}V_A' + \tfrac{2}{3}V_C'$$
$$= \tfrac{1}{3}(-j9600) + \tfrac{2}{3}(j4800 + 9530)$$
$$= -j3200 + j3200 + 6353 = 6353 \text{ V}$$

$$V_b' = \tfrac{1}{3}V_B' + \tfrac{2}{3}V_A'$$
$$= \tfrac{1}{3}(j4800 - 9530) + \tfrac{2}{3}(-j9600)$$
$$= j1600 - 3177 - j6400$$
$$= -j4800 - 3177 = 5820 \text{ V}$$

$$V_c' = \tfrac{1}{3}V_C' + V_B'$$
$$= \tfrac{1}{3}(j4800 + 9530) + \tfrac{2}{3}(j4800 - 9530)$$
$$= j1600 + 3177 + j3200 - 6353$$
$$= 4800 - 3177 = 5820 \text{ V}.$$

The voltages of T1 and T3 are, of course, the same.
All voltages throughout the circuit are summarised as follows:

Transformers T1 and T3:

All primary line voltages, 33 kV.
Secondary line to neutral voltages,

$$V_a' = 6353 \text{ V}$$
$$V_b' = -j4800 - 3177 = 5820 \text{ V}$$
$$V_c' = j4800 - 3177 = 5820 \text{ V}.$$

Secondary line voltages,

$$V_A' = -j9600 = 9600 \text{ V}$$
$$V_B' = j4800 - 9530 = 10\,650 \text{ V}$$
$$V_C' = j4800 + 9530 = 10\,650 \text{ V}.$$

Transformer T2:

Primary line voltages,

$$V_A' = -j9600 = 9600 \text{ V}$$
$$V_B' = j4800 - 9530 = 10\,650 \text{ V}$$
$$V_C' = j4800 + 9530 = 10\,650 \text{ V}.$$

Secondary line voltages,

$$V_A'' = 440 \text{ V}$$
$$V_B'' = -j110\cdot7 - 220 = 246 \text{ V}$$
$$V_C'' = j110\cdot7 - 220 = 246 \text{ V}.$$

Secondary line to neutral voltages,

$$V_a'' = 0 \text{ V}$$
$$V_b'' = -j110\cdot7 + 220 = 246 \text{ V}$$
$$V_c'' = -j110\cdot7 - 220 = 246 \text{ V}.$$

These voltages are shown in phasor form in Fig. A7.19.

If a resistance of $0\cdot25\ \Omega$ is now included in the earth fault current circuit, the sequence impedances are as given in Fig. A7.12, except that the resistance $R_0 = 3R = 0\cdot75$ is inserted in series in the zero sequence network; the overall zero sequence impedance is then,

$$Z_0 = 0\cdot75106 + j0\cdot0058 = 0\cdot752\ \Omega$$

as given before.

The total current in each sequence network is 338 A and one half this value in each of the parallel branches of the positive and negative sequence networks. The voltages throughout the circuit are then as follows.

Voltages at 440 V busbars

Phase sequence voltages, from eqns. A7.9 and A7.10 and Fig. A7.12, bearing in mind the inclusion of a series resistance $R_0 = 0\cdot75\ \Omega$ in the zero sequence network:

750

11 kV   440 V

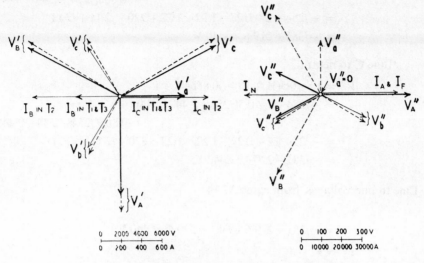

Fig. A7.19   Currents and voltages.
Case 3.   E.R. = 0. Broken lines show no-load voltages

Positive sequence voltage is,

$$V_1 = V_G - I_1 Z_1$$
$$= j254 - [(j169 \times j0\cdot00261) + (j338 \times j0\cdot0059)]$$
$$= j254 + 0\cdot44 + 2$$
$$= j254 + 2\cdot44 = 254 \text{ V}.$$

Negative sequence voltage is,

$$V_2 = V_G - I_2 Z_2$$
$$= 0 - [(j169 \times j0\cdot00261) + (j338 \times j0\cdot0059)]$$
$$= 0\cdot44 + 2 = 2\cdot44 \text{ V}.$$

Zero sequence voltage is,

$$V_0 = V_G - I_0 Z_0$$
$$= 0 - j338(0\cdot75106 + j0\cdot0058)$$
$$= -j254 + 2 = 254 \text{ V}.$$

Line to neutral voltages, from eqns. A7.11:

Line A to neutral,

$$V_A = V_a''' = V_0 + V_1 + V_2$$
$$= -j254 + 2 + j254 + 2\cdot44 + 2\cdot44 = 6\cdot88 \text{ V}.$$

Line B to neutral,

$$V_B = V_b'' = V_0 - 0\cdot5(V_1 + V_2) - j0\cdot866(V_1 - V_2)$$

751

$$= -j254 + 2 - 0{\cdot}5(j254 + 2{\cdot}44 + 2{\cdot}44)$$
$$\qquad\qquad\qquad\qquad -j0{\cdot}866(j254 + 2{\cdot}44 - 2{\cdot}44)$$
$$= -j254 + 2 - j127 - 1{\cdot}22 - 1{\cdot}22 + 220 - j2{\cdot}11 + j2{\cdot}11$$
$$= -j381 + 219{\cdot}6 = 440 \text{ V.}$$

Line C to neutral,

$$V_C = V_c'' = V_0 - 0{\cdot}5(V_1 + V_2) + j0{\cdot}866(V_1 - V_2)$$
$$= -j254 + 2 - 0{\cdot}5(j254 + 2{\cdot}44 + 2{\cdot}44)$$
$$\qquad\qquad\qquad\qquad + j0{\cdot}866(j254 + 2{\cdot}44 - 2{\cdot}44)$$
$$= -j254 + 2 - j127 - 1{\cdot}22 - 1{\cdot}22 - 220 + j2{\cdot}11 - j2{\cdot}11$$
$$= -j381 - 220{\cdot}4 = 440 \text{ V.}$$

Line to line voltages, from eqns. A7.14:

$$V_A'' = V_b'' - V_c''$$
$$= -j381 + 219{\cdot}6 + j381 + 220{\cdot}4 = 440 \text{ V}$$
$$V_B'' = V_c'' - V_a''$$
$$= -j381 - 220{\cdot}4 - 6{\cdot}88$$
$$= -j381 - 227{\cdot}3 = 444 \text{ V}$$
$$V_C'' = V_a'' - V_b''$$
$$= 6{\cdot}88 + j381 - 219{\cdot}6$$
$$= j381 - 212{\cdot}7 = 436 \text{ V.}$$

The voltage drop across the neutral earth resistance is,

$$V_R = I_0 R_0 = I_0 3R$$
$$= j338 \times 0{\cdot}75 = j254 \text{ V.}$$

Voltages at 11 kV busbars

Phase sequence voltages, from eqns. A7.9 and A7.10 and Fig. A7.12:

Positive sequence voltage is,

$$V_1 = V_G - I_1 Z_1$$
$$= j254 - (j169 \times j0{\cdot}00261)$$
$$= j254 + 0{\cdot}44 = 254 \text{ V.}$$

Negative sequence voltage is,

$$V_2 = V_G - I_2 Z_2$$
$$= 0 - (j169 \times j0{\cdot}00261) = 0{\cdot}44 \text{ V.}$$

Zero sequence voltage is $V_0 = 0$.

Line to neutral voltages, from eqns. A7.11:

Line A to neutral,

$$V_A = V_a'' = V_0 + V_1 + V_2$$
$$= 0 + j254 + 0{\cdot}44 + 0{\cdot}44$$
$$= j254 + 0{\cdot}88 = 254 \text{ V}.$$

Line B to neutral,

$$V_B = V_b'' = V_0 - 0{\cdot}5(V_1 + V_2) - j0{\cdot}866(V_1 - V_2)$$
$$= 0 - 0{\cdot}5(j254 + 0{\cdot}44 + 0{\cdot}44) - j0{\cdot}866(j254 + 0{\cdot}44 - 0{\cdot}44)$$
$$= -j127 - 0{\cdot}22 - 0{\cdot}22 + 220 - j0{\cdot}38 + j0{\cdot}38$$
$$= -j127 + 219{\cdot}6 = 254 \text{ V}.$$

Line C to neutral,

$$V_C = V_c'' = V_0 - 0{\cdot}5(V_1 + V_2) + j0{\cdot}866(V_1 - V_2)$$
$$= 0 - 0{\cdot}5(j254 + 0{\cdot}44 + 0{\cdot}44) + j0{\cdot}866(j254 + 0{\cdot}44 - 0{\cdot}44)$$
$$= -j127 - 0{\cdot}22 - 0{\cdot}22 - 220 + j0{\cdot}38 - j0{\cdot}38$$
$$= -j127 - 220{\cdot}4 = 254 \text{ V}.$$

Converting these to true 11 kV delta values, from Fig. A7.14 and eqns. A7.12:

$$V_A' = V_a''/n$$
$$= (-j254 - 0{\cdot}88)/0{\cdot}0231$$
$$= -j11\,000 - 39 = 11\,000 \text{ V}$$

$$V_B' = -V_b''/n$$
$$= (j127 - 219{\cdot}6)/0{\cdot}0231$$
$$= j5500 - 9500 = 10\,970 \text{ V}$$

$$V_C' = -V_c''/n$$
$$= (j127 + 220{\cdot}4)/0{\cdot}0231$$
$$= j5500 + 9545 = 11\,020 \text{ V}.$$

The corresponding true line to neutral voltages on the 11 kV side of T1 and T3 are, from eqns. A7.13:

$$V_a' = \tfrac{1}{3}V_A' + \tfrac{2}{3}V_C'$$
$$= \tfrac{1}{3}(-j11\,000 - 39) + \tfrac{2}{3}(j5500 + 9545)$$
$$= -j3667 - 13 + j3667 + 6363 = 6350 \text{ V}$$

$$V_b' = \tfrac{1}{3}V_B' + \tfrac{2}{3}V_A'$$
$$= \tfrac{1}{3}(j5500 - 9500) + \tfrac{2}{3}(-j11\,000 - 39)$$
$$= j1833 - 3167 - j7333 - 26$$
$$= -j5500 - 3193 = 6360 \text{ V}$$

$$V_c' = \tfrac{1}{3}V_C' + \tfrac{2}{3}V_B'$$
$$= \tfrac{1}{3}(j5500 + 9545) + \tfrac{2}{3}(j5500 - 9500)$$
$$= j1833 + 3182 + j3667 - 6333$$
$$= j5500 - 3151 = 6340 \text{ V}.$$

The voltages of T1 and T3 are the same.
All voltages throughout the circuit are summarised as follows:

Transformers T1 and T3:

All primary line voltages, 33 kV.
Secondary line to neutral voltages,

$$V'_a = 6350 \text{ V}$$
$$V'_b = -j5500 - 3193 = 6360 \text{ V}$$
$$V'_c = j5500 - 3151 = 6340 \text{ V}.$$

Secondary line voltages,

$$V'_A = -j11\,000 - 39 = 11\,000 \text{ V}$$
$$V'_B = j5500 - 9500 = 10\,970 \text{ V}$$
$$V'_C = j5500 + 9545 = 11\,020 \text{ V}.$$

11 kV   440 V

ALL H.V. VOLTAGES
PRACTICALLY
NO-LOAD
VOLTAGE

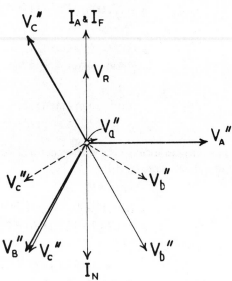

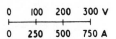

Fig. A7.20   Currents and voltages.
Case 3.   E.R. = 0·25 Ω. Broken lines show no-load voltages

Transformer T2:

Primary line voltages,

$$V_A' = -j11\,000 - 39 = 11\,000 \text{ V}$$
$$V_B' = j5500 - 9500 = 10\,970 \text{ V}$$
$$V_C' = j5500 + 9545 = 11\,020 \text{ V.}$$

Secondary line voltages,

$$V_A'' = +440 = 440 \text{ V}$$
$$V_B'' = -j381 - 227 \cdot 3 = 444 \text{ V}$$
$$V_C'' = j381 - 212 \cdot 7 = 436 \text{ V.}$$

Secondary line to neutral voltages,

$$V_a'' = +6 \cdot 88 = 6 \cdot 88 \text{ V}$$
$$V_b'' = -j381 + 219 \cdot 6 = 440 \text{ V}$$
$$V_c'' = -j381 - 220 \cdot 4 = 440 \text{ V.}$$

Voltage drop across earth resistance, $V_R = j254 = 254$ V.

The phasor diagram corresponding to these voltages is shown in Fig. A7.20.

Currents and voltages for all the three cases chosen for this investigation are scheduled in Tables A7.1, page 756 and A7.2 (pages 757 to 759). Current *magnitudes* are controlled almost entirely by the value of the earth resistances, while the *distribution* of current depends upon the impedances of the various parallel paths of the circuit.

Table A7.1
LINE CURRENTS—AMPERES

	Case 1		Case 2		Case 3	
	E.R. = 0	E.R. = 0·25 Ω	E.R. = 0	E.R. = 0·25 Ω	E.R. = 0	E.R. = 0·25 Ω
On 440 V side:						
I_F	66 498	2028	52 896	1014	37 500	1014
T2, secondary:						
I_A	33 249	1014	35 264	676	37 500	1014
I_B	0	0	8816	169	0	0
I_C	0	0	8816	169	0	0
I_N	33 249	1014	52 896	1014	37 500	1014
T4, secondary:						
I_A	33 249	1014	17 632	338	...	...
I_B	0	0	−8816	−169	...	...
I_C	0	0	−8816	−169	...	...
I_N	33 249	1014	...	...	...	...
On 11 kV side:						
T1, secondary:						
I_A	0	0	0	0	0	0
I_B	−768	−23·4	−612	−11·75	−433	−11·75
I_C	768	23·4	612	11·75	433	11·75
T2, primary:						
I_A	0	0	0	0	0	0
I_B	−768	−23·4	−612	−11·75	−866	−23·5
I_C	768	23·4	612	11·75	866	23·5
T3, secondary:						
I_A	0	0	0	0	0	0
I_B	−768	−23·4	−612	−11·75	−433	−11·75
I_C	768	23·4	612	11·75	433	11·75
T4, primary:						
I_A	0	0	0	0	...	...
I_B	−768	−23·4	−612	−11·75	...	...
I_C	768	23·4	612	11·75	...	...
On 33 kV side:						
T1, primary:						
I_A	−295	−9	−235	−4·5	−166	−4·4
I_B	147·5	4·5	117·5	2·25	83	2·25
I_C	147·5	4·5	117·5	2·25	83	2·25
T3, primary:						
I_A	−295	−9	−235	−4·5	−166	−4·5
I_B	147·5	4·5	117·5	2·25	83	2·25
I_C	147·5	4·5	117·5	2·25	83	2·25

Table A7.2 VOLTAGES

		At no-load		Case 1			
				E.R. = 0 (Fig. A7.15)		E.R. = 0.25 Ω (Fig. A7.16)	
At 11 kV bars and windings—							
Line to neutral:	V'_a	$+6351$	$= 6351$ V	$+6353$	$= 6353$ V	$+6347$	$= 6347$ V
	V'_b	$-j5500-3175$	$= 6351$ V	$-j4250-3177$	$= 5300$ V	$-j5500-3218$	$= 6380$ V
	V'_c	$+j5500-3175$	$= 6351$ V	$+j4250-3177$	$= 5300$ V	$+j5500-3180$	$= 6350$ V
Line to line:	V'_A	$-j11\,000$	$= 11\,000$ V	$-j8500$	$= 8500$ V	$-j11\,000-77$	$= 11\,000$ V
	V'_B	$+j5500-9526$	$= 11\,000$ V	$+j4250-9530$	$= 10\,420$ V	$+j5500-9500$	$= 10\,750$ V
	V'_C	$+j5500+9526$	$= 11\,000$ V	$+j4250+9530$	$= 10\,420$ V	$+j5500+9560$	$= 11\,020$ V
At 440 V bars and windings—							
Line to neutral:	V''_a	$+j254$	$= 254$ V	0	$= 0$ V	$+7.8$	$= 7.8$ V
	V''_b	$-j127+220$	$= 254$ V	$-j98.1+220$	$= 241$ V	$-j381+219.1$	$= 440$ V
	V''_c	$-j127-220$	$= 254$ V	$-j98.1-220$	$= 241$ V	$-j381-220.9$	$= 440$ V
Line to line:	V''_A	$+440$	$= 440$ V	$+440$	$= 440$ V	$+440$	$= 440$ V
	V''_B	$-j381-220$	$= 440$ V	$-j98.1-220$	$= 241$ V	$-j381-228.7$	$= 443$ V
	V''_C	$+j381-220$	$= 440$ V	$+j98.1-220$	$= 241$ V	$+j381-211.3$	$= 435$ V
Neutral to earth*:	V_{NE}	0	$= 0$ V	0	$= 0$ V	$+j254$	$= 254$ V
E.R. drop	V_R	0	$= 0$ V		...	$+j254$	$= 254$ V

*Voltage from neutral point to zero earth potential.

Table A7.2 VOLTAGES (*continued*)

		At no-load	Case 3	
			E.R. = 0 (Fig. A7.19)	E.R. = 0.25 Ω (Fig. A7.20)
At 11 kV bars and windings—				
Line to neutral: … … …	V'_a	$+6351 = 6361$ V	$+6353 = 6353$ V	$+6350 = 6350$ V
	V'_b	$-j5500 - 3175 = 6351$ V	$-j4800 - 3177 = 5820$ V	$-j5500 - 3193 = 6360$ V
	V'_c	$+j5500 - 3175 = 6351$ V	$+j4800 - 3177 = 5820$ V	$+j5500 + 3151 = 6340$ V
Line to line: … … …	V'_A	$-j11000 = 11000$ V	$-j9600 = 9600$ V	$-j11000 - 39 = 11000$ V
	V'_B	$+j5500 - 9526 = 11000$ V	$+j4800 - 9530 = 10650$ V	$+j5500 - 9500 = 10970$ V
	V'_C	$+j5500 + 9526 = 11000$ V	$+j4800 + 9530 = 10650$ V	$+j5500 + 9545 = 11020$ V
At 440 V bars and windings—				
Line to neutral: … … …	V''_a	$+j254 = 254$ V	$0 = 0$ V	$+6·9 = 6·9$ V
	V''_b	$-j127 + 220 = 254$ V	$-j110·7 + 220 = 246$ V	$-j381 + 219·6 = 440$ V
	V''_c	$-j127 - 220 = 254$ V	$-j110·7 - 220 = 246$ V	$-j381 - 220·4 = 440$ V
Line to line: … … …	V''_A	$+440 = 440$ V	$+440 = 440$ V	$+440 = 440$ V
	V''_B	$-j381 - 220 = 440$ V	$-j110·7 - 220 = 246$ V	$-j381 - 227·3 = 444$ V
	V''_C	$+j381 - 220 = 440$ V	$+j110·7 - 220 = 246$ V	$+j381 - 212·7 = 436$ V
Neutral to earth*: … …	V_{NE}	$0 = 0$ V	$0 = 0$ V	$+j254 = 254$ V
E.R. drop … … …	V_R	$0 = 0$ V	…	$+j254 = 254$ V

* Voltage from neutral point to zero earth potential.

758

Table A7.2 VOLTAGES (*continued*)

At 11 kV bars and windings—

	At no-load	Case 2 — E.R. = 0 (Fig. A7.17)	Case 2 — E.R. = 0.25 Ω (Fig. A7.18)
Line to neutral: … V'_a	$+6351$ = 6351 V	$+6347$ = 6347 V	$+6354$ = 6354 V
V'_b	$-j5500-3175$ = 6351 V	$-j4500-3173$ = 5500 V	$-j5500-3193$ = 6360 V
V'_c	$+j5500-3175$ = 6351 V	$+j4500-3174$ = 5500 V	$+j5500-3150$ = 6340 V
Line to line: … V'_A	$-j11000$ = 11 000 V	$-j9000$ = 9000 V	$-j11000-39$ = 11 000 V
V'_B	$+j5500-9526$ = 11 000 V	$+j4500-9520$ = 10 500 V	$+j5500-9500$ = 10 750 V
V'_C	$+j5500+9526$ = 11 000 V	$+j4500+9520$ = 10 500 V	$+j5500+9550$ = 11 020 V

At 440 V bars and windings— (Case 2 columns split into T2 and T4)

	At no-load	Case 2 — E.R. = 0 (Fig. A7.17)	Case 2 — E.R. = 0.25 Ω (Fig. A7.18)
Line to neutral: … V''_a	$+j254$ = 254 V	T2: 0 = 0 V; T4: $+j104$ = 104 V	T2: $+4.9$ = 4.9 V; T4: $+j254+2.9$ = 254 V
V''_b	$-j127+220$ = 254 V	T2: $-j156+220$ = 270 V; T4: $-j52+220$ = 226 V	T2: $-j381+220.55$ = 440 V; T4: $-j127+218.55$ = 252 V
V''_c	$-j127-220$ = 254 V	T2: $-j156-220$ = 270 V; T4: $-j52-220$ = 226 V	T2: $-j381-219.45$ = 440 V; T4: $-j127-221.45$ = 255 V
Line to line: … V''_A	$+440$ = 440 V	T2: $+440$ = 440 V; T4: $+440$ = 440 V	$+440$ = 440 V
V''_B	$-j381-220$ = 440 V	T2: $-j156-220$ = 270 V; T4: $-j52+220$ = 270 V	$-j381-224.35$ = 433 V
V''_C	$+j381-220$ = 440 V	T2: $+j156-220$ = 270 V; T4: $+j156-220$ = 270 V	$+j381-215.65$ = 438 V
Neutral to earth*: V_{NE}	0 = 0 V	T2: 0 = 0 V; T4: $+j104$ = 104 V	$+j254$ = 254 V
E.R. drop … V_R	0 = 0 V	…	$+j254$ = 254 V

*Voltage from neutral point to zero earth potential.

List of reports issued by the British Electrical and Allied Industries Research Association (E.R.A.) relating to transformers and to surge phenomena therein

TRANSFORMERS

Ref. Q/T101a   'Mechanical stresses in transformer windings', by M. Waters.

Ref. Q/T103   'Electrical and mechanical effects of internal faults in transformers', by E. Billig.

Ref. Q/T113   'The measurement of axial magnetic forces in transformer windings', by M. Waters.

Ref. Q/T115   'The calculation of transformer thermal data from readings taken in service', by M. R. Dickson.

Ref. Q/T116   'Generation of gases in transformers. Résumé of available information', by M. R. Dickson.

Ref. Q/T117   'Temperature gradients in transformer windings and rates of oil flow in transformer tanks. A critical review of published information', by B. L. Coleman.

Ref. Q/T118   'The operation of naturally cooled outdoor transformers as affected by weather and surroundings. Preliminary review', by M. R. Dickson.

Ref. Q/T121   'The calculation of currents due to faults between turns in transformer windings', by B. L. Coleman.

Ref. Q/T126   'The causes and effects of water in oil-immersed transformers. A critical résumé', by M. R. Dickson.

Ref. Q/T130   'Corrosion of internal tank surfaces in non-conservator transformers', by M. Waters.

Ref. Q/T134   'The measurement and calculation of axial electromagnetic forces in concentric transformer windings', by M. Waters.

Ref. Q/T139   'The effects of dissolved gases in the design and operation of oil immersed transformers', by M. R. Dickson.

Ref. Q/T141   'An adjustable ambient-temperature thermometer for use when testing transformers', by M. R. Dickson.

Ref. Q/T144	'The effect of core properties on axial electromagnetic forces in transformers with concentric windings', by M. Waters.
Ref. Q/T151	'A method based on Maxwell's equations for calculating the axial short-circuit forces in the concentric windings of an idealized transformer', by P. R. Vein.
Ref. Q/T153	'The measurement of axial displacement of transformer windings', by M. Waters.
Ref. Q/T158	'Measurement of axial forces in a transformer with multi-layer windings', by E. D. Taylor, J. Page and M. Waters.
Ref. Q/T161	'Copper for transformer windings', by J. E. Bowers and E. C. Mantle.
Ref. Q/T163	'E.R.A. researches on transformer noise 1951–59', by A. I. King, A. S. Ensus and M. Waters.
Ref. G/T130	'The effect of zero phase sequence exciting impedance of three-phase core transformers on earth fault currents', by L. Gosland.
Ref. G/T140	'Some measurements of zero phase sequence impedance of three-phase, three-limb, core-type transformers with a delta winding', by L. Gosland.
Ref. G/T313	'Measurement of overvoltages caused by switching out a 75 MVA, 132/33 kV transformer from the high voltage side', by M. P. Reece and E. L. White.
Ref. V/T123	'Application of the dispersion test to the drying of high voltage transformers', by D. C. G. Smith.
Ref. 5028	'The mechanical properties of high conductivity copper conductors for power transformers', by M. Waters.
Ref. 5081	'The mechanical properties of high conductivity aluminium conductors for power transformers', by M. Waters.
Ref. 5096	'The ventilation of transformer substations or cubicles', by M. R. Dickson.
Ref. 5149	'An exploration of some mechanical factors affecting vibration and noise of transformer cores', by L. Gosland and M. Waters.
Ref. 5152	'Transformer magnetising inrush currents,' A résumé of published information by A. A. Hudson.

SURGE PHENOMENA

Ref. S/T35	'Surge phenomena. Seven years' research for the Central Electricity Board (1933–1940)', edited by H. M. Lacey with a foreword by E. B. Wedmore.
Ref. S/T43	'Surge tests on a transformer with and without protection', by H. M. Lacey and E. W. W. Double.
Ref. S/T48	'Surge voltage distribution in a continuous-disc transformer winding', by K. L. Selig.
Ref. S/T54	'An investigation of flashovers on a low voltage busbar system', by L. Gosland and E. L. White.
Ref. S/T69	'The effects of cylindrical end rings on the distribution of surge voltages in transformer windings', by E. L. White.
Ref. S/T73	'Surge voltage distribution in transformer windings due to current chopping', by E. L. White.

761

Ref. S/T85   'Impulse-excited terminal oscillations due to no-load switching of a three-phase transformer installation', by E. L. White.

Ref. S/T95   'Transients in transformer windings', by B. L. Coleman.

Ref. S/T97   'Line and neutral currents in multi-limb transformers under impulse-test conditions', by E. L. White.

Ref. S/T98   'An experimental study of transient oscillations in windings of core-type transformers', by E. L. White.

Ref. S/T103   'A simple technique for producing test voltages across a transformer winding by current chopping', by M. P. Reece and E. L. White.

Ref. S/T109   'Transference of surges through a generator transformer with special reference to neutral earthing. Field tests on a 15·8 MVA generator', by E. L. White.

Ref. S/T111   'Switching surges on a 275/132 kV auto-transformer', by E. L. White.

Ref. S/T112   'A summary of the ERA theory of oscillations and surges in transformer windings', by R. J. Clowes and E. L. White.

Ref. S/T115   'A capacitive probe method of exploring voltage distributions in windings', by E. L. White.

Ref. S/T116   'Maximum voltages on concentric transformer windings subjected to one-, two- or three-pole impulses', by R. J. Clowes.

Ref. 5015   'Excessive voltages induced in an inner winding of a transformer during impulse tests on an outer winding', by R. J. Clowes.

Ref. 5063   'Calculation of voltages induced in an inner winding of a transformer when an impulse is applied to the outer winding', by R. J. Clowes.

Ref. 5133   'Surge transference in generator transformers', A study based on published information, by E. L. White.

Ref. 5134   'Controlled current chopping as a possible overvoltage test method for transformers on site', by E. L. White.

Ref. 5144   'Excessive surge voltages in a 33 kV earthing transformer', by E. L. White.

Ref. 5153   'Surge voltage transference in a 100 MW unit-connected generator set at Aberthaw Station', by E. L. White.

Ref. 5210   'Surge transference measurements on generator transformers connected to systems above 100 kV', by E. L. White.

Appendix 9

List of the principal specifications issued by the British Standards Institution relative to power transformers

77 Voltages for a.c. transmission and distribution systems.
148 Insulating oil for transformers and switchgear.
156 Enamelled (round) copper conductors (oleo-resinous enamel)
Part 1: Round wire.
Part 2: Rectangular conductors.
Part 3: Round wire metric units.
171 Power transformers.
205 Glossary of terms used in electrical engineering.
Part 1: (Group 4) Terms used for power transformers.
Part 2: (Group 71) Voltage fluctuations terminology.
223 High-voltage bushings.
229 Flameproof enclosure of electrical apparatus.
231 Pressboard for electrical purposes.
269 Rules for methods of declaring efficiency of electrical machinery (excluding traction motors).
355 Mining-type transformers. Part 1 Dry type; Part 2 Oil immersed.
358 Method for measurement of voltage with sphere-gaps (one sphere earthed).
542 Cable-glands and sealing boxes for association with apparatus for use at mines.
587 Motor starters and controllers.
601 Steel sheets for magnetic circuits of power electrical apparatus. Part 1: Non-oriented steel; Part 2: Oriented steel.
638 Arc welding plant, equipment and accessories.
661 Glossary of acoustical terms.
698 Papers for electrical purposes.
822 Terminal markings for electrical machinery.
923 Impulse-voltage testing.
933 Magnetic materials for use under combined a.c. and d.c. magnetization.
1432 Copper for electrical purposes. Strip with drawn or rolled edges.
1791 Cotton-covered Copper Conductors.
Part 1: Round wire

Part 2: Rectangular conductors.

Part 3: Round wire metric units.

1844   Enamelled copper conductors (enamel with vinyl acetal base).

Part 1: Round wire.

Part 2: Rectangular conductors.

Part 3: Round wire metric units.

1858   Bitumen-base filling compounds for electrical purposes.

2497   Reference zero for the calibration of the pure-tone audio meters.

2520   Barometer conventions and tables.

2538   Air-cooled flameproof single-phase lighting transformer units supplied from high-voltage systems

2562   Cable sealing boxes for oil-immersed transformers.

Part 1: Boxes for voltages up to and including 11 kV.

Part 2: Cable sealing boxes for 22 kV and 33 kV solid type cables.

2658   Guide to terms used in a.c. power system studies.

2725   Memorandum on the measurement of cooling-medium temperature when testing electrical machines, transformers and other electrical apparatus.

2757   Classification of insulating materials for electrical machinery and apparatus on the basis of thermal stability in service.

2776   Paper-covered copper conductors.

Part 1: Round wire.

Part 2: Rectangular conductors.

3045   The relation between the sone scale of loudness and the phon scale of loudness level.

3383   Normal equal-loudness contours for pure tones and normal threshold of hearing under free-field listening conditions.

3489   Sound level meters (industrial grade)

3535   Safety isolating transformers for industrial and domestic purposes.

3938   Current transformers.

3941   Voltage transformers.

4109   Copper for electrical purposes; wire for general electrical purposes, and for insulated cables and flexible cords.

4197   A precision sound level meter.

4571   On-load tap-changers for power transformers.

CP.321.102 Installations and maintenance of electrical machines, transformers, rectifiers, capacitors and associated equipment.

CP.1009   Maintenance of insulating oils.

CP.1010   Guide to loading of oil-immersed transformer to B.S.171.

These Specifications may be obtained from The British Standards Institution, Sales Dept., Newton House, 110–113 Pentonville Road, London W1Y 4AA.

Index